D1162632

SPECIALIZATION, SPECIATION, AND RADIATION

SPECIALIZATION, SPECIATION, AND RADIATION

THE EVOLUTIONARY BIOLOGY OF HERBIVOROUS INSECTS

Edited by

KELLEY JEAN TILMON

UNIVERSITY OF CALIFORNIA PRESS
Berkeley Los Angeles London

University of California Press, one of the most distinguished university presses in the United States, enriches lives around the world by advancing scholarship in the humanities, social sciences, and natural sciences. Its activities are supported by the UC Press Foundation and by philanthropic contributions from individuals and institutions. For more information, visit www.ucpress.edu.

University of California Press
Berkeley and Los Angeles, California

University of California Press, Ltd.
London, England

Library of Congress Cataloging-in-Publication Data

Specialization, speciation, and radiation : the evolutionary biology of herbivorous insects / edited by Kelley Jean Tilmon.
p. cm.
Includes bibliographical references and index.
ISBN 978-0-520-25132-8 (case : alk. paper)
1. Insects—Evolution. 2. Insects—Behavior. 3. Insect–plant relationships. I. Tilmon, Kelley Jean.
QL468.7.S64 2008
595.7138—dc22

2007032362

Manufactured in the United States of America
10 09 08
10 9 8 7 6 5 4 3 2 1

The paper used in this publication meets the minimum requirements of ANSI/NISO Z39.48-1992 (R 1997) (*Permanence of Paper*).

Cover photograph: A specialist caterpillar (Lepidoptera: Limacodidae) with brilliant warning coloration and stinging hairs feeds on a leaf of *Macaranga tsonane* (Euphorbiaceae) in the lowland rainforest of Papua New Guinea. Photo by George Weiblen.

This book is dedicated to Tom Wood, whose abiding love of the Membracidae led him to study them in all their measure, and who was all that a naturalist, a mentor, and a friend should be.

CONTENTS

PART II
Co- and Macroevolutionary Radiation

CONTRIBUTORS

WARREN G. ABRAHAMSON Department of Biology, Bucknell University, Lewisburg, Pennsylvania 17837
abrahmsn@bucknell.edu

LYNN S. ADLER Department of Plant, Soil, and Insect Science, University of Massachusetts, Amherst, Massachusetts 01003
lsadler@ent.umass.edu

ANURAG A. AGRAWAL Department of Ecology and Evolutionary Biology, Cornell University, Carson Hall, Ithaca, New York 14853
aa337@cornell.edu

MAY R. BERENBAUM Department of Entomology, University of Illinois, Urbana, Illinois 61801
maybe@life.uiuc.edu

CATHERINE P. BLAIR Department of Biology, Bucknell University, Lewisburg, Pennsylvania 17837
cblair@bucknell.edu

CAROL L. BOGGS Department of Biological Sciences, Stanford University, Stanford, California 94305-5020
cboggs@stanford.edu

MARIE BUTCHER Integrative Biology, The University of Texas at Austin, Austin, Texas 78712-0253

YVES CARRIÈRE Department of Entomology, University of Arizona, Tucson, Arizona 85721-0036
ycarriere@ag.arizona.edu

WENDY L. CLEMENT Department of Plant Biology, University of Minnesota, St. Paul, Minnesota 55108

REGINALD B. COCROFT Division of Biological Sciences, University of Missouri, Columbia, Missouri 65211
cocroftr@missouri.edu

TIMOTHY P. CRAIG Department of Biology, University of Minnesota–Duluth, Duluth, Minnesota 55812
tcraig@d.umn.edu

MARY ELLEN CZESAK Department of Biology, Vassar College, Poughkeepsie, New York 12604-0133
maczesak@vassar.edu

ROBERT F. DENNO Department of Entomology, University of Maryland, College Park, Maryland 20742
rdenno@umd.edu

PAUL R. EHRLICH Department of Biological Sciences, Center for Conservation Biology, Stanford University, Stanford, California 94305-5020

JEFFREY L. FEDER Department of Biological Sciences, University of Notre Dame, Notre Dame, Indiana 46556-0369
jeffrey.l.feder.2@nd.edu

PAUL P. FEENY Department of Ecology and Evolutionary Biology, Cornell University, Ithaca, New York 14853

ANDREW A. FORBES Department of Biological Sciences, University of Notre Dame, Notre Dame, Indiana 46556-0369
aforbes@nd.edu

ROBERT S. FRITZ Department of Biology, Vassar College, Poughkeepsie, New York 12604-0133
fritz@vassar.edu

DANIEL J. FUNK Department of Biological Sciences, Vanderbilt University, Nashville, Tennessee 37235
daniel.j.funk@vanderbilt.edu

DOUGLAS J. FUTUYMA Department of Ecology and Evolution, State University of New York at Stonybrook, Stony Brook, New York 11794-5245
futuyma@life.bio.sunysb.edu

SARA HAWKINS Integrative Biology, The University of Texas at Austin, Austin, Texas 78712-0253

DAVID HAWTHORN Department of Entomology, University of Maryland, College Park, Maryland 20742

CRIS HOCHWENDER Department of Biology, University of Evansville, Evansville, Indiana 47722
ch81@evansville.edu

RANDY E. HUNT Department of Biology, Indiana University Southeast, New Albany, Indiana 47150

JOANNE K. ITAMI Department of Biology, University of Minnesota–Duluth, Duluth, Minnesota 55812
jitami@d.umn.edu

NIKLAS JANZ Department of Zoology, Stockholm University, SE-106 91 Stockholm, Sweden
niklas.janz@zoologi.su.se

JOHN L. MARON Division of Biological Sciences, University of Montana, Missoula, Montana 59812
john.maron@mso.umt.edu

RODRIGO J. MERCADER Department of Entomology, Michigan State University, East Lansing, Michigan 48824

CHARLES MITTER Department of Entomology, University of Maryland, College Park, Maryland 20742

KAILEN A. MOONEY Department of Ecology and Evolutionary Biology, University of California, Steinhaus Hall, Irvine, California 92697
mooneyk@tritrophic.org

PATRIK NOSIL Zoology Department and Centre for Biodiversity Research, University of British Columbia, Vancouver, British Columbia V6T 1Z4 Canada

SÖREN NYLIN Department of Zoology, Stockholm University, SE-106 91 Stockholm, Sweden
soren.nylin@zoologi.su.se

GABE J. ORDING Department of Entomology, Michigan State University, East Lansing, Michigan 48824

DIANA M. PERCY UBC Botanical Garden and Centre for Plant Research, University of British Columbia, Vancouver, British Columbia V6T 1Z4 Canada

MERRILL A. PETERSON Biology Department, Weston Washington University, Bellingham, Washington 98225

PETER W. PRICE Department of Biological Sciences, Northern Arizona University, Flagstaff, Arizona 86011-5640
peter.price@nau.edu

GEORGE K. RODERICK Department of Environmental Science, Policy, & Management, University of California at Berkeley, Berkeley, California 94720-3112
roderick@nature.berkeley.edu

RAFAEL L. RODRÍGUEZ Department of Biological Sciences, University of Wisconsin–Milwaukee, Milwaukee, Wisconsin 53201

J. MARK SCRIBER Department of Entomology, Michigan State University, East Lansing, Michigan 48824
scriber@msu.edu

SUMMER I. SILVIEUS Department of Biology, Aquinas College, Grand Rapids, Michigan 49506
silvisum@aquinas.edu

MICHAEL C. SINGER Integrative Biology, The University of Texas at Austin, Austin, Texas 78712-0253
sing@mail.utexas.edu

MICHAEL S. SINGER Department of Biology, Wesleyan University, Middletown, Connecticut 06459-0170
msinger@wesleyan.edu

BRUCE E. TABASHNIK Department of Entomology, University of Arizona, Tucson, Arizona 85721-0036
brucet@ag.arizona.edu

JOHN N. THOMPSON Department of Ecology and Evolutionary Biology, University of California at Santa Cruz, Santa Cruz, California 95064
thompson@biology.ucsc.edu

KELLEY J. TILMON Plant Science Department, South Dakota State University, Brookings, South Dakota 57007-2141
kelley.tilmon@sdstate.edu

MONTSERRAT VILÀ Estacion Biologica deDonana, Avda Parque de Mana Luisa, Pabellon del Peru, 41013 Sevilla, Spain

MATTHEW R. WEAVER Department of Entomology, University of Maryland, College Park, Maryland 20742

BRIAN WEE Department of Ecology, Evolution and Behavior, The University of Texas at Austin, Austin, Texas 78712-0253

GEORGE D. WEIBLEN Department of Plant Biology, University of Minnesota, St. Paul, Minnesota 55108
gweiblen@umn.edu

ISAAC S. WINKLER Department of Entomology, University of Maryland, College Park, Maryland 20742
isw@umd.edu

PREFACE

This book examines the evolutionary biology of herbivorous insects, including their relationships with their host plants and natural enemies. It is a compendium pulling together many aspects of evolutionary study at different levels of biological organization, from individuals to clades.

The inspiration for this book was a symposium I organized for the 2003 annual meeting of the Entomological Society of America, in memory of the late Thomas K. Wood, who worked on sympatric speciation in *Enchenopa* treehoppers. My goal for the symposium was to bring together people working on insect-plant evolution from several different perspectives, including those working on ecological and population genetics, behavior, speciation, macroevolution, and systematics. In chapter 16, Thompson notes that there has historically been a "conceptual tension between fields of study that focus on current selection, such as evolutionary ecology and population genetics, and fields that focus on higher-level patterns in the diversification of life, such as systematics and paleobiology." During the symposium, there was a noticeable synergy that came from having people speak about evolutionary topics normally considered somewhat independent subdisciplines, but all united by the common theme of insect-plant interactions. These speakers were the nucleus of contributors to this volume, which doubled in chapter number after the inclusion of additional topics and authors. It is my hope that the common theme of insect-plant evolution will help readers to link ideas from micro- and macroevolutionary levels of study to overcome the conceptual tension that exists between them, and that this book will thereby contribute to a more holistic understanding of evolutionary process.

Chapters are organized into three sections: Evolution of Populations and Species; Co- and Macroevolutionary Radiation; and Evolutionary Aspects of Pests, Invasive Species, and the Environment. The first section focuses on evolutionary mechanisms that maximize the fitness of individuals and the populations they comprise, and mechanisms that drive populations to speciation. The second section concentrates on the evolutionary forces that generate and maintain insect and plant biodiversity. Chapters in the third section draw on concepts developed in the first two, but relate specifically to insect-plant systems evolutionarily influenced—whether intentionally or inadvertently—by human activity. These three sections represent a logical continuum of inference about whole-organism evolutionary study. Evolution proceeds at the population level in an ecological context; species and speciation are at the crux of the evolutionary process in that they represent both the outcome of ecological and population genetic processes and the basic units of phylogenetic pattern; macroevolutionary processes and patterns build from the evolutionary process as a whole; finally, evolutionary study at various levels contributes to our understanding of practical biological problems that face society.

Although the 23 chapters in this volume are divided, by organizational necessity, among sections, certain ideas and themes run throughout the book, occurring again and again. One such unifying theme is the importance of host-plant range and host-plant shifts in the evolution of herbivorous insects. Most insect herbivores have very intimate associations with their host plants. This characteristic is biologically and ecologically defining and has evolutionary implications at all levels. Some authors examine host-plant range at a genetically or ecologically mechanistic level. For example, Berenbaum and Feeny (chapter 1) look at the biochemical and genetic links between adult ovipositional preference and offspring performance that help determine host-plant range. Craig and Itami (chapter 2) examine adult and offspring preference and performance, and host range in light of the ecological characteristics of plant resources, and M. C. Singer, Wee, Hawkins, and Butcher (chapter 22) discuss the evolution of host range in response to ecologically changing plant communities altered by human activity. M. S. Singer (chapter 3) looks at how a broad host-plant range can enhance individual herbivore fitness through escape from natural enemies.

Most of the herbivorous insect speciation mechanisms examined in detail in section I (Cocroft, Rodriguez, and Hunt, chapter 7; Feder and Forbes, chapter 8; Funk and Nosil, chapter 9; Futuyma, chapter 10) center around host-plant shifts and host-use patterns (but see Scriber, Ording, and Mercader, chapter 6, for an exception, where life-history variation across hybrid zones is more important than host-plant use). Host shifts and host-plant range also feature prominently in chapters discussing broader patterns of evolutionary radiation. Roderick and Percy (chapter 11) use the special case of island systems to evaluate the relative importance of host shifts and geographic isolation in cladogenesis. Janz and Nylin (chapter 15) discuss how oscillations in host-plant range (with periods of wider host use followed by periods of specialization) may promote high net speciation rates in herbivorous insect clades; Thompson (chapter 16), how local patterns of rapid and shifting adaptation to novel plants are important for the long-term maintenance of broad insect-plant associations. And Winkler and Mitter (chapter 18) compile a literature review of heroic scale, providing a broad phylogenetic perspective on maintenance of and change in host-use patterns by herbivorous insects. In addition, they have made their data compilation and analyses available online as a resource for the biological community.

Host shifts are also shown to be important for generating biodiversity not just of herbivorous insects, but also of their natural enemies. Silvieus, Clement, and Weiblen (chapter 17) write about historical patterns of host-plant associations and cospeciation among three trophic levels, and Abrahamson and Blair (chapter 14) discuss how herbivore radiation in response to host-plant shifts can cascade to the next trophic level, influencing natural enemy radiation.

Other themes also appear, implicitly or explicitly, across many chapters. The evolution of resistance and defense—of either plants or their herbivores—is an important recurring concept. Adler (chapter 12) applies a fresh approach to the study insect–floral plant coevolution by examining how pollinators can influence the evolution of plant resistance traits (and, conversely, how herbivores can affect floral traits). Czesak, Fritz, and Hochwender (chapter 5) discuss the genetic architecture of resistance from the perspective of the plant, with the insect herbivore as the agent of selection. Tabashnik and Carrière (chapter 19) also deal with the genetic architecture of resistance, but from the insect perspective, where the plant is the agent of selection, in their discussion of resistance evolution of crop pests to genetically modified plants. Maron (chapter 20) also writes about the practical implications of resistance evolution, in terms of biological control and how resistance evolves or is lost in invasive plants (noxious weeds) exposed to altered herbivore landscapes. These two chapters, like many of those in section III, are examples of the emerging field of applied evolution.

The ability of organisms to express different traits in response to different environments or conditions is an important mediating factor in the evolution of adaptive responses and also the generation of community structure. The role of phenotypic plasticity in evolution is another concept developed in this book, often in the context of trade-offs between opposing selective forces. Mooney and Agrawal (chapter 4) write about this specifically, and the concept of plasticity appears in other chapters as well, for example, in those by M. C. Singer (chapter 3), Cocroft et al. (chapter 7), and Adler (chapter 12). Conversely, Price (chapter 13) discusses the impact of morphological constraints and a lack of plasticity on evolution in phylogenetic groups, and the ecological repercussions for those groups.

The recent interest in landscape factors in ecological fields of study is also apparent in evolutionary study, and several chapters in this volume deal implicitly or explicitly with the impact of spatial scale, gene flow, and dispersal on the evolution of biodiversity. For example, spatial considerations relative to cladogenesis are important in systems discussed by Thompson (chapter 16) and Roderick and Percy (chapter 11). Denno, Peterson, Weaver, and Hawthorne (chapter 21) discuss the evolution of dispersal ability in an introduced insect herbivore, in the context of variation in habitat persistence. And M. C. Singer et al. (chapter 22) and Boggs and Ehrlich (chapter 23) discuss the evolutionary implications for insects metapopulations in habitat patches altered by human activity.

Running throughout this book is the idea that the evolution of both herbivorous insects and the plants with which they interact are inexorably linked, and that understanding the ecological framework of these interactions is fundamental to understanding their evolution. Also evident in every chapter is the authors' deep regard for the pursuit of whole-organism biology and the intrinsic value they recognize and reveal in the species they study. Tom Wood, to whom this volume is dedicated, was motivated by the same love of the natural world, and it has been a privilege to work with this group of naturalists in his memory. They are an exciting combination of authors who have a whole career of perspective on their topics, and relatively new investigators at the cutting edge of this field. They have written about long-term study systems to which they and their coinvestigators have contributed a depth of knowledge that spans decades, and also about emerging systems that push our knowledge in new directions. I wish to thank here the authors of this volume for their contributions to it.

I also wish to thank the Entomological Society of America for sponsoring the symposium that inspired this work;

Deirdre Prischmann, Ana Micijevic, and other members of my lab group who helped with manuscript preparation; Jason Owens, who made my life happier as I worked on this project; Chuck Crumly, his assistant Francisco Reinking, and the production staff at University of California Press, who worked with me to make this book a reality on its long journey from symposium to shelf; and all the reviewers who have helped to improve it along the way.

Kelley Jean Tilmon
South Dakota State University
November 2006

PART I

EVOLUTION OF POPULATIONS AND SPECIES

ONE

Chemical Mediation of Host-Plant Specialization: The Papilionid Paradigm

MAY R. BERENBAUM AND PAUL P. FEENY

Understanding the physiological and behavioral mechanisms underlying host-plant specialization in holometabolous species, which undergo complete development with a pupal stage, presents a particular challenge in that the process of host-plant selection is generally carried out by the adult stage, whereas host-plant utilization is more the province of the larval stage (Thompson 1988a, 1988b). Thus, within a species, critical chemical, physical, or visual cues for host-plant identification may differ over the course of the life cycle. An organizing principle for the study of host-range evolution is the preference-performance hypothesis (Jaenike 1978). According to this hypothesis, ovipositing females should maximize their fitness by selecting plants on which offspring survival will be high; in other words, over a range of potential host plants, adult female preference should be correlated with larval performance. Evidence in support of this hypothesis has been mixed (Thompson 1988a, 1988b; Scheirs and deBruyn 2002). Poor correlations have been attributable in some cases to opposing evolutionary forces that include conflicting selection pressures exerted by trophic factors such as predation and parasitism (Camara 1997; Heisswolf et al. 2005); ecological factors such as relative rarity of most-suitable hosts (Rausher 1980), recent invasion of less-suitable hosts into an otherwise coevolved community (Wiklund 1975), or thermal constraints on voltinism patterns (Scriber 1996, 2002); phenotypic plasticity of host-plant selection (Mercader and Scriber 2005); or parent-offspring conflict in the case of species in which adults are also phytophagous (Roitberg and Mangel 1993). Although positive correlations are frequently found in other taxa (viz., Craig and Itami, this volume), such correlations are exceptional in lepidopterans.

The dichotomy between adult behavioral preference and larval physiological performance in Lepidoptera is key to longstanding debates over the evolution of specialization in general. The timing and nature of adaptation to plant chemistry throughout the life cycle are central to these debates. Almost 60 years ago, Dethier (1948) suggested that "the first barrier to be overcome in the insect-plant relationship is a behavioral one. The insect must sense and discriminate before nutritional and toxic factors become operative" (p. 98). Thus, Dethier argued for the primacy of adult preference, or detection and response to kairomonal cues, in host-plant shifts. In contrast, Ehrlich and Raven (1964) reasoned that "after the restriction of certain groups of insects to a narrow range of food plants, the formerly repellent substances of these plants might . . . become chemical attractants" (p. 602), arguing for the primacy of shared allomonal phytochemistry and larval detoxification in initiating host shifts (with the evolution of kairomonal responses following). Oviposition "mistakes," in which females lay eggs on host plants that do not support larval growth (e.g., Berenbaum 1981b), are cited as evidence for the importance of behavioral cues in initiating host shifts; colonization of novel host plants that share the same range of defense compounds is claimed as evidence of the primacy of larval performance in the process (see Berenbaum 1990 for a review).

In view of the fact that the physiological and behavioral changes necessary to effect a host shift arise by random mutation, primacy is in a sense irrelevant; host shifts are completed when the necessary behavioral and physiological traits are in place, irrespective of the order in which they occur (Berenbaum 1990). Complicating the resolution of this issue are recent experimental studies indicating that preference and performance traits involved in host shifts may not be entirely under heritable genetic control and are instead influenced by epigenetic and even nongenetic factors. Evidence suggests that larval feeding experience in at least some taxa may influence adult oviposition preference either via "chemical legacy" (retained chemical signals) or via some form of retained memory (Barron 2001; Rietdorf and Steidle 2002; Akhtar and Isman 2003; but see van Emden et al. 1996). No real resolution of such debates can take place until the biochemical and genetic bases of preference and performance are elucidated for a broad array of species.

How Lepidopterans Prefer

A priori prediction of which chemicals influence oviposition preference and larval performance and whether linkages in chemical perception or processing between developmental stages exist is hampered by the fact that, within the Lepidoptera, caterpillars and adults experience very different chemical environments, with very different attendant behavioral and physiological consequences. Adult females, with relatively few exceptions, search for host plants while flying; complementing visual cues for identifying host plants are volatile odorant cues. Long-distance orientation to host plants by female butterflies is generally mediated by volatile signals detected by antennae (Feeny et al. 1989); in that this process involves upwind anemotaxis, it might be expected to resemble pheromone plume-following behavior and rapid antennal processing in males. Male moths orienting in an odor plume benefit from the ability to assess changes in ambient pheromone concentrations and accordingly rely on rapid processing of signals to facilitate rapid assessment (Wang et al. 1999). How female butterflies process volatile kairomonal cues is virtually unexplored.

Once host plants are encountered, suitability assessment generally involves contact chemoreception, mediated by tarsal and in some cases ovipositor receptors. Oviposition kairomonal cues tend to be contact chemical cues restricted to plant surfaces or tissues close to the surface (Heinz and Feeny 2005). In contrast with volatile cues, rapid clearance may not be of such paramount importance in the absence of upwind anemotaxis, nor is directionality of the signal at issue. As well, plant chemicals present on the surface constitute a small subset of the total allelochemical inventory of the plant, and these chemicals tend to be present on the plant surface in concentrations orders of magnitude smaller than those in internal tissue (Brooks et al. 1996; Brooks and Feeny 2004).

Signal sorting differs qualitatively and quantitatively between immature and adult stages of Lepidoptera. The ability to fly exposes ovipositing females to a far more chemically complex environment than that experienced by the vast majority of lepidopteran larvae. Adult females must locate and identify particular host plants within a community of dozens or even hundreds of potential nonhosts; the ability to differentiate among complex chemical signals is at a premium. In contrast, caterpillars, with few exceptions, are restricted to walking and thus "sample" a vastly smaller subsection of the plant community. In most species, caterpillars remain on the plant on which they hatch and are rarely if ever called upon to differentiate among potential host-plant species. Host-plant acceptance is mediated primarily by contact chemoreception. Yet there is a fundamental difference even in contact chemoreception between adult and larval lepidopterans; whereas "tasting" by ovipositing females is intermittent and brief, foraging larval lepidopterans may be in constant contact with surface chemicals. Moreover, for species hatching from eggs laid within plant tissues, neonates are essentially immersed in a chemical cauldron and are potentially exposed to locally high concentrations of phytochemical cues (Kreher et al. 2005).

The neurological burden imposed by the necessity of differentiating among host and nonhost signals has been suggested to represent a constraint on the evolution of host preferences in insects (Bernays 2001). In what was purportedly a test of the so-called information processing hypothesis, Janz and Nylin (1997) compared the discriminatory capacity of two monophagous nymphalids, *Polygonia satyrus* and *Vanessa indica*, with that of the polyphagous nymphalids *P. c-album* and *Cynthia cardui*. As adults, the specialists were better able to distinguish and discriminate against "bad-quality" nettle than were the generalists, consistent with the idea that generalists are well equipped to sort through a large number of signals but less capable of evaluating subtle differences in particular signals. In this study, the fact that larvae of the generalist species displayed no growth impairment on the "poor-quality leaves" makes interpreting these results more ambiguous in that so-called poor quality did not manifest itself in performance differences.

How Lepidopteran Larvae Perform

Allomones, or deterrent compounds, as well as kairomones, or attractants, play a role in determining preference and performance in both larval and adult Lepidoptera. Toxicological consequences of exposure to plant chemicals, however, differ dramatically between larval and adult stages, due to qualitative differences in composition of toxins and quantitative differences in toxicological loads. Whereas the exposure of an ovipositing female to particular host-plant chemicals may be on the order of seconds or minutes, caterpillars are exposed to host-plant chemicals more or less continuously for the duration of larval life, which can extend for days or weeks. Caterpillars that are capable of ingesting their body weight in plant material are potentially exposed to enormous quantities of compounds that often possess detrimental biological activities at high concentrations. Moreover, by virtue of consuming leaf or other plant tissue in its entirety, in all but a few cases caterpillars are exposed to a much broader inventory of host-plant chemicals. Skeletonizers and leaf miners may avoid indigestible plant material, but, even so, their intake of plant allomones is extremely likely to exceed that of ovipositing adults.

Although chemical complexity has been thought to act as a constraint on ovipositing adults via limits on neural information processing, the idea that chemical complexity acts as a constraint on larvae via limits on detoxification processing has received considerably less attention of late. There has been a longstanding assumption that the principal detoxification enzymes in Lepidoptera (e.g., cytochrome P450 monooxygenases) are broadly substrate specific, as they generally are in vertebrates (Berenbaum 1999), and thus well

TABLE 1.1

Phenotypic Correlations Between Leaf and Fruit Furanocoumarin Content of *Pastinaca Sativa* (From Berenbaum et al. [1986], Zangerl et al. [1997], and Unpublished Data). $N = 124$, Values are Correlation Coefficients, and Significant Correlations ($p < 0.05$) are in Bold

	Seed Content				
Leaf Content	Imperatorin	Bergapten	Isopimpinellin	Xanthotoxin	Sphondin
Imperatorin	−0.022	0.016	0.033	0.032	**0.258**
Bergapten	0.016	0.045	−0.068	0	−0.003
Isopimpinellin	0.046	−0.051	−0.157	0.01	0.067
Xanthotoxin	−0.031	−0.106	−0.142	−0.077	0.035
sphondin	−0.061	0.006	−0.032	−0.156	0.134

designed to accommodate novel substrates; indeed, this idea is the basis for the hypothesis that enzymes that process plant allelochemicals are preadaptations for the evolution of insecticide resistance (Gordon 1961). In fact, as more detoxification enzymes are characterized at the biochemical and molecular level, it has become abundantly clear that broad substrate specificity is the exception, particularly in oligophagous species (Berenbaum 2002; Mao et al. 2005).

Thus, chemically mediated interactions between host plant and lepidopteran differ profoundly over the course of development. In addition, by virtue of the fact that in essentially all plants chemistry changes with time and development, ovipositing females must make choices based on chemicals that may not be predictive of the chemistry that will be experienced by their progeny. The time between egg deposition and egg hatch can be on a scale of hours, in which case chemical changes in host plants may be trivial. However, for eggs that undergo seasonal diapause, the time displacement between oviposition and hatch can be on a scale of months, in which case chemical changes in host plants may be dramatic. The parsnip webworm *Depressaria pastinacella*, for example, feeds as a caterpillar exclusively on the developing buds, flowers, and fruits of the biennial forb *Pastinaca sativa*. Larval performance in this species is profoundly affected by furanocoumarin content of the reproductive tissues of its host plants (Berenbaum and Zangerl 1993), yet the number of eggs laid by ovipositing females is independent of the furanocoumarin chemistry of foliage (Zangerl and Berenbaum 1992). Adult females oviposit in late spring before this biennial plant has produced a flowering stalk; the only tissue available at the time of oviposition for assessment is foliage. Furanocoumarin content of foliage present before production of the flowering stalk is in virtually no case significantly correlated with furanocoumarin chemistry of the reproductive structures in this plant (Table 1.1). Larvae of this species do respond behaviorally to octyl esters, which are present in reproductive tissues that serve as food and absent in foliage, even though, upon ingestion, these compounds are toxic (Carroll et al. 2000). The within-plant distribution of these compounds may allow larvae to move from oviposition sites on the leaves to feeding sites within the reproductive structures. This example also illustrates the difficulty of classifying plant infochemicals as either kairomones (host-recognition cues) or allomones (plant-defense compounds).

Preference-Performance Relationships in Lepidoptera

An enormous amount of information exists about preference and performance in lepidopterans (Table 1.2); over a dozen studies have been conducted to determine the relationship between oviposition preference and larval performance. In over half of these studies, no correlation could be found. The lack of correlation may result from any number of possible explanations, not the least of which is the difficulty in measuring these attributes. Thompson's (1988a, 1988b) definition of preference as the "hierarchical ordering of plant species by ovipositing females when the plants are presented in equal abundance and availability" (p. 4) is widely but not universally accepted (Singer et al. 1992). As such, "preference" is in reality a series of behaviors, including orientation in flight, decision to land, and decision to oviposit. "Performance" is even more nebulous experimentally, generally measured by survival but also by such fitness proxies as development time, growth rate, or pupal weight (Moreau et al. 2006).

Operational problems notwithstanding, the underlying genetics of preference and performance characterized in numerous systems offer little support for a mechanistic link between these attributes. In general, interspecific differences in female preference are associated with the X chromosome (Thompson 1988a, 1988b; Thompson and Pellmyr 1991); in Lepidoptera the female is the heterogametic sex, and in reciprocal crosses oviposition preferences map onto the male source population (Scriber, Giebink, and Snider, 1991; Thompson and Pellmyr 1991). X-linked traits are thought to evolve rapidly if mutations are recessive because such traits are exposed to selection in the heterogametic sex (Charlesworth et al. 1987); this idea has lent support to the idea that adult preference is

TABLE 1.2
Tests for Significant Positive Correlations Between Preference (pref) and Performance (perf) Involving Lepidopterans

Family	*Taxon*	*Pref/perf Correlation*	*Author*
Papilionidae	*Papilio machaon*	No	Wiklund 1974
	Papilio glaucus	Yes[a]	
		No[b]	Bossart 2003
			Scriber 1996, 2002
	Battus philenor	No	Rausher 1980
	Eurytides marcellus	Yes	Damman and Feeny 1988
Pieridae	*Ascia monuste*	No	Carta-Preta and Zucoloto 2003
	Eucheira socialis	No	Underwood 1994
	Pieris rapae crucivorae	Yes	Chen et al. 2004
	Pieris rapae melete,napi	Yes	Ohsaki and Sato 1999
Nymphalidae	*Polygonia c-album*	Yes	Nylin and Janz 1996
	Polygonia c-album	No	Nylin and Janz 1996
	Junonia coenia	No	Prudic et al. 2005
	Junonia coenia	No	Camara 1997
	Melitoea cinxia	No	Van Nouhuys et al. 2003
	Danaus plexippus	Yes	Mattila and Otis 2003
	Danaus plexippus	Yes	DiTommaso and Losey 2003
	Danaus plexippus	No	Ladner and Altizer 2005
Lycaenidae	*Polyommatus icarus*	Yes	Bergstrom et al. 2004
	Mitoura spp.	Yes	Forister 2004
	Glaucopsyche lygdamus		Carey 1994
Noctuidae	*Spodoptera exigua*	No	Berdegue et al. 1998

[a]Ohio population.
[b]Florida population.

key to initiating host shifts. In addition, the disproportionate influence of X-linked traits on oviposition preferences may facilitate the evolution of adaptive gene complexes due to reduced recombination (Jaenike 1989). Within-species variation in preference, however, does not appear to be predominantly sex-linked (Janz 1998). Low levels of additive genetic variance for within-species variation in oviposition preference argue for involvement of major genes. In the rare cases in which larval preference has been genetically mapped, it is not necessarily genetically correlated with adult preference, although it can be X-linked (Nylin and Janz 1996).

Chemical Mediation of Preference and Performance: Papilionids as Paradigm

Despite the fact that butterflies have been the subject of chemical ecology studies for over a century, the chemistry of host-plant preference and performance has been elucidated for only a handful of groups. One group in which host-utilization patterns are closely associated with particular plant allelochemicals is the family Papilionidae, the swallowtail butterflies (Fig. 1.1). The swallowtails have been a paradigm group for the study of chemical coevolution for over 50 years (Feeny 1991, 1992; Scriber et al. 1995). In this cosmopolitan family with over 500 species, host-plant use patterns are remarkably conservative. Only 21% of the 281 species with known host associations use more than a single plant family (Scriber, Lederhouse, and Hagen 1991). Of the plant families utilized by papilionids, five dominate; these are the Annonaceae, Apiaceae, Aristolochiaceae, Lauraceae, and Rutaceae. Two tribes, Troidini and Zerynthiini, are restricted to Aristolochiaceae; caterpillars in these groups sequester toxic aristolochic acids from their host plants and acquire chemical defense against predators that is carried over to the adult stage. In contrast, over 75% of species in the tribe Papilionini, comprising the genus *Papilio*, utilize host plants in the families Rutaceae and Apiaceae, with the North American *P. machaon* complex of approximately a dozen species account-

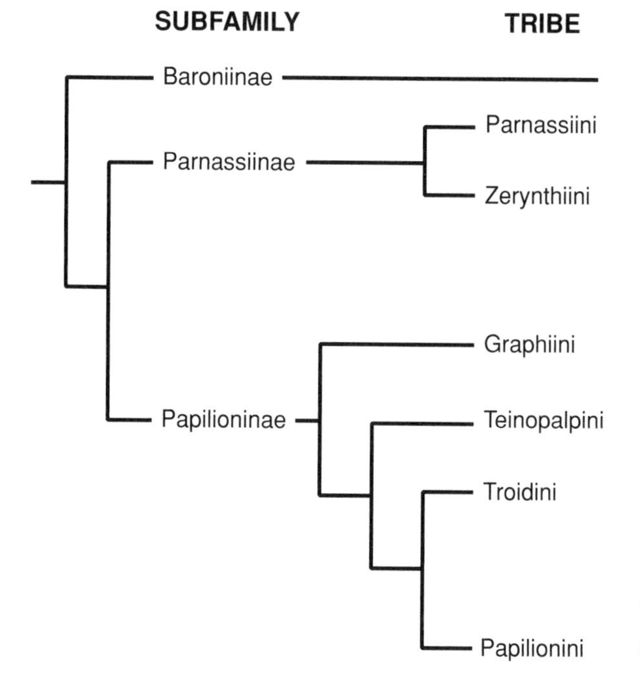

FIGURE 1.1. Phylogeny of the family Papilionidae, based on Hancock (1983), Miller (1987), Caterino and Sperling (1999), and Caterino et al. (2001). Figures in parentheses indicate approximate numbers of species in each genus. Not listed are several small genera, each containing one or two species. Recent molecular evidence (Caterino et al. 2001) suggests that the subfamily Parnassiinae is not monophyletic and that the tribe Zerynthiini, with or without *Luehdorfia* (cf. Zakharov et al. 2004), and the subfamily Papilioninae may be sister taxa.

ing for most of the specialists on Apiaceae (Sperling 1987). The genus *Papilio* also contains the most polyphagous of papilionids: species in the *P. glaucus* complex. *Papilio glaucus* and *P. canadensis* have distinct but overlapping host ranges encompassing several families of early successional trees, including species in Rosaceae, Oleaceae, and Tiliaceae.

Host Search and Oviposition

Long-distance orientation to host plants by papilionids involves the processing of visual cues. Females of the zebra swallowtail, *Eurytides marcellus*, approach understory shrubs of various nonhost species as well as shrubs of their larval host plants (*Asimina* spp.). After close-range inspection by a slowly fluttering female, a nonhost shrub is about twice as likely as is a host shrub to be rejected by a female without landing. Nevertheless, about one-third of nonhosts approached are not rejected until after a female landed and had access to contact chemical cues (Damman and Feeny 1988). Although *E. marcellus* females are strongly attracted to host-plant volatiles, at least at short range (Haribal and Feeny 1998), the initial process of filtering potential oviposition sites from surrounding vegetation is strongly dependent on responses to visual cues such as the ratio of young to mature leaves and the presence of leaf buds (Damman and Feeny 1988).

Females of the pipevine swallowtail, *Battus philenor*, in southeast Texas lay small batches of eggs on two herbaceous *Aristolochia* species in the forest understory. Individual females can be found landing preferentially either on narrow-leaved or on broad-leaved plants, the large majority of which are nonhosts (Rausher 1978). A female's leaf shape preference is determined by her recent contact experience with the narrow-leaved host, *A. serpentaria*, or the broad-leaved host, *A. reticulata*. Contact alone, even without oviposition, is sufficient to bring about this associative learning (Papaj 1986). Since the relative availability of the two host species as suitable oviposition substrates changes seasonally, short-term learning of leaf shapes can result in greater searching efficiency (Rausher 1995).

Papilio species possess six classes of spectral receptors, including UV, violet, blue, green, red, and broadband (Arikawa 2003). Physiological receptivity to colors is reinforced by chemical cues; *B. philenor* displays the capacity to learn to associate different color cues with sucrose (representative of nectar plant chemistry) and leaf extracts (representative of oviposition substrate chemistry). Landings and probes on nontrained colors were biased toward the training alternatives, suggesting neural limitations on processing visual information (Weiss and Papaj 2003), akin to that proposed for processing chemical information (Kelber and Pfaff 1999).

The role of olfaction in the oviposition behavior of swallowtails has remained less clear despite the early demonstration by Saxena and Goyal (1978) that volatile extracts of the host plant *Citrus limettioides* (lime) elicited significant oviposition by *P. demoleus* females. In laboratory bioassays with model plants, females of the black swallowtail, *P. polyxenes*, were significantly more active when host volatiles were added to artificial leaves treated with contact stimulant extracts. They also landed more frequently and laid more eggs (Feeny et al. 1989). Activity levels and landing rates were significantly reduced, however, in the presence of volatiles from cabbage, a nonhost (Feeny et al. 1989). In comparable experiments with *E. marcellus*, females were highly responsive to model leaves treated with both contact stimulants and volatiles but volatiles alone were sufficient (in combination with visual cues) to stimulate some oviposition (Haribal and Feeny 1998). Recently, Heinz (2002) has shown that responses to volatiles by *P. polyxenes* females are largely innate but that the females are capable of learning combinations of volatile and leaf-shape cues.

Once a female has landed on a leaf, contact stimulants are of crucial importance in determining whether a potential oviposition site is accepted or rejected (Städler 1992; Renwick and Chew 1994). The stimuli are perceived by tarsal contact chemoreceptors during drumming behavior, in which the females "taste" the leaf surface with their forelegs (Ilse 1956; Feeny et al. 1983). In the pierid butterfly *Pieris brassicae*, Ma and Schoonhoven (1973) showed that each taste sensillum on the foretarsi of females contains five receptor cells, including at least one cell that responds to the glucosinolates that characterize the crucifer host plants. Among swallowtails, cells responding to contact stimulants from carrot, *Daucus carota*, have been detected in the foretarsi of *Papilio polyxenes* females (Roessingh et al. 1991). The stimulant activity of host-plant extracts typically results from synergistic interactions between several ingredients that are seldom active alone (Honda 1986; Feeny et al. 1988; Ohsugi et al. 1991; Sachdev-Gupta et al. 1993; Nishida 1995). Contact receptors in butterflies, including swallowtails, can also respond to compounds that inhibit rather than stimulate oviposition (Nishida et al. 1990; Sachdev-Gupta et al. 1990). Although a particular compound or class of compounds may contribute the major sensory cue, it is the total complex that forms the basis for perception and behavioral response (cf. Dethier 1976; Renwick and Chew 1994).

Responses of *P. polyxenes* females to contact chemical cues are innate and not altered by specific host-plant experience. Females responded to extracts of two host plants, carrot *(D. carota)* and poison hemlock *(Conium maculatum)* differentially but without regard to the plant previously experienced (Heinz and Feeny 2005). The postalighting response is unaffected by host-plant experience; however, Heinz (2002) has shown that the females are capable of learning all three major cue types (contact stimulants, volatile stimulants, and visual cues) in free-flight bioassays, especially when the cues were presented in pairs. Females also demonstrated learning in whole-plant bioassays, being most likely to approach and land on the same species with which they had prior oviposition experience (Heinz 2002). Such chemically mediated learning is likely to enhance the rate of egg-laying in an oligophagous species such as *P. polyxenes*, successive generations of which may encounter shifting arrays of potential host-plant species that differ significantly in chemistry and leaf shape (Heinz 2002).

Identifying the Chemical Cues: Kairomones and Allomones

Oviposition Kairomones: Volatile Attractants

Little is known about the identities of volatile compounds used as oviposition kairomones by swallowtails, probably because bioassays of volatiles are much more time-consuming than are those for contact stimulants and concentrations of active material are typically much smaller. Electroantennogram recordings revealed that the antennae of females of *P. polyxenes*, *P. machaon*, and *P. troilus* respond selectively to several compounds in the volatiles of their host plants (Baur and Feeny 1995). *Cis*- and *trans*-sabinene hydrate, 4-terpineol, and *cis*-3-hexenyl acetate are among the dozen or so components of carrot volatiles that evoke responses in *P. polyxenes* antennae, but their role in oviposition behavior is not yet known; several of the compounds evoke antennal responses in males as well as females (Baur et al. 1993). More likely candidates for volatile oviposition kairomones should perhaps be sought among the volatile phenylpropanoids and other compound classes that are more characteristic of the Apiaceae, Rutaceae, and other typical swallowtail host families.

Although they have to date eluded chemical characterization, it is becoming clear that volatile stimulants may be no less significant than contact kairomones in the evolution of host preference in swallowtails. First, there is increasing experimental evidence that host volatiles represent a major component of the oviposition-stimulant profiles to which many swallowtails respond. Second, the work of Heinz (2002) indicates that responses to volatile stimulants are heritable and hence potentially more conserved during swallowtail evolution than previously suspected. And third, the circumstantial evidence is compelling; the host plants of swallowtails are commonly aromatic, and both terpenoid and phenylpropanoid volatiles have been reported from virtually all the major food-plant families (Feeny et al. 1983 and references therein). Moreover, several reports of oviposition "mistakes" by swallowtails are suggestive of a role for a common theme in the volatile profiles of swallowtail food plants. Within the genus *Papilio*, in particular, there have been numerous observations of females of one species laying eggs on host plants used more typically by larvae of other species (references in Feeny et al. 1983). Stride and Straatman (1962), for example, found that females of *P. aegeus*, a Rutaceae feeder, would oviposit on *Cinnamomum camphora* (Lauraceae), a larval food plant of *Graphium* species. Berenbaum (1981b) found two eggs of *P. glaucus* on leaves of *Angelica atropurpurea*, a most unusual food plant for this tree-feeding generalist, but a potential food plant of *P. polyxenes* and other umbellifer-feeding *Papilio* species. D. A. West (1981, personal communication) found an egg of *P. troilus*, a Lauraceae-feeding species, on *Aristolochia macrophylla*, a host plant of the pipevine swallowtail, *Battus philenor*. In such examples, it seems more plausible to invoke a role for shared volatiles, rather than similar contact stimulants or visual cues, in triggering the mistakes.

Oviposition Kairomones: Contact Stimulants

Contact oviposition stimulants have been at least partially characterized for eight species (genus *Papilio*) in the tribe Papilionini, two species *(B. philenor* and *Atrophaneura alcinous)* in the tribe Troidini, and one species each in the

teristic of ancestral plants could arise in at least two ways: ancestral detoxification enzymes may be retained unmodified as new enzymes evolve by gene duplication events; alternatively, gene duplication events may lead to the acquisition of novel substrates without a loss of the ability to metabolize ancestral allomones because the genetic changes facilitating novel substrate acquisition occur in places in the protein that are not involved in ancestral allomone metabolism.

Preference and Performance Genes

Even less well understood than the chemical cues mediating preference and performance are the actual genes influencing the processing of chemical signals. Although the process by which host-plant shifts occur is widely recognized to involve genetic changes, documenting such genetic change has proved difficult. Examples are vanishingly rare, even among the well-studied butterflies. The idea that independent gene complexes regulate preference systems in larval and adult swallowtails has been embraced for over a quarter-century (Wiklund 1975), at least in part because of the abundance of evidence that host-plant ranges of larvae and adults are generally not congruent.

In swallowtails, what little genetic work has been done with respect to mapping preference and performance genes provides support for the idea of independent gene complexes. The adult oviposition preference hierarchy in *P. zelicaon* is controlled by at least one locus on the X chromosome and at least one locus that is on an autosome (Thompson 1988c, 1993; Thompson et al. 1990). Indications are that preference strength is not regulated by a locus on the X chromosome. In contrast, hybridization studies demonstrate no X-chromosome contribution to any component of larval performance; thus, X-linkage of preference and performance is eliminated as a possibility. Pupal mass and development time and survival on host and nonhosts may be Y-linked or the result of nongenetic maternal effects (Thompson et al. 1990).

In the polyphagous tiger swallowtail *P. glaucus*, Scriber, Giebink, and Snider (1991) demonstrated X-chromosome influence over oviposition preference. Bossart and Scriber (1995) evaluated variation in oviposition preference and found differences not only in relative fidelity to specific host plants, but also differences in the preference rank of particular host plants. Heritable variation was detectable in relative fidelity toward less-preferred hosts but not in preference rank. That host-plant choice is influenced by past experience and physiological state in this species argues against strict genetic correlation between adult preference and larval performance. Bossart (2003) examined covariance between preference and performance (as evidenced in the proxy measures of larval development time, relative growth rate, pupal mass, and mortality) in strains of *P. glaucus* that differ in fidelity—a locally monophagous population from Florida and a locally polyphagous population in Ohio. Covariance was detected independently within each population, although relative performance was consistent across host plants, independent of adult preferences. Although preference-performance correlations were in the expected direction in the Ohio population (whereby performance measures were correlated with preference for the typical host), in Florida, the correlation was in a direction opposite to that expected. Ovipositing females with the strongest preference for the unusual host plant *(Liriodendron tulipifera)* produced progeny attaining the greatest pupal weight. Based on her findings, "multiple genetic control mechanisms (e.g., pleiotropy and coaptation)" (p. 477), were invoked to account for preference-performance relationships (Bossart 2003).

Candidates

In terms of characterizing specific genes mediating host-plant specialization, one gene superfamily—the cytochrome P450 monooxygenases (P450s)—appears to play a role in both preference and performance in lepidopterans in general and swallowtails in particular. P450s catalyze the NADPH-associated reductive cleavage of oxygen to produce a functionalized product and water. The genes encoding these enzymes constitute one of the largest superfamilies known (Nelson et al. 1996; http://drnelson.utmem.edu/CytochromeP450.html). The enormous proliferation of these proteins is in part a reflection of their functional versatility. Reactions catalyzed by P450s include such oxidative transformations as monooxygenations, dehydrogenations, and peroxidase-type oxidations (Mansuy 1998), so P450s participate in a wide variety of biosynthetic and detoxification reactions. In insects, P450s are involved in the biosynthesis of pheromones, cuticular hydrocarbons, and hormones and are also responsible for metabolism and detoxification of exogenous substrates such as insecticides and plant allomones (Feyereisen 1999).

Gonzalez and Nebert (1990) suggested that the P450 superfamily began to diversify over 400 million years ago, concomitant with the colonization of terrestrial habitats by plants and herbivorous animals. Rapid diversification via a series of gene duplication events may have been a consequence of reciprocal selection pressures whereby plants evolved biosynthetically novel defense compounds, and insects (and other herbivores) overcame the toxins with novel detoxification pathways. Multiple gene duplication events allowed P450s to acquire new functions in the presence of new environmental stress factors while retaining ancestral metabolic capabilities. Protein diversification is, however, only one of several possible evolutionary mechanisms; evolution may also proceed by alterations in gene expression. In insects, P450 expression is regulated in response to a wide range of factors, including developmental stage and exposure to both endogenous and exogenous substances. Examining P450-mediated metabolism of host-plant chemicals within a group of closely related species with well-characterized host-utilization patterns can provide insights into how P450s evolve and diversify in response to environmental selective agents.

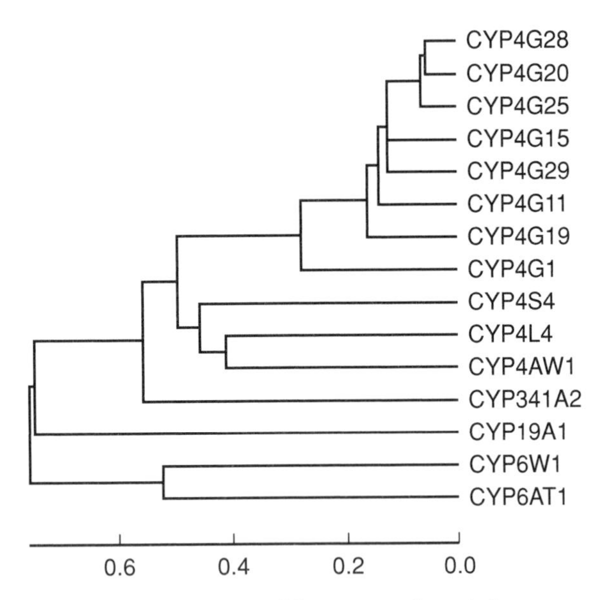

FIGURE 1.3. Simple amino acid distance tree of putatively chemosensory P450s (CYP4G28 from *Papilio polyxenes* [Mao et al. in preparation]; sequences other than *P. polyxenes* from Genbank).

P450s contribute to preference and performance in herbivorous insects in two ways. Along with glutathione-S-transferases and carboxylesterases, they are the principal detoxification system for metabolizing plant allelochemicals in caterpillar midguts (Berenbaum 2002). They are also key enzymes in chemosensory systems of adults, functioning as degradative enzymes that break down plant and other odorants bound to receptors, thereby regenerating these receptors (Vogt 2003). How P450s that function as degradative enzymes in insect chemoreception relate to those involved in larval midgut detoxification is subject to speculation; in no one species has the full inventory of P450s been characterized functionally. Perireceptor phenomena are not involved in the signal transduction cascade *per se* but rather are involved with the disposition of the chemical signal itself—uptake, delivery to neuronal receptors, and eventual release of chemicals to regenerate the system. In both vertebrate and invertebrate systems, proteins mediate these processes. Whereas odorant-binding proteins play a role in transporting relatively lipophilic signal substances through the aqueous medium bathing the sensory neuron to receptor proteins in the membrane, degradative enzymes may act extracellularly to facilitate termination of signals by breaking down the odorant. Among the odorant-degrading enzymes, including esterases, glutathione-S-transferases, epoxide hydrolases, and aldehyde dehydrogenases, P450s contribute to chemical signal clearance in both vertebrates and invertebrates (Gu et al. 1999) (Fig. 1.3). While the specificity of P450s involved in degrading pheromones has not been widely assessed, it is assumed that activity and specificity levels are high. With respect to insects, cytochrome P450 and NADPH-P450 reductase genes have been identified that are preferentially expressed in antennae of *Drosophila melanogaster* (Hovemann et al. 1997), suggesting a function in odorant clearance. Maibeche-Coisne et al. (2002) report that CYP4L4 is expressed preferentially in the antennae of the moth *Mamestra brassicae,* and CYP4S4 expression is limited to the sensilla trichodea in the antennae responsible for odor detection.

In the most compelling study to date implicating P450s in odorant degradation, Leal (2001) showed that enzymatic degradation of the sex pheromone of the scarabaeid beetle *Phyllopertha diversa* is mediated by P450s (Wojtasek and Leal 1999). Subsequent work (Maibeche-Coisne et al. 2004) led to the cloning of three P450s from male antenna cDNA: CYP4AW1, which is antenna specific, CYP4AW2, and CYP6AT1, which is antenna rich. To date, there have been no reports on female-specific expression of P450s associated with sensory organs involved in perceiving chemical cues from plants. For Lepidoptera, then, chemical mediation of preference and performance should converge when adults and larvae assess the same host-plant tissues—when oviposition is directly on tissues eventually consumed by the larvae and when the time between egg hatch and larval development is short (so that chemical changes associated with plant maturation are minimal). Kairomone-degrading P450s, by virtue of exposure to lower and less toxic concentrations of plant chemicals than larval midgut allomone-degrading P450s, should have broader substrate specificity. As well, kairomone-degrading P450s, by virtue of encountering only a subset of the chemical content of host-plant foliage, should be fewer in number than allomone-degrading P450s.

The nature of P450s contributing to allelochemical processing in both larval and adult preference remains a subject for speculation. To date, one P450, CYP4G28, has been characterized from female *Papilio polyxenes* adults that has highest identity (ca. 85%) with two other lepidopteran P450s in the CYP4 subfamily that have been implicated in chemosensory processing (Fig. 1.4). Expression patterns of CYP4G28 are consistent with a chemosensory function. Reverse transcription polymerase chain reaction analysis with head, thorax, abdomen, wing, antennae, and thorax transcripts showing highest levels of expression in tarsi and antennae, structures that in adult females are associated with detection of plant kairomones. The precise substrate specificity of CYP4G28 and, indeed, other CYP4 P450s expressed in tarsi awaits characterization.

P450s IN LARVAL SWALLOWTAIL PERFORMANCE

P450s in *Papilio* display differences both in constitutive activity against furanocoumarins and in inducibility in response to furanocoumarin ingestion (see Berenbaum 1995 for a review). In general, these differences correspond to the frequency of ecological exposure to furanocoumarins. The genus *Papilio* has historically been divided into five sections (Munroe 1961). Of these, sections II, III, and IV have some association with furanocoumarin-containing plants. Constitutive activity against xanthotoxin, a linear furanocoumarin, is high in *P. cresphontes*, a section IV specialist on linear furanocoumarin-containing Rutaceae, the putative ancestral host-plant family of the group. Activity is

FIGURE 1.4. Rooted tree of lepidopteran P450 proteins generated with the CLUSTALW program in Biology Workbench (ver. 3.2), employing defaults for all settings except that the "Blosum series" was employed for the weight matrix. Known or putatively xenobiotic-metabolizing P450s (involved in larval performance) are represented by dots, known or putatively kairomone-processing P450s (involved in adult preference) by long dashes, and P450s involved in processing endogenous substrates (e.g., hormone or pheromone synthesis) by short dashes.

high as well in *P. polyxenes* and *P. brevicauda*, section II specialists on Apiaceae containing furanocoumarins. In contrast, P450-mediated metabolism of xanthotoxin is undetectable in *P. troilus*, a section III specialist on Lauraceae, which lacks furanocoumarins (Cohen et al. 1992).

The molecular basis of furanocoumarin resistance has been investigated extensively in *P. polyxenes*. CYP6B1 transcripts are inducible by xanthotoxin (Cohen et al. 1992), and the protein it encodes metabolizes linear and to a lesser degree angular furanocoumarins (Ma et al. 1994).

Transcripts of another P450, CYP6B3 (Hung et al. 1995), are induced by a wider range of linear and angular furanocoumarins but encode a P450 capable of only low levels of furanocoumarin metabolism (Wen et al. 2003). Consistent with a function of allelochemical detoxification, both of these genes are expressed only in feeding stages of the life cycle; no expression is detectable in eggs, pupae, or adults, which do not consume furanocoumarin-containing host-plant tissue (Harrison et al. 2001).

Coexpression of CYP6B1 and P450 reductase (Wen et al. 2003) resulted in high turnover rates for the linear furanocoumarins xanthotoxin and psoralen; in contrast, substrate binding and turnover rates for angular furanocoumarins are low to undetectable in heterologous expression systems (Ma et al. 1994; Wen et al. 2003). This relative ranking is consistent with the relative frequency of occurrence of these furanocoumarins in host plants; linear furanocoumarins such as xanthotoxin and bergapten are widespread among apiaceous as well as rutaceous host plants, whereas angular furanocoumarins are restricted to a handful of genera in the Apiaceae (Berenbaum 1983).

Comparing two closely related CYP6B4 and CYP6B17 groups in the polyphagous congeners *P. glaucus* and *P. canadensis*, Li et al. (2003) found that, generally, P450s from *P. glaucus*, which feeds occasionally on furanocoumarin-containing host plants, displays higher activities against furanocoumarins than those from *P. canadensis*, which normally does not encounter furanocoumarins. These P450s in turn catalyze a larger range of furanocoumarins at lower efficiency than CYP6B1 from *P. polyxenes*, a specialist on furanocoumarin-containing host plants. Reconstruction of the ancestral CYP6B sequences using maximum likelihood predictions and comparisons of the sequence and geometry of their active sites to those of contemporary CYP6B proteins indicate that substrate specificity is related to host-plant diversity; P450s of oligophagous swallowtails have a narrower range of substrates, but higher activity toward those substrates, than do P450s of polyphagous swallowtails.

The broadening of P450 substrate specificity that appears to have accompanied the evolution of polyphagous feeding within the genus *Papilio* is associated with remarkably few genetic changes. Substitution of aliphatic amino acids for the aromatic Phe484 and Phe371 of CYP6B1 results in an enzyme with a more open and accessible catalytic pocket and, as a consequence, a broader range of substrates in these polyphagous species. Along the lineage leading to *Papilio* P450s, the ancestral, highly versatile CYP6B protein presumed to exist in a polyphagous ancestor evolved through time into a more efficient and specialized CYP6B1-like protein in *Papilio* species with continual exposure to furanocoumarins. Further diversification of *Papilio* CYP6Bs has likely involved interspersed events of positive selection in oligophagous species and relaxation of functional constraints in polyphagous species (Li et al. 2003).

Determining whether changes in the specificity of regulatory pathways accompanies change in substrate specificity within a lineage requires comparative analysis of promoter regions of these genes (Hung et al. 1996). The CYP6B1 promoter possesses a xanthotoxin-responsive regulatory element required for basal transcription and xanthotoxin inducibility (Prapaipong et al. 1994; Petersen et al. 2003). XRE-xan appears to be conserved in all swallowtail CYP6B genes and may account for the furanocoumarin inducibility of these genes even in species that rarely encounter furanocoumarins in their host plants (McDonnell et al. 2004).

In addition to these elements, the CYP6B1 promoters also contain putative XRE-AhR elements identical to the aryl hydrocarbon–response elements present in mammalian phase I detoxification genes. Transfections of CYP6B4 and CYP6B1 promoters containing EcRE/ARE/XRE-xan and XRE-AhR elements indicate that both are induced significantly by benzo(α)pyrene, an aryl hydrocarbon widespread in the environment, as well as by xanthotoxin, an allelochemical encountered in the host plants (McDonnell et al. 2004). In the mammalian CYP1A1 gene, the benzylisoquinoline alkaloid berberine interacts with XRE-AhR (Vrzal et al. 2005); if the promoter elements function similarly in the CYP6B genes, the presence of XRE-AhR may reflect conservation from ancestral species within the genus associated with magnoliaceous host plants, possibly contributing to the ability of other section II Rutaceae-feeding swallowtails (e.g., *P. aegeus*) to feed and develop on such ancestral host plants as *Magnolia virginica, Michellia champaca,* and *Cinnamomum camphora* (J. M. Scriber, 2006, personal communication) and may share regulatory elements responsive to ancestral toxins.

P450s and Host-Use Evolution

The potential involvement of *P. polyxenes* cytochrome P450s in both host-plant selection by adults and host-plant utilization by larvae provides an extraordinary opportunity to examine within a single gene superfamily some of the mechanistic bases for host-use evolution. Although progress is being made in understanding the genetic architecture of host-race formation within the *Rhagoletis pomonella* complex with respect to odor perception (Linn et al. 2003), comparable studies have not yet been done within the Lepidoptera, and to date there have been no studies examining the involvement of allelochemical metabolism in host-plant perception and host-plant consumption within the same species. Whether P450s recruited as degradative enzymes in chemoreception are similar in function to those recruited as degradative enzymes in allelochemical detoxification remains an open question. Rapid evolution of host shifts may be facilitated if in fact these two groups of P450s share similar substrate specificities or regulatory pathways. Chlorogenic acid, a component of the oviposition stimulant for *P. polyxenes*, is in fact a substrate for CYP6B8, a P450 in the midgut of the generalist *Helicoverpa zea*. Extensive background information on the chemical mediation of oviposition behavior and larval performance in *P. polyxenes* coupled with the unique inventory of cytochrome P450s characterized from the genome of this species provide a unique and timely opportunity to investigate the mechanistic basis of host-use patterns in Lepidoptera.

Among the other groups of environmental response genes (Berenbaum 2002) that are likely to influence both preference and performance are olfactory and gustatory receptor genes. The absence of a *Papilio* genome project presents a tremendous obstacle to characterizing these genes. In *Drosophila melanogaster*, for example, odor perception depends on activation of a family of receptors sharing a seven-transmembrane domain (a region of a protein that crosses the cell membrane seven times—characterizes many olfactory and gustatory receptors); 60 Or genes have been identified in the *D. melanogaster* genome (Clyne et al. 1999; http://flybase.bio.indiana.edu). Only a subset of these genes expressed in larvae are also expressed in adults: 13 in adult antennae and 1 in the maxillary palp (Kreher et al. 2005). The intriguing possibility exists that genes such as these may provide a mechanistic link between larval and adult preference. Gustatory receptors are less well characterized, but these genes may also have conserved function throughout development. There are 68 gustatory receptor genes *(Gr)* (Amrein and Thorne 2005). Little is known of their function, and expression levels are low, even in known taste neurons. A subset of these receptors is expressed in both larvae and adults and may mediate perception of compounds that are deterrent in both life stages.

The fact that certain chemicals are deterrent to larval and adult stages of certain swallowtail species (e.g., hydroxybenzoic acid and *P. xuthus* [Ono et al. 2004]) is suggestive of a "congruent sensory mechanism between the tarsal chemoreceptors of adults and the gustatory receptors of larvae" (p. 294) (Nishida 2005). Specific gustatory receptors expressed throughout the life cycle may thus represent a genetic link between nonpreference in butterflies and nonperformance in caterpillars, as it were. Similarly, certain compounds function as host-recognition cues in both life stages (e.g., aristolochic acids in troidines such as *Atrophaneura alcinous* [Nishida and Fukami 1989a]), and specific receptors may also link preference and performance genetically.

Conclusions

With respect to the relationship between preference and performance in the evolution of host-plant specialization, whether *Papilio* is in fact representative of the 120,000 species of Lepidoptera (or, indeed, of just the 20,000 species of butterflies) is an open question. If in fact the genus is typical, then the expectation of a tight correlation between these two important aspects of host-plant utilization is perhaps unrealistic. Among the purported selective advantages of a holometabolous lifestyle is the opportunity to allow for more efficient exploitation of different environments, and with those environments come vastly different selective pressures. Nonetheless, the host plant is more than just a source of caterpillar food in the life of most butterflies. Over and above serving as a substrate for depositing eggs, it can be a rendezvous site for mating, and a source of chemical protection for eggs and adults, among other things (Nishida 2002). Because of these commonalities, it is a reasonable expectation that certain sensory and physiological traits will be shared across life stages. Just how many are held in common is likely a function of the specific ecological association. As both preference and performance are more precisely defined at the genetic level, and as more Lepidoptera genomes become available, the specific linkages are likely to be uncovered, to shed new light on questions that have bedeviled investigators for the better part of a century.

Acknowledgments

This work was supported in part by NSF grants IOB0614726 (to MRB) and IBN-9986250 (to PF). We thank Mark Scriber and an anonymous reviewer for helpful comments on our manuscript, and Kelley Tilmon for her extraordinary editorial patience.

References Cited

Akhtar, Y., and M.B. Isman. 2003. Larval exposure to oviposition deterrents alters subsequent oviposition behavior in generalist, *Trichoplusia ni*, and specialist, *Plutella xylostella*, moths. J. Chem. Ecol. 29: 1853–1870.

Amrein, H., and N. Thorne. 2005. Gustatory perception and behavior in *Drosophila melanogaster*. Curr. Biol. 15: R673–R684.

Arikawa, K. 2003. Spectral organization of the eye of a butterfly, *Papilio*. J. Comp. Physiol. A 189: 791–800.

Barron, A.B. 2001. The life and death of Hopkins' host-selection principle. J. Insect Behav. 14: 725–737.

Baur, R., and P. Feeny. 1995. Comparative electrophysiological analysis of plant odor perception in females of three *Papilio* species. Chemoecology 5/6: 26–36.

Baur, R., P. Feeny, and E. Städler. 1993. Oviposition stimulants for the black swallowtail butterfly: identification of electrophysiologically active compounds in carrot volatiles. J. Chem. Ecol. 19: 919–937.

Berdegue, M., S.R. Reitz, and J.T. Trumble. 1998. Host plant selection and development in *Spodoptera exigua*: do mother and offspring know best? Entomol. Exp. Appl. 89: 57–64.

Berenbaum, M. 1981a. Effects of linear furanocoumarins on an adapted specialist insect *(Papilio polyxenes)*. Ecol. Entomol. 6: 345–351.

Berenbaum, M. 1981b. An oviposition "mistake" by *Papilio glaucus* (Papilionidae). J. Lepid. Soc. 35: 75.

Berenbaum, M.R. 1983. Coumarins and caterpillars: a case for coevolution. Evolution 37: 163–179.

Berenbaum, M.R. 1990. Plant consumers and plant secondary metabolites: past, present, and future. In J. Antonovics and D. Futuyma (eds.), Oxford Rev. Evol. Biol. 7: 285–307 (Oxford University Press).

Berenbaum, M.R. 1991. Coumarins, pp. 221–249. In G. Rosenthal and M.R. Berenbaum (eds.), Herbivores: their interactions with secondary plant metabolites. Academic Press, New York.

Berenbaum, M.R. 1995. Phototoxicity of plant secondary metabolites: insect and mammalian perspectives. Arch. Insect Biochem. Physiol. 29: 119–134.

Berenbaum, M. 1999. Animal-plant warfare: molecular basis for cytochrome P450–mediated natural adaptation, pp. 553–571. In A. Puga and K.B. Wallace (eds.), Molecular biology of the toxic response. Taylor and Francis, New York.

Berenbaum, M. R. 2002. Post-genomic chemical ecology: from genetic code to ecological interactions. J. Chem. Ecol. 28: 873–896.

Berenbaum, M. R., and A. R. Zangerl. 1993. Furanocoumarin metabolism in *Papilio polyxenes*: genetic variability, biochemistry, and ecological significance. Oecologia 95: 370–375.

Berenbaum, M. R., A. R. Zangerl, and J. K. Nitao. 1986. Constraints on chemical coevolution: wild parsnips and the parsnip webworm. Evolution 40: 1215–1228.

Bergstrom, A., S. Nylin, and G. H. Nygren. 2004. Conservative resource utilization in the common blue butterfly: evidence for low costs of accepting absent host plants? Oikos 107: 345–351.

Bernays, E. A. 2001. Neural limitations of phytophagous insects: implications for diet breadth and host affiliation. Annu. Rev. Entomol. 46: 703–727.

Bossart J. L. 2003. Covariance of preference and performance on normal and novel hosts in a locally monophagous and locally polyphagous butterfly population. Oecologia 135: 477–486.

Bossart, J. L., and J. M. Scriber. 1995. Maintenance of ecologically significant genetic variation in the tiger swallowtail butterfly through differential selection and gene flow. Evolution 49: 1163–1171.

Brooks, J. S., and P. Feeny. 2004. Seasonal variation in *Daucus carota* leaf-surface and leaf-tissue profiles. Biochem. Syst. Ecol. 32: 769–782.

Brooks, J. S., E. H. Williams, and P. Feeny. 1996. Quantification of contact oviposition stimulants for black swallowtail butterfly, *Papilio polyxenes*, on the leaf surfaces of wild carrot, *Daucus carota*. J. Chem. Ecol. 22: 2341–2357.

Camara, M. D. 1997. Predator responses to sequestered plant toxins in buckeye caterpillars: are tritrophic interactions locally variable? J. Chem. Ecol. 23: 1573–1561.

Carey, D. B. 1994. Patch dynamics of *Glaucopsyche lygdamus* (Lycaenidae): correlations between butterfly density and host species diversity. Oecologia 99: 337–342.

Carroll, M. J., A. R. Zangerl, and M. R. Berenbaum. 2000. Octyl acetate and octyl butyrate in the mature fruits of the wild parsnip, *Pastinaca sativa* (Apiaceae). J. Heredity 91: 68–71.

Carta-Preta, P. D., and F. S. Zucoloto. 2003. Oviposition behavior and performance aspets of *Ascia monuste* (Godart, 1919) (Lepidoptera, Pieridae) on kale *(Brassica oleracea* var. *acephala)*. Rev. Brasil. Entomol. 47: 169–174.

Carter, M., K. Sachdev-Gupta, and P. Feeny. 1998. Tyramine in the leaves of wild parsnip: a stimulant and synergist for oviposition by the black swallowtail butterfly. Physiol. Entomol. 23: 303–312.

Carter, M., P. Feeny, and M. Haribal. 1999. An oviposition stimulant for the spicebush swallowtail butterfly, *Papilio troilus*, from the leaves of *Sassafras albidum*. J. Chem. Ecol. 25: 1233–1245.

Caterino, M. S., R. D. Reed, M. M. Kuo, and F. A. H. Sperling. 2001. A partitioned likelihood analysis of swallowtail butterfly phylogeny (Lepidoptera: Papilionidae), Syst. Biol. 50:106–127.

Charlesworth, B., J. A. Coyne, and N. H. Barton. 1987. The relative rates of evolution of sex chromosomes and autosomes. Am. Nat. 130: 113–146.

Chen, Y.-Z., L. Lin, C.-W. Wang, C.-C. Yeh, and S.-Y. Hwang. 2004. Response of two *Pieris* (Lepidoptera: Pieridae) species to fertilization of a host plant. Zool. Stud. 43: 778–786.

Clyne, P. J., C. G. Warr, M. R. Freeman, D. Lessing, J. H. Kim, and J. R. Carlson. 1999. A novel family of divergent seven-transmembrane proteins: candidate odorant receptors in *Drosophila*. Neuron 22: 327–338.

Cohen, M. B., M. A. Schuler, and M. R. Berenbaum. 1992. A host-inducible cytochrome P450 from a host-specific caterpillar: molecular cloning and evolution. Proc. Natl. Acad. Sci. USA 89: 10920–10924.

Damman, H., and P. Feeny. 1988. Mechanisms and consequences of selective oviposition by the zebra swallowtail butterfly. Anim. Behav. 36: 563–573.

Dethier, V. G. 1970. Chemical interactions between plants and insects, pp. 83–102. In E. Sondheimer and J. B. Simeone, eds. Chemical Ecology. Academic Press, New York.

Dethier, V. G. 1976. The importance of stimulus patterns for host plant recognition and acceptance, pp. 67–70. In T. Jermy (ed.), The host plant in relation to insect behavior and reproduction. Plenum Press, New York.

DiTommaso, A. and J. E. Losey. 2003. Oviposition preference and larval performance of monarch butterflies *(Danaus plexippus)* on two invasive swallow-wort species. Entomol. Exp. Appl. 108: 205–209.

Ehrlich, P. R. and P. H. Raven. 1964. Butterflies and plants: a study in coevolution. Evolution 18: 586–608.

Feeny, P. 1991. Chemical constraints on the evolution of swallowtail butterflies, pp. 315–340. In P. W. Price, T. M. Lewinsohn, G. W. Fernandes, and W. W. Benson (eds.), Plant-animal interactions: evolutionary ecology in tropical and temperate regions. John Wiley and Sons, New York.

Feeny, P. 1992. The evolution of chemical ecology: contributions from the study of herbivorous insects, pp. 1–44. In G. A. Rosenthal and M. Berenbaum (eds.), Herbivores: their interactions with secondary plant metabolites, 2nd edition, vol. 2: Evolutionary and ecological processes. Academic Press, San Diego.

Feeny, P., L. Rosenberry, and M. Carter. 1983. Chemical aspects of oviposition behavior in butterflies, pp. 27–76. In S. Ahmad (ed.), Herbivorous insects: host-seeking behavior and mechanisms. Academic Press, New York.

Feeny, P., K. Sachdev, L. Rosenberry, and M. Carter. 1988. Luteolin 7-O-(6"-O-malonyl)-β-D-glucoside and *trans*-chlorogenic acid: oviposition stimulants for the black swallowtail butterfly. Phytochemistry 27: 3439–3448.

Feeny, P., E. Städler, I. Åhman, and M. Carter. 1989. Effects of plant odor on oviposition by the black swallowtail butterfly, *Papilio polyxenes* (Lepidoptera: Papilionidae). J. Insect Behav. 2: 803–827.

Feyereisen, R. 1999. Insect P450 enzymes. Annu. Rev. Entomol. 44: 507–533.

Forister, M. L. 2004. Oviposition preference and larval performance within a diverging lineage of lycaenid butterflies. Ecol. Entomol. 29: 264–272.

Gonzalez, F. J., and D. W. Nebert. 1990. Evolution of the P450 gene superfamily: animal-plant "warfare," molecular drive, and human genetic differences in drug oxidation. Trends Genet. 6: 182–186.

Gordon, H. T. 1961. Nutritional factors in insect resistance to chemicals. Annu. Rev. Entomol. 6: 27–54.

Gu, J., C. Dudley, T. Su, D. C. Spink, Q. Y. Zhang, R. L. Moss, and X. X. Ding. 1999. Cytochrome P450 and steroid hydroxylase activity in mouse olfactory and vomeronasal mucosa. Biochem. Biophys. Res. Commun. 266: 262–267.

Hancock, D. L. 1983. Classification of the Papilionidae (Lepidoptera): a phylogenetic approach. Smithersia 2: 1–48.

Harborne, J. B. , V. H Heywood, and C. A. Williams. 1969. Distribution of myristicin in seeds of the Umbelliferae. Phytochemistry 8: 1729–1732.

Haribal, M., and P. Feeny, 1998. Oviposition stimulant for the zebra swallowtail butterfly, *Eurytides marcellus*, from the foliage of pawpaw, *Asimina triloba*. Chemoecology 8: 99–110.

Haribal, M., and P. Feeny. 2003. Combined roles of contact stimulant and deterrents in assessment of host-plant quality by ovipositing zebra swallowtail butterflies. J. Chem. Ecol. 29: 653–670.

Haribal, M., P. Feeny, C. C. Lester. 1998. Caffeoylcyclohexane-1-carboxylic acid derivative from *Asimina triloba*. Phytochemistry 49: 103–108.

Harrison, T., A. R. Zangerl, M. A. Schuler, and M. R. Berenbaum. 2001. Developmental variation in cytochrome P450 expression in *Papilio polyxenes*. Arch. Insect Biochem. Physiol. 48: 179–189.

Heininger, E. A. 1989. Effects of furocoumarin and furoquinoline allelochemicals on host-plant utilization by Papilionidae. Doctoral thesis, University of Illinois at Urbana-Champaign.

Heinz, C. A. 2002. Host finding and recognition by *Papilio polyxenes* (Lepidoptera: Papilionidae): the effects of three host cues and of host-plant experience on oviposition behavior. PhD. Thesis, Cornell University, Ithaca, NY.

Heinz, C. A., and P. Feeny. 2005. Effects of contact chemistry and host plant experience in the oviposition behaviour of the eastern black swallowtail butterfly. Anim. Behav. 69: 107–115.

Heisswolf, A., E. Obermaier, and H. J. Poethke. 2005. Selection of large host plants for oviposition by a monophagous leaf beetle: nutritional quality or enemy-free space? Ecol. Entomol. 30: 299–306.

Honda, K. 1986. Flavanone glycosides as oviposition stimulants in a papilionid butterfly, *Papilio protenor*. J. Chem. Ecol. 12: 1999–2010.

Honda, K. 1990. Identification of host plant chemicals stimulating oviposition by swallowtail butterfly, *Papilio protenor*. J. Chem. Ecol. 16: 325–337.

Honda, K. 1995. Chemical basis of differential oviposition by lepidopterous insects. Arch. Insect Biochem. Physiol. 30: 1–23.

Honda, K., S. Kawano, and N. Hayashi. 1997. Flavonoids as oviposition stimulants or deterrents in host selection by swallowtail butterflies *(Papilio)*, p. 259. In Abstracts of the Third Asia-Pacific Conference of Entomology (APCEIII), 16–22 November 1997, Taichung, Taiwan.

Hovemann, B. T., F. Sehlmeyer, and J. Malz. 1997. *Drosophila melanogaster* NADPH-cytochrome P450 oxidoreductase: pronounced expression in antennae may be related to odorant clearance. Gene 189: 213–219.

Hung, C. F., T. L. Harrison, M. R. Berenbaum, and M. A. Schuler. 1995. CYP6B3: a second furanocoumarin-inducible cytochrome P450 expressed in *Papilio polyxenes*. Insect Mol. Biol. 4: 149–160.

Hung, C. F., R. Holzmacher, E. Connolly, M. R. Berenbaum, and M. A. Schuler. 1996. Conserved promoter elements in the *CYP6B* gene family suggest common ancestry for cytochrome P450 monooxygenases mediating furanocoumarin detoxification. Proc. Natl. Acad. Sci. USA 93: 12200–12205.

Ilse, D. 1956. Behaviour of butterflies before oviposition. J. Bombay Nat. Hist. Soc. 53: 486–488.

Jaenike J. 1978. On optimal oviposition behaviour in phytophagous insects. Theoret. Pop. Biol. 14: 350–356.

Jaenike, J. 1989. Genetics of butterfly-hostplant associations. Trends Ecol. Evol. 4: 34–35.

Janz, N. 1998. Sex-linked inheritance of host-plant specialization in a polyphagous butterfly. Proc. R. Soc. Lond. Ser. B 265: 1675–1678.

Janz, N., and S. Nylin. 1997. The role of female search behaviour in determining host plant range in plant feeding insects: a test of the information processing hypothesis. Proc. R. Soc. Lond. Ser. B 264: 701–707.

Janz, N., S. Nylin, and N. Wedell. 1994. Host plant utilization in the comma butterfly: sources of variation and evolutionary implications. Oecologia 99: 132–140.

Kelber, A., and M. Pfaff. 1999. True colour vision in the orchard butterfly, *Papilio aegeus*. Naturwissenschaften 86: 221–224.

Kreher, S. A., J. Y. Kwon, and J. R. Carlson. 2005. The molecular basis of odor coding in the *Drosophila* larva. Neuron 46: 445–456.

Ladner, D., and S. Altizer. 2005. Oviposition preference and larval performance of North American monarch butterflies on four *Asclepias* species. Entomol. Exp. Appl. 116: 9–20.

Leal, W. S. 2001. Molecules and macromolecules involved in chemical communication of scarab beetles. Pure Appl. Chem. 73: 613–616.

Li, W., M. R. Berenbaum, and M. A. Schuler. 2001. Molecular analysis of multiple CYP6B genes from polyphagous *Papilio* species. Insect Biochem. Mol. 31: 999–1011.

Li, W. M., M. A. Schuler, and M. R. Berenbaum. 2003. Diversification of furanocoumarin-metabolizing cytochrome P450 monooxygenases in two papilionids: specificity and substrate encounter rate. Proc. Natl. Acad. Sci. USA 100: 14593–14595.

Linn, C., J. L. Feder, S. Nojima, H. R. Dambroski, S. H. Berlocher, and W. Roelofs. 2003. Fruit odor discrimination and sympatric host race formation in *Rhagoletis*. Proc. Natl. Acad. Sci. USA 100: 11490–11493.

Ma, R., M. B. Cohen, M. R. Berenbaum, and M. A. Schuler. 1994. Black swallowtail *(Papilio polyxenes)* alleles encode cytochrome P450s that selectively metabolize linear furanocoumarins. Arch. Biochem. Biophys. 310: 332–340.

Ma, W. C., and L. M. Schoonhoven. 1973. Tarsal contact chemosensory hairs of the large white butterfly *Pieris brassicae* and their possible role in oviposition behavior. Entomol. Exp. Appl. 16: 343–357.

Maibeche-Coisne, M., E. Jacquin-Joly, M. C. Francois, and P. Nagnan-LeMeillour. 2002. cDNA cloning of biotransformation enzymes belonging to the cytochrome P450 family in the antennae of the noctuid moth *Mamestra brassicae*. Insect Mol. Biol. 11: 273–281.

Maibeche-Coisne, M., A. A. Nikonov, Y. Ishida, E. Jacquin-Joly, and W. S. Leal. 2004. Pheromone anosmia in a scarab beetle induced by in vivo inhibition of a pheromone-degrading enzyme. Proc. Natl. Acad. Sci. USA 101: 11459–11464.

Mansuy, D. 1998. The great diversity of reactions catalyzed by cytochromes P450. Comp. Biochem. Physiol. C: Pharmacol. Toxicol. Endocrinol. 121: 5–14.

Mao, W., S. Rupasinghe, R. Zangerl, M. A. Schuler, and M. R. Berenbaum. 2006. Remarkable substrate-specificity of CYP6AB3 in *Depressaria pastinacella*, a highly specialized caterpillar. Insect Mol. Biol. 15: 169–179.

Mattila, H. R., and G. W. Otis. 2003. A comparison of the host preference of monarch butterflies *(Danaus plexippus)* for milkweed *(Asclepias syriaca)* over dog-strangler vine *(Vincetoxicum rossicum)*. Entomol. Exp. Appl. 107: 193–199.

McDonnell, C. M., R. P. Brown, M. R. Berenbaum, and M. A. Schuler. 2004. Conserved regulatory elements in the promoters of two allelochemical-inducible cytochrome P450 genes differentially regulate transcription. Insect Biochem. Mol. Biol. 34: 1129–1139.

Mercador, R. J., and J. M. Scriber. 2005. Phenotypic plasticity of host selection in adult tiger swallowtails; *Papilio glaucus* (L.), pp. 25–57. In T. N. Ananthakrishnan and D. Whitman (eds.), Insects plasticity: Diversity of responses. Science Publishers, Enfield, NH.

Miller, J. S. 1987. Phylogenetic studies in the Papilioninae (Lepidoptera: Papilionidae). Bull. Amer. Mus. Nat. Hist. 186: 365–512.

Miller, J. S., and P. Feeny. 1983. Effects of benzylisoquinoline alkaloids on the larvae of polyphagous Lepidoptera. Oecologia 58: 332–33.

Miller, J. S., and P. Feeny. 1989. Interspecific differences among swallowtail larvae (Lepidoptera: Papilionidae) in susceptibility to aristolochic acids and berberine. Ecol. Entomol. 14: 287–296.

Moreau, J., B. Benrey, and D. Thiery. 2006. Assessing larval food quality for phytophagous insects: are the facts as simple as they appear? Funct. Ecol. 20: 592–600.

Munroe, E. 1961. The generic classification of the Papilionidae. Can. Entomol. Suppl. 17: 1–51.

Murray, R.D. H., J. Mendez, and S.A. Brown. 1982. The natural coumarins: occurrence, chemistry, and biochemistry. John Wiley and Sons, Bristol, UK.

Nakayama, T., and K. Honda. 2004. Chemical basis for differential acceptance of two sympatric rutaceous plants by ovipositing females of a swallowtail butterfly, *Papilio polytes* (Lepidoptera, Papilionidae). Chemoecology 14: 199–205.

Nakayama, T., K. Honda, and N. Hayashi. 2002. Chemical mediation of differential oviposition and larval survival on rutaceous plants in a swallowtail butterfly, *Papilio polytes*. Entomol. Exp. Appl. 105: 35–42.

National Center for Biotechnology Information, 2007. Genbank, http://www.ncbi.nlm.nih.gov/.

Nakayama, T., K. Honda, H. Omura, and N. Hayashi. 2003. Oviposition stimulants for the tropical swallowtail butterfly, *Papilio polytes*, feeding on a rutaceous plant, *Toddalia asiatica*. J. Chem. Ecol. 29: 1621–1634.

Neal, J.J., and D. Wu. 1994. Inhibition of insect cytochromes P450 by furanocoumarins. Pestic. Biochem. Phys. 50: 43–50.

Nelson, D.R., L. Koymans, T. Kamataki, J.J. Stegeman, R. Feyereisen, D.J. Waxman, M.R. Waterman, O. Gotoh, M.J. Coon, R.W. Estabrook, I.C. Gunsalus, and D.W. Nebert. 1996. P450 superfamily: update on new sequences, gene mapping, accession numbers and nomenclature. Pharmacogenetics 6: 1–42.

Nishida, R. 1994. Oviposition stimulant of a zerynthiine swallowtail butterfly, *Luehdorfia japonica*. Phytochemistry 36: 873–877.

Nishida, R. 1995. Oviposition stimulants of swallowtail butterflies, pp. 17–26. In M. Scriber, Y. Tsubaki , and R.C. Lederhouse (eds.), Swallowtail butterflies: their ecology and evolutionary biology. Scientific Publishers, Gainesville, FL.

Nishida, R. 2002. Sequestration of defensive substances from plants by Lepidoptera. Annu. Rev. Entomol. 47: 57–92.

Nishida, R. 2005. Chemosensory basis of host recognition in butterflies: multi-component system of oviposition stimulants and deterrents. Chem. Senses. 30 (Suppl 1): 1293–1294.

Nishida, R., and H. Fukami. 1989a. Oviposition stimulants of an Aristolochiaceae feeding swallowtail butterfly, *Atrophaneura alcinous*. J. Chem. Ecol. 15: 2565–2575.

Nishida, R., and H. Fukami. 1989b. Ecological adaptation of an Aristolochiaceae-feeding swallowtail butterfly, *Atrophaneura alcinous*, to aristolochic acids. J. Chem. Ecol. 15: 2549–2563.

Nishida, R., T. Ohsugi, H. Fukami, and S. Nakajima. 1990. Oviposition deterrent of a Rutaceae-feeding swallowtail butterfly, *Papilio xuthus*, from a non-host rutaceous plant, *Orixa japonica*. Agric. Biol. Chem. 54: 1265–1270.

Nishida, R., S. Kokubo, Y. Kuwahara, and H. Fukami. 1994. Oviposition stimulant of *Papilio macilentus* from *Orixa japonica*. (In Japanese). Nippon Nogeikagaku Kaishi 68: 394.

Nylin, S., and N. Janz. 1996. Host plant preferences in the comma butterfly *(Polygonia c-album)*: do parents and offspring agree? Ecoscience 3: 285–289.

Ohsaki, N., and Y. Sato. 1999. The role of parasitoids in evolution of habitat and larval food plant preference by three *Pieris* butterflies. Res. Pop. Ecol. 41: 107–119.

Ohsugi, T., R. Nishida, and H. Fukami. 1991. Multi-component system of oviposition stimulants for a Rutaceae-feeding swallowtail butterfly, *Papilio xuthus* (Lepidoptera: Papilionidae). Appl. Entomol. Zool. 26: 29–40.

Ono, H., and H. Yoshikawa. 2004. Identification of amine receptors from a swallowtail butterfly, *Papilio xuthus* L.: cloning and mRNA localization in foreleg chemosensory organ for recognition of host plants. Insect Biochem. Mol. 34: 1247–1256.

Ono, H., R. Nishida, and Y. Kuwahara. 2000a. Oviposition stimulant for a Rutacae-feeding swallowtail butterfly, *Papilio bianor* (Lepidoptera: Papilionidae): hydroxycinnamic acid derivative from *Orixa japonica*. Appl. Entomol. Zool. 35: 119–123.

Ono, H., R. Nishida, and Y. Kuwahara. 2000b. A dihydroxy-γ-lactone as an oviposition stimulant for the swallowtail butterfly, *Papilio bianor*, from the rutaceous plant, *Orixa japonica*. Biosci. Biotechnol. Biochem. 64: 1970–1973.

Ono, H., Y. Kuwahara, and R. Nishida. 2004. Hydroxybenzoic acid derivatives in a nonhost rutaceous plant, *Orixa japonica*, deter both oviposition and larval feeding in a Rutaceae-feeding swallowtail butterfly, *Papilio xuthus* L. J. Chem. Ecol. 30: 287–301.

Papaj, D.R. 1986. Conditioning of leaf-shape discrimination by chemical cues in the butterfly, *Battus philenor*. Anim. Behav. 34: 1281–1288.

Papaj, D.R., P. Feeny, K. Sachdev-Gupta, and L. Rosenberry. 1992. D-(+)-pinitol, an oviposition stimulant for the pipevine swallowtail butterfly, *Battus philenor*. J. Chem. Ecol. 18: 799–815.

Petersen, R.A., H. Niamsup, M.R. Berenbaum, and M.A. Schuler. 2003. Transcriptional response elements in the promoter of CYP6B1, an insect P450 gene regulated by plant chemicals. Biochem. Biophys. Acta 1619: 269–282.

Prapaipong, H., M.R. Berenbaum, and M.A. Schuler. 1994. Transcriptional regulation of the *Papilio polyxenes CYP6B1* gene. Nucleic Acids Res. 22: 3210–3217.

Prudic K.L., J.C. Oliver, and M.D. Bowers. 2005. Soil nutrient effects on oviposition preference, larval performance, and chemical defense of a specialist insect herbivore. Oecologia 143: 578–587.

Rausher, M.D. 1978. Search image for leaf shape in a butterfly. Science 200: 1071–1073.

Rausher, M.D. 1980. Host abundance, juvenile survival, and oviposition preference in *Battus philenor*. Evolution 34: 342–355.

Rausher, M.D. 1995. Behavioral ecology of oviposition in the pipevine swallowtail, *Battus philenor*, pp. 53–62. In J.M. Scriber, Y. Tsubaki, and R.C. Lederhouse (eds.), Swallowtail butterflies: their ecology and evolutionary biology. Scientific Publishers, Gainesville, FL.

Renwick, J.A. A., and F.S. Chew. 1994. Oviposition behavior in Lepidoptera. Annu. Rev. Entomol. 39: 377–400.

Rietdorf, K., and J.L. M. Steidle. 2002. Was Hopkins right? Influence of larval and early adult experience on the olfactory response in the granary weevil *Sitophilus granarius* (Coleoptera, Curculionidae). Physiol. Entomol. 27: 223–227.

Roessingh, P., E. Städler, R. Schöni, and P. Feeny. 1991. Tarsal contact chemoreceptors of the black swallowtail butterfly, *Papilio polyxenes*: responses to phytochemicals from host- and nonhost plants. Physiol. Entomol. 16: 485–495.

Roitberg, B.D., and M. Mangel. 1993. Parent-offspring conflict and life-history consequences in herbivorous insects. Am. Nat. 142: 443–456.

Sachdev-Gupta, K., J.A. A. Renwick, and C.D. Radke. 1990. Isolation and identification of oviposition deterrents to cabbage butterfly, *Pieris rapae*, from *Erysimum cheiranthoides*. J. Chem. Ecol. 16: 1059–1067.

Sachdev-Gupta, K., P. Feeny, and M. Carter. 1993. Oviposition stimulants for the pipevine swallowtail butterfly, *Battus philenor*, from an *Aristolochia* host plant: synergism between inositols, aristolochic acids and a monogalactosyl diglyceride. Chemoecology 4: 19–28.

San Diego Supercomputer Center, 2007. Biology Workbench 3.2. http://workbench.sdsc.edu/.

Saxena, K.N., and S. Goyal. 1978. Host-plant relations of the citrus butterfly *Papilio demoleus* L.: orientational and ovipositional responses. Entomol. Exp. Appl. 24: 1–10.

Scheirs, J., and L. DeBruyn. 2002. Integrating optimal foraging and optimal oviposition theory in plant-insect research. Oikos 96: 187–191.

Scriber, J.M. 1996. A new cold pocket hypothesis to explain local host preference shifts in *Papilio canadensis*. Entomol. Exp. Appl. 80: 315–319.

Scriber, J.M. 2002. Latitudinal and local geographic mosaics in host plant preferences as shaped by thermal units and voltinism. Eur. J. Entomol. 99: 225–39.

Scriber J.M., B.L. Giebink, and D. Snider. 1991. Reciprocal latitudinal clines in oviposition. Behavior of *Papilio glaucus* and *P. canadensis* across the Great Lakes hybrid zone: possible sex-linkage of oviposition preferences. Oecologia 87: 360–368.

Scriber, J.M., R. Lederhouse, and R. Hagen. 1991. Foodplant and evolution within *P. glaucus* and *P. troilus* species groups (Lepidoptera: Papilionidae), pp. 341–373. In P.W. Price, T.M. Lewinsohn, G.W. Fernandes, and W.W. Benson (eds.), Plant-animal interactions: evolutionary ecology in tropical and temperate regions. J. Wiley, New York.

Scriber, J.M., Y. Tsubaki, and R.C. Lederhouse (eds.). 1995. Swallowtail butterflies: their ecology and evolutionary biology. Scientific Publishers, Gainesville, FL.

Singer, M.C., D. Vasco, C. Parmesan, and C.D. Thomas. 1992. Distinguishing between "preference" and "motivation" in food choice: an example from insect oviposition. Anim. Behav. 44: 463–471.

Sperling, F.A. H. 1987. Evolution of the *Papilio machaon* species group in western Canada (Lepidoptera: Papilionidae). Quaestiones Entomologicae 23: 198–319.

Städler, E. 1992. Behavioral responses of insects to plant secondary compounds, pp. 45–88. In G.A. Rosenthal and M. Berenbaum (eds.), Herbivores: their interactions with secondary plant metabolites, 2nd ed., vol. 2: evolutionary and ecological processes. Academic Press, San Diego.

Stride, G.O., and R. Straatman. 1962. The host plant relationship of an Australian swallowtail, *Papilio aegeus*, and its significance in the evolution of host plant selection. Proc. Linn. Soc. NSW 87: 69–78.

Thompson, J.N. 1988a. Evolutionary ecology of the relationship between oviposition preference and performance of offspring in phytophagous insects. Entomol. Exp. Appl. 47: 3–14.

Thompson, J.N. 1988b. Variation in preference and specificity in monophagous and oligophagous swallowtail butterflies. Evolution 42: 118–128.

Thompson, J.N. 1988c. Evolutionary genetics of oviposition preference in swallowtail butterflies. Evolution 42:1223–1234.

Thompson, J.N. 1993. Preference heirarchies and the origin of geographic specialization in host use in swallowtail butterflies. Evolution 47: 1585–1594.

Thompson, J.N., and O. Pellmyr. 1991. Evolution of oviposition behavior and host preference in Lepidoptera. Annu. Rev. Entomol. 36: 65–89.

Thompson, J.N., W. Wehling, and R. Podolsky. 1990. Evolutionary genetics of host use in swallowtail butterflies. Nature 344: 148–150.

Thompson, J.D., D.G. Higgins, and T.J. Gibson. 1994. CLUSTAL W: improving the sensitivity of progressive multiple sequence alignment through sequence weighting, position-specific gap penalties and weight matrix choice. Nucleic Acids Res. 22: 4673–4680.

Underwood, D.L. A. 1994. Intraspecific variability in host plant quality and ovipositional preferences in *Eucheira socialis* (Pieridae: Lepidoptera). Ecol. Entomol. 19: 245–256.

van Emden, H.F., V. Sponagl, E. Wagner, T. Baker, S. Ganguly, and S. Douloumpaka. 1996. Hopkins "host selection principle," another nail in its coffin. Physiol. Entomol. 21: 325–328.

Van Nouhys, S., M.C. Singer, and M. Nieminen. 2003. Spatial and temporal patterns of caterpillar performance and the suitability of two host plant species. Ecol. Entomol. 28: 193–202.

Vogt, R.G. 2003. Biochemical diversity of odor detections: OBPs, ODEs and SNMPs, pp. 391–445. In G.J. Blomquist and R.G. Vogt (eds.), Insect pheromone biochemistry and molecular biology: the biosynthesis and detection of pheromones and plant volatiles. Elsevier Academic, London.

Vrzal, R., A. Zdarilova, J. Ulrichova, L. Blaha, J.P. Giesy, and Z. Dvorak. 2005. Activation of the aryl hydrocarbon receptor by berberine in HepG2 and H4IIE cells: biphasic effect on CYP1A1. Biochem. Pharmacol. 70: 925–936.

Wang, Q., G. Hasan, and C.W. Pikielny. 1999. Preferential expression of biotransformation enzymes in the olfactory organs of *Drosophila melanogaster*, the antennae. J. Biol. Chem. 274: 10309–10315.

Wehling, W.F., and J.N. Thompson. 1997. Evolutionary conservatism of oviposition preference in a widespread polyphagous insect herbivore, *Papilio zelicaon*. Oecologia 111: 209–215.

Weiss, M.R., and D.R. Papaj. 2003. Colour learning in two behavioural contexts: how much can a butterfly keep in mind? Anim. Behav. 65: 425–434.

Wen, Z., L. Pan, M.R. Berenbaum, and M.A. Schuler. 2003. Metabolism of linear and angular furanocoumarins by *Papilio polyxenes* CYP6B1 coexpressed with NADPH cytochrome P450 reductase. Insect Biochem. Mol. Biol. 333: 937–947.

Wiklund, C. 1973. Host plant suitability and the mechanism of host selection in larvae of *Papilio machaon*. Entomol. Exp. Appl. 16: 232–242.

Wiklund, C. 1975. The evolutionary relationship between adult oviposition preferences and larval host plant range in *Papilio machaon* L. Oecologia 18: 185–197.

Wojtasek H., and W.S. Leal. 1999. Degradation of an alkaloid pheromone from the pale-brown chafer, *Phyllopertha diversa* (Coleoptera: Scarabaeidae), by an insect olfactory cytochrome P450. FEBS Lett. 458: 333–336.

Zakharov, E.V., M.S. Caterino, and F.A. H. Sperling. 2004. Molecular phylogeny, historical biogeography, and divergence time estimates for swallowtail butterflies of the genus *Papilio* (Lepidoptera: Papilionidae). Syst. Biol. 53: 193–215.

Zangerl, A.R., and M.R. Berenbaum. 1992. Oviposition patterns and hostplant suitability: parsnip webworms and wild parsnip. Am. Midl. Nat. 128: 292–298.

Zangerl, A.R., E.S. Green, R.L. Lampman, and M.R. Berenbaum. 1997. Phenological changes in primary and secondary chemistry of reproductive parts of *Pastinaca sativa*. Phytochemistry 44: 825–831.

TWO

Evolution of Preference and Performance Relationships

TIMOTHY P. CRAIG AND JOANNE K. ITAMI

Natural selection should favor female phytophagous insects that have a preference for ovipositing on resources where their offspring will have the highest fitness (Dethier 1959a, 1959b; Singer 1972; Jaenike 1978). This assertion has been termed the naïve adaptationist hypothesis (Courtney and Kibota 1990). This hypothesis has been tested by measuring oviposition preference and offspring performance in a wide range of interactions, and contrary to initial expectations a wide range of preference-performance relationships have been found (Thompson 1988; Courtney and Kibota 1990; Mayhew 1998, 2001). General hypotheses explaining broad patterns in preference-performance relationships have been largely lacking, with the exception of the phylogenetic constraints hypothesis described below. We propose the "feeding niche constraints" hypothesis that adaptations to the characteristics of an herbivore's feeding niche will determine the evolution of the relationship between preference and performance.

In this chapter, to determine whether eggs are oviposited where offspring fitness is highest we looked at the correlation between preference and performance. Preference is nonrandom oviposition on resources offered simultaneously or sequentially (Singer 1986). Performance is any measure of offspring survival, growth, or reproduction that is presumed to be correlated with fitness (Thompson 1988). These indirect measures do not always correlate with ultimate fitness (Thompson 1988), but they are frequently used because of the difficulty of directly measuring fitness.

Preference-performance relationships have important implications for ecological and evolutionary interactions. Insect population dynamics are strongly influenced by the nature of these relationships. Price (2003) showed that species that have highly selective oviposition behavior with a strong preference-performance correlation have stable or latent population dynamics, while species with low correlations between preference and performance have eruptive population dynamics. Preference-performance relationships also have evolutionary implications: species with strong relationships tend to be specialists, and those with weak relationships tend to be generalists. As a result an understanding of the forces responsible for the evolution of preference-performance relationships will allow us to understand the evolution of host range.

To explain patterns in preference and performance, the feeding niche constraints hypothesis posits that the feeding niche of a species influences the evolution of the preference-performance relationship. The feeding niche of an herbivore determines the kinds of plant resource variation it will encounter. We hypothesize that insects share certain fundamental constraints that limit their ability to respond to this variation and that therefore impact the evolution of a positive preference-performance relationship. If all insects share these constraints, then the evolution of the preference-performance relationships depends on the resources insects encounter and not on their particular phylogenetic background.

A testable prediction of this hypothesis is that insects with similar feeding niches will evolve similar preference-performance relationships. For example, stem-gallers on woody plants will have similar types of relationships, and free-feeding leaf-feeders on herbaceous plants will have relationships that differ from those of the stem-gallers. An alternative is the phylogenetic constraints hypothesis (Price 2003), which proposes that taxonomic groups differ from each other in preference-performance relationships because they have specific characters that constrain the evolution of these relationships. This hypothesis predicts that closely related species will evolve similar preference-performance relationships. The null hypothesis for both hypotheses is that there will be no relationship between either the taxonomic group or the feeding niche and the strength of the preference-performance relationship.

To develop predictions from the feeding niche hypothesis, we proceed through three steps. First, we identify the insect constraint: for example, an insect's limited neural capacity to integrate information. Second, we identify resource variation that will cause this constraint to limit the evolution of the preference-performance relationship: for example, variation in a plant's resource quality that differs in complexity. Third, we compare the preference-performance relationships in insects with different types of feeding niches: for example, monophagous species with less resource complexity versus polyphagous species.

We argue that all insect herbivores are subject to at least three types of constraints that limit the kind of resources where strong preference-performance relationships can evolve.

Limiting Constraints

Complexity Constraints

Bernays (1998, 2001) has advanced the neural limitation hypothesis, that there are inherent limitations of the insect nervous system such that specialists processing a limited set of information make more efficient and accurate choices than generalist insects faced with complex choices. This hypothesis also implies that as the complexity of resource variation in the feeding niche increases, the preference-performance relationship will weaken because of the limitations on herbivore ability to integrate the complex information required to choose the best oviposition sites. It has previously been hypothesized that when resource variation is complex it may limit the evolution of strong preference-performance relationships (Singer 1972; Craig et al. 1989).

We predict that the following niche characteristics would increase resource complexity and decrease the strength of the preference-performance relationship.

Polyphagy Increases Resource Complexity and Weakens the Preference-Performance Correlation An herbivore with a polyphagous feeding niche where it must evaluate both intraspecific and interspecific host variation will have a weaker preference-performance relationship than a monophagous herbivore that has only to evaluate intraspecific host variation.

Feeding on Multiple Plant Modules Increases Resource Complexity and Weakens the Preference-Performance Correlation A feeding niche where multiple host modules must be utilized during development will increase complexity, because determining the resource quality for offspring requires a female to evaluate the quality of multiple host modules rather than the single oviposition site. Plant resources could select for larval movement in several ways. The initial vegetation module may be too small for completion of development, and the larva must find new resources. Alternately, the larva may move because it requires different resources, or because higher-quality resources become available as the larva matures.

Natural Enemies Increase Resource Complexity, and This Weakens the Preference-Performance Correlation The feeding niche is also influenced by the availability of enemy-free space. If a female must choose an oviposition site based not only on plant resource quality, but also on the susceptibility of her offspring at the site to natural enemies, then complexity will increase and the preference-performance relationship will weaken.

Sensory Constraints

Insects have sensory limitations that constrain their ability to detect cues about which resources would be best for offspring development. We propose that feeding niches will differ in how sensory limitations constrain the evolution of preference and performance. The poorer the ability of a female to predict the quality of resources for larval development at the time of oviposition, the weaker the preference-performance relationship will be (Craig et al. 1999).

Changes in Plant Resources Between Oviposition Period and Insect Maturation Will Weaken the Preference-Performance Relationship The greater the change in the feeding niche between oviposition and the completion of immature development, the less likely that the insect sensory system will be able to detect cues at oviposition that forecast resource quality for the complete period of immature development. Insect sensory systems are constrained from evolving the capacity to detect future stochastic changes in resources. Large differences between resource quality at oviposition and adult maturation can occur for several reasons. Increased insect development time will create the opportunity for plant resources to change, and plant characteristics can strongly influence herbivore development time. Plants that are low in nutrition and high in defenses can slow insect development (Strong et al. 1984). If the feeding niche requires that the plant tissues mature before the insects can complete development (as with gall-formers), this may also increase development time. The life-history strategy of the plant can also influence the change in resources between oviposition and maturation. For example, an herbaceous plant changes much more completely during a growing season than a woody plant; as a result, cues about future resource quality are easier to detect on a woody plant than on a weedy species.

Movement Between Plant Resources Between Oviposition and Offspring Maturation Will Weaken the Preference-Performance Relationship If a larva must move to complete development, then sensory cues available at the oviposition site may have very little predictive value about the quality of resources when development is completed. The sensory system of the female is constrained from being able to sense resource quality at all the sites to which immature stages could potentially disperse. The greater the disparity in resource quality at the oviposition site and at the location where insect development is completed, the weaker the preference-performance relationship.

Sensory Biases Will Influence the Evolution of Preference-Performance Relationships Insect sensory systems are limited in the information they can detect. A change in the range of host-plant species an insect accepts for oviposition (i.e., the gain or loss of a host species) requires a change in the sensory system, such as a change in chemoreceptors (Bernays and Chapman 1994). Such changes are probably also required to recognize intraspecific variation in host-plant susceptibility due to its secondary chemistry. Some changes in sensory systems may be more likely to occur than others. For example, many herbivores have oviposition preference for rapidly growing plant modules (Price 1991, 2003), indicating that herbivores can readily evolve the sensory ability to detect differences in plant growth rate. There are several potential reasons why this ability might evolve so readily. First, plant growth rate cues may be relatively universal, unlike secondary plant chemicals that can be specific to a given host species or even to a genotype within a plant species. Thus, once a phytophagous insect lineage evolves the ability to detect plant growth cues, this could be maintained in the lineage because it is also adaptive for descendant species using a variety of host plants. Alternatively, there might be many different cues indicating plant growth, and thus many different mutations that would allow insects the ability to detect plant growth rate. In contrast, the evolutionary steps needed to assess many other secondary chemicals that influence performance may be difficult or impossible (Craig et al. 1999). Different feeding niches will have different cues indicating quality, and biases in which sensory abilities can most readily evolve will influence the evolution of a preference-performance relationship.

Searching Constraints

The feeding niche of a species will also determine how an insect must search for oviposition sites; this in turn will influence the evolution of the preference-performance relationship. A female searching for oviposition sites has a limited amount of time and resources to invest in the effort. The spatial and temporal distribution and the relative proportion of high- or variable-quality oviposition sites can all constrain the evolution of a strong preference-performance relationship.

High Variation in Resource Quality Selects for a Strong Preference-Performance Relationship If a feeding niche has highly variable resource quality, then the benefits gained through increased offspring performance at a high-quality oviposition site will outweigh the costs of searching; under these conditions a strong preference-performance correlation will evolve. Many tenthredinid sawflies with very strong positive preference-performance relationships follow this trend (Price 2003). Conversely, Price et al. (2004) showed that in one tenthredinid sawfly inhabiting a niche with low resource-quality variation, there was no link between preference and performance.

Rarity of High-Quality Resources Will Select for a Strong Preference-Performance Relationship If the high-quality resources in a feeding niche are rare, then few eggs deposited through random oviposition will end up where offspring can survive. Strong preference-performance relationships have been repeatedly found where suitable oviposition sites represent a small proportion of all possible oviposition sites (Price 2003).

The Narrower the Range of Resources Used, the Stronger the Preference-Performance Relationship If the feeding niche is broad, for example, in a highly polyphagous species, then searching the entire range of potential plant resources will be energetically expensive. This will weaken the preference-performance relationship because high-quality resources may not be located.

The More Time Available for Oviposition, the Stronger the Preference-Performance Relationship If the feeding niche is a resource with a short window of availability for oviposition, then time limitations will constrain the ability of an ovipositing female to assess resources. This will inhibit the evolution of a positive preference-performance relationship.

Preference and Performance in Three Well-Studied Interactions

Here we compare the preference-performance relationships in three herbivores that differ in their feeding niches, and in the strength of their preference-performance relationships. Their feeding niches, and therefore their interactions with plant resource variation, differ in crucial ways. The herbivores come from three different orders, so any similarities are not due to recent shared common ancestry.

Euura lasiolepis and Willows

Feeding Niche *Euura lasiolepis* (Hymenoptera: Tenthredinidae) is a monophagous shoot-galler on a woody plant, the arroyo willow, *Salix lasiolepis*, that completes development in the oviposition site chosen by the female. Natural enemies have little impact on sawfly larval survival.

Preference-Performance Relationship *Euura lasiolepis* has one of the strongest positive correlations between oviposition preference and offspring performance that has been reported (Price 2003). It has a strong preference for rapidly growing shoots, where offspring performance is highest (Craig et al. 1986, 1989).

Feeding Niche Resource Variation The feeding niche of the sawfly has low complexity, facilitating the development of a strong preference-performance relationship (prediction 1). The sawfly is highly specialized: it oviposits only on shoots of one host species, and it does not have to assess multiple axes of resource variation (prediction 1a). The sawfly has only to evaluate variation in shoot growth in order to determine the oviposition site quality, as variation in shoot quality explains a high proportion of variation in larval survival (Craig et al. 1986, 1989). Intraspecific variation among plants in larval survival is also largely explained by shoot length, and so the same cues can be used to assess resource quality variation within and among plants (Craig et al. 1986, 1989). Larvae complete development in the galls and do not utilize any other resources other than those chosen by the female (prediction 1b). Since larvae use only one resource module, females have only to assess the quality of a single plant node to assess the probability of larval survival. Natural enemies are not an important source of mortality (reviewed in Price 2003), and so there is no complexity added by the need to assess enemy-free space (prediction 1c). Intraspecific competition is the only other source of variation that a female must assess, as females must avoid leaf nodes that have already received an oviposition (Craig et al. 1988). Sawfly survival increases slightly if there are other galls on the shoot (Craig et al. 1990), and oviposition preference increases for a previously galled shoot because gall induction increases shoot growth. However, this does not increase complexity of the oviposition decision because shoot growth is again the oviposition cue.

Sensory Constraints Sensory constraints do not limit the ability of *E. lasiolepis* to accurately assess resource quality, facilitating the evolution of a strong preference-performance relationship (prediction 2). Resource quality does not change radically throughout larval development, so the cues available at oviposition are good predictors of the resource quality for the entire period of this development (prediction 2a). Larvae complete development and overwinter in the galls, and this extended development period is predicted to weaken the preference-performance relationship. The assumption, however, that extended development leads to changes in resource quality through time is not supported. *Salix lasioplepis* is a woody plant, so the resources that will be devoted to shoot growth are largely determined at the time of oviposition; and shoot length at oviposition is highly correlated with the shoot length at larval maturity (Craig et al. 1989). The potentially important causes of resource variation—shoot growth, intraspecific or interspecific competition, and natural enemy attack—do not change much during larval development. The sawfly uses tremulacin, a phenolic glycoside, to detect plant growth rate (Roininen et al. 1999) to choose the most rapidly growing shoot for oviposition. Above, we argue that there is sensory bias in the evolution of sensitivity to chemical cues related to plant growth rate (prediction 2c).

The *E. lasiolepis* feeding niche has a distribution of resources that facilitates the evolution of a strong preference-performance relationship (prediction 3). A large variability in oviposition site quality produces strong selection for selective oviposition (prediction 3a). Survival to adulthood ranges from zero on the shortest shoots, which abscise and fall to the ground, to nearly 90% on the longest shoots. Oviposition sites on rapidly growing shoots, where survival is high, are rare, strengthening the selection for selective oviposition (prediction 3b).

The spatial and temporal distribution of *S. lasiolepis* oviposition sites produces few searching constraints on the sawfly, facilitating the evolution of a strong preference-performance relationship (predictions 3a, 3b, and 3c). *Salix lasiolepis* are distributed contiguously along streams and springs so that the complete range of resource variation is within easy dispersal range of the sawfly (Stein et al. 1994). Female *E. lasiolepis* oviposit in rapidly growing shoots of *S. lasiolepis* in May and early June, when growth is at its peak (Price and Craig 1984), and there is an adequate window of time for adults to emerge and assess resources, so that resource phenology does not constrain the preference-performance relationship (prediction 3d).

Aphrophora pectoralis and Willows

Feeding Niche The spittlebug, *A. pectoralis* (Hemiptera: Cercopidae), is a polyphagous, xylem-feeder on woody plants (willows, *Salix* spp.). In Japan, it feeds on at least seven willow species (Komatsu 1997). The spittlebug utilizes multiple shoots during its development. It does not suffer natural enemy mortality.

Preference-Performance Relationship *Aphrophora pectoralis* has a positive preference-performance relationship, but it is weaker than in *E. lasiolepis*. The spittlebug has a preference for ovipositing on rapidly growing shoots, and larval performance there is high. It also has an ovipositional preference hierarchy among willow species that is significantly, but imperfectly, correlated with offspring performance on these species.

Feeding Niche Resource Variation The feeding niche of *A. pectoralis* has a higher resource complexity than is encountered by *E. lasiolepis,* and the preference-performance relationship is positive but lower (prediction 1) than in the sawfly. The spittlebug female must assess the resource variation on at least three axes: shoot growth, number of other egg masses, and willow species. As was found in *E. lasiolepis*, the spittlebug has a strong preference for rapidly growing willow shoots, where performance is high. However, shoot quality is not determined by shoot growth alone, because survival is facilitated when different females aggregate eggs together. Unlike the sawfly interaction, shoot growth is not facilitated by egg mass deposition, so the spittlebug female must evaluate additional cues to determine the egg load of the shoot.

The spittlebug is polyphagous, feeding on at least seven willow species in Japan (Komatsu 1997), and this adds to the complexity of the feeding niche (prediction 1a). Craig and Ohgushi (2002) studied spittlebug preference and performance on four willow species and found a preference ranking among willow species that correlated with performance. However, the correlation was not perfect, as the highest-ranking host species for performance was not the most preferred species.

The spittlebug nymphs move as they develop, utilizing multiple shoots. This potentially increases the complexity of the oviposition decision and decreases the preference-performance relationship (prediction 1b). It would be difficult for females to assess the quality of all of the resources used during nymphal development because nymphs move to multiple shoots as far as a meter away from the oviposition site. Direct observation of oviposition behavior shows no evidence that females systematically assess the quality of shoots adjacent to the oviposition site that offspring might likely encounter (T.P.C., 1997, personal observation).

A strong preference-performance relationship in shoot choice is maintained in spite of nymphal movement because it does not greatly alter the resource quality encountered. Early instar nymphs usually feed at the base of the old shoot where eggs were oviposited, or on an immediately adjacent shoot. Shoot diameter of the immediately adjacent shoots is strongly correlated with the shoot chosen for oviposition (Craig and Ohgushi 2002). If a nymph emerges on a good shoot, it does not move far, so that it develops on resources similar to that chosen by the female for oviposition. It usually moves only to the adjacent shoots of similar quality, or later in the season to new shoots that are also of similar quality. In the rare cases where eggs are oviposited on very poor quality shoots, nymphs will move over a meter because adjacent shoots are also of poor quality. In observations, these nymphs often disappear and may frequently die while searching for a better feeding site. Natural enemy attack is also virtually absent, and so this does not influence the complexity of searching for an oviposition site (prediction 1c).

Sensory Constraints Females of *A. pectoralis* encounter sensory constraints that potentially weaken the preference-performance relationship, but other factors mitigate these constraints (prediction 2). An extended developmental period creates the potential for resource change between oviposition and the adult stage (prediction 2a). Females oviposit into rapidly growing shoots over an extended period in the late summer and early fall (Craig and Ohgushi 2002), and nymphs do not emerge and develop on those resources until the following midsummer. Shoot breakage during the winter on shoots with heavy egg loads alters oviposition site quality by reducing performance in highly preferred sites (Nozawa and Ohgushi 2002). Also, the nymphal movement discussed above reduces the connection between the cues available at the oviposition site and the quality of sites where nymphs complete development. Two factors may somewhat increase the ability of females to sense the quality of sites where nymphs develop. First, there is a high correlation between shoot quality chosen for oviposition and the quality of both the immediately adjacent shoots and all shoots on the willow. Second, resource quality on woody plants is correlated among years (Craig et al. 1986), and this contributes to the predictability of resource quality (prediction 2b).

As with *E. lasiolepis*, shoot growth is again the cue used by *A. pectoralis* females in oviposition decisions. This similarity between distantly related insects supports our previously stated hypothesis that insect sensory systems are generally adapted to detect shoot growth cues (prediction 2c).

As seen in *E. lasiolepis*, the wide range in oviposition site quality (with the best sites being very rare) exerts the same selection on the spittlebug for the evolution of a strong preference-performance relationship (prediction 3). Resource quality varies widely, as nymphal survival rates are strongly influenced by three traits: shoot size, the number of previously oviposited egg masses, and willow species (prediction 3a). High-quality oviposition sites are extremely rare (prediction 3b) because the best ones are a combination of these traits: large-diameter shoots with previous egg masses on *S. hultenii* (Craig and Ohgushi 2002).

Aphrophora pectoralis is polyphagous, and the oviposition sites on different host species are widely spatially dispersed. This complicates the search process and weakens the preference-performance relationship (prediction 3c). The willow species where performance is best, *S. hultenii,* is small and tended to be isolated from the other species (Craig and Ohgushi 2002). This apparently makes locating this species difficult for females searching for oviposition sites, and weakens the preference-performance relationship.

Eurosta solidaginis and Goldenrods

Feeding Niche *Eurosta solidaginis* (Diptera: Tephritidae) is a monophagous or narrowly oligophagous shoot-galler on an herbaceous plant. It completes development in the oviposition site chosen by the female. The tephritid fly has host races that induce galls on two species of goldenrod, *Solidago altissima* and *S. gigantea*. *Eurosta solidaginis* survival is strongly influenced by natural enemy mortality.

Preference-Performance Relationship At the level of host species there is a strong preference-performance relationship: flies reared from *S. altissima* show a strong preference for oviposition on their own host plants when given a choice of *S. gigantea*, where survival is low (Craig et al. 1993). *Eurosta solidaginis* from *S. altissima* show no preference-performance relationship when choosing among *S. altissima* plants. Flies show a preference for rapidly growing plants, but plant growth has no significant relationship to larval survival (Craig et al. 1999). Variation among plant genotypes has a strong impact on larval survival, but there is no significant preference for genotypes where survival is high (Craig et al. 1999, 2000; Cronin and Abrahamson 2001; Cronin et al. 2001). At times, the most preferred genotype has a zero survival rate (Craig et al. 1999, 2000). Intraspecific competition also affects performance. A developing bud may receive multiple ovipositions, but only one larva can survive per node (Craig et al. 1999). Females, however, show no avoidance of previously used oviposition sites, despite the negative impact of larval competition (Craig et al. 1999).

Feeding Niche Resource Variation Resource complexity is high for a *Eurosta* female searching for an oviposition site because host-plant species, plant genotype, intraspecific competitors, interspecific interactions, and natural enemy densities all vary within and among sites (prediction 1). Within their dispersal range, most flies encounter not only a range of plant species but also a range of intraspecific plant genotypes with different potential for offspring performance. Natural enemy attack, especially bird predation, is spatially sporadic and temporally variable, creating variation in enemy-free space (prediction 1c). Birds will consume more than 50% of all larvae in some sites (Weis et al. 1992; Craig 2006), and this is strongly influenced by proximity to trees. It is probably difficult for ovipositing females to assess the number of trees in an area that are suitable for bird perches. Competition from other larvae also influences oviposition site quality, and it is unpredictable since previous oviposition neither deters nor facilitates further oviposition. Conspecific interactions with other herbivores can influence oviposition preference and larval performance (Cronin and Abrahamson 1999, 2001). The complexity of evaluating resource quality is, however, simplified somewhat because as a gall-inducer, *E. solidaginis* larvae develop within the oviposition site, and they use only one resource module (prediction 2b).

Sensory Constraints Sensory constraints are high for *E. solidaginis,* and we hypothesize that this has contributed to the weak preference-performance relationship (prediction 2). Cues available at the time of oviposition are poor predictors of site quality over the entire period of larval development because during the extended larval development period, resource quality changes (prediction 2a). Stem growth in this herbaceous plant subsequent to oviposition is highly unpredictable. Future weather can strongly influence larval survival, as drought stress can cause high mortality in dry areas, with high survival occurring only in wet sites (Sumerford et al. 2000).

The quality of a plant for larval development can be altered due to natural enemy attack (for reviews, see Abrahamson and Weis [1997] and Craig [2006]), and this is difficult to predict at the time of oviposition. Bird predation is an important mortality factor that is highly unpredictable both spatially and temporally (Weis et al. 1992; Craig 2006). Bird predation can be very heavy as transient bird flocks happen to roost in trees near the goldenrod. Gall size determines susceptibility to natural enemy attack, and gall size is difficult to predict at the time of oviposition because it is influenced by the interaction of plant genotype, insect genotype, and the environment (Weis and Abrahamson 1986). Larvae in small galls are susceptible to parasitism by an inquiline (Craig 2007) and a parasitoid (Weis and Abrahamson 1986), and large galls are susceptible to bird predation (Weis and Abrahamson 1986). Gall size is highly variable within plants of the same genotype even when grown in the same experimental pot under standardized experimental conditions (T.P.C., 1994, personal observation). Precipitation subsequent to oviposition can greatly alter gall size, changing the susceptibility to natural enemy attack (Sumerford et al. 2000).

We hypothesize that sensory bias limits the ability of *E. solidaginis* to evolve a strong preference-performance relationship (prediction 2c). While it does respond to variation in plant growth rate that has no detectable impact on performance, *E. solidaginis* shows no evidence of being able to detect intraspecific genotypic and phenotypic differences among plants that would increase performance. Evolving to detect cues to intraspecific host-plant variation may be difficult because of the complexity of this variation. The strong interaction of plant genotype, fly genotype, and environment in determining fitness suggests that cues predicting performance at a particular oviposition site would be complex and harder to detect than cues for plant growth rate.

Variation in resource quality is high, with large variation in survival among offspring at different oviposition sites (prediction 3) due to the variation among genotypes, environments, and intraspecific and interspecific competitors described above. This suggests the potential exists for a strong preference-performance relationship to evolve (prediction 3a). Oviposition sites where survival is high are rare in some years, but in some years more than 30% of the stems produce survivors. The wide availability of suitable oviposition sites may decrease the selection for selective oviposition behavior (prediction 3b).

Eurosta solidaginis females may also face large problems in assessing resource quality due to the dispersal of resources. Goldenrod is a midsuccessional plant that grows patchily in isolated disturbed sites, and once a goldenrod clone is established it can spread by rhizomes to cover large areas. *Eurosta solidaginis* has very limited dispersal ability (Cronin and Abrahamson 1999). This fact combined with the distribution of goldenrod genotypic and phenotypic variation means that an individual female can assess only a very limited range of the variation in resource quality (prediction 3c). The window of time when successful oviposition is possible does not appear to limit the ability of females to assess resource quality (Horner et al. 1999). However, there is evidence that females may be time constrained for reasons that are not understood. Females are active for only very brief periods in a narrow range of environmental conditions (Uhler 1951; Craig et al. 1993), and they live for only a few days in outdoor cages. They emerge with a full complement of eggs (Uhler 1951) and oviposit rapidly when environment conditions are suitable (Craig et al. 1999).

The Three Systems

Differences in niche complexity and sensory, resource, and searching constraints all seem to play a role in the decreasing correlation between preference and performance, with *Euura lasiolepis* > *Aphrophora pectoralis* > *Eurosta solidaginis*. Simple grouping by such factors as being gall-formers or endophytic feeders is not sufficient to predict the preference-performance relationship. For example, all three are adapted for oviposition into plant tissue, a trait that Price (2003) predicts would result in a positive preference-performance relationship, and yet they differ in this relationship. Two of the species are gall-inducers, but they had very different preference-performance relationships. We conclude that a more complete understanding of the resource variation in the feeding niche is required to predict preference-performance correlations.

Increasing resource complexity is a key difference predicting the differences in the preference-performance relationships. *Euura lasiolepis* has a very simple environment in which to make choices: if it chooses the most rapidly growing shoot it will have the highest performance. The next ranking species, *A. pectoralis*, also increases performance by choosing the most rapidly growing shoot, but it also must choose among host species, and it faces a more spatially complex environment. *Eurosta solidaginis* has the weakest preference-performance link, and it faces a great deal of resource complexity, particularly in the form of host-plant intraspecific genotypic and phenotypic complexity.

The ability to predict oviposition site quality, which we view as a sensory constraint, appears critical in the evolution of the preference-performance relationship. The factors that limit the ability to detect future resource quality are not those that we originally predicted. We assumed that gall-makers, because they do not move from the oviposition site, would have a better ability to detect cues predicting the quality of the resources for offspring than insects whose offspring disperse from the oviposition site. Resource quality changes least for the gall-maker *Euura lasiolepis*, where the preference-performance relationship is strongest. We have found, however, that the gall-maker *Eurosta solidaginis* has a weaker preference-performance relationship than *A. pectoralis*, which disperses from the oviposition site. This is because resource quality changes much more for the gall-maker that stays in one place than it does for the spittlebug nymphs, which disperse. An important factor in explaining this increased resource variability is predictable development on woody plants versus the unpredictability of herbaceous plants.

All three species encounter a wide range of resource qualities selecting for the evolution of strong preference-performance relationships. High-quality resources are rare for the spittlebug and sawfly, which we also predicted would result in the evolution of selective oviposition and strong preference-performance correlations. This contrasts with the resource variation for *E. solidaginis*, where in some years a large proportion of oviposition sites are suitable for larval development, while in other years almost none are. This should reduce selection for selectivity in oviposition site choice, decreasing the probability that a positive preference-performance relationship would evolve. We think that the combination of high resource complexity and the low predictability of resource quality for offspring at the time of oviposition in *E. solidaginis* results in a weak preference-performance relationship. *Eurosta solidaginis* has to hedge it bets by ovipositing large number of eggs because it lacks cues that could indicate whether oviposition site quality is high, and it has a difficult time dealing with the complex variation in resources.

The two species with more dispersed resources, the spittlebug and the gall-fly, have weaker preference-performance relationships, indicating that barriers to assessing resource quality due to searching difficulties may decrease the preference-performance relationship.

Alternative Hypotheses

A multitude of specific explanations have been advanced for why the predicted strong preference-performance relationship does not occur in specific interactions (Thompson 1988; Courtney and Kibota 1990; Mayhew 1998, 2001). The evolution of oviposition behavior is only one aspect of the evolution of optimal behavior, and other factors such adult feeding preferences can influence oviposition preferences (Scheirs et al. 2004). In addition, idiosyncratic features in the history of the plant-insect interaction may influence oviposition preference. For example, herbivores may not have enough time to evolve adaptive oviposition responses to recently introduced plants (Chew 1977). Such idiosyncratic characteristics of interactions may influence, and somewhat obscure, the general patterns in the evolution of preference-performance relationships we have hypothesized.

Another hypothesis explaining overall preference-performance patterns is the phylogenetic constraints hypothesis (Price 1994; Price et al. 1995, 1998a, 2003; Price and Carr 2000; Price, this volume) that proposes that a phylogenetic constraint is "a critical plesiomorphic character, or set of characters common to a major taxon, that limits the major adaptive developments in a lineage and thus the ecological options for that taxon" (Price, 2003, p. 5). Preference-performance relations can evolve based on these characters. For example, Price has argued that tenthredinid sawflies, such as *Euura lasiolepis*, all have a sawlike ovipositor that forces them to oviposit into plant tissue. This ovipositor allows them to detect host-plant variation and to evolve positive preference performance relationships. In contrast, a group of macrolepidopteran moths that cause outbreaks in forests lack an ovipositor that can provide information about plant quality, and they have not evolved a strong preference-performance relationship. The phylogenetic constraints hypothesis proposes that shared preference-performance patterns will depend on shared ancestry, and not necessarily on shared feeding niches.

In many cases, the phylogenetic constraints hypothesis and the feeding niche hypothesis predict the same patterns. For example, all tenthredinid sawflies oviposit into plant tissue, and this leads to them occupying a similar feeding niche and encountering similar resource variation, such as initial endophytic feeding. Almost of all of these species have a positive preference-performance relationship, and they have a number of other feeding niche characters that we have hypothesized to favor the evolution of a positive preference-performance pattern. The phylogenetic and feeding niche constraints hypotheses differ in their predictions. If species share a common adaptation that is hypothesized to constrain the evolution of the preference-performance relationship, but they have different preference-performance relationships, this would support the feeding niche hypothesis (but not the phylogenetic constraints hypothesis). For example, the three species discussed in this chapter share the trait of ovipositors adapted to oviposit into plant tissue, but they differ in their preference-performance relationships. On the other hand, if species were observed to have different feeding niches, but similar phylogenetic constraints and similar preference-performance relationships, this would support the phylogenetic constraints hypothesis (but not the feeding niche hypothesis).

Testing Hypotheses on Preference and Performance

A broader, systematic approach is needed to test general hypotheses about the evolution of preference-performance relationships. Certainly no general conclusions can be drawn from a comparison of only three interactions, as we have done in this chapter. We reviewed over 100 papers to determine if we could test the ecological niche constraints hypothesis in other species, but we found that the data did not exist to rigorously test the hypothesis. Studies typically examine preference-performance relationships only in response to single factors, making it difficult to assess the overall preference-performance relationships. Most studies do not describe feeding niches in sufficient detail to evaluate our hypothesis. Differences in the methods used to measure preference and performance also make comparisons difficult.

To test the feeding niche constraints hypothesis, detailed comparative studies are needed. Two approaches would be most fruitful. The first would be to continue the approach taken in this study to choose insects from a variety of different taxa and predict, a priori, their preference-performance relationships based on their feeding niche characteristics. The second approach would be to take a closely related group of species that have radiated into different feeding niches, to detail the characteristics of their feeding niches and precisely measure their preference-performance relationships. The tenthredinid sawflies, for which a great deal of information has already been collected (Price 2003), would be a promising group on which to work.

Conclusion

In retrospect, it is no surprise that there has been little support for the naïve adaptationist hypothesis. The unstated assumption is that an ovipositing female with a perfect knowledge of resource quality, unlimited resources, and no other trade-offs in her life-history strategy would oviposit her eggs where her offspring would have the highest fitness. These sound like preliminary assumptions a modeler would make for a first stab at an overly simplified and admittedly unrealistic model. It is remarkable that such simplified systems have been found in nature, such as in *Euura lasiolepis*, where preference and performance are highly correlated. We now need to determine what other factors need to be added to this simplified model to make it more realistic and applicable to a broader number of interactions. We suggested some of these factors in this chapter, and we hope that it generates more focused and systematic study of preference-performance relationships.

Acknowledgments

We thank Peter Price for his review of an earlier version of this chapter, and for many insightful discussions over the years on the ideas in this chapter. We also thank Kelley Tilmon, Mark Scriber, and an anonymous reviewer for very helpful reviews.

References Cited

Abrahamson, W.G., and E.A. Weis. 1997. Evolutionary ecology across three trophic levels: goldenrods, gallmakers, and natural enemies. Princeton University Press, Princeton.

Bernays, E.A. and R.F. Chapman. 1994. Host-plant selection by phytophagous insects. Chapman and Hall, New York.

Bernays, E.A. 1998. The value of being a resource specialist: behavioral support for a neural hypothesis. Am. Nat. 151: 451–464.

Bernays, E.A. 2001. Neural limitations in phytophagous insects: implications for diet breadth and evolution of host affiliation. Annu. Rev. Entomol. 46: 703–727.

Chew, F.S. 1977. Coevolution of pierid butterflies and their cruciferous foodplants. II. The distribution of eggs of potential food plants. Evolution 31: 568–579.

Courtney, S.P., and T.T. Kibota. 1990. Mother doesn't know best: host selection by ovipositing insects, pp. 161–188. In E.A. Bernays (ed.), Insect-plant relationships, vol. 2. CRC Press, Boca Raton, FL.

Craig, T.P. 2007. Evolution of plant mediated interactions. In T. Ohgushi, T.P. Craig, and P.W. Price (eds.), Ecological communities: plant mediation in indirect interaction webs. Cambridge University Press, Cambridge, UK.

Craig, T.P., and T. Ohgushi. 2002. Preference and performance in the spittlebug, *Aphrophora pectoralis,* on four species of willow. Ecol. Entomol. 27: 529–540

Craig, T.P., P.W. Price, and J.K. Itami. 1986. Resource regulation by a stem-galling sawfly, on the arroyo willow. Ecology 67: 419–425.

Craig, T.P., P.W. Price, K.M. Clancy, G.L. Waring, and C.F. Sacchi. 1988. Forces preventing coevolution in the three-trophic-level system: willow, a gall-forming herbivore, and parasitoid, pp. 57–80. In K.C. Spencer (ed.), Chemical mediation of coevolution. Academic Press, New York.

Craig, T.P., J.K. Itami, and P.W. Price. 1989. A strong relationship between oviposition preference and larval performance in a shoot-galling sawfly. Ecology 70: 1691–1699.

Craig, T.P., J.K. Itami, and P.W. Price. 1990. Intraspecific competition and facilitation by a shoot-galling sawfly. J. Anim. Ecol. 59: 147–159.

Craig, T.P., J.K. Itami, W.G. Abrahamson, and J.D. Horner. 1993. Behavioral evidence of host race formation in *Eurosta solidaginis*. Evolution 47: 1696–1710.

Craig, T.P., J.K. Itami, W.G. Abrahamson, J.D. Horner, and J.V. Craig. 1999. Oviposition preference and offspring performance of *Eurosta solidaginis* on genotypes of *Solidago altissima*. Oikos 86: 119–126.

Craig, T.P., J.K. Itami, C. Schantz, W.G. Abrahamson, J.D. Horner, and J.V. Craig. 2000. The influence of host plant variation and intraspecific competition on oviposition preference in the host races of *Eurosta solidaginis*. Ecol. Entomol. 25: 7–18.

Cronin, J.T., and W.G. Abrahamson. 1999. Host-plant genotype and other herbivores influence goldenrod stem galler preference and performance. Oecologia 121: 392–404.

Cronin, J.T., and W.G. Abrahamson. 2001. Goldenrod stem galler preference and performance: effects of multiple herbivores and plant genotypes. Oecologia 127: 87–96.

Cronin, J.T., W.G. Abrahamson, and T.P. Craig. 2001. Temporal variation in host-plant preference and offspring performance: constraints on host-plant specialization. Oikos 93: 312–320.

Dethier, V.G. 1959a. Foodplant distribution and larval dispersal as factors affecting insect populations. Can. Entomol. 91: 581–596.

Dethier, V.G. 1959b. Egg-laying habits of Lepidoptera in relation to available food. Can. Entomol. 91: 554–561.

Horner, J.D., T.P. Craig, and J.K. Itami. 1999. The influence of oviposition phenology on survival in the host races of *Eurosta solidaginis*. Entomol. Exp. Appl. 93: 121–129.

Jaenike, J. 1978. On optimal oviposition behavior in phytophagous insects. Theor. Popul. Biol. 14: 350–356.

Komatsu, T. 1997. A revision of the froghopper genus *Aphrophora* Gemar (Homoptera, Cercopoidea, Aphrophoridae) from Japan, part 2. Japanese J. Entomol. 65: 369–383.

Mayhew, P.J. 1998. Testing the preference-performance hypothesis in phytophagous insects: lessons from chrysanthemum leafminer (Diptera: Agromyzidae). Environ. Entomol. 27: 45–52.

Mayhew, P.J. 2001. Herbivore host choice and optimal bad motherhood. Trends Ecol. Evol. 16: 165–167.

Nozawa, A., and T. Ohgushi. 2002. Indirect effects mediated by compensatory shoot growth on subsequent generations of a willow spittlebug. Popul. Ecol. 44: 235–239.

Price, P.W. 1991. The plant vigor hypothesis and herbivore attack. Oikos 62: 244–251.

Price, P.W. 1994. Phylogenetic constraints, adaptive syndromes, and emergent properties: from individuals to population dynamics. Res. Popul. Ecol. 36: 3–14.

Price, P.W., T.P. Craig, and M.D. Hunter 1998. Population ecology of a gall-inducing sawfly, *Euura lasiolepis*, and relatives, pp. 323–340. In J.P. Dempster and I.F.G. McLean (eds.), Insect populations in theory and practice. Kluwer, Dordrecht, The Netherlands.

Price, P.W. 2003. Macroevolutionary theory on macroecologidal patterns. Cambridge University Press, Cambridge, UK.

Price, P.W., and T.G. Carr. 2000. Comparative ecology of membracids and tenthredinids in a macroevolutionary context. Evol. Ecol. Res. 2: 645–665.

Price, P.W., and T.P. Craig. 1984. Life history, phenology, and survivorship of a stem-galling sawfly, *Euura lasiolepis* (Hymenoptera: Tenthredinidae), on the arroyo willow, *Salix lasiolepis*, in Northern Arizona. Ann. Entomol. Soc. Am. 77: 712–719.

Price, P.W., T.P. Craig, and H. Roininen. 1995. Working toward theory on galling sawfly population dynamics, pp. 321–338. In N. Cappuccino and P.W. Price (eds.), Population dynamics: new approaches and synthesis. Academic Press, San Diego, CA.

Roininen, H., P.W. Price, R. Julkunen-Tiitto, J. Tahvanainen, A. Ikonen. 1999. Oviposition stimulant for a gall-inducing sawfly, *Euura lasiolepis*, on willow is a phenolic glucoside. J. Chem. Ecol. 25: 943–953.

Scheirs, J., T.G. Zoebisch, D.J. Schuster, and L. De Bruyn. 2004. Optimal foraging shapes host preference of a polyphagous leafminer. Ecol. Entomol. 29: 375–379.

Singer, M.C. 1972. Complex components of habitat suitability within a butterfly colony. Science 176: 75–77.

Singer, M.C. 1986. The definition and measurement of oviposition preference in plant feeding insects, pp. 66–94. In J. Miller and T.A. Miller (eds.), Insect-plant relations. Springer-Verlag, New York.

Stein, S.J., P.W. Price, T.P. Craig, and J.K. Itami. 1994. Dispersal of a galling sawfly: implications for studies of insect population dynamics. J. Anim. Ecol. 63: 666–676.

Strong, D.R., J.H. Lawton, and T.R.E. Southwood. 1984. Insects on plants community patterns and mechanisms. Blackwell Scientific, Oxford, UK.

Sumerford, D.V., W.G. Abrahamson, and A.E. Weis. 2000. The effects of drought on the *Solidago altissima–Eurosta solidaginis*–natural enemy complex: population dynamics, local extirpations, and measures of selection intensity on gall size. Oecologia 122: 240–248.

Thompson, J.N. 1988. Evolutionary ecology of the relationship between oviposition preference and performance of offspring in phytophagous insects. Entomol. Exp. Appl. 47: 3–14.

Uhler, L.D. 1951. Biology and ecology of the goldenrod gall fly, *Eurosta solidaginis* (Fitch). Cornell University Agricultural Station Memoir 300: 1–51.

Weis, A.E., and W.G. Abrahamson. 1986. Evolution of host plant manipulation by gallmakers: ecological and genetic factors in the *Solidago-Eurosta* system. Am. Nat. 127: 681–695.

Weis, A.E., W.G. Abrahamson, and M.C. Andersen. 1992. Variable selection on *Eurosta*'s gall size. I. The extent and nature of variation in phenotypic selection. Evolution 46: 1674–1697.

THREE

Evolutionary Ecology of Polyphagy

MICHAEL S. SINGER

The evolutionary ecology of polyphagy by phytophagous insects has been overshadowed by an intense focus on the evolutionary ecology of their host specificity. This bias reflects the preponderance of host specificity in phytophagous insects (reviewed by Weis and Berenbaum 1989 and Novotny and Basset 2005) and its fascinating consequences for community structure and evolutionary diversification. Truly, the study of host-specific herbivores has provided many key insights and motivated the conceptual side of the study of plant-insect interactions (e.g., Brues 1924; Dethier 1954; Fraenkel 1959; Ehrlich and Raven 1964; Feeny 1976; Rhodes and Cates 1976) as well as the broader issues of ecological specialization (e.g., Futuyma and Moreno 1988; Novotny et al. 2002), coevolution (e.g., Thompson 1994, 1999; Becerra 1997; Berenbaum and Zangerl 1998), and speciation (e.g., Schluter 2000; Nosil et al. 2002; Stireman et al. 2005). However, I believe the development of evolutionary-ecology theory and the strength of its empirical tests require further attention to the full range of host-plant use observed in insect herbivores. After all, specialist and generalist herbivore species co-occur in the same communities (e.g., Novotny and Basset 2005), have overlapping host-plant ranges (e.g., Novotny et al. 2004), and have frequently sprung from the same phylogenetic lineages (Janz et al. 2001; Nosil 2002; Morse and Farrell 2005). Therefore, any complete theory of evolutionary ecology must ultimately aim to explain the existence of variation in diet breadth, in addition to the preponderance of host specificity seen in most phytophagous insect lineages. I argue here that the ultimate causes of host specificity may become clearer by greater consideration of the polyphagous exceptions to the rule. The study of dietary generalists, of course, also illuminates the causes of polyphagy, which has both conceptual and practical value of its own, as many invertebrate and vertebrate herbivores and many pestiferous insect herbivores are polyphagous.

In my approach to the evolutionary ecology of polyphagy, I will assume that ultimate explanations for host-plant use by phytophagous insects are generally historical and functional (i.e., adaptive). That is, I contend that functional explanations of microevolutionary processes are the most important cause of macroevolutionary patterns. I raise this point because there is a widespread awareness in evolutionary biology that natural selection operates within various genetic-developmental constraints, and the role of such constraints in determining macroevolutionary patterns is unresolved (e.g., Pigliucci and Kaplan 2000; Arthur 2004). Accordingly, some authors have recently invoked such constraints as ultimate causes of the macroevolutionary patterns of host use either in particular instances or more generally. These genetic or developmental constraints are thought to limit the intrinsic evolutionary potential of particular phylogenetic lineages, creating phylogenetically conservative patterns of host-plant use (Futuyma et al. 1995; Janz et al. 2001; Price 2003). Although such constraints on host use may be important in certain cases and in ecological time (but see Radtkey and Singer [1995] and Singer et al. [this volume] for counterexamples), alone they offer limited explanatory power for macroevolutionary patterns of host-plant use when one considers polyphagy as well as stenophagy at the level of herbivore clades. Host-specific herbivore species nested within lineages or clades that use a phylogenetically disparate set of host-plant species (case B in Table 3.1) indicate the widespread past occurrence of shifts between phylogenetically disparate hosts. Although the frequency of phylogenetic lability in host use has not been quantified among randomly selected lineages, several known examples indicate that this pattern is not uncommon in nature (Janz and Nylin 1998; Powell et al. 1998; Lopez-Vaamonde et al. 2003; Farrell and Sequeira 2004). Unlike explanations that focus on constraints as properties of lineages, functional explanations based on

TABLE 3.1

Theoretical Combinations of Host Specificity and Polyphagy at the Levels of Species Nested within a Larger Clade or Lineage

	Herbivore Clade or Lineage	
Herbivore Species	Specialist	Generalist
Specialist	Related specialists use phylogenetically restricted hosts (case A)	Related specialists use phylogenetically disparate hosts (case B)
Generalist	Polyphagous species use phylogenetically restricted hosts (case C)	Polyphagous species use phylogenetically disparate hosts (case D)

NOTE: In practice, relative specialists and generalists are defined in the context of the study system.

ecological constraints can explain host specificity *or* polyphagy at the species (or lower) level, nested within a clade with either broad *or* restricted host-plant use (cases A–D in Table 3.1). The following discussion of host-plant use will focus on host range at the level of herbivore species or below (population, individual).

General Explanations for Host Specificity

Theoretical discussions of host-plant range by phytophagous insects reveal several plausible adaptive explanations with scattered empirical support, a complex state of affairs that is poorly resolved (e.g., Jaenike 1990; Mayhew 1997). For context, I will first focus on the familiar ground of host-plant specificity. The strongest general hypotheses advanced to explain the predominance of host specificity in insect herbivores include the physiological-efficiency, enemy-free-space, optimal-foraging, and neural-constraints hypotheses.

Physiological-Efficiency Hypothesis

The physiological-efficiency hypothesis is implicit in much of the literature and was most seminally articulated by Dethier (1954). It states that herbivores have evolved host specificity in adapting physiologically to the food quality of their host plants. There is a large amount of evidence for the notion that host-specific herbivores are physiologically adapted to the nutritional and secondary chemistry of their host plants (reviewed by Slansky 1993 and Cornell and Hawkins 2003), but limited support for the prediction that specialists perform better than generalists on shared host plants (reviewed by Strauss and Zangerl 2002). The physiological-efficiency hypothesis also assumes a physiological trade-off in an herbivore's ability to efficiently utilize alternative host-plant species as food. That is, by adapting its physiology to the chemical and physical characteristics of one host-plant species (or related set of host species), an herbivore will generally become less adapted to using other host-plant species. This part of the hypothesis has also received mixed support (reviewed by Strauss and Zangerl 2002). Finally, the physiological-efficiency hypothesis assumes that the evolution of adult oviposition preference will reflect that of immature performance, the commonly discussed preference-performance correlation. As pointed out by Scheirs et al. (2004) and various other authors, this assumption is not always well founded.

In a community setting with intra- and interspecific competition, the expected outcome of physiological trade-offs in host utilization is niche partitioning among many herbivore species, each specialized to eat particular plant species in a particular adaptive way. In a phylogenetic context, the expected outcome of this process is varied: some herbivore lineages may track particular host-plant lineages via strict or loose co-cladogenesis; other herbivore lineages may be associated with a variety of host-plant lineages. The pattern depends on the process of coevolution between plants and herbivores (Thompson 1999). In pure form, the physiological-efficiency hypothesis is bitrophic, strictly focused on the interaction between plant and herbivore. Interactions with higher trophic levels are not part of the conceptual or methodological framework.

Enemy-Free-Space Hypothesis

The enemy-free-space hypothesis states that herbivores have evolved host specificity via specialized ways of using their host plants as a refuge or defense against carnivores. Enemy-free space thus includes defensive strategies such as the use of plant toxins for resistance to enemies, physical or chemical crypsis to avoid detection from enemies, and many others (Singer and Stireman 2005). This hypothesis has gained support in recent years, with increasing investigations of multitrophic interactions by ecologists. Brower (1958), Gilbert (1979), Price et al. (1980), and Janzen (1985) provided some of the first synthetic articulations of the idea that interactions among plants, herbivores, and carnivores may be essential to understanding the evolutionary ecology of plant-insect interactions. The concept of enemy-free space was forcefully developed by Jeffries and Lawton (1984), then famously used by Bernays and Graham (1988) to challenge the primacy of plant chemical defense (within the framework of the physiological-efficiency hypothesis) as an explanation for host specificity. Bernays and Graham (1988) argued that mortality from generalist predators is likely to be the strongest agent of natural selection causing host specificity. Several experimental studies show that selection from generalist predators favors host specialists over generalists (Bernays 1988; Bernays and Cornelius 1989; Dyer 1995, 1997; Vencl et al. 2005), and others show that enemy-free space can be more important than food quality in determining host-plant preference (Damman 1987; Denno et al. 1990; Baur and Rank 1996; Camara 1997a, 1997b, 1997c; Murphy 2004). Bernays and Graham (1988) also posited the enforcement of host specificity via ecological trade-offs involving predation and other factors, rather than physiological trade-offs in the efficient utilization of particular foods.

Extended to community ecology, the expected outcome of the enemy-free-space hypothesis is very similar to that expected from the physiological-efficiency hypothesis: niche partitioning among herbivore species that are specialized to eat and defensively use particular plant species in particular ways. The phylogenetic patterns expected from the enemy-free-space hypothesis are variable, like those expected from the physiological-efficiency hypothesis. It is important to be clear about one captivating phylogenetic pattern that may be predicted by either hypothesis: constrained phylogenetic or phytochemical associations between plants and herbivores (e.g., Farrell and Mitter 1990, 1998; Becerra 1997, 2003; Janz et al. 2001; Farrell and Sequiera 2004). Herbivores may gain a selective advantage from eating particular host plants with particular toxins because of physiological efficiency or enemy-free space, or both. Either selective advantage could cause a long-term historical association between plant and herbivore lineages.

Optimal-Foraging Hypothesis

The optimal-foraging hypothesis focuses on components of adult fitness such as mating success and realized fecundity (Scheirs and De Bruyn 2002). These fitness components are likely to be influenced by insect traits, such as longevity, dispersal ability, and the use of host plants for mating sites, as well as host-plant traits, such as host availability and proximity to adult food. In theory, herbivores may maximize fitness by specializing on host plants on which females oviposit the maximum number of eggs via the combined effects of these insect and plant traits. This hypothesis is not necessarily at odds with the physiological-efficiency or enemy-free-space hypotheses; but, in pure form, it ranks the fitness contributions of adults over those of immatures. The empirical work by Scheirs, DeBruyn, and colleagues on grass-mining agromyzid flies most directly addresses this issue, and demonstrates the importance of adult fitness components in host-plant use (e.g., Scheirs et al. 2000). This hypothesis is underexplored for analyzing host specificity of insects that mate and feed as adults on their larval host plants (e.g., many chrysomelid beetles). Indeed, in such organisms, it might be fruitful to further subdivide the issues of adult feeding and mating as factors favoring host specificity. Like the physiological-efficiency and enemy-free-space hypotheses, the optimal-foraging hypothesis predicts the variable, but largely specialized, patterns of insect herbivore community structure and phylogenetic associations with host plants that are observed.

Neural-Constraints Hypothesis

The neural-constraints (or information-processing) hypothesis is another recent addition to the literature. Bernays and Wcislo (1994) offered it as a general explanation for specialized resource use, and Bernays (1998, 1999) and colleagues (e.g., Bernays and Funk 1999) have most actively applied it in theory and practice to the issue of host specificity by phytophagous insects. The neural-constraints hypothesis states that the limited information processing ability of insects will restrict their ability to make efficient or high-quality decisions in the recognition and acceptance of potential host plants (Bernays 1998). Inefficient or poor decision making will be evolutionarily penalized by reduced fitness from choosing poor-quality food, an increased risk of predation, or squandered time for oviposition. Because the fitness costs and benefits of this hypothesis are meted out in terms of food quality, natural enemies, or realized fecundity costs, this hypothesis is not an alternative to the other three, but rather a proximate mechanism by which they may operate. Its status as a proximate mechanism does not diminish the importance of the neural-constraints hypothesis in providing a comprehensive understanding of host specificity by phytophagous insects.

What about Polyphagy?

Understanding dietary generalism is not a separate issue from understanding host-plant specificity. The framework I will present here covers the full range of observed patterns of host range, without a clear distinction between "generalists" and "specialists." I develop this point by first considering the different hierarchical levels at which an herbivore may be polyphagous. The general hypothetical explanations for polyphagy apply to all of these levels, as well as to cases of stenophagy. Polyphagy at the individual level (i.e., food mixing or grazing) occurs in a minority of phytophagous insect species, with many grasshoppers and arctiid caterpillars being notable examples. Polyphagy at the population level (composite generalists; Fox and Morrow 1981), with individuals feeding on a single plant species, is probably more common (e.g., relatively immobile larvae and many immature bugs). Lastly, some currently accepted species appear to be polyphagous at the species level, with different populations feeding on a single plant species. The saturniid moth *Eupackardia calleta*, for example, has three geographically distinct populations in the southwestern USA and adjacent Mexico that each feeds specifically on a phylogenetically distinct host-plant species: *Leucophyllum frutescens* (Scrophulariaceae) in western Texas, *Fouquieria splendens* (Fouquieriaceae) in southeastern Arizona, and *Sapium biloculare* (Euphorbiaceae) in southwestern Arizona (Tuskes et al. 1996). What is the functional distinction between the monophagous populations of the polyphagous species, *E. calleta* (Table 3.1, case B), and the stenophagous checkerspot butterfly species, *Euphydryas editha* (Table 3.1, case C), which uses a more phylogenetically restricted range of host-plant species across its geographic range but multiple host-plant species within most local populations (Singer et al., this volume)? In principle, these kinds of cases are not functionally distinct because the functional explanations for host-plant use (described below) may apply to polyphagy at the species or population levels, as well as to stenophagy at the species or population levels.

The greatest functional distinction among herbivore feeding strategies is between grazers (food-mixing individuals) and parasites (monophagous individuals) (Thompson 1994), rather than between "specialists" and "generalists." The lack of functional distinctness between monophagous and polyphagous parasites is due in large part to a new appreciation for the importance of adaptive phenotypic and behavioral plasticity that allows generalist genotypes to phenotypically match their local environment (e.g., Agrawal 2001; West-Eberhard 2003). Therefore, some phenotypic traits of monophagous and polyphagous parasites might similarly match those of their host plants under natural conditions. However, these polyphages ("polyspecialists" sensu West-Eberhard [2003]) and monophages achieve functional similarity via different genetic-developmental mechanisms. By contrast, the functional constraints on grazers and parasites differ because the ecological consequences of small herbivores moving through a complex vegetative environment to take a mixed diet differ considerably from those of small herbivores remaining stationary or moving short distances on a single plant. The greatest differences are the kinds of ecological risks faced by individuals of each type of generalist: vulnerability to induced plant defenses, nutritive imbalances of host plants, predators, and parasitoids, as well as abiotic stressors. The direction and magnitude of differences in these risk factors will depend on local ecological details (i.e., natural history).

Trade-offs as Explanations for Host-Plant Use

The simplest general, functional explanation for host-plant use is that the set of host-plant species used by the herbivore have, on average, equivalent fitness consequences for the herbivore, and that the uniformly high fitness on this set of plants is superior to that on potential host plants that are not used. However, a set of host plants is unlikely to be uniform in all aspects that influence herbivore fitness, for example, food quality and availability, enemy-free space, and exposure to abiotic factors. Therefore, polyphagous herbivores may commonly achieve equivalent fitness across a set of host-plant species when different host-plant species offer different costs and benefits to fitness, and these factors trade off against one another. Similar fitness trade-offs may also occur between different life-history stages (e.g., larva versus adult). Testing the existence and relative importance of these various kinds of ecological, physiological, or life-history trade-offs, described below, is thus critical to understanding the evolutionary ecology of polyphagy, and perhaps host specificity as well.

Physiological Efficiency Hypothesis

Although the physiological-efficiency hypothesis would frequently favor host specificity, its broader application hypothesizes that polyphagy is adaptive when a general diet enhances physiological performance, hence fitness, over that gained from a specific diet. For habitual food-mixers, this fitness gain (in terms of survival, reproduction, or both) is most likely to occur through a more favorable balance of nutrients (nutrient balancing) or a less harmful intake of secondary metabolites (toxin dilution) afforded by a mixed diet (e.g., Singer et al. 2002). The trade-off here is in terms of complementarity of different host-plant species for an herbivore's physiological needs. For example, different host-plant species offer different and complementary aspects of beneficial food quality (abundance of a particular limiting nutrient, rarity of a particular toxin). One adaptive solution to this trade-off is to mix complementary foods, whereas another solution is physiological adaptation to the nutritional and allelochemical content of a particular food (host specificity). For either composite generalists or food-mixers that leave a host plant only once they have consumed it, the advantage of polyphagy stems from benefits of food quantity over quality. There may be a trade-off between host quality versus availability, particularly when the availability of high-quality host plants is unpredictably variable (Jaenike 1978; Courtney et al. 1989; Mangel 1989; Doak et al. 2006). For example, if the highest-quality host-plant species is unpredictably rare amidst numerous lower-quality host plants, an ovipositing female could lay a greater number of eggs by adopting a polyphagous strategy, offsetting the performance cost to individual offspring on lower-quality hosts. A feeding herbivore seeking a high-quality host plant could similarly reduce the risk of starvation or desiccation with increasing search time by using inferior, alternative host plants that are immediately available. For generalists at the species level, the physiological-efficiency hypothesis is the same explanation as that given for host specialists: superior physiological advantages to using a particular host-plant species in a local area. The assumption in this case is that the host-plant species of highest food quality differs among populations or geographic areas.

Enemy-Free-Space Hypothesis

The enemy-free-space hypothesis could explain cases of polyphagy if the use of multiple host-plant species can improve survival through greater refuge from or defense against natural enemies, with the assumption that the enemy-free space associated with particular host plants is variable in time or space. For example, habitual food-mixers could enhance enemy-free space by dividing their feeding time between host plants that offer the best refuge or defense from enemies at different times (e.g., morning versus afternoon) or places (e.g., exposed versus sheltered microsites). This strategy is most likely to be viable for relatively mobile herbivores with intrinsic defenses (e.g., armature, escape behavior) against abundant, opportunistic predators such as birds, ants, ground beetles, and spiders. Hairy caterpillars and agile grasshoppers are good candidates. Here the costs (risk of exposure to enemies and the elements) and benefits (safer new food) of moving between host plants trade off against the costs (risk of exposure to enemies associated with the host plant) and benefits (safety from opportunistic predators) of staying put. Again, food mixing is one solution to this trade-off, whereas behavioral, physiological, and morphological adaptation to maximize enemy-free space on a particular host plant is another (Janzen 1985; Bernays and Graham 1988). Composite generalists might use multiple host-plant species for the same reason as food-mixers, but over a larger scale of time or space. For example, a certain host-plant species may be safest early in the season or in a particular microhabitat type, whereas another is safest later in the season or in a different microhabitat type (Lill 2001). If these differences in enemy-free space are predictable, the spatial or temporal variation in host-plant use by individual females is expected to adaptively track this variation. If variation in enemy-free space is unpredictable, the herbivore is likely to employ a bet-hedging strategy, with individual ovipositing females spreading risk among their offspring by distributing eggs over multiple host-plant species (reviewed by Hopper 1999). Polyphagy at the species level could be favored by geographic variation in enemy-free space provided by different host-plant species. Monophagy would result from a single host-plant species offering superior enemy-free space in all geographic locations.

Food Quality versus Enemy-Free Space

Several additional kinds of trade-offs involving combinations of ecological factors could explain polyphagy. One of the best supported within the field of plant-insect interactions and more generally is the trade-off between food quality and enemy-free space. That is, some host-plant species offer relatively high-quality food but relatively little enemy-free space, whereas other host-plant species offer the reverse (e.g., Denno et al. 1990; Camara 1997a, 1997b, 1997c; Ballabeni et al. 2001; Mira and Bernays 2002; Schmitz 2003; Singer et al. 2004b). This kind of trade-off could occur via several possible mechanisms. First, it might result if the highest-quality host-plant species harbor the highest densities of herbivores, coupled with density-dependent mortality imposed by natural enemies. Second, it might result if the highest-quality host-plant species support the highest-quality herbivores (as food for natural enemies), coupled with selective attack of high-quality prey or hosts by natural enemies. Third, it might happen if the highest-quality host-plant species either offer herbivores relatively little defense or refuge from enemies (e.g., toxins, places to hide) or attract enemies of the herbivores by various means (e.g., extrafloral nectar, domatia, releasing volatile organic compounds).

Trade-offs and Competition

In situations in which interspecific competition is important, competition may trade off against other factors to promote polyphagy. For example, polyphagy may be favored when food quality (in the absence of competitors) and competition intensity positively covary among different host-plant species. Here, an herbivore that competes poorly with others may gain a performance advantage by switching to alternative host plants under highly competitive circumstances (induced polyphagy), or may routinely use multiple host-plant species because highly competitive circumstances are unpredictable (constitutive polyphagy). This trade-off could result from a limitation of food quantity due to intense exploitative competition; or it could be the outcome of reduced food quality due to feeding by competitors. Recent work has shown the latter to be mediated by induced chemical responses of plants to particular herbivores (e.g., Redman and Scriber 2000; Van Zandt and Agrawal 2004) and may be more important than currently recognized.

It is also possible that polyphagy may be favored by the positive covariance of enemy-free space and food competition across host-plant species. Hypothetically, a particular host-plant species may offer enemy-free space for an assemblage of herbivore species, allowing them to survive and reach high densities, resulting in relatively intense competition. Herbivores may achieve equivalent fitness on alternative hosts for which the risk of mortality from natural enemies is higher, but dietary performance is also higher due to reduced competition.

Fitness of Adults versus Immatures

Finally, several recent studies have provided evidence for opposing benefits and costs to adult versus immature fitness components (Scheirs and De Bruyn 2002; Scheirs et al. 2004). In theory, such trade-offs could enforce polyphagy or, alternatively, promote host specificity when the fitness effects from one life stage (e.g., adults) supersede those of another life stage. Other kinds of trade-offs are, of course, possible and may play a role in the evolutionary ecology of polyphagy. However, the theoretical possibilities discussed above have both plausibility and some support from ecological studies showing that the hypothesized mechanisms or factors exist in nature.

Testing Theory with Polyphagous Woolly Bear Caterpillars

To illustrate the application of these ideas, I principally recount my collaborative, empirical work on the evolutionary ecology of host-plant use by two polyphagous woolly bear caterpillars, *Grammia geneura* (Strecker) and *Estigmene acrea* (Drury) (Lepidoptera: Arctiidae). Note that Ferguson and Opler (2006) have recently changed the name of the taxon I refer to as *G. geneura* to *G. incorrupta* (Henry Edwards), but I use *G. geneura* here to avoid confusion. In practice, I have restricted empirical tests to only those hypotheses relevant and experimentally tractable in light of the natural history of these insects in southeastern Arizona, USA, where my colleagues and I conducted this work. The following account of their natural history provides observations relevant to empirical tests described subsequently.

Natural History

In southeastern Arizona, both *G. geneura* and *E. acrea* predominantly occur in semidesert and subtropical grassland and savanna plant communities, typically 1200 to 1800 m in elevation (Brown 1994). These plant communities mainly consist of grasses, forbs, and, to a lesser degree, woody plants (Brown 1994; Singer and Stireman 2001). Two major growing seasons are dictated by a bimodal pattern of precipitation: late winter/early spring and summer/early autumn. *Grammia geneura* has two generations per year in which its larval stage coincides with these two growing seasons, whereas *E. acrea* has but a single summer/early autumn generation. Neither moth species feeds during the adult stage (they have no functional mouthparts), but they have different oviposition habits. Like most Lepidoptera, *E. acrea* females oviposit on host plants, placing variably sized clusters of eggs on the leaves of various plant species. I have mainly observed *E. acrea* egg clusters and young larvae on the underside of large, mature leaves of sunflowers *(Helianthus annua)* and close relatives (*Viguiera* spp.). By contrast, *G. geneura* females cannot adhere eggs to a substrate; instead, they apparently deposit their eggs on the ground, in loose groups most likely at the base of bunch grasses on which I have found adult females resting. I have never observed the eggs or young larvae of *G. geneura* in nature, though I have inferred this bit of natural history from behavior in the laboratory, as no other information exists in the literature.

The consequence of these contrasting oviposition behaviors is that *E. acrea* larvae start life on a particular host plant selected by their mother and feed specifically on it for the first three instars. *Estigmene acrea* caterpillars leave their natal host plant as early as the fourth instar and become polyphagous at the individual level (food-mixers). *Grammia geneura* larvae must initiate host-plant selection without direct parental assistance, and probably express some degree of food-mixing behavior from the outset. I have observed first-instar larvae moving between host plants in the laboratory. *Grammia geneura* neonates can also await the germination of annual host plants, living for days without food or water by slowing down their metabolism (Woods and Singer 2001). The late instars of both species have very similar food-mixing behavior; they feed selectively on forbs (annual and perennial herbaceous dicots) (Singer and Stireman 2001; M.S.S. personal observation) and customarily switch among host plants over minutes or hours (Singer et al. 2002; M.S.S., unpublished data for *E. acrea*). My colleagues and I have recorded *G. geneura* populations feeding on nearly 80 plant species in over 50 plant families (Singer and Stireman 2001) and *E. acrea* populations on at least 88 plant species in 33 families (Singer et al. 2004b), with much overlap in host-plant species used by these herbivores.

Despite their polyphagy, both caterpillar species prefer to eat host plants that contain pyrrolizidine alkaloids (PAs) (e.g., *Senecio longilobus* [Asteraceae], *Crotalaria pumila* [Fabaceae], *Plagiobothrys arizonicus* [Boraginaceae]). PAs are widely deterrent and toxic compounds to nonadapted herbivores and carnivores (Hartmann 1999). *Grammia geneura* additionally prefers plants containing iridoid glycosides (IGs) (e.g., *Plantago* spp. [Plantaginaceae]), which, when sequestered by specialist herbivores, have been shown to be deterrent or toxic to invertebrate and vertebrate predators (reviewed by Nishida 2002). These host-plant preference measures are based on observations of caterpillars in nature, as well as direct and indirect observations from laboratory experiments. We have quantified the acceptability of different host-plant species from six-hour continuous observations of final-instar caterpillars in nature (Singer and Stireman 2001; Singer et al. 2004b). We found two measures of host preference to be particularly informative: the probability that caterpillars initiate feeding on different host-plant species and the average duration of their feeding events on various host-plant species (Singer and Stireman 2001; Singer et al. 2004b). Electrophysiological and behavioral experiments in the laboratory show that PAs are strong feeding stimulants for both caterpillar species (Bernays et al. 2002), and IGs are important feeding stimulants for *G. geneura* (Bernays et al. 2000). Both caterpillar species sequester PAs from their host plants in the blood and integument (Hartmann et al. 2004, 2005a, 2005b), and *G. geneura* sequesters IGs in addition to PAs (M.S.S. and J.O. Stireman, unpublished; M.D. Bowers, unpublished). Preliminary evidence indicates that *E. acrea* also sequesters IGs (M.D. Bowers, 2005, personal communication).

Like other phytophagous insects, both *G. geneura* and *E. acrea* caterpillars are subject to frequent attack by carnivores. However, they appear to experience infrequent attack from generalist invertebrate and vertebrate predators (e.g., ants, predatory wasps, birds) relative to other caterpillars in the same community (M.S.S., personal observations), probably by virtue of intrinsic defenses such as their long, dense setae (hairiness) and considerable locomotory speed by caterpillar standards. It is likely that the chemical defenses of *G. geneura* and *E. acrea* deter predators as well, particularly

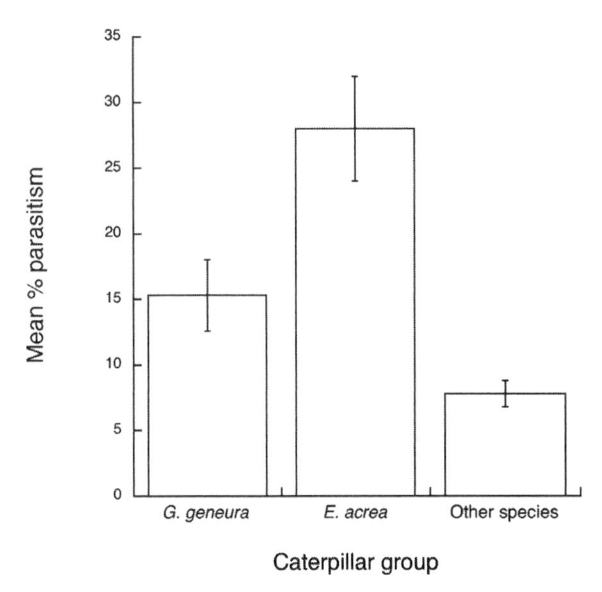

FIGURE 3.1. Comparison of the frequency of mortality from parasitoids among natural populations of *Grammia geneura, Estigmene acrea*, and 168 other species of macrolepidoptera in the same ecological community, based on field collections and rearings between 1996 and 2001 (Singer and Stireman 2003; Stireman and Singer 2003; Singer et al. 2004b). Error bars represent standard errors of the means among collections of *G. geneura* (N = 45) and *E. acrea* (N = 14), and standard errors of the mean among means among all other caterpillar species (N = 168).

because both species place sequestered PAs in the integument as well as in the blood (Hartmann et al. 2004). Laboratory experiments with captive colonies of *Crematogaster opuntiae* ants indicate that *G. geneura* eggs contain a hydrophobic, chemical deterrent, even in *G. geneura* lab colonies reared on a synthetic, toxin-free diet (M.S.S. and M. T. Jones, unpublished data). In contrast to their apparently limited vulnerability to generalist predators, both caterpillar species often experience high frequencies of attack from parasitoids (particularly tachinid flies) and host large assemblages of parasitoid species in relation to other caterpillars in the same community (Fig. 3.1) (Stireman and Singer 2002, 2003). Despite their extremely similar ecological habits, *G. geneura* and *E. acrea* host distinct sets of their most important parasitoid species. *Carcelia reclinata* (Tachinidae; locally oligophagous), *Cotesia* nr. *phobetri* (Braconidae; locally oligophagous), and *Exorista mella* (Tachinidae; locally oligophagous) account for 80% of the total mortality from parasitism in *G. geneura* (Stireman and Singer 2002). *Carcelia languida* (Tachinidae; locally host specific), *Leschenaultia adusta* (Tachinidae; seasonally host specific), and *Lespesia aletiae* (Tachinidae; locally polyphagous) cause 85% of the total mortality from parasitism in *Estigmene acrea* (Singer et al. 2004b).

Hypotheses to Explain Food Mixing

This natural history information helps reduce the list of general theoretical explanations for polyphagy to a set of the most plausible, hypothetical explanations for polyphagy by *G. geneura* and *E. acrea*. It is clear that customary food mixing is the type of polyphagy in need of explanation for both species. The six-hour continuous observations of *G. geneura* showed that caterpillars typically switch host plants prior to consuming them entirely (Singer et al. 2002). These switches were proximate responses to secondary metabolites, as indicated by laboratory experiments with manipulated concentrations of macronutrients and secondary metabolites in synthetic diets (Singer et al. 2002). Thus, these switches are behavioral decisions, presumably adaptive, in response to food quality, not food quantity. As such, the physiological-efficiency hypothesis would be supported if these caterpillars generally performed better on mixed-species diets than on single-species diets.

The preference for highly toxic host-plant species coupled with frequent parasitism suggests that chemically mediated enemy-free space with respect to parasitoids may be important by itself or in combination with other factors. Chemically mediated enemy-free space with respect to generalist predators is also likely to be important in this system, but predation is difficult to study when it is infrequently observed. Therefore, tests of the enemy-free-space hypothesis have involved experiments with parasitoids. The enemy-free-space hypothesis would be supported if these caterpillars gained better defense against parasitoids by food mixing than by feeding specifically.

Field observations also suggest that interspecific competition is unlikely to be important in these study systems because most individual host plants encountered by these caterpillars are abundant, annual forbs lacking herbivores and aboveground herbivore damage. Furthermore, these caterpillars show no aversion to host plants with herbivore damage (M.S.S., personal observations), suggesting little, if any, competition mediated by induced changes in food quality. Finally, adult fitness components (e.g., feeding habits, mating sites, special oviposition microsites) are unlikely to determine host-plant use, and thus trade-offs between adult and immature fitness are unlikely to exist in the system. A possible exception is the use of PA-derived male pheromones by adult *E. acrea* moths (Davenport and Conner 2003; Jordan et al. 2005), but this is unlikely to pose a conflict with immature defensive needs derived from PAs, as discussed below.

TESTS OF THE PHYSIOLOGICAL EFFICIENCY HYPOTHESIS

Experimental work contradicts the notion that food mixing by woolly bear caterpillars may be simply explained by the physiological-efficiency hypothesis. The evidence against the physiological-efficiency hypothesis for woolly bear caterpillars comes mainly from laboratory experiments in which caterpillars were reared on different diets, including pure (single plant species) and mixed (multiple plant species) treatments composed of the plant species offered as pure diets. A series of three such experiments (four mixed treatments in total) with *G. geneura* showed the mixed diets to be superior to pure diets in terms of survival, developmental rate, or pupal mass in only one of the four comparisons with multiple pure diets (Singer 2001). In the single case matching the prediction of the physiological-efficiency hypothesis, caterpillars survived better and females attained greater pupal mass when they ate a mixed diet of two individually poor host-plant species (*Ambrosia confertiflora* [Asteraceae] and *Machaeranthera gracilis* [Asteraceae]). However, the majority of cases showed that food mixing offered no performance advantage if a single high-quality host-plant species (e.g., *Malva parviflora* [Malvaceae]) was available. Consequently, the physiological-efficiency hypothesis, at best, provides a limited explanation of food-mixing behavior in *G. geneura*.

The limited explanatory power of the physiological-efficiency hypothesis appears to be general among food-mixing arctiid caterpillars. In their study testing the performance of *E. acrea* on various diets of cultivated plants, Bernays and Minkenberg (1997) found no survival or weight gain advantage to mixed over the best pure diets in three separate experiments. Likewise, *E. acrea* experienced reduced pupal mass on a mixed diet of natural host plants, *Viguiera dentata* (Asteraceae) and *Senecio longilobus* (Asteraceae), in relation to *V. dentata* alone (Singer et al. 2004b). Haegele and Rowell-Rahier (1999) found no difference in the final mass of *Callimorpha dominula* (Arctiidae) reared on a mixed diet and the best pure diet (*Adenostyles alliariae* [Asteraceae], a PA plant). Similarly, Adler (2004) reported that *Platyprepia virginalis* (Arctiidae) caterpillars grew no better on a natural, mixed diet than on a pure diet of *Lupinus arboreus* in two different field experiments, each testing early and late instars. Food mixing by *Spilosoma virginica* (Arctiidae) caterpillars reared on all possible combinations of pure and mixed diets including dandelion (*Taraxacum officiale* [Asteraceae]), white clover (*Trifolium repens* [Fabaceae]), and plantain (*Plantago major* [Plantaginaceae]) never provided superior survival, developmental rate, or adult mass over the best pure diet (dandelion) (M.S.S., unpublished data). As a conservative summation of these studies, only 1 in 12 experimental comparisons supported the prediction that food mixing by woolly bear caterpillars enhances performance over pure diets available in a mixed treatment. Although these diets were not chosen at random from the huge array of possibilities and therefore do not represent an unbiased statistical sample, the uniformity of these results nonetheless raises the question, why do woolly bear caterpillars choose to leave high-quality host-plant species to eat a mixed diet when they could perform equally well or better by feeding specifically on such plants?

As further evidence against the physiological-efficiency hypothesis, I turn to two other, more mechanistic, predictions. Namely, caterpillars faced with a choice would be expected to choose high-quality over low-quality food. Furthermore, this self-selected diet should enhance the physiological efficiency of growth. The evidence from *G. geneura* does not match these predictions. As evidence against the behavioral prediction, *G. geneura* caterpillars showed no clear preference between *Malva parviflora* and *Ambrosia confertiflora* despite the greater suitability of the former plant species (Singer 2001; Singer and Stireman 2003). In another choice experiment, *G. geneura* spent more time feeding on *Tithonia fruticosum* (Asteraceae) than on *M. parviflora*, despite the latter plant's superior food quality (Singer 2001). In a no-choice test, the increase in feeding time on *T. fruticosum* would indicate compensatory feeding, but not for caterpillars in the choice experiment. In five comparisons of growth efficiency on mixed and pure diets, *G. geneura* also never grew more efficiently on a mixed diet (Singer 2001; Singer et al. 2004a).

TESTS OF THE ENEMY-FREE-SPACE HYPOTHESIS

Experimental work supports the notion that enemy-free space plays a critical role in the evolutionary ecology of host-plant use by *G. geneura* and *E. acrea* caterpillars. Several tritrophic experiments involving various host-plant species, the two caterpillar species, and parasitoids provide evidence that these caterpillars gain resistance against parasitoids by feeding on certain toxic plant species. These toxic plant species are mainly those containing either PAs or IGs.

Previous work on the polyphagous arctiid caterpillar *Platyprepia virginalis* showed that a caterpillar's diet could enhance its survival after hosting the successful development and emergence of a tachinid fly (English-Loeb et al. 1990, 1993; Karban and English-Loeb 1997). This work suggested the related possibility that an arctiid caterpillar's diet could modify its resistance to parasitoids. In this context, resistance means an ability to deter parasitoid attack or prevent successful parasitoid development in the event of attack. In contrast, the interaction between *Platyprepia* and the tachinid fly *Thelaira americana* is more akin to the concept of tolerance in the plant-defense literature (Strauss and Agrawal 1999).

Initial evidence for resistance against parasitoids came from laboratory experiments with *G. geneura*, three different host-plant species (*Malva parviflora, Plantago insularis,* and *Ambrosia confertiflora*), and its important natural parasitoid, *Exorista mella* (Tachinidae) (Singer and Stireman 2003). In these experiments, *G. geneura* caterpillars were reared on a wheat-germ-based synthetic diet until they reached either the penultimate or ultimate larval stadium (depending on the experiment) were subjected to attack as final instars by lab-reared, female parasitoids in a standardized way, and divided into different pure diet treatment groups. Although these experiments had limited sample sizes and surprisingly high survival rates of caterpillars that received parasitoid eggs, they generated several important findings. First, the diets of late-instar caterpillars modifed the outcome of the caterpillar-parasitoid interaction. In particular, parasitized late-instar caterpillars survived at a higher frequency when given *P. insularis* or *A. confertiflora* than when given *M. parviflora* or the same synthetic diet they ate as early instars.

Caterpillar survival means successful development into an adult moth without a sign of having been parasitized (except the tachinid egg on the larval exuvia). We concluded that the parasitoid died at some unknown stage, and thus considered the frequency of this outcome as a measure of resistance against the parasitoid. Unlike studies with *Platyprepia virginalis*, J. Stireman and I never observed cases in which both host and parasitoid survived to adulthood. This difference is most likely due to the fact that the parasitoids of *G. geneura* wait until the prepupal or pupal stage to emerge from the host. In contrast, the tachinid *Thelaira americana* frequently emerges from *Platyprepia* while the caterpillar host is still active and feeding, and thus able to heal and recover through dietary and perhaps other behavioral and physiological means (Karban 1998). The second important result from these tritrophic experiments with *G. geneura* is that the diet during the final larval stage alone could modify its resistance against *E. mella*. Not only was this result somewhat surprising, but it also suggested possibilities for future experiments with field-collected caterpillars.

My colleagues and I used the information from the laboratory experiments to design new experiments intended to test, with more ecological realism, the enemy-free-space hypothesis as an explanation for food mixing. In these experiments, we used field-collected *G. geneura* (Singer et al. 2004a) and *Estigmene acrea* (Singer et al. 2004b) caterpillars to determine if caterpillars gained enemy-free space from their parasitoids by eating mixed-species diets that included a PA plant, from which they could presumably gain chemically mediated resistance.

Although these experiments were less controlled than the strict laboratory experiments using *Exorista mella*, they were more realistic in several ways. Late-instar caterpillars were collected from the field late in the season when we knew from past study (Stireman and Singer 2002) that many were likely to have already been parasitized by flies or wasps. The fact that parasitism had occurred under natural conditions, rather than in a laboratory terrarium, discounted the possibility (present in laboratory experiments with *E. mella*) that parasitoid performance was somehow weakened or egg viability was low due to laboratory-rearing conditions. The field experiment also included other species of parasitoids that could not be easily cultured in the laboratory.

A summary of the methods and results follows. After field collection, we randomly placed the caterpillars in laboratory diet treatments that included a mixed-species treatment, representative of their natural diets. We included the PA plant *Senecio longilobus* (Asteraceae) in these experiments because we had by this time observed its frequent use by both *G. geneura* and *Estigmene acrea* caterpillars in nature. We again found differences in the survival and frequency of mortality from parasitoids of caterpillars in the different diet treatment groups. *Grammia geneura* caterpillars given a mixed diet *(S. longilobus, M. parviflora,* and *A. confertiflora)* had higher survival to adulthood (68% versus 50%) due to reduced mortality from parasitoids than caterpillars given only *M. parviflora* in the laboratory. Of the three parasitoid species that developed successfully from these field-collected caterpillars *(Carcelia reclinata, Cotesia* nr. *phobetri*, and *Chetogena tachinomoides),* all three emerged in lower frequency from the caterpillars in the mixed treatment. In the *E. acrea* experiment, the total survival of caterpillars was highest in the mixed treatment *(Viguiera dentata* + *Senecio longilobus)* in relation to either *V. dentata* or *S. longilobus* alone. In this case, the improved experimental design (comparing all combinations of *V. dentata* and *S. longilobus* diets) showed that the survival benefit of the mixed diet resulted not only from reduced mortality from the parasitoid, *Carcelia languida* (mixed diet versus *V. dentata*), but also from reduced mortality from unknown causes that were most likely related to host-plant quality (mixed diet versus *S. longilobus*). It appears that food mixing allows *E. acrea* caterpillars to gain both enemy-free space from *S. longilobus* and relatively high food quality from *V. dentata*, with an overall fitness advantage over pure diets (Singer et al. 2004b). For both caterpillar species, concentrations of sequestered PAs from particular diets were positively correlated with resistance against parasitoids on those diets (*G. geneura*: Singer et al. 2004a; *E. acrea*: T. Hartmann, D. Rodrigues, and M. S. Singer, unpublished) (Fig. 3.2).

Food Quality and Enemy-Free Space

My overall conclusion from these experiments is that food mixing by *G. geneura* and *E. acrea* caterpillars is an adaptive response to a trade-off between food quality and enemy-free space from various host plants. This is significant for at least two reasons. First, it reveals the importance of a pluralistic approach to studying the complex issue of host-plant use by phytophagous insects. Food quality of host plants is clearly important in combination with enemy-free space, exposing the limitations of considering the physiological-efficiency and enemy-free-space hypotheses strictly as alternatives. Indeed, the relationship between food quality and enemy-free space may be a prime factor dictating the evolutionary ecology of host-plant use. Whereas a trade-off between host-plant species with respect to food quality and enemy-free space may favor the use of multiple host plants in ecological and evolutionary time, a coupling of relatively high food quality *and* enemy-free space on particular host plants may drive host specificity in ecological and evolutionary time. One consequence may be phylogenetically conservative host-plant use (Table 3.1, case A), characteristic of certain herbivore lineages such as particular butterfly lineages (reviewed by Janz and Nylin 1998) and various chrysomelid beetle clades (Farrell and Mitter 1990, 1998; Becerra 1997; Termonia et al. 2001; Farrell and Sequeira 2004).

Second, this ecological trade-off points to the possibility that an individual caterpillar may improve its prospects for survival and reproduction by changing its diet in relation to its ecological circumstance. Do these caterpillars show

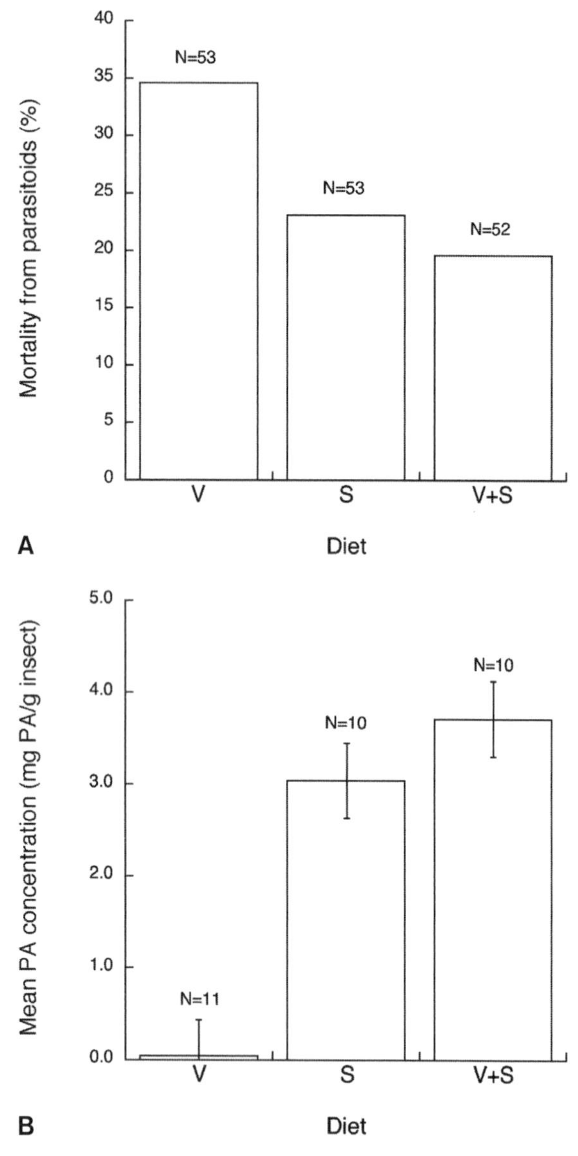

FIGURE 3.2. Effect of experimental diets fed to *Estigmene acrea* on mortality from parasitoids (A) and concentration of sequestered pyrrolizidine alkaloids (PA) (B). Experimental diets included *Viguiera dentata* (V), *Senecio longilobus* (S), and both plant species in combination (V + S). Percentage mortality from parasitoids was calculated as (frequency of caterpillars collected from which adult parasitoids emerged) × 100, based on the field/laboratory experiment described in Singer et al. 2004b. Concentrations of sequestered PAs are shown as means with standard error bars and measured in the pupae of survivors in the laboratory performance experiment described in Singer et al. 2004b, using the same methods described in Singer et al. 2004a.

adaptive plasticity in diet choice, induced by parasitism? Preliminary data suggest that this is likely. Bernays and Singer (2005) investigated the responses of taste sensory cells in parasitized and unparasitized caterpillars' mouthparts to several ecologically important chemicals: seneciphylline N-oxide (PA, feeding stimulant), catalpol (IG, feeding stimulant), caffeine or protocatechuic acid (representative feeding deterrents), and sucrose (nutrient, feeding stimulant). Field-collected *G. geneura* and *E. acrea* caterpillars showed specific differences in taste sensation (rate of action potentials fired from specific taste cell types) depending on their parasitism status (parasitized versus unparasitized), determined by dissection following electrophysiological recordings. Namely, parasitized caterpillars of both species had stronger taste responses to the PA and weaker responses to the deterrent. This suggests that caterpillars harboring parasitoids would have a stronger feeding preference for PA plants and would be less deterred by plant chemicals that normally inhibit feeding. These deterrents would include secondary metabolites characteristic of toxic plants normally avoided by the caterpillars, but also deterrent secondary metabolites found in combination with PAs in PA plants. Either way, the parasitized caterpillars would be expected to feed on toxic plants to a greater degree than unparasitized caterpillars. Parasitized *G. geneura* also had a stronger taste response to the IG, suggesting an enhanced preference for another type of defensive plant species (e.g., *Plantago* spp.). These changes in taste sensation, however, were specifically in response to secondary metabolites: both caterpillar species showed no consistent directional change in their taste responses to sucrose in relation to parasitism. This indicates that the taste system of parasitized caterpillars did not simply reflect a generalized need to increase the quantity of feeding due to nutritional losses from internal parasitoids. Further work is needed to determine if these specific and seemingly adaptive changes in taste sensation actually translate into self-medicative feeding behavior. Preliminary data from a similar behavioral experiment parallels the electrophysiological results in supporting the self-medication hypothesis (M.S.S. and E. A. Bernays, unpublished data), and further study is underway. The study of feeding behavioral changes in response to parasitism should reveal the magnitude of changes in preference (if any) for high-quality host-plant species in addition to those for defensive host-plant species.

These differences in taste physiology in *G. geneura* and *E. acrea* caterpillars indicate putatively adaptive feeding responses to parasitoids per se. The ecological experiments testing the enemy-free space-hypothesis could not precisely address this question because, without testing predators, they left open the possibility that the use of toxic plant species for chemical defense is an adaptation to predators, with the effect on parasitoids being an incidental byproduct. Yet the current evidence suggests that parasitoids have acted as agents of natural selection on the taste physiology and, in all likelihood, host-plant selection of these caterpillars. It is noteworthy that both parasitized and unparasitized *G. geneura* and *E. acrea* caterpillars feed preferentially on defensive host plants over host plants offering high-quality food, suggesting that enemy-free space (with respect to predators as well as parasitoids) has superseded food quality in its selective impact on diet choice (Singer et al. 2004a, 2004b).

Conclusions

On theoretical grounds, ecological trade-offs are central to understanding the use of multiple host-plant species by individuals, populations, and species of phytophagous insects. Different host-plant species are likely to offer different combinations of fitness benefits and costs to particular herbivore species. Consequently, an herbivore species may gain equivalent fitness, on average, by using a set of available host plants that is likely to pose ecological trade-offs for the herbivore. The particular ecological factors that determine these fitness components (e.g., food quality, enemy-free space, competition, adult resources, thermal stress) are likely to vary from case to case. In the empirical cases of southern Arizona populations of *G. geneura* and *E. acrea*, the trade-off between food quality and resistance against parasitoids can explain the polyphagy of individual caterpillars. These caterpillars principally gain high food quality from some host-plant species and chemical defense from others. The viability of this strategy probably depends on the existence of intrinsic defenses possessed by these caterpillars, including hairiness and locomotory speed, allowing individual caterpillars to conspicuously switch among host plants with relatively low chances of being attacked and killed by generalist predators. The most recent evidence suggests that these caterpillar species can adaptively shift the balance of their nutritional and defensive intake in accordance with their risk of death from parasitoids.

This framework of ecological trade-offs offers the view that host-plant specificity may be an alternative solution in which herbivores adaptively overcome this or other possible ecological trade-offs. That is, herbivore phenotypes adapted to their host plants in terms of both food quality and enemy-free space gain greater benefits of survival and reproduction over other phenotypes. Over evolutionary time, continued adaptation on these fronts would be expected to cause the sustained use of particular host-plant lineages (phylogenetic conservatism of host use), a relatively common observation (Farrell and Sequeira 2004). This result would be especially likely if the set of traits providing adaptation to food quality and enemy-free space were the same (e.g., sequestration of plant secondary metabolites, gall induction). It would be interesting to know if phylogenetically conserved patterns of host-plant use were statistically more likely in these cases. This adaptive process, repeated in many cases over large scales of space and time, would accumulate host-specific herbivores in ecological communities. Natural selection would also intensify this process in communities in which plant defense and enemy pressure are greatest, causing the greatest predominance of host-plant specificity in such areas. Comparisons of temperate and tropical communities are consistent with this scenario (Dyer and Coley 2002), yet empirical tests of these ideas and the important natural-history information needed to properly apply these tests are still needed.

Acknowledgments

I thank Kelley Tilmon for inviting me to contribute to this volume. Critical comments from Warren Abrahamson, Catherine Blair, an anonymous reviewer, and Kelley Tilmon improved the manuscript. The empirical work described was conducted with important collaborative contributions from Liz Bernays, Yves Carrière, the late Reg Chapman, Thomas Hartmann, Daniela Rodrigues, and John Stireman. Thanks to the Evolution Journal Club in the Biology Department at Wesleyan University for discussions that helped spur and clarify some of my ideas.

References Cited

Adler, L.S. 2004. Size-mediated performance of a generalist herbivore feeding on mixed diets. Southwest. Nat. 49: 189–196.

Agrawal, A.A. 2001. Phenotypic plasticity in the interactions and evolution of species. Science 294: 321–326.

Arthur, W. 2004. The effect of development on the direction of evolution: toward a twenty-first century consensus. Evol. Develop. 6: 282–288.

Ballabeni, P., M. Wlodarczyk, and M. Rahier. 2001. Does enemy-free space for eggs contribute to a leaf beetle's oviposition preference for a nutritionally inferior host plant? Func. Ecol. 15: 318–324.

Baur, R., and N.E. Rank. 1996. Influence of host quality and natural enemies on the life history of the alder leaf beetles *Agelastica alni* and *Linaeidea aenea*, pp. 173–194. In P.H. Jolivet and M.L. Cox (eds.), Chrysomelidae biology, vol. 2. Ecological studies. SPB, Amsterdam.

Becerra, J.X. 1997. Insects on plants: macroevolutionary chemical trends in host use. Science 276: 253–256.

Becerra, J.X. 2003. Synchronous coadaptation in an ancient case of herbivory. Proc. Natl. Acad. Sci. USA 100: 12804–12807.

Berenbaum, M.R., and A.R. Zangerl. 1998. Chemical phenotype matching between a plant and its insect herbivore. Proc. Natl. Acad. Sci. USA 95: 13743–13748.

Bernays, E.A. 1988. Host specificity in phytophagous insects: selection pressure from generalist predators. Entomol. Exp. Appl. 49: 131–140.

Bernays, E.A. 1998. The value of being a resource specialist: behavioral support for a neural hypothesis. Am. Nat. 151: 451–464.

Bernays, E.A. 1999. When host choice is a problem for a generalist herbivore: experiments with the whitefly, *Bemesia tabaci*. Ecol. Entomol. 24: 260–267.

Bernays, E.A., and M.L. Cornelius. 1989. Generalist caterpillar prey are more palatable than specialists for the generalist predator *Iridomyrmex humilis*. Oecologia 79: 427–430.

Bernays, E.A., and D.J. Funk. 1999. Specialists make faster decisions than generalists: experiments with aphids. Proc. R. Soc. Lond. B 266: 151–156.

Bernays, E., and M. Graham. 1988. On the evolution of host specificity in phytophagous arthropods. Ecology 69: 886–892.

Bernays, E.A., and O.P.J.M. Minkenberg. 1997. Insect herbivores: different reasons for being a generalist. Ecology 78: 1157–1169.

Bernays, E.A., and M.S. Singer. 2005. Taste alteration and endoparasites. Nature 436: 476.

Bernays, E.A., and W.T. Wcislo. 1994. Sensory capabilities, information processing and resource specialization. Q. Rev. Biol. 69: 187–204.

Bernays, E.A., R.F. Chapman, and M.S. Singer. 2000. Sensitivity to chemically diverse phagostimulants in a single gustatory neuron of a polyphagous caterpillar. J. Comp. Phys. A 186: 13–19.

Bernays, E.A., R.F. Chapman, and T. Hartmann. 2002. A taste receptor neurone dedicated to the perception of pyrrolizidine alkaloids in the medial galeal sensillum of two polyphagous arctiid caterpillars. Physiol. Entomol. 27: 1–10.

Brower, L.P. 1958. Bird predation and foodplant specificity in closely related procryptic insects. Am. Nat. 92: 183–187.

Brown, D.E. 1994. Biotic communities of the southwestern United States and northwestern Mexico. University of Utah Press, Salt Lake City.

Brues, C.T. 1924. The specificity of food-plants in the evolution of phytophagous insects. Am. Nat. 58: 127–144.

Camara, M.D. 1997a. A recent host range expansion in *Junonia coenia* Hübner (Nymphalidae): oviposition preference, survival, growth, and chemical defense. Evolution 51: 873–884.

Camara, M.D. 1997b. Physiological mechanisms underlying the costs of chemical defence in *Junonia coenia* Hübner (Nymphalidae): a gravimetric and quantitative genetic analysis. Evol. Ecol. 11: 451–469.

Camara, M.D. 1997c. Predator responses to sequestered plant toxins in buckeye caterpillars: are tritrophic interactions locally variable? J. Chem. Ecol. 23: 2093–2106.

Cornell, H.V., and B.A. Hawkins. 2003. Herbivore responses to plant secondary compounds: a test of phytochemical coevolution theory. Am. Nat. 161: 507–522.

Courtney, S.P., G.K. Chen, and A. Gardner. 1989. A general model for individual host selection. Oikos 55: 55–65.

Damman, H. 1987. Leaf quality and enemy avoidance by larvae of a pyralid moth. Ecology 68: 87–97.

Davenport, J.W., and W.E. Conner. 2003. Dietary alkaloids and the development of androconial organs in *Estigmene acrea*. J. Insect Sci. 3: 3. Available at http://insectscience.org/3.3.

Denno, R.F., S. Larsson, and K.L. Olmstead. 1990. Role of enemy-free space and plant quality in host-plant selection by willow beetles. Ecology 71: 124–137.

Dethier, V.G. 1954. Evolution of feeding preferences in phytophagous insects. Evolution 8: 33–54.

Doak, P., P. Kareiva, and J. Kingsolver. 2006. Fitness consequences of choosy oviposition for a time-limited butterfly. Ecology 87: 395–408.

Dyer, L.A. 1995. Tasty generalists and nasty specialists? A comparative study of antipredator mechanisms in tropical lepidopteran larvae. Ecology 76: 1483–1496.

Dyer, L.A. 1997. Effectiveness of caterpillar defenses against three species of invertebrate predators. J. Res. Lepidop. 34: 48–68.

Dyer, L.A., and P.D. Coley. 2002. Latitudinal gradients in tri-trophic interactions, pp. 67–88. In T. Tscharntke, and B.A. Hawkins (eds.), Multitrophic level interactions. Cambridge University Press, Cambridge, UK.

Ehrlich, P.R., and P.H. Raven. 1964. Butterflies and plants: a study in coevolution. Evolution 18: 586–608.

English-Loeb, G.M., R. Karban, and A.K. Brody. 1990. Arctiid larvae survive attack by a tachinid parasitoid and produce viable offspring. Ecol. Entomol. 15: 361–362.

English-Loeb, G.M., A.K. Brody, and R. Karban. 1993. Host-plant-mediated interactions between a generalist folivore and its tachinid parasitoid. J. Anim. Ecol. 62: 465–471.

Farrell, B.D., and C. Mitter. 1990. Phylogenesis of insect/plant interactions: have *Phyllobrotica* and the Lamiales diversified in parallel? Evolution 44: 1389–1403.

Farrell, B.D., and C. Mitter. 1998. The timing of insect/plant diversification: might *Tetraopes* (Coleoptera: Cerambycidae) and *Asclepias* (Asclepiadaceae) have co-evolved? Biol. J. Linn. Soc. 63: 553–577.

Farrell, B.D., and A. Sequeira. 2004. Evolutionary rates in the adaptive radiation of beetles on plants. Evolution 58: 1984–2001.

Feeny, P. 1976. Plant apparency and chemical defense. Rec. Adv. Phytochem. 10: 1–40.

Ferguson, D.C., and P.A. Opler. 2006. Checklist of the Arctiidae (Lepidoptera: Insecta) of the continental United States and Canada. Zootaxa 1299: 1–33.

Fox, L.R., and P.A. Morrow. 1981. Specialization: species property or local phenomenon? Science 211: 887–893.

Fraenkel, G. 1959. The raison d'etre of secondary plant substances. Science 129: 1466–1470.

Futuyma, D.J., and G. Moreno. 1988. The evolution of ecological specialization. Annu. Rev. Ecol. Syst. 19: 207–233.

Futuyma, D.J., M.C. Keese, and D.J. Funk. 1995. Genetic constraints on macroevolution: the evolution of host affiliation in the leaf beetle genus *Ophraella*. Evolution 49: 797–809.

Gilbert, L.E. 1979. Development of theory in the analysis of insect-plant interactions, pp. 117–154. In H. Horn, R. Mitchel, and G. Stairs (eds.), Analysis of ecological systems. Ohio State University Press, Columbus.

Haegele, B.F., and M. Rowell-Rahier. 1999. Dietary mixing in three generalist herbivores: nutrient complementation or toxin dilution? Oecologia 119: 521–533.

Hartmann, T. 1999. Chemical ecology of pyrrolizidine alkaloids. Planta 207: 483–495.

Hartmann, T., C. Theuring, T. Beuerle, L. Ernst, M.S. Singer, and E.A. Bernays. 2004. Acquired and partially de novo synthesized pyrrolizidine alkaloids in two polyphagous arctiids and the alkaloid profiles of their larval food-plants. J. Chem. Ecol. 30: 229–254.

Hartmann, T., C. Theuring, T. Beuerle, N. Klewer, S. Schulz, M.S. Singer, and E.A. Bernays. 2005a. Specific recognition, detoxification and metabolism of pyrrolizidine alkaloids by the polyphagous arctiid *Estigmene acrea*. Insect Biochem. Mol. Biol. 35: 391–411.

Hartmann, T., C. Theuring, T. Beuerle, E.A. Bernays, and M.S. Singer. 2005b. Acquisition, transformation and maintenance of plant pyrrolizidine alkaloids by the polyphagous arctiid *Grammia geneura*. Insect Biochem. Mol. Biol. 35: 1083–1099.

Hopper, K.R. 1999. Risk-spreading and bet-hedging in insect population biology. Annu. Rev. Entomol. 44: 535–560.

Janz, N., and S. Nylin. 1998. Butterflies and plants: a phylogenetic study. Evolution 52: 486–502.

Janz, N., K, Nyblom, and S. Nylin. 2001. Evolutionary dynamics of host-plant specialization: a case study of the tribe Nymphalini. Evolution 55: 783–796.

Janzen, D.H. 1985. A host plant is more than its chemistry. Ill. Nat. Hist. Surv. Bull. 33: 141–174.

Jaenike, J.J. 1978. On optimal oviposition behavior in phytophagous insects. Theor. Popul. Biol. 14: 350–356.

Jaenike, J. 1990. Host specialization in phytophagous insects. Annu. Rev. Ecol. Syst. 21: 243–273.

Jeffries, M.J., and J.H. Lawton. 1984. Enemy free space and the structure of ecological communities. Biol. J. Linn. Soc. 23: 269–286.

Jordan, A.T., T.H. Jones, and W.E. Conner. 2005. If you've got it, flaunt it: ingested alkaloids affect corematal display behavior in

TABLE 4.1
Plasticity in Plants Induced by Interactions with Herbivores

Stimulus	*Notes*	*References*
Oviposition (egg deposition in or on the plant)	Remarkable plant responses, at least sometimes mediated by the plant horomone jasmonic acid, and shown to increase levels of egg parasitism.	Meiners and Hilker 2000, Hilker and Meiners 2002
Walking on the leaf surface	Unlikely to be adaptive.	Bown et al. 2002
Folivory (caterpillars, beetles, etc.)	The best studied category; often mediated by jasmonic acid, shown to be adaptive.	Karban and Baldwin 1997, Agrawal 1998
Sap sucking (e.g., aphids)	Well-studied responses, chemically distinct from folivory and often similar to plant responses to disease; mixed results for the impact on subsequent herbivores.	Stout et al. 1998, Walling 2000
Cell-content feeding (e.g., spider mites)	Tends to cause plant responses with attributes similar to both folivory and piercing sucking damage.	Ozawa et al. 2000
Leaf mining	Not well studied.	Faeth 1991, Karban 1993
Galling insects	Responses appear to be more adaptive for galler than for plants.	Larson and Whitham 1991, Stone and Schonrogge 2003

reduced growth of caterpillars on damaged compared to undamaged birch trees (Haukioja and Hakala 1975) were responsible for popularizing the study of induced responses. The study of plant morphological responses to herbivory has a more recent history, with studies demonstrating induced plant production of thorns (Young 1987; Young and Okello 1998) and trichomes (Baur et al. 1991; Agrawal 1999). There are now several hundred examples of traits in plants that are induced by herbivore feeding (see examples of these in Table 4.1). For the best-studied systems, those where insects chew the leaves of plants, there is a high level of evolutionary conservation in the hormonal regulation of induced plant responses. For example, responses including plant toxins (Bodnaryk 1994; Baldwin 1996), trichomes (Traw and Bergelson 2003), volatiles (Boland et al. 1999; Thaler 1999), and extrafloral nectar (Heil et al. 2001) all appear to be, at least in part, regulated by jasmonic acid.

Phenotypic plasticity occurs not only as a relatively immediate response to an environmental change, but also across growing seasons and across generations. For example, many woody plant species, when defoliated, show an increase in the resistance of foliage in the following growing season (Haukioja 1991). The adaptive nature of this response is believed to be based on the correlations between herbivore densities across years. Maternal effects occur when the environment of an organism in one generation affects the phenotype expressed by its offspring (Rossiter 1996). In order to be adaptive, the environment in one generation must predict the environmental conditions experienced by offspring. We have found that damage by *Pieris rapae* caterpillars to *Raphanus raphanistrum* plants, even before they are reproductive, induces a response that makes progeny (seedlings) more resistant to caterpillars than progeny from undamaged plants (Agrawal et al. 1999b). Applying jasmonic acid to plants can induce this same response (Agrawal 2002). Finally, there is genetic variation for this maternal effect, and some evidence for selection acting on maternal effects from a field experiment (Agrawal 2001a).

Many of the traits induced by herbivores have been shown to reduce herbivory, but these studies do not document the adaptive value of plasticity in these traits. For plasticity to be adaptive, these same traits must carry a fitness cost in the absence of herbivores (Fig. 4.1B), because it is the avoidance of these costs that makes plasticity adaptive. Only two systems have been studied to determine whether the ecological conditions favor plasticity, and for both systems this was the case. Wild tobacco plants that were induced to produce alkaloids with jasmonates enjoyed benefits of this phenotype in the presence of herbivores and costs in the absence of herbivores (Baldwin 1998). Inducing plants with herbivores as well as jasmonates, similar benefits and costs were demonstrated for wild radish plants (Agrawal 1998, 1999; Agrawal et al. 1999a).

The source of trait costs can be a matter of trade-offs due to resources allocated to those traits. At the same time, trait costs may also be more ecologically complex (Agrawal and Karban 1999). For example, with the induced responses of plants to herbivore damage, a given "defensive" trait may have a negative effect on most herbivores, but also a beneficial effect on some herbivores (i.e., in host location) (Giamoustaris and Mithen 1995; Agrawal and Sherriffs 2001). Consequently, inducible expression may protect the plant

TABLE 4.2
Plasticity in Herbivores Induced by Interactions with Plants

Stimulus	*Notes*	*References*
Plant part consumed/ chemical defense	Mimicry of plant part (catkin vs. twig)	Greene 1989, 1996
Physical plant defense	Behavioral deactivation	Dussourd and Eisner 1987; Malcolm 1995
Antidigestive plant defense	Induced alternate digestive enzymes	Broadway 1997; Cloutier et al. 2000
Plant toxins	Enzymatic detoxification	Krieger et al. 1971; Feyereisen 1999

against being a continuously attractive target to adapted insect species (Agrawal and Karban 1999).

Studies by Baldwin (1998) and Agrawal and colleagues (1998, 1999; Agrawal et al. 1999a) measured the benefits and costs of the full repertoires of multifaceted, coordinated induced responses. In contrast, we know relatively little about the adaptive value or costs of the distinct components of such induced responses that involve multiple traits. In wild tobacco, trypsin proteinase inhibitors function as a defense that directly deters herbivory. The costs and benefits of this single induced defense have been studied, but this has been restricted to laboratory environments (Zavala et al. 2004a, 2004b). For responses that involve damaged plants attracting predators or parasitoids of herbivores (i.e., induced indirect defenses), a single study has estimated benefits to plants (Kessler and Baldwin 2001). No studies have been successful in measuring the costs of induced predator attractants, though some calculations have been attempted (Dicke and Sabelis 1992). There is as yet little evidence that induction of extrafloral nectar to attract ants is costly in the absence of herbivores (Heil et al. 2001). Consequently, while predator recruitment itself has proven fitness benefits, there is little evidence to suggest that plasticity in predator recruitment is adaptive.

Attempts to study individual components of herbivore-induced responses are made difficult by the fact that the trait of interest cannot necessarily be separated from other components of the plant response. For instance, most studies of costs or benefits of induced predator-attracting volatiles confound the induction of direct foliar defenses and indirect volatile defenses (Heil 2004; Hoballah et al. 2004). Similarly, benefits of induced extrafloral nectar production have not been separated from other induced effects (Heil et al. 2001).

Although it is part of our reductionist nature as scientists to want to decompose each of the induced response systems of a plant into adaptive plasticity (or not), perhaps this is not the most fruitful path. The very fact that multiple plant responses in a single individual are induced by herbivory and co-regulated by the same plant hormones indicates that "plastic responses" are correlated in nature and may not decompose into adaptive parts. Indeed, the summed effects of multiple plant responses may be greater (or less) than additive in their effects on herbivores. As an example, consider the fact that the benefits of a "direct" induced plant response (i.e., slowed herbivore growth) may depend on the "indirect" attraction of natural enemies of herbivores (i.e., the slow growth–high mortality hypothesis) (Clancy and Price 1987; Benrey and Denno 1997). When suites of plant traits function synergistically with respect to herbivore resistance, we predict that selection should favor their co-occurrence and that such "defense syndromes" should evolve repeatedly and independently in a diversity of plant taxa (Kursar and Coley 2003; Agrawal and Fishbein 2006). Thus, care needs to be taken when trying to decompose the independent adaptive value of coregulated traits.

HERBIVORES IN RESPONSE TO PLANTS

Herbivorous insects, like plants, exhibit phenotypic plasticity in morphology and physiology, but they also have plastic behaviors. Because herbivores are caught in the trophic sandwich between plant defenses and predators and parasitoids, and because there are so many varied lifestyles among herbivorous insects, herbivores face a greater diversity of challenges. Perhaps for this reason, they employ a greater diversity of plastic phenotypes than plants (Table 4.2).

While it was recognized in the 1960s that cytochrome P-450 enzymes in insects detoxified pesticides and plant secondary compounds (Brattsten et al. 1977; Estabrook 1996), it was not until the 1970s that it was determined they were inducible. Brattsen et al. (1977) determined that their production in *Spodoptera eridania* was induced by exposure to a variety of plant secondary compounds. It is now known that the production of proteases that are insensitive to plant-produced protease inhibitors are inducible (Broadway 1997). With respect to induced changes in herbivore morphology, Bernays (1986) showed that caterpillar head size responded plastically to the toughness of foods, and this apparently provides the appropriate musculature for individuals feeding on hard versus soft leaves. Similarly, Thompson (1992) demonstrated that plant-induced changes in mandible morphology increased feeding efficiency.

Herbivores also respond to plant defense behaviorally. Insect behaviors to circumvent plant defenses had been noted for decades, but the first work to rigorously document simply the consequence of plasticity ("Unifying Ecological and Evolutionary Concepts"), and evaluate the evidence for the ecological and evolutionary importance of phenotypic plasticity ("Is Phenotypic Plasticity Important in General?").

Ecological Consequences

Prior to the mid-1990s, both population and community ecology were based on models in which species interactions occurred via changes in population densities. In this type of interaction the net result is a change in the density of at least one species. The theoretical work of Abrams (1995) suggested that phenotypic plasticity allows organisms to interact not only via changes in density, but also by altering each other's phenotypes (Fig. 4.2). Abrams called these "density-mediated interactions" (DMIs) and "trait-mediated interactions" (TMIs), respectively. Schmitz et al. (1998) provided one of the first empirical tests of these ideas with the tritrophic interactions between spiders, grasshoppers, and old-field plants. Surprisingly, predatory spiders that did not feed upon grasshoppers (i.e., no DMI) nevertheless had a TMI when their presence induced predator-avoidance behaviors.

Abrams (1995) went on to argue that the principal significance of phenotypic plasticity occurs when a change in a trait has subsequent consequences for the same individual that induced the phenotypic change, or for other individuals of the same or a different species (Fig. 4.2). He termed such events "trait-mediated indirect interactions" (TMIIs) (Abrams 1995), as they are the indirect consequence of one species inducing trait changes in another. While the importance of indirect density effects (so-called density-mediated indirect interactions [DMIIs]) had been recognized for decades in plant-herbivore interactions (i.e., trophic cascades) (Hairston et al. 1960), the importance TMIIs first became apparent through these theoretical arguments. The study by Schmitz et al. (1998) provided empirical evidence for the importance of these dynamics: the indirect effects of spiders on plants via induction of predator-avoidance behaviors was equal in magnitude to their effects as consumers of spiders (i.e., TMIIs and DMIIs were equal in magnitude). Subsequent work has documented the importance of TMIIs in numerous contexts.

There are ecological implications for TMIIs strictly within the context of plant-herbivore interactions. For instance, Van Zandt and Agrawal (2004a, 2004b) documented that feeding by different species of herbivores on milkweed *(Asclepias syriaca)* induced distinct responses in plants that influenced the composition of the herbivore community later in the growing season. Cyclical population dynamics in some herbivores may be attributable to induced plant resistance, where feeding by one herbivore generation reduces the growth rates of subsequent generations (Haukioja and Hakala 1975; Haukioja 1980; Klemola et al. 2004). These examples demonstrate how phenotypic plasticity provides a mode of interaction among herbivores—both within and across species—that use the same plant,

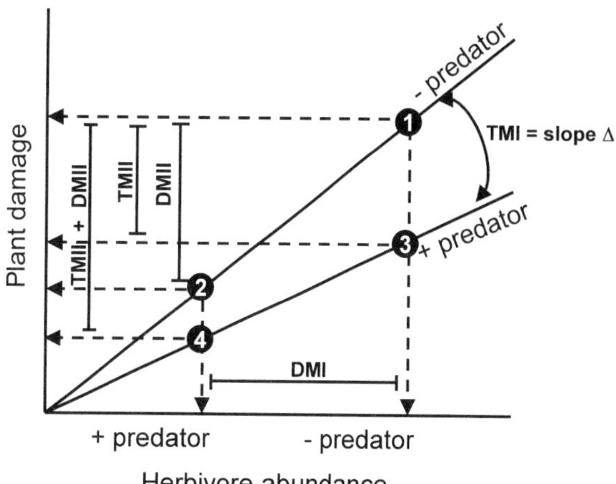

FIGURE 4.2. Graphical depiction of trait-mediated and density-mediated direct interactions (TMI and DMI) and indirect interactions (TMII and DMII) using the tri-trophic interactions among predators, herbivores, and plants as an example. The DMI between predators and herbivores is the difference in herbivore abundance as a function of predator presence versus absence (change on *x* axis from point 1 to point 2). The TMI between predators and herbivores is the change in per capita effects of herbivores on plants, namely, the change in the slope of plant damage regressed on herbivore abundance as a function of predator presence versus absence. The DMII between predators and plants is the difference in plant damage as a function of predator presence versus absence when controlling for the per capita effects of herbivores on plants (change on *y* axis from point 1 to point 2). The TMII between predators and plants is the change in plant damage due to changes in the per capita effects of herbivores on plants as a function of predator presence versus absence, when controlling for herbivore abundance (change in *y* axis from point 1 to point 3). The total effect of predators on plants is the difference in plant damage as a function of predator presence versus absence due both to changes in herbivore abundance and to changes in the per capita effects of herbivores on plants, namely, the sum of the TMII and the DMII (change on *y* axis from point 1 to point 4).

but not necessarily at the same time. Though these interactions are often competitive (Denno et al. 1995), facilitative effects are also common (Carroll and Hoffman 1980; Haukioja et al. 1990; Karban and Baldwin 1997).

Phenotypic plasticity can also be relevant to plant-herbivore interactions when a third party (predator, mutualist, or competitor) induces changes in a plant or herbivore phenotype that then indirectly affects herbivores or plants, respectively. This is what Schmitz et al. (1998) documented when they showed that a predator caused a change in herbivore behavior that resulted in reduced damage to plants. Rudgers et al. (2003) showed not only that ants *(Forelius pruinosus)* reduced caterpillar *(Bucculatrix thuberiella)* feeding on wild cotton *(Gossypium therberi)*—a result parallel to that of Schmitz et al. (1998)—but also that ants altered the spatial distribution of both caterpillars and plant damage; ants caused caterpillars to spend relatively less time on the undersurfaces of leaves and caused feeding damage to be closer to the leaf margin. Thus, in addition to the tritrophic impact of predator-induced reduction in herbivore feeding rates, if patterns of damage are important for plant fitness

(Marquis 1992) this TMII may also have evolutionary implications for herbivore selection on plant traits.

Predators may also indirectly affect plants by altering the dynamics between herbivores and their mutualists. Mooney investigated the effects of insectivorous birds on the mutualism between *Cinara* spp. aphids and the ant *Formica podzolica* (Mooney 2006), and the consequent indirect effects of birds on ponderosa pine *(Pinus ponderosa)* growth and secondary chemistry (Mooney 2007). These studies suggested that the negative effect of birds on aphids was largely due to bird-induced changes in aphid and ant behavior that caused the mutualism to break down, and that birds increased pine growth and the concentration of several phloem monoterpenes by disrupting the ant-aphid mutualism.

Another third-party effect can be seen when one plant alters another's interaction with herbivores via competition-induced plasticity. Agrawal and Van Zandt (2003) demonstrated that milkweeds *(A. syriaca)* grown in competition with grass were taller and thinner than plants grown in the absence of competition. As a result of these induced morphological changes, plants received less herbivory by the specialist weevil *Rhyssomatus lineaticollis* that preferentially oviposited on thicker stems. Perhaps more common is the situation where competition induces changes that increase plant susceptibility to herbivores. For instance, where competition reduces resources available to plants, induced plant responses to herbivores may be subsequently lowered, indirectly leading to higher levels of herbivory in the field (Karban et al. 1989; Karban 1993; Cipollini and Bergelson 2001, 2002).

Mutualists of plants and herbivores also induce traits that affect plant-herbivore interactions. Gange and West (1994) showed that *Plantago lanceolata* grown with mutualist arbuscular mycorrhizal fungi responded to caterpillar feeding damage with increased production of secondary compounds (iridoid glycosides) compared to fungicide-treated plants, and that mutualist fungi thus indirectly reduced herbivore feeding. A similar pattern has been reported for several grass-endophyte-herbivore interactions (Eichenseer et al. 1991; Bultman and Ganey 1995; Bultman et al. 2004). Mutualist *Lasius niger* ants directly affected the aphid *Metopeurum fuscoviride* (i.e., in the absence of predators) by increasing aphid life span, development, and reproduction (Flatt and Weisser 2000), possibly because of increased feeding rate by ant-attended individuals (Herzig 1937). While the indirect effects of ant attendance on plant growth were not shown, it is likely that under some circumstances such effects occur.

Unifying Ecological and Evolutionary Concepts

The terminology and conceptual model first proposed by Abrams (1995) are now thoroughly integrated into studies of community-level interactions, and there are hundreds of studies that have identified nonconsumptive, trait-mediated interactions in a wide variety of systems (Werner and Peacor 2003; Preisser et al. 2005). These plasticity-mediated interactions have typically been interpreted in an ecological context, but of course ecological interactions frequently have important evolutionary implications. In most cases the TMIIs are *linear* indirect effects in that one organism indirectly affects another by inducing trait changes in an intermediate species. For example, in Schmitz (1998) spiders indirectly affected old-field grasses by inducing predator-avoidance behaviors in grasshoppers that reduced feeding. Just as TMIIs provide a mode of indirect interaction between two species, TMIIs likewise represent a mode for indirect selection. In the work by Schmitz (1998), one could imagine that grass fitness is increased by the TMII between spiders and grasshoppers; this TMII could reasonably be expected to lead to selection for grass traits that increased the ability of spiders to intimidate grasshoppers. Predators can also induce traits that alter the patterns of selection herbivores impose on plants. For instance, as was discussed above, Rudgers et al. (2003) showed that ants changed not only the amount of caterpillar feeding on wild cotton, but also the location of feeding. Such changes in herbivore feeding location could alter selection on plants if, for instance, certain traits that provide resistance to herbivores are present in one feeding location but not on another. Parallel scenarios can be envisioned for TMIIs in other contexts as well (Miller and Travis 1996).

Evolutionary ecologists have typically considered plasticity with respect to its adaptive value and not within the theoretical constructs of Abrams (1995) that are central to modern community ecology. Yet adaptive plasticity also functions by TMII. Instead of these indirect interactions being linear (see above), these indirect effects *loop* such that the end consequence of an induced effect is to alter the fitness of the plastic organism. For example, consider the interaction dynamics of induced defense by plants to herbivores using the ecological lexicon. Induced defense occurs when herbivore damage of plants induces plant phenotypes that reduce future herbivory. This is a TMII, where one herbivore indirectly affects another by induced changes in a plant. This TMII in itself does not affect plant fitness. The putative adaptive explanation of the induced plant phenotype only occurs when this TMII continues to propagate as an additional indirect effect to increase plant fitness, thus forming a complete interaction loop. Induced defense is thus a three-step indirect interaction (herbivore → plant → herbivore → plant) initiated and propagated by TMIIs. Indirect induced defense occurs when plants respond to herbivores by attracting herbivore enemies, increasing predation or parasitism of herbivores, which in turn reduces plant damage. This interaction is thus a four-step indirect interaction (herbivore → plant → predator → herbivore → plant), again initiated and propagated by TMIIs. Plant induction (herbivore plasticity similarly can be considered in this light: plant secondary compounds induce changes in herbivores (such as the production of P-450 detoxification enzymes) that subsequently allow for increased herbivore feeding (plant → herbivore → plant). Looping and linear plasticity-based interactions (TMIIs) are thus analogous with respect to the interaction mechanism and only differ in that the former, but not the latter, provides putative explanations for the evolution of that plasticity.

The ecological and evolutionary consequences of phenotypic plasticity are also united through the parallel issues of ecological contingency and diffuse selection. Ecological contingency comes from the fact that the structure of any given pairwise interaction may not be fixed but rather may depend on the context in which the interaction takes place. Such "interaction modification" (Wootton 1994) is essentially a result of phenotypic plasticity, where the pairwise interaction between two species is altered because the phenotype of one or both interactors is modified by third parties. It is also now recognized that selection imposed by one species on another may be dependent on the ecological context, or "diffuse" (Iwao and Rausher 1997; Linhart et al. 2005; Strauss et al. 2005). While diffuse selection can be a result of genetic correlations, another principal cause is the change in effect of one species on another due to phenotypic plasticity. In other words, diffuse selection occurs when ecological interactions that affect fitness are context dependent (Inouye and Stinchcombe 2001). Consequently, phenotypic plasticity is one of the root causes for diffuse selection and ecological contingency.

Is Plasticity Important?

There are strong arguments and some good empirical evidence that phenotypic plasticity is an adaptive strategy in many traits of taxonomically disparate organisms with varied life histories. Consequently, opportunities exist for interactions to be trait mediated in every ecological and evolutionary context. There is a growing appreciation for the ecological and evolutionary consequences of organisms being plastic. One indication comes from the explicit comparison of TMIIs and DMIIs when they act on the same dependent variable. Many studies have compared the effects of predators on plants, when predators function as consumers of herbivores versus their effects on herbivore behavior. Preisser et al. (2005) synthesized the results of many such studies. Their meta-analysis showed that the TMIIs between predators and plants were as strong or stronger than the DMIIs. In this scenario, trait-mediated effects provided interaction pathways within food webs that supplemented the consumption-based pathways upon which most thought and theory on community ecology has historically been based. In other scenarios TMIIs may have effects of opposite sign from DMIIs. But whether TMIIs and DMIIs work in unison or against each other, an understanding is emerging that plasticity-based interactions play a central role in community structure and function and in trait evolution in species.

Evidence for the strength of TMIIs also comes from the commonness of induced resistance in plants. Induced resistance (direct and indirect) is adaptive only because of remarkably long chains of TMIIs (see above). These chains are as long or longer than those typically reported from the ecological literature. While there are arguments for why DMIIs are likely to attenuate, this may not be the case for TMIIs (Preisser et al. 2005). The fact that three-step (direct induced resistance) and four-step (indirect induced resistance) interactions are sufficiently strong and consistent to shape the evolution of plant traits is circumstantial evidence for both the ecological and evolutionary importance of interactions based on organism plasticity.

Future Directions

Individual and Combined Effects of TMIIs and DMIIs

A major focus of empirical studies on TMIIs has been to compare the strength of TMIIs, the more recently recognized mode of interaction, with DMIIs, the mode by which most interactions were historically assumed to occur. In almost all studies to date, the experimenters have assumed, either explicitly or implicitly, that TMIIs and DMIIs function additively. In other words, the assumption is that the strength of TMIIs when they operate alone is neither stronger nor weaker than their effects when they operate simultaneously with DMIIs.

In fact, no a priori reason exists to assume that TMIIs and DMIIs are additive and do not interact with each other. Where they do not function additively, estimates of the relative strengths of TMIIs and DMIIs are inaccurate because of the experimental designs typically used. The standard design employs three treatments: a control treatment without any indirect interaction, a full-effect treatment where TMIIs and DMIIs operate simultaneously, and a treatment where TMIIs operate but DMIIs are prevented from occurring. For example, in studies investigating the TMIIs and DMIIs between predators, herbivores, and plants, these treatments have been herbivores and plants alone (no TMIIs or DMIIs occurring), herbivores and plants with predators capable of consuming herbivores (lethal predators with TMIIs and DMIIs), and predator cues in the absence of herbivore consumption (risk predators with TMIIs alone). The DMII has then been calculated—under this questionable assumption of additivity—by subtracting the TMII from the full predator effect. If TMIIs and DMIIs do not function additively, the strength of TMIIs (relative to DMIIs) will have been overestimated or underestimated based on whether the combined effects of the two are synergistic (i.e., superadditive) or antagonistic (i.e., subadditive), respectively.

Given the lack of additivity in the combined effects of many other types of interactions, this assumption should be explicitly tested. Two studies have done so and outline appropriate methodologies for disentangling the independent and combined effects of TMIIs and DMIIs (Peacor and Werner 2001; Griffin and Thaler 2006): in a fourth treatment, DMIIs are imposed in the absence of TMIIs. In the tritrophic example outlined above, herbivores can be hand-removed at a rate that mimics that of the predator in the lethal predator treatment. This added treatment thus creates a fully crossed two-by-two factorial design where DMIIs and TMIIs operate alone and in combination, thus allowing for an explicit statistical test for the interaction between these two factors (simply, the interaction term in a two-way ANOVA).

Future studies should adopt this or similar approaches that explicitly test for additivity in TMIIs and DMIIs. Indeed, both studies report nonadditive effects (Peacor and Werner 2001; Griffin and Thaler 2006). Such studies will better inform us as to the relative strengths of TMIIs and DMIIs. Perhaps of greater importance, this approach will allow us to predict under in what circumstances TMIIs and DMIIs are likely to work synergistically and increase total interaction strength, versus antagonistically and decrease total interaction strength.

Expanding the Types of Interactions

The ecological literature on TMIIs has been dominated by studies where effects are transmitted by induced changes in herbivore or predator behavior (Werner and Peacor 2003; Preisser et al. 2005). While evolutionary studies have investigated the consequences of plasticity in plants, this work has often been conducted under relatively controlled conditions with the goal of providing proof of principle rather than testing for relative ecological importance. For example, while volatiles from one plant have been shown to induce defenses and reduce herbivory in neighboring plants, the broader importance of such dynamics in a field setting is still being evaluated (Karban et al. 2004). Future studies should investigate the broad significance of plasticity in plants, and compare the strength such effects with other forms of ecological interaction (Agrawal and Van Zandt 2003; Callaway et al. 2003).

Broad Patterns?

It is presently unclear whether the commonness, magnitude, and ecological importance of phenotypic plasticity show broad patterns in terms of habitats (terrestrial verus aquatic, tropical versus temperate), trophic roles (plants, herbivores, predators, parasites), or phylogeny (angiosperms versus nonangiosperm plants, vascular versus nonvascular plants, vertebrate versus nonvertebrate animals, etc.). When contrasting habitats and trophic roles, predictions for patterns should be based on differences in the expected benefits and costs. Inherent limitations to plasticity should also be considered with respect to phylogenetic patterns.

An example of this approach is the contrast between terrestrial and aquatic systems. Though herbivore morphological and behavioral plasticity in response to predators has been more widely documented in aquatic compared to terrestrial systems (Kats and Dill 1998), it is unclear whether this is simply an artifact of research traditions. If this distinction is based in a true difference between the levels of plasticity across habitats, it may be due to the fact that chemical cues from predators persist longer and are more easily detected in water than in air. Indeed, the meta-analysis of Preisser et al. (2005) found that TMIIs were stronger in aquatic than terrestrial systems. The total effect of predators on reducing herbivory via both DMIIs and TMIIs may thus be greater in wet than dry systems, perhaps explaining in part the greater strength of trophic cascades in aquatic communities (Strong 1992).

The relative importance of plasticity may also vary predictably within communities. For instance, herbivores may be expected to respond differently to predator cues based on the foraging strategy of the predator. Schmitz and Suttle (2001) compared the effects of three species of hunting spider on grasshoppers. Grasshopper behavior changed in response to the sit-and-wait and sit-and-pursue predators, but not to the species that employed an active hunting strategy. Such predator-specific responses by herbivores would be expected if cues from active (as compared to less mobile) hunters provide unreliable information about the likelihood of future predator encounters. Plants also demonstrate specificity in their responses to herbivores (Agrawal 2000; Van Zandt and Agrawal 2004a, 2004b). One factor leading to variable responses may be differences in the information contained in feeding by different herbivore species: plants would be expected to respond differently to herbivore damage when current feeding is indicative of future feeding compared to when the information content of current feeding is relatively low (Karban et al. 1999).

Phenotypic plasticity may also be a predictive axis to explain variation in the strategies that herbivores employ in consuming plants (Karban and Agrawal 2002). Herbivores' strategies vary from the relatively passive, such as host choice (choosing the best foods to eat), to more active strategies, such as detoxification of defense chemicals. Given that generalist herbivores are more likely to employ plastic strategies, presumably because they encounter more variable hosts than specialists, we predict that one benefit of specialization is that costs of plasticity may be avoided. Circumstantially it appears that specialists that do not invest in plastic choices benefit from this lack of plasticity, especially in complex environments (Bernays 1999, 2001).

While phenotypic plasticity is often assumed to be an evolutionarily derived trait evolved from fixed strategies, this has been little tested. The data we have for induced plant responses are ambiguous in terms of the ancestral state (Thaler and Karban 1997; Heil et al. 2004). Karban and Baldwin (1997) have further argued that there is no a priori reason why we should expect inducibliity to be the derived state. Furthermore, plasticity may in turn shape the macroevolution of species. According to a somewhat controversial hypothesis, phenotypic plasticity may lead to ecological success in a novel habitat that ultimately leads to evolutionary divergence and speciation (West-Eberhard 1989; Robinson and Dukas 1999). Ironically enough, evolutionary biologists have historically viewed phenotypic

Miller, T.E., and J. Travis. 1996. The evolutionary role of indirect effects in communities. Ecology 77: 1329–1335.

Mitton, J.B., and K.B. Sturgeon. 1982. Bark beetles in North American conifers: a system for the study of evolutionary biology. University of Texas Press, Austin.

Mooney, K.A. 2006. The effects of birds and ants on pine herbivore load are contingent on whether birds disrupt an ant-aphid mutualism. Ecology 87: 1805–1815.

Mooney, K.A. In press. Tritrophic effects of birds and ants on a canopy foodweb, tree growth and phytochemistry. Ecology 88(8): 2005–2014.

Mullin, C., and B. Croft. 1983. Host-related alterations of detoxification enzymes in *Tetranychus urticae* (Acari: Tetranychidae). Environ. Entomol. 12: 1278–1282.

Neal, J.J. 1987. Metabolic costs of mixed-function oxidase induction in *Heliothis zea*. Entomol. Exp. Appl. 43: 175–180.

Nguyen, T.N.M., Q.G. Phan, L.P. Duong, K.P. Bertrand, and R.E. Lenski. 1989. Effects of carriage and expression of the Tn10 tetracycline-resistance operon on the fitness of *Escherichia coli* K12. Mol. Biol. Evol. 6: 213–226.

Ozawa, R., G. Arimura, J. Takabayashi, T. Shimoda, and T. Nishioka. 2000. Involvement of jasmonate- and salicylate-related signaling pathways for the production of specific herbivore-induced volatiles in plants. Plant Cell Phys. 41: 391–398.

Peacor, S.D., and E.E. Werner. 2001. The contribution of trait-mediated indirect effects to the net effects of a predator. Proc. Nat. Acad. Sci. USA 98: 3904–3908.

Preisser, E.L., D.I. Bolnick, and M.F. Benard. 2005. Scared to death? The effects of intimidation and consumption in predator-prey interactions. Ecology 86: 501–509.

Relyea, R.A. 2002. Costs of phenotypic plasticity. Am. Nat. 159: 272–282.

Robinson, B.W., and R. Dukas. 1999. The influence of phenotypic modifications on evolution: the Baldwin effect and modern perspectives. Oikos 85: 582–589.

Rossiter, M. 1996. Incidence and consequences of inherited environmental effects. Annu. Rev. Ecol. Syst. 27: 451–476.

Rotem, K., A.A. Agrawal, and L. Kott. 2003. Parental effects in *Pieris rapae* in response to variation in food quality: adaptive plasticity across generations? Ecol. Entomol. 28: 211–218.

Rudgers, J.A., J.G. Hodgen, and J.W. White. 2003. Behavioral mechanisms underlie an ant-plant mutualism. Oecologia 135: 51–59.

Scheiner, S.M., and D. Berrigan. 1998. The genetics of phenotypic plasticity. VIII. The cost of plasticity in *Daphnia pulex*. Evolution 52: 368–378.

Schmitz, O.J. 1998. Direct and indirect effects of predation and predation risk in old-field interaction webs. Am. Nat. 151: 327–342.

Schmitz, O.J., and K.B. Suttle. 2001. Effects of top predator species on direct and indirect interactions in a food web. Ecology 82: 2072–2081.

Steinger, T., B.A. Roy, and M.L. Stanton. 2003. Evolution in stressful environments. II. Adaptive value and costs of plasticity in response to low light in *Sinapis arvensis*. J. Evol. Biol. 16: 313–323.

Stone, G.N., and K. Schonrogge. 2003. The adaptive significance of insect gall morphology. Trends Ecol. Evol. 18: 512–522.

Stout, M.J., K.V. Workman, R.M. Bostock, and S.S. Duffey. 1998. Specificity of induced resistance in the tomato, *Lycopersicon esculentum*. Oecologia 113: 74–81.

Strauss, S.Y., H. Sahli, and J.K. Conner. 2005. Toward a more trait-centered approach to diffuse (co)evolution. New Phytol. 165: 81–89.

Strong, D.R. 1992. Are trophic cascades all wet? Differentiation and donor-control in speciose ecosystems. Ecology 73: 747–754.

Thaler, J.S. 1999. Jasmonate-inducible plant defences cause increased parasitism of herbivores. Nature 399: 686–688.

Thaler, J.S., and R. Karban. 1997. A phylogenetic reconstruction of constitutive and induced resistance in *Gossypium*. Am. Nat. 149: 1139–1146.

Thompson, D.B. 1992. Consumption rates and the evolution of diet-induced plasticity in the head morphology of *Melanoplus femurrubrum* (Orthoptera, Acrididae). Oecologia 89: 204–213.

Traw, M.B., and J. Bergelson. 2003. Interactive effects of jasmonic acid, salicylic acid, and gibberellin on induction of trichomes in *Arabidopsis*. Plant Physiol. 133: 1367–1375.

van Kleunen, M., M. Fischer, and B. Schmid. 2000. Costs of plasticity in foraging characteristics of the clonal plant *Ranunculus reptans*. Evolution 54: 1947–1955.

Van Tienderen, P.H. 1991. Evolution of generalists and specialists in spatially heterogeneous environments. Evolution 45: 1317–1331.

Van Zandt, P.A., and A.A. Agrawal. 2004a. Specificity of induced plant responses to specialist herbivores of the common milkweed, *Asclepias syriaca*. Oikos 104: 401–409.

Van Zandt, P.A., and A.A. Agrawal. 2004b. Community-wide impacts of herbivore-induced plant responses in milkweed (*Asclepias syriaca*). Ecology 85: 2616–2629.

Walling, L.L. 2000. The myriad plant responses to herbivores. J. Plant Growth Reg. 19: 195–216.

Werner, E.E., and S.D. Peacor. 2003. A review of trait-mediated indirect interactions in ecological communities. Ecology 84: 1083–1100.

West-Eberhard, M.J. 1989. Phenotypic plasticity and the origins of diversity. Annu. Rev. Ecol. Syst. 20: 249–278.

Wootton, J.T. 1994. The nature and consequences of indirect effects in ecological communities. Annu. Rev. Ecol. Syst. 25: 443–466.

Yao, I., and S. Akimoto. 2001. Ant attendance changes the sugar composition of the honeydew of the drepanosiphid aphid *Tuberculatus quercicola*. Oecologia 128: 36–43.

Yao, I., and S.I. Akimoto. 2002. Flexibility in the composition and concentration of amino acids in honeydew of the drepanosiphid aphid *Tuberculatus quercicola*. Ecol. Entomol. 27: 745–752.

Yao, I., H. Shibao, and S. Akimato. 2000. Costs and benefits of ant attendance to the drepanosiphid aphid *Tuberculatus quercicola*. Oikos 89: 3–10.

Young, T.P. 1987. Increased thorn length in *Acacia drepanolobium*—an induced response to browsing. Oecologia 71: 436–438.

Young, T.P., and B.D. Okello. 1998. Relaxation of an induced defense after exclusion of herbivores: spines on *Acacia drepanolobium*. Oecologia 115: 508–513.

Zavala, J.A., A.G. Patankar, K. Gase, and I.T. Baldwin. 2004a. Constitutive and inducible trypsin proteinase inhibitor production incurs large fitness costs in *Nicotiana attenuata*. Proc. Nat. Acad. Sci. USA 101: 1607–1612.

Zavala, J.A., A.G. Patankar, K. Gase, D.Q. Hui, and I.T. Baldwin. 2004b. Manipulation of endogenous trypsin proteinase inhibitor production in *Nicotiana attenuata* demonstrates their function as antiherbivore defenses. Plant Physiol. 134: 1181–1190.

FIVE

Selection and Genetic Architecture of Plant Resistance

MARY ELLEN CZESAK, ROBERT S. FRITZ, AND CRIS HOCHWENDER

Basic and applied research programs can both benefit by approaching concerns regarding resistance to herbivores from a perspective centering on natural selection and genetic architecture of resistance. In natural systems, quantification of selection, determination of genetic correlations with other traits, and evaluation of genetic architecture (i.e., estimation of additive and nonadditive genetic effects) can enhance our ability to predict the evolutionary trajectory of plant resistance. Evaluation of genetic architecture has become increasingly emphasized because hybridization is widespread in plant species. Moreover, genetic architecture studies involving interpopulational or interspecific hybrids can give insight into the processes of population differentiation, speciation, and the formation and persistence of hybrid zones. From an applied perspective, programs proposing to use biological controls (e.g., insect herbivores and pathogens carried by insects) to regulate introduced weed species can benefit by considering the direct and indirect selective effects that those herbivores may have on the weedy species. Moreover, programs using integrated pest management strategies are increasingly utilizing plants bred or genetically engineered for increased resistance to pests to reduce pesticide use and its negative impact on beneficial organisms. Integrated pest management strategies require a fundamental understanding of selection pressure, correlational selection, and ecological costs in order to plan long-term management strategies.

This chapter explores the selective agents that influence plant resistance to insect herbivores, the strength of selection, and the genetic architecture of resistance traits. Additionally, this chapter examines the relationship between plant hybrid resistance and the resistance of parental populations or species, which have implications for hybrid zone dynamics and for introgression of resistance traits between populations or species (see Scriber et al., this volume).

Selection on Resistance within Populations

Genetic Variation in Traits

Although resistance traits can be controlled by one or a few loci (e.g., Daday 1954), it is more common for resistance traits to be controlled by many loci. Often, broad-sense heritability (h^2) of resistance or resistance traits has been estimated in natural populations and agricultural species (e.g., significant effect of genotype in *Piper arieianum* C.D.C. [Marquis 1990]; cultivars of chrysanthemum *Dendranthema grandiflora* Tzvelev [deJager et al. 1995]; common milkweed *Asclepias syriaca* L. [Agarwal 2005]). Narrow-sense heritability is considered more valuable when quantifying genetic variation in resistance because a response to selection can occur only if significant additive genetic variance exists in a plant population. Additive genetic variation in resistance or resistance traits has been detected in a wide variety of plant species. These include leaf furanocoumarins in wild parsnip *Pastinaca sativa* L. (for most furanocoumarins, h^2 = 0.33–1.17) (Berenbaum et al. 1986); proportion of damaged leaves in common morning glory *Ipomoea purpurea* L. (Roth) (Simms and Rausher 1989); leaf trichome number in *Brassica rapa* L. ($h^2 = 0.56 \pm 0.14$) (Ågren and Schemske 1994); glucosinolates in Brussels sprout hybrids *B. oleracea* L. (h^2 = 0.72 or 1.09, sinigrin and progoitrin, respectively) (van Doorn et al. 1999); leaf trichome number in *B. nigra* L. (Koch) ($h^2 = 0.54 \pm 0.21$) (Traw 2002); and phenolic glycoside concentration in *Salix sericea* Marshall leaves (h^2 = 0.20–0.59) (Orians and Fritz 1995). For the wild radish, *Raphanus raphanistrum* L., additive genetic variance in induced glucosinolate production was detected in response to damage by caterpillars of the white cabbage butterfly (*Pieris rapae* L.). Induced responses varied among half-sib families, ranging from a 7% to 140% increase from constitutive levels (Agrawal et al. 2002b). Realized heritability of resistance has been demonstrated through artificial selection experiments. For example, iridoid glycoside concentrations in leaves of the plantain *Plantago lanceolata* L. respond to selection and have a significant realized heritability estimate (0.23 ± 0.07) (Marak et al. 2000).

Natural selection within a population (or species) can give rise to epistatic genetic effects. Coadapted complexes arise when selection acts to favor the coevolution of alleles at different loci independent of the environment (Waser 1993), whereas local adaptation occurs when selection is contingent upon the environment (e.g., Sork et al. 1993). Epistatic genetic effects can create evolutionary trajectories that can be difficult to predict and can cause asymmetrical responses to selection (Merilä and Sheldon 1999; Galloway and Fenster 2001). For example, as the genetic composition of a population changes, nonadditivity can change the additive effects of alleles. Because additive effects of alleles are averaged across all the genotypes in the population, those averages can change as epistatic interactions among alleles change. This can potentially change which alleles are favored by selection (Barton and Keightley 2002) and influence the outcome of selection. Additionally, epistatic variance can be converted to additive genetic variance, causing genetic variance-covariance matrices to change in response to selection (Agrawal et al. 2002a).

Differences in resistance between inbred and outcrossed individuals can reveal nonadditive genetic effects. Inbreeding causes an increase in homozygosity, thereby increasing the expression of recessive alleles. When recessive alleles are deleterious, inbreeding depression can occur (Charlesworth and Charlesworth 1987). For example, selfed individuals of yellow monkey flower (*Mimulus guttatus* Fisch. ex D.C.) are negatively affected by spittlebug nymphs (*Philaenus spumarius* L.), whereas outcrossed individuals are not negatively affected (Carr and Eubanks 2002). Selfed individuals with nymphs produced fewer flowers and had lower aboveground biomass compared to selfed control plants without nymphs, whereas outcrossed individuals with nymphs produced more flowers and had a similar aboveground biomass compared to outcrossed control plants without nymphs.

Herbivores as Selective Agents

Plant defense theory and plant-herbivore coevolutionary theory both make the assumption that plant resistance traits have evolved as adaptations to reduce herbivory (Ehrlich and Raven 1964; Feeny 1976; Rhoades and Cates 1976). Still, the selective agent(s) for some resistance traits remains unclear. For example, cyanogenic glycosides provide effective resistance to generalist herbivores (e.g., Horrill and Richards 1986), but these compounds may have other functions including nitrogen storage and control of germination. Similarly, glucosinolates may benefit mustard plants by reducing the harmful effects of UV light. Therefore, in addition to documenting genetic variation in resistance traits, research must also document selection by insect herbivores favoring resistant phenotypes over susceptible phenotypes.

Natural selection by insect herbivores on plant resistant traits can be inferred from a signification additive covariance between a resistance trait and fitness, for example, negative additive covariance between corn earworm *Helicoverpa zea* (Broddie) damage and seed number in common morning glory *Ipomoea purpurea* (Roth) (Simms and Rausher 1989). Natural selection by insect herbivores can also be measured by quantifying selection differentials (S) or by determining linear selection gradients (β_i), which estimate directional phenotypic selection as the average slope of the relationship between a trait and relative fitness (Lande and Arnold 1983). Selection gradients are preferred because correlational selection for multiple traits can be considered (i.e., the simultaneous selection on a combination of two traits). Several studies have demonstrated that herbivores are strong selective agents on plant resistance or resistance traits using selection gradients. For example, significant stabilizing selection was detected on total glucosinolate concentration in *Arabidopsis thaliana* L. (Heynh.) when herbivores were experimentally removed but not when herbivores were present ($\gamma = -0.00329$) (Mauricio and Rausher 1997). Additionally, significant directional selection for increased resistance to herbivores was detected in three out of six populations of *Datura stramonium* L. (β = 0.529–0.564) (Valverde et al. 2001).

Plant resistance traits can evolve in a relatively short period of time in response to insect herbivores. For example, the defoliating hemlock moth (*Agonopterix alstroemeriana* Clerck), a specialist leaf-rolling caterpillar, appears to have acted as a selective agent on the defensive chemistry of poison hemlock (*Conium maculatum* L.) (Castells et al. 2005). Poison hemlock was introduced to the United States from Europe and remained relatively free of herbivory until 30 years ago when the defoliating hemlock moth was introduced. Populations of poison hemlock experiencing high caterpillar herbivory have much higher alkaloid concentrations than populations with low caterpillar herbivory. Moreover, caterpillars exhibit preference for genotypes within a population that have lower alkaloid concentrations. The differences in alkaloid concentrations among populations and preference by caterpillars for low alkaloid producing genotypes suggest that the caterpillar has acted as a selective agent on alkaloid concentration.

Resistance can directly affect fitness by reducing herbivore damage, but resistance traits can also indirectly affect fitness through correlated effects (Lande and Arnold 1983). By using multivariate analysis, relative fitness can be partitioned into direct and indirect selection. Correlational selection can be estimated as a quadratic selection gradient (γ_{ij}) (Lande and Arnold 1983). Significant correlational selection was detected in *Brassica rapa* L. for resistance to a flea beetle (*Phyllotreta cruciferae* Goeze) and the diamondback moth (*Plutella xylostella* L.); at low levels of moth damage, there is negative directional selection on damage by flea beetles, whereas at high levels of moth damage there is positive directional selection on damage by flea beetles (Pilson 1996).

Because multivariate analysis estimates phenotypic selection gradients, the trait will not necessarily respond to selection unless the trait also expresses additive genetic variation. Additionally, estimates of selection can be biased if environmental factors cause the observed covariance between resistance traits and relative fitness (Stinchcombe et al. 2002). To avoid this bias and to determine the genetic basis for traits, regression coefficients of breeding values can be used to create genotypic selection gradients (Rausher 1992). Path analysis, which allows for consideration of known causal relationships between traits in a selection model, can also be used to separate the importance of direct and indirect selection on resistance traits (Scheiner et al. 2000). For example in the ivyleaf morning glory (*Ipomoea hederacea* L.), no direct effect of cotyledon damage by insects was detected on plant fitness, although indirect effects were detected through the effect of cotyledon damage on plant size (Stinchcombe 2002).

Constraints on the Evolution of Resistance

Strong selection can reduce genetic variation through the fixation of alleles (Fisher 1958), but genetic variation in resistance may be maintained within a population through constraints. Plant resistance can be constrained by (1) costs related to resource limitation (allocational cost), (2) trade-offs between resistance and tolerance, (3) resistance trade-offs among herbivores (or among herbivores and pollinators; ecological cost), and (4) constraints related to the environment.

ALLOCATIONAL COSTS

Allocation costs to resistance should cause plant fitness to be reduced in the absence of herbivores. In the presence of herbivores, though, the costs are outweighed by the benefits of reducing leaf tissue loss (reviewed in Strauss et al. 2002). Constraints in resource allocation would generate negative genetic correlations between resistance and fitness-enhancing traits, causing a genetic trade-off between these traits. When the same alleles affect the expression of both the resistance traits and the other fitness-enhancing traits, these pleiotropic effects will fundamentally constrain the expression of an optimal phenotype. In contrast, when the negative genetic correlation is due to linkage disequilibrium (i.e., the association of alleles at differing loci), selection will be slowed, but independent assortment and recombination during meiosis will lead to the disruption of linkage, allowing for the eventual selection of an optimal phenotype.

Allocation costs have been detected in some studies. For example, in *Arabidopsis thaliana* L. (Heynh.), negative genetic correlations were detected between fruit number and both trichome density (genetic correlation $r_G = -0.46$) and total glucosinolate concentration ($r_G = -0.27$) (Mauricio 1998). Thus, selection for increased trichome density or total glucosinolate concentration should lead to reduced fruit number. However, allocation costs are not always observed. In the wild radish, *Raphanus raphanistrum* L., neither genetic correlations between glucosinolate concentration and early season flower number nor between glucosinolate concentration and lifetime fruit mass were significant (additive genetic correlation r_A ranged from -0.32 to 0.10, all $p > 0.10$) (Agrawal et al. 2002a). Similarly, no negative genetic correlation was detected between iridoid glycoside concentration growth in *Plantago lanceolata* L., despite a high biosynthetic cost of these compounds (Marak et al. 2003).

TRADE-OFF BETWEEN RESISTANCE AND TOLERANCE

Trade-offs between plant genotypes that produce defenses against herbivores and genotypes that tolerate herbivore damage (i.e., that reduce the fitness consequences of herbivores through regrowth and/or the production of additional flowers) have been predicted (sensu Simms and Triplett 1994) for resistance and tolerance to fungal disease. Models that evaluate these two alternate strategies of defense assume that resistance and tolerance draw upon the same pool of resources, so that a plant genotype possessing both high resistance and high tolerance would suffer from the fitness costs of both. These models suggest that a plant population might evolve either high resistance or high tolerance, but not both; disruptive correlational selection on resistance and tolerance is thus predicted (i.e., simultaneous selection on a combination of high resistance and low tolerance, or vice versa). A few studies have detected a trade-off between resistance and tolerance (e.g., Stowe 1998; Pilson 2000). Other empirical studies, however, have found high levels of both resistance and tolerance in plant populations (e.g., Valverde et al. 2003); these findings could reflect either a transitory evolutionary stage or indicate that intermediate levels can evolve. A recent model by Fornoni et al. (2004a) predicts that intermediate levels of resistance and tolerance will evolve if resistance and tolerance do not have equal fitness costs. This assumption is plausible because the severity of fitness costs for resistance and tolerance can be independent and can depend on the environment (Fornoni et al. 2004b).

ECOLOGICAL COSTS

Ecological costs to resistance can also constrain resistance evolution. The degree of resistance can be specific to herbivore species (e.g., Berenbaum et al. 1986), so increased resistance to one herbivore species could lead to reduced resistance to another herbivore species. Similarly, herbivory by one insect species may change the quality of the host plant for other insect species (Pilson 1992), thereby affecting resistance to other insect species.

Herbivores and pollinators can both affect plant fitness; when herbivores and pollinators exert opposing selection pressures, ecological costs may maintain genetic variation in traits. For example, a flower color polymorphism in wild radish (*Raphanus sativus* L.) is maintained by pollinators and herbivores (Irwin et al. 2003). Pollinators strongly prefer anthocyanin-recessive color morphs (yellow and white) compared to anthocyanin-dominant color morphs (pink and bronze). For plants in bloom, the specialist white cabbage butterfly (*Pieris rapae* L.) and generalist beet armyworm (*Spodoptera exigua* Hübner) preferred to oviposit on the yellow and white morphs versus pink and bronze morphs. This herbivore preference may be due to higher concentrations of indole glucosinolates in the leaves of pink and bronze morphs compared to the leaves of yellow and white morphs. Thus, floral pigments and secondary compounds in *R. sativus* may be genetically linked.

ENVIRONMENTAL EFFECTS

Environmental factors, such as soil chemistry (Brown et al. 1984), water availability (Berneys and Lewis 1986), light availability (Collinge and Louda 1988), and competitive interactions can also alter the effect of herbivores on plant fitness. Costs of resistance are predicted to be more severe in stressful environments because the pool of resources shared by plant functions is severely limited (Herms and Mattson 1992). Some evidence supports this viewpoint; the cost of resistance to the lettuce root aphid (*Pemphigus bursarius* L.) was more severe for lettuce (*Lactuca sativa* L. var. *avoncrisp*) grown under nutrient limitation (Bergelson 1994). A trade-off between resistance and other plant functions also may result from increased allocation to resistance in stressful environments. In the case of the cape periwinkle, *Catharanthus roseus* (Don.), alkaloid production is increased in stressful environments, reducing resources available to other fitness-enhancing functions (Hirata et al. 1993). In contrast, other studies have detected higher costs in resource-abundant environments (reviewed by Bergelson and Purrington 1996; for competitive environment, see Siemens et al. 2003). For carbon-based defensive chemicals, production may not incur a significant cost if nutrient limitation leads to greater availability of fixed carbon within the plant (Herms and Mattson 1992).

Because environmental factors can alter the expression of alleles (Falconer and Mackay 1996), selection on resistance traits can vary in magnitude and/or direction among different environments. For example, an altitudinal cline (i.e., change in allele frequency over geographic area) evolved very rapidly for populations of introduced white clover (*Trifolium repens* L.) near Vancouver, Canada, such that populations with a high percentage of cyanogenic individuals were found at lower altitudes, while populations with a low percentage of cyanogenic individuals were found at higher altitudes (Ganders 1990). This cline may have evolved due to differences in herbivore abundance, soil moisture stress, and temperature across elevations (Ganders 1990). As another example, selection gradients (δ) for resistance of *Datura stramonium* L. differ for genotypes originating from two different sites in central Mexico; positive directional selection on resistance was observed within a tropical dry forest population ($\delta = 0.304$), but no selection on resistance was detected within a pine-oak forest population ($\delta = -0.008$, not significant; plants grown in common environment) (Fornoni et al. 2004b).

Genetic Architecture of Resistance Traits between Populations and Species

Related species of plants, distinct populations of the same species, and cultivars/varieties can differ in their resistance to insects, either through natural or artificial selection or genetic drift. Studies investigating the genetic architecture between populations or species can give insight into (1) the process of population differentiation, (2) the process of speciation, and (3) the formation and persistence of hybrid zones. From an applied perspective, such studies are necessary for the prediction of hybrid resistance in breeding and conservation programs (Linder et al. 1998). Hybridization between populations/species can affect resistance evolution by causing hybrid breakdown in novel genotypes, hybrid superiority, or introgression of resistance traits.

Potential Impacts of Hybridization

HYBRID BREAKDOWN

Hybrids can exhibit outbreeding depression or hybrid breakdown by having lower resistance or higher susceptibility than either parental population or species (Christensen et al. 1995; Fritz et al. 2003; Czesak et al. 2004). During allopatric population differentiation or speciation, there can be an accumulation of allele substitutions that differs between populations/species as a result of natural selection or genetic drift. These allele substitutions are not predicted to be deleterious within each population/species, but hybridization can bring together unique combinations of alleles that are deleterious (negative epistatic interactions in the Dobzhansky-Muller model). Natural selection can also create coadapted gene complexes or result in local adaptation to an environment, and hybridization can disrupt such gene complexes or create hybrids that express genes that are not locally adapted (adverse genotype × environment interactions) (Templeton 1986). Hybrid breakdown may not be evident until recombination occurs in the F_2 generation or later (but see Dungey et al. 2000). Although selection can act against hybrids, they can persist if such selection is balanced by gene flow (Barton and Hewitt 1989). In addition, hybrid breakdown of resistance traits may influence the population dynamics of insect herbivores, with hybrids acting as sinks for insect pests (Whitham 1989), allowing for the expansion of host range for insect herbivores (the "hybrid bridge" hypothesis of Floate and Whitham [1993; see also Pilson 1999]).

HYBRID SUPERIORITY

Hybridization between two populations/species can lead to hybrid superiority, the increased resistance compared to one or both parents. When hybrids are superior, hybrid populations can persist (Barton and Hewitt 1989) and ultimately may become new species (Rieseberg and Carney 1998). Hybrids exhibit heterosis if they have higher resistance than either parent. Heterosis for traits has been found in novel environments outside the parental population or species range, possibly contributing to hybrid zone formation. Heterosis also has been detected by crossing small populations with low to moderate levels of gene flow (Whitlock et al. 2000).

In F_1 hybrids, heterosis in resistance can be attributed to the masking of deleterious recessive alleles from one parent by the alleles present at the same locus of the other parent. Alternatively, F_1 hybrids can exhibit heterosis for resistance if such loci have increased heterozygosity. Dominance interactions that contribute to heterosis can be detected if half the heterosis in F_1 hybrids is lost in F_2 hybrids, due to a loss of half the heterozygosity from the segregation of chromosomes during meiosis (Lynch and Walsh 1998). Epistatic interactions that contribute to heterosis can be detected if the resistance of F_2 hybrids is not intermediate to the average between the F_1 hybrid mean and the parental mean. Ultimately, the resulting fitness of hybrids depends on a balance between the loss of favorable and the loss of unfavorable interactions within the parental genomes (Arnold 1997).

INTROGRESSION

Lastly, hybridization can allow for introgression of resistance traits through repeated backcrossing and movement of beneficial alleles from one population/species to the other. Evidence suggests that hybrids may act as selective filters, allowing the introgression of some genes while not allowing the introgression of others (Martinsen et al. 2001).

Hybrid Zone Dynamics

Because environmental factors can alter the effect of insect herbivores on host plants, hybrid resistance can be dependent upon the environment. If so, the environment can be an important factor in the maintenance of hybrid zones. Models of hybrid zone dynamics differ in the importance of the environment on hybrid fitness. For example, the tension zone model predicts that hybrid zones are maintained by a balance between dispersal and endogenous selection against hybrids, which is independent of the environment (Barton and Hewitt 1985; Arnold 1997). The mosaic hybrid zone model predicts endogeneous selection against hybrids but also predicts exogeneous selection against hybrids in parental habitats (Howard 1986). The bounded hybrid superiority model predicts hybrid superiority in intermediate or novel habitats (Moore 1977; Arnold 1997). Although the environment can clearly change the pattern of hybrid fitness relative to parental fitness, relatively few experimental studies in natural systems have examined hybrid fitness across different environments using hybrids of known genetic background (Campbell 2004).

The relative fitness of hybrids versus parental populations or species can also vary throughout the growing season. For example, hybrids of the skyrockets *Ipomopsis aggregata* (Grant) and *I. tenuituba* (Grant) often have a high proportion of flowers with anthomyiid fly eggs throughout the season, while on *I. aggregata*, this proportion is reduced later in the season. This pattern is not likely due to differences in flowering phenology between *I. aggregata* and the hybrids or to phenological differences of the fly, but could be due to phenological differences in defensive chemical production or plant stress levels (Campbell et al. 2002).

Architecture of Resistance in a Willow Hybrid System

Studying the genetic architecture of resistance between two populations or species in natural systems can be challenging. Hybrid genotypes (whether F_1, F_2, or backcross hybrids) can differ in expression of resistance (e.g., Dungey et al. 2000), yet the genetic identity of natural hybrids often remains unknown. Additionally, hybrids may not co-occur with parental population or species in the field, so differences in resistance between hybrids and parents could be due to a patchy insect distribution or other environmental factors. Moreover, resistance can change over time because of changes in insect abundance, plant development and phenological changes, or environmental variation. To overcome these challenges, identification of pure species and the creation of controlled crosses to create hybrids of known identity can be implemented. Resistance studies in a common garden setting or through reciprocal transplanting are able to quantify hybrid and parental resistance in relevant environmental contexts (e.g., Campbell et al. 2002; Fritz et al. 2003).

The willows *Salix eriocephala* Michx. and *S. sericea* Marshall are shrubs that occur in swampy habitats and along streams in the eastern United States and Canada. Various insect species attack these willows, including *Chaitophorus* sp. aphids, leaf-mining moths *(Phyllonorycter salicifoliella* Chambers and *Phyllocnistis salicifolia),* leaf-folding moths (*Caloptilia* sp. and two unidentified species), stem gall–inducing flies (*Rabdophaga rigidae* Osten Sacken and *R. salicisbrassicoides* Packard), leaf gall–inducing sawflies (*Euponta-nia s-gracilis, Phyllocolpa eleanorae* Smith and Fritz, *Phyllocolpa nigrita* Marlatt, *P. terminalis* Marlatt, and an unidentified *Phyllocolpa* sp.), and leaf-eating beetles (*Calligrapha multipunctata bigsbyana* Kirby, *Plagiodera versicolora* Laicharting, *Chrysomela scripta* F., *C. knabi* Brown, Japanese beetles *Popillia japonica* Newman). The two willow species differ in various traits that can potentially affect resistance to herbivores. *Salix eriocephala* has glabrous leaves that contain high concentrations of condensed tannins (Orians and Fritz 1995). *Salix sericea* has leaves that are densely covered with trichomes on the lower surface and contain high concentrations of phenolic glycosides (higher concentrations of salicortin versus 2′-cinnamoyl salicortin) and low concentrations of condensed tannins (Orians and Fritz 1995). Additive genetic variation is present within *S. sericea* for phenolic glycoside concentrations, and thus these traits have the potential to evolve (h^2 salicortin = 0.20, h^2 2′-cinnamoyl salicortin = 0.59) (Orians et al. 1996). It is unknown at this time if trichome density in *S. sericea* or condensed tannin concentration in *S. eriocephala* exhibits additive genetic variation.

Throughout their geographic range, *S. eriocephala* and *S. sericea* hybridize and co-occur with their hybrids. Introgression has occurred between these two species as revealed through molecular analyses of chloroplast DNA, 20 diagnostic random amplified polymorphic DNA markers and secondary chemical analyses (Hardig et al. 2000). At a field site in central New York state, F_2-type and backcross hybrids are present, with F_2-type hybrids being the most common hybrid class (Hardig et al. 2000).

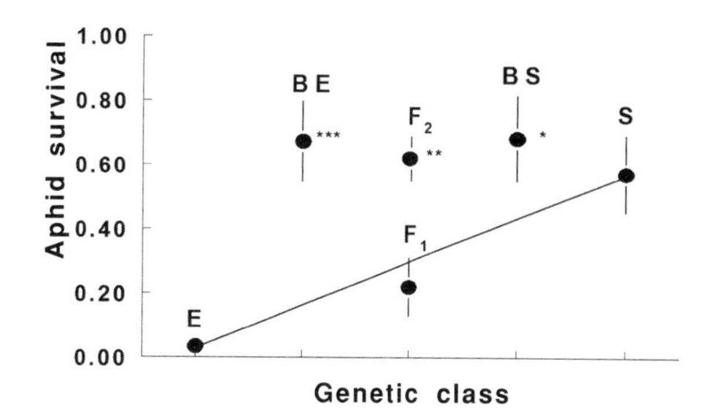

FIGURE 5.1. Mean survival of *Chaitophorus* sp. aphids (±SE) on *Salix eriocephala* (E), *S. sericea* (S), and their interspecific hybrids (F_1, F_2, backcross to *S. eriocephala* [BE], and backcross to *S. sericea* [BS]). The line connecting mean aphid survival on E and S represents where means should fall if additive effects alone explained variation in resistance to aphids. *, Significantly different from line of additivity at $p < 0.05$, **$p < 0.01$, ***$p < 0.001$. Data from Czesak et al. (2004).

Hybrid Resistance

Traits that confer resistance to insect herbivores vary among the parental willow species and their hybrids. F_1 hybrids have leaves with a dense covering of trichomes on the lower surface (*S. sericea* trait) and concentrations of condensed tannins and phenolic glycosides that are intermediate between the concentrations of the two parental species (Orians and Fritz 1995). The concentration of these secondary chemicals could affect insect herbivore preference for the species and their hybrids. The beetles *Calligrapha multipunctata bigsbyana* and *P. versicolora* prefer hybrids over either species, while the beetles *Chrysomela scripta* and *C. knabi* prefer hybrids and *S. sericea* over *S. eriocephala*, and Japanese beetles prefer *S. eriocephala*. Tests with purified salicortin indicate that this chemical contributes to the pattern of preference among beetle species. Although untested, trichome density or condensed tannin concentration could also contribute to this pattern. Thus, hybrid resistance to different herbivores varies from that of the parental species, such that unique combinations of herbivores are present on hybrids (Hochwender and Fritz 2004), as found in other study systems (e.g., Wimp and Whitham 2001). For some herbivores in this study system, hybrids can support populations of multiple herbivores, while each parental species is highly resistant to certain herbivores.

Because F_1 hybrids have intermediate concentrations of condensed tannins and phenolic glycosides, additive genetic effects are important in explaining the difference in these concentrations between *S. eriocephala* and *S. sericea.* Dominance and epistatic effects could potentially contribute to the differences in secondary chemical concentration between species as well. The genetic architecture of resistance or susceptibility to various insects has been determined through line cross analysis. For example, resistance to *Chaitophorus* aphids is much lower than predicted from a purely additive model of inheritance for some recombinant hybrids (measured as aphid survival) (Fig. 5.1) (Czesak et al. 2004). This could occur through disruption of gene combinations in the *S. eriocephala* genome that confer resistance to aphids. Alternatively, low resistance of hybrids can be explained by the creation of unfavourable gene interactions or underdominance (Schierup and Christiansen 1996). Line cross analysis revealed a complex genetic architecture, such that additive, dominance, and epistatic genetic effects explained the difference between *S. eriocephala* and *S. sericea* in resistance to aphids (Table 5.1).

Complex genetic architecture was found for susceptibility to 14 herbivores, measured as mean density per plant (Table 5.1) (Fritz et al. 2003). Significant epistatic variation was found for susceptibility to each of these herbivores, but the patterns of response to hybrid plant differed markedly among herbivore species. Three prevalent patterns of the expression of plant resistance emerged from the evaluation of genetic architecture. One pattern exhibited by four species was of hybrid breakdown of resistance in F_2 hybrids. A second pattern found for six herbivores was that of greater resistance of F_2 hybrids, which suggested that hybridization led to the breakdown of host-recognition traits. The last pattern was that of dominance of either resistance or susceptibility in recombinant hybrid genotypes. Species showing similar patterns of response to hybridization were often in different families and genera. Even species in the same genus (e.g., *Phyllonorycter*) responded differently to hybrid plants, suggesting that different plant resistance genes affected the abundances of these species on the willow species. The highly variable responses of herbivores in this

TABLE 5.1

Composite Genetic Effects Contributing to Differences Between *Salix eriocephala* and *S. sericea* in Mean Survival of *Chaitophorus* sp. aphids and Mean Densities of 11 Other Insect Herbivores

	Mean Composite Effects[a]					
	Mean	*a*	*d*	*aa*	*ad*	*dd*
Chaitophorus sp.	0.28	−0.24	1.52	—[b]	—	−1.58
Phyllonorycter salicifoliella	89.5	12.1	152.9	102.5	122.1	—
Phyllocnistis salicifolia	0.476	1.061	0.147	0.700	0.409	—
Caloptilia sp.	0.051	0.027	1.186	0.076	−0.375	—
Rabdophaga rigidae	0.722	0.161	−0.055	−0.257	0.628	—
Rabdophaga salicisbrassicoides	1.869	−0.017	−3.916	−1.728	—	2.157
Eupontania s-gracilis	0.241	0.101	—	−0.166	−0.094	−0.108
Phyllocolpa eleanorae	0.882	2.905	5.763	9.035	9.132	—
Phyllocolpa nigrita	0.371	1.024	1.446	2.414	—	—
Phyllocolpa terminalis	−0.483	0.185	2.421	0.886	1.898	—
Phyllocolpa sp.	0.437	0.230	2.11	—	1.924	—
Popillia japonica	1.282	−0.805	−0.934	0.393	−0.919	—

NOTE: Data from Fritz et al. (2003) and Czesak et al. (2004).

[a]Model parameters: *a*, additive; *d*, dominance; *aa*, additive-additive epistasis; *ad*, additive-dominance epistasis; *dd*, dominance-dominance epistasis.

[b]—, a term that is not included in the model or did not contribute significantly to the model.

study highlights not only the breadth of genetic effects on plant resistance, but also that the fitness effects of herbivores on hybrid plants are not easily predictable from one or a few case studies.

Herbivore Community Structure

Using the same data as above, Hochwender and Fritz (2004) investigated how interspecific hybridization affected the community structure of herbivores on plants. Discriminant function analysis grouped herbivore responses to variation in hybridization according to the patterns found above. When these discriminant variables were plotted they revealed that community structure varied as much among the hybrids (e.g., F_1 versus F_2, and backcross to *S. eriocephala* versus backcross to *S. sericea*) as it did between the parental species. This study suggests that genetic variation within plant species plays a substantial role in structuring herbivore communities.

Environmental Effects

Interestingly, the genetic architecture of resistance or susceptibility varies among environments in this study system. Within the field site, there are microhabitats that vary in soil moisture content and nutrient content (Fritz et al. 2006). The genetic architecture of susceptibility to Japanese beetles and five galling insects was determined for six common gardens within the field site in which *S. eriocephala*, *S. sericea*, and their interspecific hybrids were planted. For Japanese beetle density, additive and epistatic genetic effects were detected in all six gardens, but dominance was detected in only one garden (Fig. 5.2). The type of epistatic genetic effect varies among gardens as well. Additive × dominance epistatic effects were detected in all gardens, but additive × additive epistasis was detected in one garden, and dominance × dominance epistasis was detected in five gardens (Fig. 5.2). Thus, environmental differences in the composite effects over a small spatial scale explained the difference between *S. eriocephala* and *S. sericea* in susceptibility to Japanese beetles. Although Japanese beetle densities also can vary over small spatial scale, it was the varying pattern of susceptibility among genetic classes that resulted in the detection of differences in genetic architecture.

Selection by Slugs

In our studies of herbivory on willows, we have found that nonnative slugs, particularly *Arion subfuscus* Draparnaud, is a major source of mortality of willows seedlings (Fritz et al. 2001). Slugs discriminate between *S. eriocephala* and *S. sericea* seedlings, with *S. sericea* being largely resistant due to the presence of phenolic glycosides (Fritz et al. 2001). Because recombination produces substantial variation in phenolic glycoside concentration among F_2 hybrids, we predicted that slug herbivory would result in selection for higher phenolic glycoside concentrations in surviving hybrids. This prediction was partly informed by the presence of phenolic glycosides in backcross hybrids that are not morphologically distinguishable from *S. eriocephala*

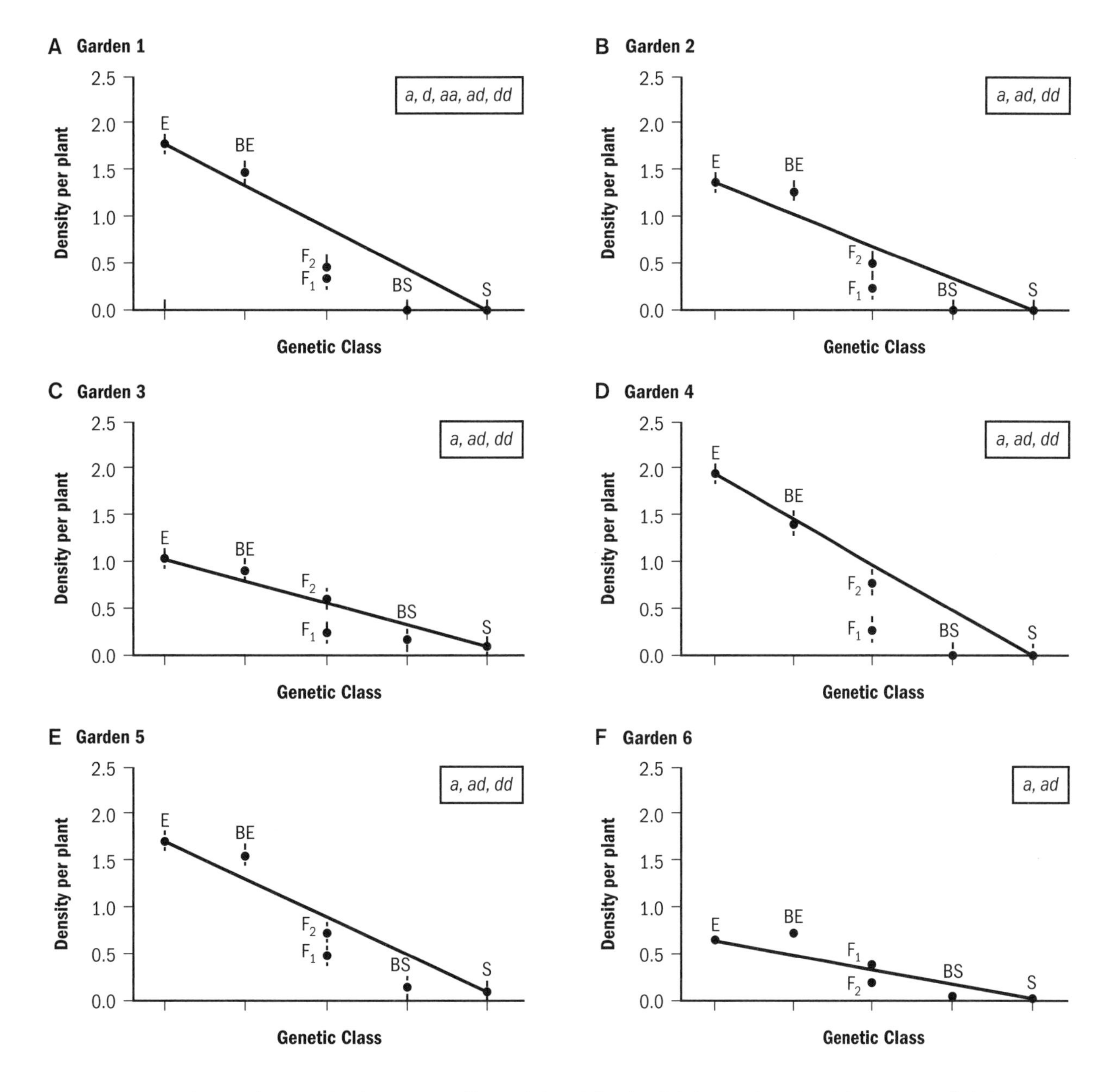

FIGURE 5.2. Mean density of Japanese beetles (± SE) *(Popillia japonica)* on *Salix eriocephala* (E), *S. sericea* (S), and their interspecific hybrids (F_1, F_2, backcross to *S. eriocephala* [BE], and backcross to *S. sericea* [BS]). Letters in the box indicate the composite genetic effects contributing to differences between *S. eriocephala* and *S. sericea* (*a*, additive; *d*, dominance; *aa*, additive-additive epistasis; *ad*, additive-dominance epistasis; *dd*, dominance-dominance epistasis).

and do not express phenolic glycosides in their leaves, suggesting that natural selection had favored introgression of *S. sericea* resistance genes (Hardig et al. 2000). We performed a field selection experiment using more than 5000 F_2 willow seedlings (R. S. Fritz, C. Hochwender, and C. M. Orians, unpublished). Slugs were allowed to consume (and kill) about 85% of the seedlings, while control seedlings were maintained slug free in the field. When we compared characteristics of seedlings exposed to selection by slugs ("slug-selected") versus control seedlings at one and two years of age, we found that slug-selected willows were more resistant to slugs, had higher growth rates, had higher nitrogen and carbon in their leaves, but did not differ in concentrations of phenolic glycosides. Thus, selection on phenolic glycosides had not occurred, but correlational selection on other plant traits was evident. Further, resistance to the native beetle *(Chrysomela knabi)* and the introduced willow leaf beetle *(P. versicolora)* had declined in slug-selected plants, and resistance to the native rust fungus *Melampsora epitea* Thum. had decreased. This study illustrated that selection studies can be useful ways to reveal some of the complicated responses of plants to selection by herbivores and the constraints on that selection caused by linkage or pleiotropy.

Summary

For resistance to evolve, there must be additive genetic variation in resistance traits. Nonadditive genetic effects are important to estimate because they can greatly complicate the evolutionary predictions based on heritability estimates. Several studies have demonstrated that resistance traits are under strong selection by insect herbivores and can evolve in a relatively short period of time. Some of this selection can be indirect, caused by indirect effects of resistance on fitness through correlations with other traits.

Genetic variation in resistance can be maintained within a population in part by allocational costs to resistance, which can generate negative genetic correlations between resistance and fitness-enhancing traits. Disruptive correlational selection may occur between resistance and tolerance, such that a plant population evolves either high resistance *or* high tolerance, although some studies find intermediate levels of both resistance and tolerance. Opposing selection pressure can also maintain genetic variation in traits, as can occur for herbivores and pollinators.

Selection on resistance traits can vary in magnitude and/or direction among different environments. Phenotypic plasticity can evolve for resistance traits, such that there is a constitutive level expressed in the absence of herbivores and an induction response expressed in the presence of herbivores. A genotype × environment interaction indicates additive genetic variation in the plastic resistance trait and the potential for evolution.

Hybridization between populations or species can influence resistance evolution. Novel genotypes that exhibit hybrid superiority or hybrid breakdown in resistance can arise through hybridization, resulting in differential survival between parental populations or species and their hybrids within hybrid zones. Introgression can occur for resistance genes between the parental populations or species. Hybrid populations can persist or become new species. The environment can be an important factor in the maintenance of hybrid zones. Studies that compare hybrid and parental resistance, especially in different environments, can be useful in elucidating the genetic effects that explain the difference in resistance between two plant populations or species.

References Cited

Agrawal, A. A. 2005. Natural selection on common milkweed *(Asclepias syriaca)* by a community of specialized herbivores. Evol. Ecol. Res. 7: 651–667.

Agrawal, A. A., E. D. Brodie III, and L. H. Rieseberg. 2002a. Possible consequences of genes of major effect: transient changes in the G-matrix. Genetica 112–113: 33–43.

Agrawal, A. A., J. K. Conner, M. T. J. Johnson, and R. Wallsgrove. 2002b. Ecological genetics of an induced plant defense against herbivores: additive genetic variance and costs of phenotypic plasticity. Evolution 56: 2206–2213.

Ågren, J., and D. W. Schemske. 1994. Evolution of trichome number in a naturalized population of *Brassica rapa*. Am. Nat. 143: 1–13.

Arnold, M. L. 1997. Natural hybridization and evolution. Oxford University Press, New York. Barton, N. H., and G. M. Hewitt. 1985. Analysis of hybrid zones. Annu. Rev. Ecol. Syst. 16: 113–148.

Barton, N. H., and G. M. Hewitt. 1989. Adaptation, speciation, and hybrid zones. Nature 341: 497–503.

Barton, N. H., and P. D. Keightley. 2002. Understanding quantitative genetic variation. Nat. Rev. Genet. 3: 11–21.

Berenbaum, M. R., A. R. Zangerl, and J. K. Nitao. 1986. Constraints on chemical coevolution: wild parsnips and the parsnip webworm. Evolution 40: 1215–1228.

Bergelson, J. 1994. The effects of genotype and the environment on costs of resistance in lettuce. Am. Nat. 143: 349–359.

Bergelson, J., and C. B. Purrington. 1996. Surveying patterns in the cost of resistance in plants. Am. Nat. 148: 536–558.

Bernays, E. A., and A. C. Lewis. 1986. The effect of wilting on palatability of plants to *Schistocerca gregaria,* the desert locust. Oecologia 70: 132–135.

Brown, P. H., R. D. Graham, and D. J. D. Nicholas. 1984. The effects of manganese and nitrate supply on the levels of phenolics and lignin in young wheat plants. Plant Soil 81: 437–440.

Campbell, D. R. 2004. Natural selection in *Ipomopsis* hybrid zones: implications for ecological speciation. New Phytol. 161: 83–90.

Campbell, D. R., M. Crawford, A. K. Brody, and T. A. Forbis. 2002. Resistance to pre-dispersal seed predators in a natural hybrid zone. Oecologia 131: 436–443.

Carr, D. E., and M. D. Eubanks. 2002. Inbreeding alters resistance to insect herbivory and host plant quality in *Mimulus guttatus* (Scrophulariaceae). Evolution 56: 22–30.

Castells, E., M. A. Berhow, S. F. Vaughn, and M. R. Berenbaum. 2005. Geographic variation in alkaloid production in Conium maculatum populations experiencing differential herbivory by *Agonopterix alstroemeriana*. J. Chem. Ecol. 31: 1693–1709.

Charlesworth, D., and B. Charlesworth. 1987. Inbreeding depression and its evolutionary consequences. Annu. Rev. Ecol. Syst. 18: 237–268.

Christensen, K. M., T. G. Whitham, and P. Keim. 1995. Herbivory and tree mortality across a pinyon pine hybrid zone. Oecologia 101: 29–36.

Collinge, S. K., and S. M. Louda. 1988. Herbivory by leaf miners in response to experimental shading of a native crucifer. Oecologia 75: 559–566.

Czesak, M. E., M. J. Knee, R. G. Gale, S. J. Bodach, and R. S. Fritz. 2004. Genetic architecture of resistance to aphids and mites in a hybrid system of willows. Heredity 96: 619–626.

Daday, H. 1954. Gene frequencies in wild populations of *Trifolium repens*. I. Distribution by latitude. Heredity 28: 61–78.

deJager, C. M., R. P. T. Butot, P. G. L. Klinkhamer, T. J. deJong, K. Wolff, and E. van der Meijden. 1995. Genetic variation in chrysanthemum for resistance to *Frankliniella occidentalis*. Entomol. Exp. Appl. 77: 277–287.

Dungey, H. S., B. M. Potts, T. G. Whitham, and H.-F. Li. 2000. Plant genetics affects arthropod community richness and composition: evidence from a synthetic eucalypt hybrid population. Evolution 54: 1938–1946.

Ehrlich, P. R., and P. H. Raven. 1964. Butterflies and plants: a study in coevolution. Evolution 18: 586–608.

Falconer, D. S., and T. F. C. Mackay. 1996. Introduction to quantitative genetics. Longman Group Ltd., Essex, UK.

Feeny, P. P. 1976. Plant apparency and chemical defense. Recent Adv. Phytochem. 10: 1–40.

Fisher, R. A. 1958. The genetical theory of natural selection, 2nd edition. Dover Publications, New York.

Floate, K. D., and T. G. Whitham. 1993. The "hybrid bridge" hypothesis: host shifting in plant hybrid swarms. Am. Nat. 141: 651–662.

Fornoni, J., J. Núñez-Farfán, P. L. Valverde, and M. D. Rausher. 2004a. Evolution of mixed strategies of plant defense allocation against natural enemies. Evolution 58: 1685–1695.

Fornoni, J., P. L. Valverde, and J. Núñez-Farfán. 2004b. Population variation in the cost and benefit of tolerance and resistance against herbivory in *Datura stramonium*. Evolution 58: 1696–1704.

Fritz, R. S., Hochwender, C. G., Lewkiewicz, D. A., Bothwell, S., and C. M. Orians. 2001. Seedling herbivory by slugs in a willow hybrid system: developmental changes in damage, chemical defense, and plant performance. Oecologia 129: 87–97.

Fritz, R. S., C. G. Hochwender, S. J. Brunsfeld, and B. M. Roche. 2003. Genetic architecture of susceptibility to herbivores in hybrid willow. J. Evol. Biol. 16: 1115–1126.

Fritz, R. S., C. G. Hochwender, B. R. Albrectsen, and M. E. Czesak. 2006. Fitness and genetic architecture of parent and hybrid willows in common gardens. Evolution 60: 1215–1257.

Galloway, L. F., and C. B. Fenster. 2001. Nuclear and cytoplasmic contributions to intraspecific divergence in an annual legume. Evolution 55: 488–497.

Ganders, F. R. 1990. Altitudinal clines for cyanogenesis in introduced populations of white clover near Vancouver, Canada. Heredity 64: 387–390.

Hardig, T. M., S. J. Brunsfeld, R. S. Fritz, M. Morgan, C. M. Orians. 2000. Morphological and molecular evidence for hybridization and introgression in a willow *(Salix)* hybrid zone. Mol. Ecol. 9: 9–24.

Herms, D. A., and W. J. Mattson. 1992. The dilemma of plants: to grow or defend. Q. Rev. Biol. 67: 283–335.

Hirata, K., M. Asada, E. Yatani, K. Miyamoto, and Y. Miura. 1993. Effects of near-ultraviolet light on alkaloid production in Catharanthus roseus plants. Planta Med. 59: 46–50.

Hochwender, C. G., and R. S. Fritz. 2004. Plant genetic differences influence herbivore community structure: evidence from a hybrid willow system. Oecologia 138: 547–557.

Horrill, J. C., and A. J. Richards. 1986. Differential grazing by the mollusk *Arion hortensis* Fér. on cyanogenic and acyanogenic seedlings of the white clover, Trifolium repens L. Heredity 56: 277–281.

Howard, D. J. 1986. A zone of overlap and hybridization between two ground cricket species. Evolution 40: 34–43.

Irwin, R. E., S. Y. Strauss, S. Storz, A. Emerson, and G. Guibert. 2003. The role of herbivores in the maintenance of a flower color polymorphism in wild radish. Ecology 84: 1733–1743.

Lande, R., and S. J. Arnold. 1983. The measurement of selection on correlated characters. Evolution 37: 1210–1226.

Linder, C. R., I. Taha, G. J. Seiler, A. A. Snow, and L. H. Rieseberg. 1998. Long-term introgression of crop genes into wild sunflower populations. Theor. Appl. Genet. 96: 339–347.

Lynch, M., and B. Walsh. 1998. Genetics and analysis of quantitative traits. Sinauer, Sunderland, MA.

Marak, H. B., A. Biere, and J. M. M. Van Damme. 2000. Direct and correlated responses to selection on iridoid glycosides in *Plantago lanceolate* L. J. Evol. Biol. 13: 985–996.

Marak, H. B., A. Biere, and J. M. M. Van Damme. 2003. Fitness costs of chemical defense in *Plantago lanceolate* L.: effects of nutrient and competition stress. Evolution 57: 2519–2530.

Marquis, R. J. 1990. Genotypic variation in leaf damage in *Piper arieianum* (Piperaceae) by a multi-species assemblage of herbivores. Evolution 44: 104–120.

Martinsen, G. D., T. G. Whitham, R. J. Turek, and P. Keim. 2001. Hybrid populations selectively filter gene introgression between species. Evolution 55: 1325–1335.

Mauricio, R. 1998. Cost of resistance to natural enemies in field populations of the annual plant *Arabidopsis thaliana*. Am. Nat. 151: 20–28.

Mauricio, R., and M. D. Rausher. 1997. Experimental manipulation of putative selective agents provides evidence for the role of natural enemies in the evolution of plant defense. Evolution 51: 1435–1444.

Merilä, J., and B. C. Sheldon. 1999. Genetic architecture of fitness and nonfitness traits: empirical patterns and development of ideas. Heredity 83: 103–109.

Moore, W. S. 1977. An evaluation of narrow hybrid zones in vertebrates. Q. Rev. Biol. 52: 263–277.

Orians, C. M., and R. S. Fritz. 1995. Secondary chemistry of hybrid and parental willows: phenolic glycosides and condensed tannins in *Salix sericea, S. eriocephala,* and their hybrids. J. Chem. Ecol. 21: 1245–1253.

Orians, C. M., B. M. Roche, and R. S. Fritz. 1996. The genetic basis for variation in the concentration of phenolic glycosides in *Salix sericea*: an analysis of heritability. Biochem. Syst. Ecol. 24: 719–724.

Pilson, D. 1992. Relative resistance of goldenrod to aphid attack: changes through the growing season. Evolution 46: 1230–1236.

Pilson, D. 1996. Two herbivores and constraints on selection for resistance in *Brassica rapa*. Evolution 50: 1492–1500.

Pilson, D. 1999. Plant hybrid zones and insect host range expansion. Ecology 80: 407–415.

Pilson, D. 2000. Herbivory and natural selection on flowering phenology in wild sunflower, *Helianthus annuus*. Oecologia 122: 72–82.

Rausher, M. D. 1992. The measurement of selection on quantitative traits: biases due to the environmental covariances between traits and fitness. Evolution 46: 616–626.

Rhoades, D. F., and R. G. Cates. 1976. Toward a general theory of plant antiherbivore chemistry. Recent Adv. Phytochem. 10: 168–213.

Rieseberg, L. H, and S. E. Carney. 1998. Plant hybridization. New. Phytol. 140: 599–624.

Scheiner, S. M., R. J. Mitchell, and H. S. Callahan. 2000. Using path analysis to measure natural selection. J. Evol. Biol. 13: 423–433.

Schierup, M. H., and F. B. Christiansen. 1996. Inbreeding depression and outbreeding depression in plants. Heredity 77: 461–468.

Siemens, D. H., H. Lischke, N. Maggiulli, S. Schürch, and B. A. Roy. 2003. Cost of resistance and tolerance under competition: the defense-stress benefit hypothesis. Evol. Ecol. 17: 247–263.

Simms, E. L., and M. D. Rausher. 1989. The evolution of resistance to herbivory in *Ipomoea purpurea*. II. Natural selection by insects and costs of resistance. Evolution 43: 573–585.

Simms, E. L., and J. Triplett. 1994. Costs and benefits of plant responses to disease: resistance and tolerance. Evolution 48: 1973–1985.

Sork, V. L., K. A. Stowe, and C. Hochwender. 1993. Evidence for local adaptation in closely adjacent subpopulations of northern red oak (*Quercus rubra* L.) expressed as resistance to leaf herbivores. Am. Nat. 142: 928–936.

Stinchcombe, J. R. 2002. Fitness consequences of cotyledon and mature-leaf damage in the ivyleaf morning glory. Oecologia 131: 220–226.

Stinchcombe, J. R., M. T. Rutter, D. S. Burdick, P. Tiffin, M. D. Rausher, and R. Mauricio. 2002. Testing for environmentally induced bias in phenotypic estimates of natural selection: theory and practice. Am. Nat. 160: 511–523.

Strauss, S. Y., J. A. Rudgers, J. A. Lau, and R. E. Irwin. 2002. Direct and ecological costs of resistance to herbivory. Trends. Ecol. Evol. 17: 278–285.

Stowe, K. A. 1998. Experimental evolution of resistance in *Brassica rapa*: Correlated response of tolerance in lines selected for glucosinolate content. Evolution 52: 703–712.

Templeton, A. R. 1986. Coadaptation and outbreeding depression, pp. 105–116. In M. E. Soulé (ed.), Conservation biology: the science of scarcity and diversity. Sinauer, Sunderland, MA.

Traw, M. B. 2002. Is induction response negatively correlated with constitutive resistance in black mustard? Evolution 56: 2196–2205.

Valverde, P. L., J. Fornoni, and J. Núñez-Farfán. 2001. Defensive role of leaf trichomes in resistance to herbivorous insects in *Datura stramonium*. J. Evol. Biol.14: 424–432.

Valverde P. L., J. Fornoni, and J. Núñez-Farfán. 2003. Evolutionary ecology of *Datura stramonium*: equivalent plant fitness benefits of growth and resistance to herbivory. J. Evol. Biol. 16: 127–137.

van Doorn, J. E., G. C. van der Kruk, G. J. van Holst, M. Schoofs, J. B. Broer, and J. J. M. deNijs. 1999. Quantitative inheritance of the progoitrin and sinigrin content in Brussels sprouts. Euphytica 108: 41–52.

Waser, N. M. 1993. Population structure, optimal outbreeding, and assortative mating in angiosperms, pp. 173–199. In N. W. Thornhill (ed.), Natural history of inbreeding and outbreeding. University of Chicago Press, Chicago.

Whitham, T. G. 1989. Plant hybrid zones as sinks for pests. Science 244: 1490–1493.

Whitlock, M. C., P. K. Ingvarsson, and T. Hatfield. 2000. Local drift load and the heterosis of interconnected populations. Heredity 84: 452–457.

Wimp, G. M., and T. G. Whitham. 2001. Biodiversity consequences of predation and host plant hybridization on an aphid-ant mutualism. Ecology 82: 440–452.

SIX

Introgression and Parapatric Speciation in a Hybrid Zone

J. MARK SCRIBER, GABE J. ORDING,
AND RODRIGO J. MERCADER

Hybridization has been recognized by some as a potent evolutionary force that rapidly can generate new (novel) gene combinations for adaptive evolution and speciation (Arnold 1997; Burke and Arnold 2001; Schluter 2001; McKinnon et al. 2004). However, others have historically viewed it as a minor evolutionary force (barring allopolyploids in plants) or simply as a local or transient type of evolutionary noise or dead end (Rhymer and Simberloff 1996; Schemske 2000; Barton 2001). While definitive proof is generally lacking, especially for animals, diploid hybrid recombinant speciation may represent a mechanism of evolution of new species (Dowling and Secor 1997), especially if it occurs rapidly (Coyne and Orr 2004; Schwarz et al. 2005). The rarity of animal hybrid speciation may be partly due to the difficulty in detection of hybrids (until the use of recent technological tools).

Hybrid Zones, "Evolutionary Novelties," and Isolation

Populations at the species borders that are under stress often show increased recombination (Hoffmann and Hercus 2000) and may exhibit increased dispersal tendencies or become more polyphagous (Thomas et al. 2001). When the species edge is a hybrid zone, some species-related traits, including X-linked traits, may move into different populations independently of others due to extensive interspecific introgression and recombination (Martinson et al. 2001; Scriber 2002a, 2002b; Scriber and Ording 2005)—all of which can result in genetically novel populations. Unfortunately, for most organisms there remains little information on geographic variation in specific traits known to be associated with range margins (Hoffmann and Blows 1994; Endler 1995).

Homoploid hybrid speciation requires hybrids to be fertile, "fit," and reproductively isolated from the parental species types (Buerkle et al. 2000; Coyne and Orr 2004). Hybrids may be more fit than parental species inside a hybrid zone (bounded superiority model), or less fit (tension zone model), or they may display variable fitness in a "mosaic zone" model (see review by Arnold 1997). Rarely have more than one or two fitness traits been measured in assessing hybrid fitness relative to parental types (Arnold and Hodges 1995; Barton 2001). However, hybrid fitness may in some cases give rise to "evolutionary novelties" that are not explained by the tension zone model or mosaic models (since hybrid genotypes may be very fit). The "novelty" model also differs from the bounded superiority model since hybrids are not necessarily restricted to the ecotone in which they arose and may spread outside the historical hybrid zone (Arnold 1997). In addition, although they may arise infrequently, these hybrid populations with genetic novelties may give rise to new, long-lived evolutionary lineages (Arnold and Emms 1998), adaptive radiations (Seehausen 2004; Bell and Travis 2005), or new species via recombinant hybrid speciation (Rieseberg 2001; Coyne and Orr 2004; Schwarz et al. 2005; Gompert et al. 2006).

Rapid chromosomal repatterning, ecological divergence, and/or spatial separation have been invoked to explain the reproductive isolation between hybrid lineages and parental gene flow (Rieseberg 2001). Necessarily sympatric or parapatric in origin, homoploid hybrid speciation is assumed to require a unique ecological niche for recombinant hybrids to avoid being outcompeted or gene-swamped by introgression from parental types before the establishment of an independent lineage from hybrid origins (Coyne and Orr 2004; Seehausen 2004; Schwarz et al. 2005). New host-plant-associated races could provide hybrids with such a competition-free habitat (Emalianov et al. 2003; Schwarz et al. 2005). However, temporal reproductive isolation can provide a different but equally effective mechanism to avoid introgression or competition from parental types (Scriber and Ording 2005). Changes in voltinism patterns or other phenological constraints in insect populations may affect their temporal isolation from other populations. Climate change is one mechanism that might cause such changes.

Climate Warming, Thermal Constraints, and Voltinism

The impact of recent regional climate warming on geographic distribution limits, abundance, and population dynamics has been significant for a variety of herbivorous insects. While many species may extend or expand their ranges poleward (Hellmann 2002; Parmesan and Yohe 2003), some contractions or extinctions of some local populations might also result from warming (Thomas et al. 2004). Phenological coordination between parasites and herbivores or specialized pollinators and plants (Thomas and Blanford 2003), altered susceptibility of different insect herbivores to insect pathogens (Altizer et al. 2003), sexual synchrony, size variation and mating coordination (e.g., with protandry) (Nylin et al. 1993; Zonneveld 1996; Nylin and Gotthard 1998) may prove to be even more evolutionarily significant than geographic range shifts.

Constraints on the length of the available growing season exist at both ends of the summer for herbivorous insects. A late spring start in adult emergence may be safer than one that risks death from a late spring freeze (Tesar and Scriber 2003). Alternatively, it might be worth risking early fall freezes to squeeze in a second generation if possible. Although safer, stopping at a single generation where two would be possible would presumably lower fitness greatly.

Increased rate of insect growth, longer growing seasons, and milder winters may all be manifestations of warming climates that could release local herbivore populations from their previous phenological constraints and historical voltinism patterns. Species can emerge earlier in the spring season (Roy and Sparks 2000) or stay in place and "adapt" (Coope 1995; Crozier 2003). Examples of such generational telescoping include semivoltine (two-year) to univoltine shifts (e.g., in pine beetles; Logan and Powell 2001), univoltine to bivoltine, or bivoltine to trivoltine (Danks 1994; Nygren 2005). This may largely account for patterns of reduced size with latitude in some insects (i.e., Bergmann's rule) (Blanckenhorn and Demont 2004). A "saw-toothed pattern" in the shape of wing size clines of individuals at latitudes with a "tooth" where voltinism shifts occur is not uncommon and is presumably due to trade-offs between larger size for univoltine and smaller size for bivoltine populations (Mousseau and Roff 1989; Scriber 1994; Nygren 2005).

In areas where thermal constraints exist on completion of development to the winter diapausing stage, selection of highest-quality host plants for fastest larval growth rates may be one way to increase voltinism potential (Nylin 1988; Scriber and Lederhouse 1992; Scriber 1996, 2002b). For example, oviposition preferences for the hosts permitting fastest larval growth rates are narrower where latitudinal (Scriber 2002b) or local (Scriber 1996) seasonal thermal constraints exist, but the range of hosts utilized for oviposition broadens elsewhere (leaving alternating latitudinal bands and "cold pocket" mosaics of host range for two *Papilio* species). The voltinism/suitability hypothesis has had mixed data support (Scriber 2002b; Nygren 2005), as has the prevalence of adult oviposition "preference" and larval survival or growth "performance" correlations (Thompson 1988), perhaps due to phenotypic plasticity (Agrawal 2001; Mercader and Scriber 2005).

Other biochemical, physiological, and behavioral adaptations to thermal constraints exist along latitudinal gradients (e.g., Kukal et al. 1991; Ayres and Scriber 1994). Adaptations to adjust insect life histories and voltinism patterns to local environmental conditions can evolve rapidly (Holzapfel and Bradshaw 2002; Carroll et al. 2005). Diapause regulation may in some cases be obligate and require some sort of "breaking" cues. Alternatively, it may be facultative, depending on environmental conditions (such as host quality and photoperiod) (Tauber et al. 1986). Genetic and biochemical mechanisms involved in environmental cue recognition for diapause largely remain poorly understood (Denlinger 2002).

The Tiger Swallowtail Butterflies

Hybridization is common in Lepidoptera, and occurs in more than 15% of all species (Sperling 1990; Presgraves 2002). The tiger swallowtail butterflies of eastern North America in particular provide an excellent study system to examine the role of hybridization in evolution. *Papilio canadensis* and *P. glaucus* (Lepidoptera: Papilionidae; these *Papilio* = *Pterourus*) have allopatric distributions separated by a narrow but extensive hybrid zone running from Minnesota to southern New England and southward in the Appalachian Mountains possibly to northern Georgia (Fig. 6.1) (Luebke et al. 1988; Hagen et al. 1991; Scriber et al. 2003; Stump et al. 2003). The historical boundaries of this hybrid zone are closely delineated by different mean annual thermal accumulations (degree-days above a base 50°F, or 10°C, noted here as $°D_F$).

We are fortunate to have a 20-year multitrait analysis of the historical hybrid zone between *P. canadensis* and *P. glaucus* (1975–1997) (Scriber et al. 2003), which provides an excellent foundation for analyses of multitrait geographical variation (Arnold and Emms 1998; Endler 1998). In addition, we have studied the multitrait interactions of these taxa during the recent eight years (1998–2005) of climate warming across the hybrid zone (Scriber and Ording 2005).

Our most recent evidence on patterns of hybridization and introgression between *P. glaucus* and *P. canadensis* suggests an even more complex interaction between the two, which has possibly resulted in the formation of a new species, Pavulaan and Wright's (2002) *Pterourus* (= *Papilio*) *appalachiensis*. This recently described hybridlike *Papilio appalachiensis*, with its delayed postdiapause emergence,

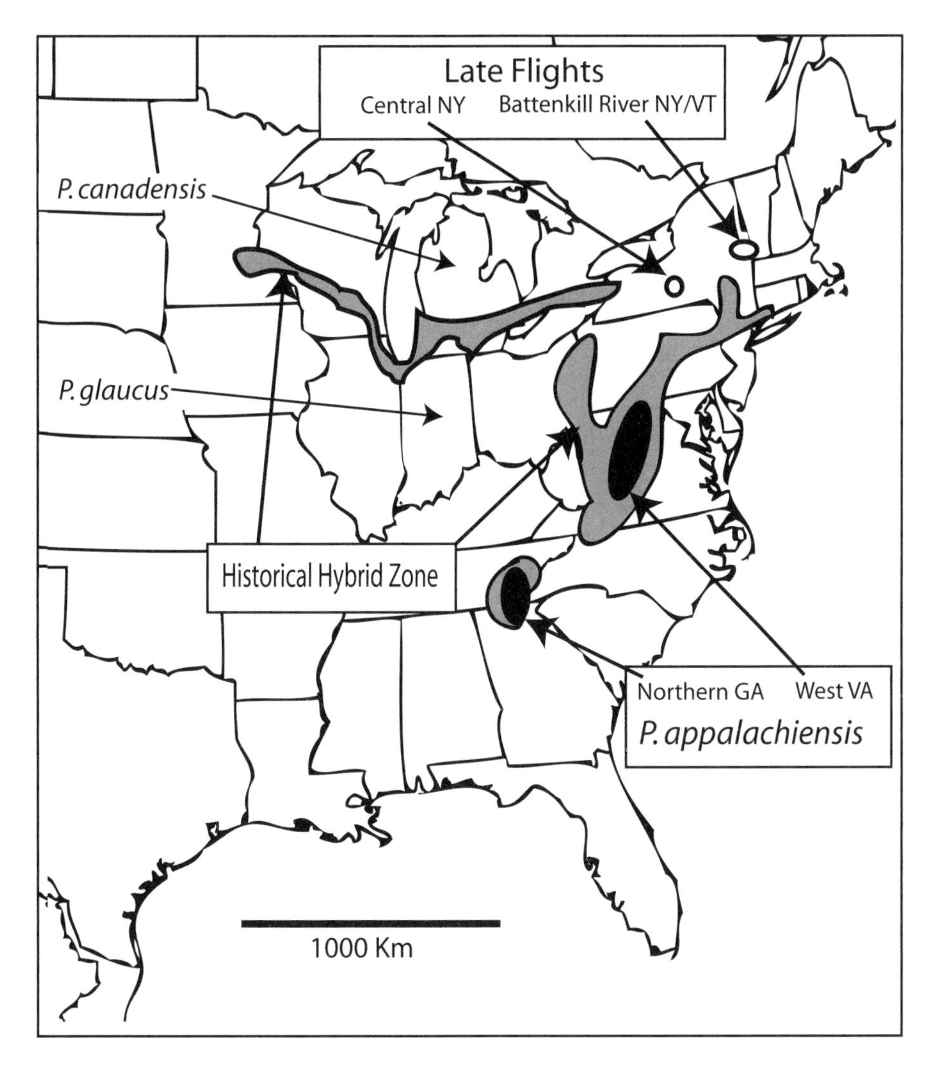

FIGURE 6.1. The geographic distributions of *Papilio glaucus, P. canadensis, P. appalachiensis*, and the late flight hybrid swarm populations of central New York and at the Battenkill River Valley border of New York and Vermont. The historical (pre-1998) hybrid zone (shaded bands) had concordant, coincidental, and rather sharply defined limits of morphological, biochemical, behavioral, and physiological species-diagnostic traits (Table 6.1) (Scriber et al. 2003). However, this has changed since 1998 with regional climate warming, and species-diagnostic traits have shown extensive interspecific introgression, recombination, and independent geographic movement northward and upward into the mountains (Scriber 2002a; Scriber and Ording 2005).

distinctive X-linked allozymes, and unique thermal zone niche (2600–2850°D_F), appears to be ecologically differentiated from both parapatric parental species and apparently stable in the hybrid zone between them. In addition, we have recently found another recombinant, temporally segregated "hybrid swarm" population on the cooler side of the hybrid zone, which emerges later (late flight [LF] population) than the parental *P. canadensis* population found in the same location (early flight [EF] population).

In the following sections, we present evidence for a model of *Papilio* divergence in the hybrid zone between *P. canadensis* and *P. glaucus*, characterized by fertile hybrid recombinant genotypes and rapid temporal isolation from both parental species (via shifts in postdiapause flight times). Divergence in this system would seem to fit the criteria for homoploid hybrid speciation (Coyne and Orr 2004; Schwarz et al. 2005). We present evidence for the role of thermal constraints in shaping geographical patterns of voltinism and reproductive isolation in eastern North America and evaluate the likely mechanisms leading to the postdiapause delays in the LF hybrid swarms and *P. appalachiensis*. Unlike the case with many other herbivorous insects, this *Papilio* divergence appears to involve neither sex pheromones nor host-plant races.

Note: Although *P. glaucus, P. canadensis*, and the recently described *P. appalachiensis* form distinct genotypes, their status as species is controversial and, depending on the species definition selected, could be considered subspecies of each other. In this chapter we are primarily concerned with the mechanisms separating and maintaining these distinct genotypes and use their current status as separate species for simplicity.

Species Distribution and Traits

The tree-feeding, polyphagous *P. glaucus* and *P. canadensis* are allopatrically distributed in eastern North America, with *P. canadensis* occurring to the north, and *P. glaucus* to the south (below 42°N latitude), with a long, narrow hybrid border zone between them (at 41.5–43°N latitude) (Fig. 6.1). This *Papilio* hybrid zone has historically, and rather precisely, been delineated by summer growing thermal accumulations of 2500–2700°D_F (see Scriber et al. 2003), corresponding closely with the boreal forest/temperate deciduous forest ecotone. The average thermal landscape has retained largely the same pattern from 1960 to 1997. However, considerable regional climate warming has subsequently been documented in the eastern United States from 1998 to 2005 (Scriber and Ording 2005). This warming trend has increased the total seasonal thermal unit accumulations of the areas surrounding the historical hybrid zone (Scriber and Ording 2005). The genetic implications of this recent warming for *Papilio* populations are discussed below.

Since the early 1980s, we have annually monitored various species-diagnostic morphological, physiological, biochemical, behavioral, and genetic adaptations of these *Papilio* butterfly species (e.g., Scriber 1982, 1994; Scriber et al. 2003; Sperling 2003) and geographically delineated them across the parapatric distribution. The species-specific diagnostic traits of *P. glaucus* and *P. canadensis*, as well as *P. appalachiensis* and an experimental lab population (discussed below), are outlined in Table 6.1. We have historically found little parental overlap between *P. glaucus and P. canadensis*, and little meaningful introgression had occurred in either direction across the hybrid zone during the 1978–1997 period (Luebke et al. 1988; Hagen et al. 1991; Scriber et al. 2003). Introgression patterns have changed, however, during the post-1998 warming period (see "Genetic Introgression of Traits into the Hybrid Zone," below).

That X-linked allozymes diverge to fixation inside the hybrid zone suggests the possibility of inversion loops that prevent recombination within parts of the *Papilio* genome, but we do not really have much evidence for this yet (Hoffmann et al. 2004; Jiggins and Bridle 2004; Mallet 2005; Schwarz et al. 2005). Such loops might also explain the close linkage of the *Ldh*-100 and the nondiapause (od^-) traits. As with the *Lonicera* fly (Schwarz et al. 2005), we have been virtually unable to locate any primary hybrids between *P. glaucus* and *P. canadensis* in the hybrid zone populations (but see Donovan and Scriber 2003).

Voltinism and Diapause

THERMAL CONSTRAINTS ON VOLTINISM

Papilio canadensis, the northern species, is univoltine, while the southern *P. glaucus* is bivoltine (or even trivoltine in the far south). The northern limits of direct bivoltine development for *P. glaucus* populations end roughly at the hybrid zone near *P. canadensis*. We have shown that thermal constraints on larval growth that prevent successful completion of the second larval generation correspond to a seasonal thermal accumulation of 2800 to 2900°D_F (Scriber and Lederhouse 1992; Scriber 1996). Below this thermal

TABLE 6.1
Papilio Species Diagnostic Traits

Diagnostic Trait	*P. canadensis*	*P. glaucus*	Lab hybrids[a]	*P. appalachiensis*	Late Flight
Thermal zone (°D_F)	<2300	>2900	N/A	2600–2850	2250–2550
Adult flight	May/June	June and August	N/A	Mid-July	Mid-July
Voltinism	Univoltine	Bivoltine	(X-linked)	Univoltine	Univoltine
Diapause	od^+	od–	(X-linked)	od^+?	od^+
Hindwing bands	55–90%	10–40%	40–55%	40–55%	25–65%
Forewing length (mm)	35–44	55–75	40–60	52–62	45–55
Tails	Short, narrow, none	Long, wide, bulb	Intermediate	Long, intermediate	Long, intermediate
Tulip tree survival	0–2%	70–100%	80–100%	>90%	>85%
Quaking aspen survival	60–100%	0–5%	70–100%	>80%	>80%
Ldh allozymes	−80 or −40	−100	X-linked (heterozygous)[b]	−80 or −40	−80 or −40
Pgd allozymes	−125/−80/−150	−100 or −50	X-linked (heterozygous)[b]	-100 or -50	all five
Dark color suppressor gene	s^+	s^-	X-linked	s^+ (some s^-)	s^+ and s^-
Dark female color gene	b^- 0%	b^+ 30–98%	Y-linked 0, 50, or 100%	b^+ 10–25%	b^- 0%

NOTE: Obligate (b^+) and facultative (b^-) diapause; relative (%) width of hindwing black bands of anal cell; tail shapes may be bulbed, wide, long, or otherwise; X-linked melanic black female morph color suppressor (s^+) or enabler (s^-); Y-linked dark color (b^+) or yellow (b^-) trait. See text for additional discussion of sex-linked traits.

[a] In lab hybrids, the direction of the cross determines various X-linked traits.

[b] Hemizygous in females.

accumulation, second-generation larvae fail to reach the overwintering pupal stage before the decline in quality and the abscission of fall leaves (their food) and winter freezing. These constraints are observed even for the larvae on the best-quality hosts in the area. This "voltinism/ suitability" phenomenon (preferences for "fast host" facilitating bivoltinism) also extends to univoltine populations at high latitudes in Alaska, where, despite many life-history adaptations for faster completion of development (e.g., large eggs, faster molts at cool temperatures, pupation at smaller sizes) (Ayres and Scriber 1994), females have evolved to still select the most nutritious host-plant species (*Populus tremuloides, P. balsamifera, Alnus tenuifolia,* etc.) (Scriber and Lederhouse 1992; Scriber 2002b). In the cold pockets of northern Michigan and Wisconsin, similar local preferences for the best-quality host for fast growth (e.g., *Fraxinus* spp. [ash]) were observed in the thermally constrained univoltine *P. canadensis* during the pre-1998 period (Scriber 1996). During the period from 1951 to 1980 the average seasonal thermal unit accumulation in the cold pocket center (Vanderbilt) was 1709°D_F. The probability of exceeding 2000°D_F was only 11% (Michigan State University Agweather). Smaller females and selection of the very best hosts for rapid larval growth were observed (Scriber 1994, 1996).

DIAPAUSE INDUCTION

Photoperiod plays the major role in diapause induction in *P. glaucus* and related *Papilio* sister species (e.g., *P. troilus*) (Valella and Scriber 2005). If photoperiod cues to induce diapause are lacking or go unrecognized, "suicidal" generations of the bivoltine *P. glaucus* may occur in areas where seasonal degree-days constrain successful completion of a second generation. This would be most likely to occur near the cool side of the hybrid zone where summer daylengths are greater than 16 hours.

Papilio glaucus and *P. canadensis* exhibit both facultative (od^-) and obligate (od^+) diapause regulation, respectively (Table 6.1), which is controlled by factors on the X chromosome (Rockey et al. 1987a, 1987b). *Papilio canadensis* (the northern species) has only univoltine populations with obligate diapause requiring 8 to 12 weeks of chilling to terminate (J.M.S. and P. Valella unpublished data). On the other hand, the southern *P. glaucus* appears to have the potential for bivoltinism everywhere but is influenced by other factors such as photoperiod and host quality. The *P. glaucus* diapause is induced at shorter daylengths, and the local photoperiodic threshold inducing 50% diapause varies latitudinally (15.5 hours in southern Michigan at 41.5°N latitude; 14.8 hours in southern Ohio at 38.5°N; and 12.5 hours in southern Florida at 27°N latitude) (P. Valella and J.M.S., unpublished data). Similar thresholds are involved in the sister group *P. troilus* over the same geographic expanse (Valella and Scriber 2005).

Based on this information, most *P. glaucus* individuals are predicted to develop directly at photoperiods greater than 16:8 hours light:dark with nondiapause eclosion of adults from pupae (Scriber 1982; Hagen and Lederhouse 1985; Rockey et al. 1987a, 1987b). Emergence data from many additional populations from Florida to Canada suggest that the northern photoperiod limits to bivoltine potential in tiger swallowtails would be 42 to 43°N latitude (where the longest days are roughly 15.8 to 16 hours). Direct development frequencies (0% direct development = 100% diapause) are presented for more than 80 populations of *P. glaucus* and *P. canadensis* across eastern North America (16 states and 3 Canadian provinces) (Fig. 6.2). While photoperiod is directly correlated with latitude, the local frequencies of direct development (and bivoltine potential) show considerable variation. Several undetermined factors may be involved, including the altitude (and other local history of thermal constraints during the growing season), host-plant quality, thermoperiods, moisture, and crowding (Tauber et al. 1986; Denlinger 2002).

Experimental hybridizations between *P. canadensis* and *P. glaucus* have shown that under long-day conditions the obligate/facultative diapause trait is regulated by X-chromosome genes (Table 6.2) (Rockey et al. 1987a, 1987b) (see "Genetics of Diapause Regulation in *Papilio*," below). Females are the heterogametic sex in Lepidoptera. Primary hybrids with a *P. glaucus* father (58 families) produce offspring that all develop directly (nondiapause, as with *P. glaucus*). Primary hybrids with a *P. canadensis* father (166 families) produce mostly nondiapause sons that are heterozygous for the obligate diapause trait (od^+) and diapausing daughters that are hemizygous for the od^+ trait. Backcrosses ($N = 43$) involving hybrid fathers and/or hybrid mothers further support the X-linkage of diapause trait inheritance (Table 6.2).

POSTDIAPAUSE DEVELOPMENT

Postdiapause development and adult emergence from the pupae appear to be controlled primarily by temperature. This thermal dependence of postdiapause development has been observed for *P. glaucus, P. canadensis,* and the sister group *P. troilus* (Deering et al. 2005). Earlier adult postdiapause emergences might be expected to evolve unless spring freezes, mating asynchrony, unopened leaf buds, or temperatures too cool for flight select otherwise (Ayres and Scriber 1994; Lederhouse et al. 1995). Based on many years of field observations and lab studies we have been able to delineate the typical adult flight periods, the number of generations possible, and the number of generations found under different temperature regimes for these *Papilio* spp. A visual representation of this information for *P. glaucus, P. canadensis,* and the allochronically separated hybrid genotypes "late flight" and *P. appalachiensis* is shown as a function of calendar date with different thermal accumulations (Fig. 6.3). In the mountains (1500–4500 feet above sea level) the specific flights of *P. glaucus* and *P. appalachiensis* may be slightly different than at lower elevations.

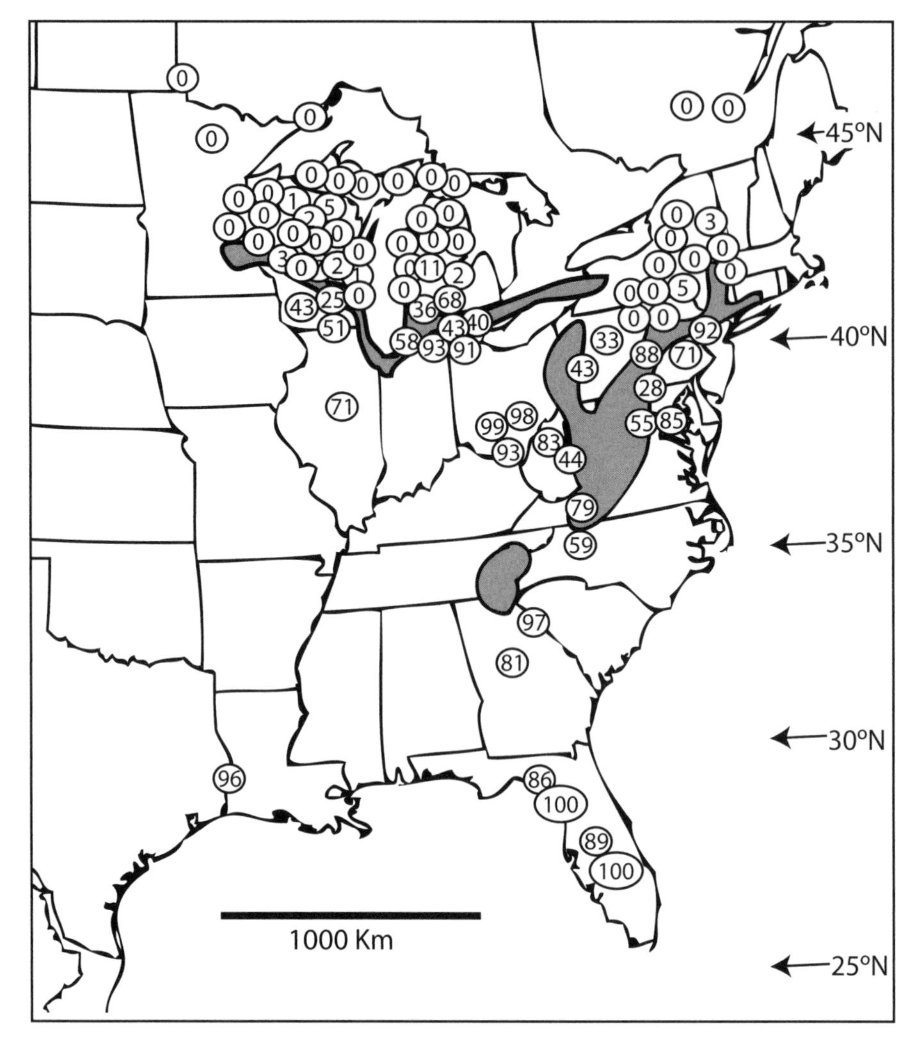

FIGURE 6.2. Direct development without pupal diapause (percent of lab-reared pupae) under long days. Families of wild females were reared in the lab under long-day conditions (some at 16:8 hours light:dark, most at 18:6 hours). The shaded areas correspond to the historical hybrid zone (cf. fig. 6.1) and the southern limits of *Papilio canadensis* distribution (ca. 2500–2700°D_F). It also represents the northern limits for successful bivoltinism and for the dark-morph genotype of *P. glaucus*. The critical photophase for diapause induction is 15.5 to 16 hours in Michigan (42°N), 14.5 to 15 hours in Ohio (39°N) , and 12 to 13 hours in Florida (27.5°N). Sample sizes varied: northern Wisconsin and Michigan, $N > 2300$; southern Wisconsin, $N = 185$; southern Michigan, $N > 280$; Illinois, $N = 31$; Ohio, $N > 120$; Virginia, $N = 33$, Pennsylvania, $N = 160$; Kentucky, $N = 95$; Georgia, $N > 220$; Louisiana/Texas, $N > 480$; Florida, $N = 339$; and Vermont, $N = 198$. In addition, data from Hagen and Lederhouse (1985) are also included here in eastern Canada and northeastern Unites States.

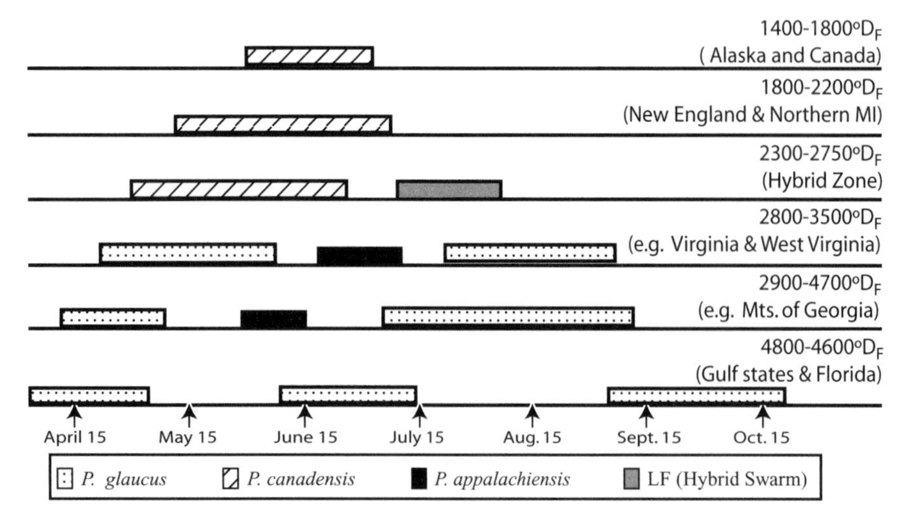

FIGURE 6.3. Flight times of *Papilio* species and populations as a function of thermal landscape. Spotted bars are *P. glaucus* (trivoltine in Florida and the Gulf States, bivoltine elsewhere). Striped bars are the univoltine *P. canadensis* (with May to June flight times from the hybrid zone to Alaska and the mountains of West Virginia). The gray bar indicates delayed (mid-July) late flight recombinant hybrid swarms in the hybrid zone (2300–2750°D_F,). Black bars are *P. appalachiensis* at the warmer side of the hybrid zone (2800–3300°D_F) in the mountains of West Virginia, Maryland, Pennsylvania, and south to Tennessee, North Carolina, South Carolina, and northern Georgia. See text for further details.

TABLE 6.2
Adult Emergence Years for *Papilio* Pupae under Long Day Summers (6 Months) and Dark Winter Periods (3–4°C for 6 months)

Genotype[a]	N families (N pupae)	% pupal mortality[b]	Sons			Daughters		
			Year 1	Year 2	Year 3	Year 1	Year 2	Year 3
Pg	132 (2570)	16	828	265	0	764	279	15
Pc	190 (2031)	10	15	871	1	7	941	1
Pc × *Pg*	58 (1227)	25	396	27	0	466	22	0
Pg × *Pc*	166 (4180)	40	1246	530	5	48	529	291
PgPc × *Pg*	10 (287)	21	59	28	0	106	12	0
PgPc × *Pc*	16 (668)	15	1	282	0	0	285	0
Pg × *PgPc*	11 (295)	32	60	61	0	45	35	2
PgPc × *PgPc*	8 (189)	30	84	19	4	63	14	4

NOTE: These assays were from 1982 to 1986 and 1996 to 2002. The successful adult emergers are shown.
[a] *Pg, P. glaucus; Pc, P. canadensis*. The pairing mother is listed first in primary, backcross, and F_2 hybrids.
[b] Pupae that failed to produce an emerged adult.

Hybrid *Papilio* Populations and Species

We are fortunate to have a three-decade (1975–present) multitrait dataset for *P. canadensis*, *P. glaucus*, and the historical hybrid zone between them (Scriber et al. 2003). This historical perspective has allowed us to detect genetic trends that might otherwise go unnoticed in a shorter frame of reference. Our more recent data (1998–present) from a period of regional warming points to more extensive hybridization and introgression of traits than in the previous period (1975–1997) (Scriber 2002a, 2002b; Scriber and Ording 2005). These data suggest a more complex interaction between the two species, which has led to an allochronically separated hybrid LF population distinct from the parental species (located at the Battenkill River Valley at the New York–Vermont border), and even possibly the formation of a hybrid species, *Pterourus* (= *Papilio*) *appalachiensis*. In the following sections, we discuss these populations and present evidence for both ecological and genetic barriers that may have limited gene flow and promoted divergence between these recombinant hybrids and the parental *Papilio canadensis* and *P. glaucus*. Such divergence may reflect a process of parapatric hybrid speciation.

Introgression in the Hybrid Zone

Recently, extensive northward interspecific introgression of *P. glaucus*–diagnostic traits (Table 6.1) has been observed at many locations near and north of the cooler (northern) side of the historical hybrid zone (such as sites in central Wisconsin, central Michigan, and southwestern Vermont) (see Fig. 6.1). These *P. glaucus*–like traits found at greater frequency include wing-band characteristics (narrow hindwing band width) and other color patterns, size (greater average forewing length), the ability to feed on tulip tree leaves, and Hk-100 allozymes—all autosomally encoded (Scriber and Ording 2005).

However, there has been little northward introgression of certain other *P. glaucus* traits, such as the X-linked Ldh-100 allozyme (which has close linkage to the facultative diapause

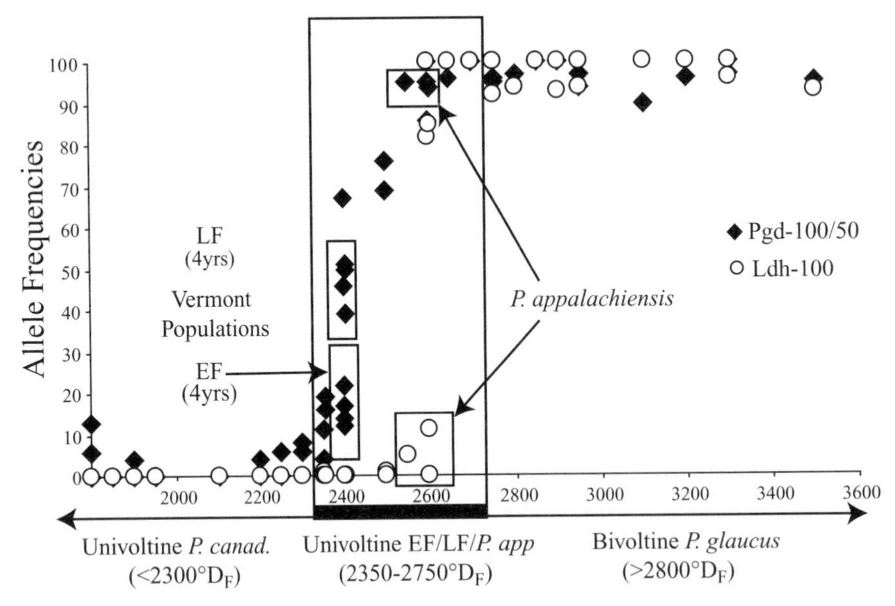

FIGURE 6.4. X-linked allele frequencies (*Pgd*-100 and *Ldh*-100 allozymes) along the thermal landscape (degree-days above a base of 50°F [°D_F]) across the eastern United States. Note the sharp decline in *Ldh*-100 at 2700°D_F or less, and the nonconcordant but steep increase in *Pgd*-100 frequencies between 2300 and 2800°D_F. These 70 different populations (each represented by 10 to 125 males) are from Vermont and northern Michigan to southern Ohio and North Carolina (data from Scriber and Ording 2005; and JMS unpublished data). The bivoltine *P. glaucus* are at the right (>2800°D_F), while everything else (*P. canadensis*, EF [early flight], LF [late flight], and *P. appalachiensis* [= *Pa*]) is univoltine. Hybrid zone increases of *P. glaucus*–type *Pgd* allozyme alleles may be due to intrinsic genetic selection against female hybrid recombinants with one or both *P. canadensis* allozyme profiles. Reduced survival of female pupae relative to male pupae has been reported in seven backcross families (Hagen and Scriber 1989, 1995).

trait), and the Y-linked melanic mimicry gene in females (b^+). In the past there has also been little indication of northern introgression of the X-linked facultative diapause (od^-) and bivoltinism genes. However, in 2005 we recovered 16 out of 300 individuals (and 8 of 500 individuals in 2006) from a population (Battenkill River) at the New York–Vermont border that demonstrated nondiapausal direct development from pupae to adult. All individuals previously recorded in 2003 and 2004 (N = 1000) were obligate diapausers (od^+) (J.M.S. unpublished data). This suggests some recent introgression of the X-linked od^- gene.

Interspecific recombination of the X chromosome has evidently occurred, as shown by discordant patterns of X-linked allozyme markers (Scriber and Ording 2005). The X-linked *P. glaucus Pgd*-100 and *Pgd*-50 alleles have introgressed 200 to 400 km north of the historical hybrid zone, yet the *P. glaucus Ldh*-100 allele has not. The allele frequency shift for both genes is more closely related to thermal landscape (i.e., mean seasonal total degree-days above a base threshold of 50°F) than to latitude (Fig. 6.4).

ROLE OF CLIMATE WARMING IN TRAIT MOVEMENT

Recent warming trends have increased the total seasonal thermal unit accumulations of the areas surrounding the historical hybrid zone (Scriber and Ording 2005). For example, the Battenkill population (at the New York–Vermont border) averaged only 1934°D_F from 1963 to 1997 but has averaged 2375°D_F during 1998–2003. This warming across the historical hybrid zone has led to extensive (but independent) trait cline movements north of the historical hybrid zone (discussed above). We believe this increase in trait movement and local increases in thermal units may have catalyzed the formation of an allochronically separated hybrid LF population. For example, when degree-days are mapped on a thermal landscape they prove to be excellent predictors of locations where the LF population as well as *P. appalachiensis* occur (Scriber and Ording 2005).

Papilio appalachiensis Characteristics

The newly described Mountain Swallowtail, *P. appalachiensis*, is possibly a mountain version of a LF population and appears to be a recombinant hybrid species (Scriber and Ording 2005). This southern species has been described from type specimens, all thus far collected from within the hybrid zone (Fig. 6.1). This species exhibits a unique combination of hybridlike morphological and behavioral traits (Table 6.1) (Scriber and Ording 2005). It has obligate diapause and is able to use host plants from both the Salicaceae (as *P. canadensis*) and the Magnoliaceae (as *P. glaucus*). It has a unique combination of the *Ldh*-80 or *Ldh*-40 alleles (as for *P. canadensis*) and the *Pgd*-100 or *Pgd*-50 alleles (as for *P. glaucus*). *Papilio appalachiensis* is univoltine and has a delayed reproductive flight period relative to the earlier univoltine *P.*

canadensis (which is an EF species). Steep altitudinal gradients put *P. appalachiensis* in geographic parapatry (i.e., spatial cruising distance) with both parental species. However, its postdiapause emergence is allochronic to both *P. canadensis* and *P. glaucus*. This, combined with its particular thermal zone affiliation, appear to facilitate reproductive isolation from both parental species—the univoltine *P. canadensis* at the cooler side of the hybrid zone (2300–2600°D_F), and the bivoltine *P. glaucus* at the warmer side of the hybrid zone (2600–2900°D_F).

The LF Population

On the cooler side of the hybrid zone at 2250–2600°D_F, a naturally occurring, univoltine LF population emerges from diapause in mid-July at the Battenkill River Valley at the New York–Vermont border (Fig. 6.1). This emergence is delayed, but bisexually synchronized (Scriber and Ording 2005). This LF appears reproductively isolated (via nonoverlapping emergences [Scriber and Ording 2005]) from the EF univoltine *P. canadensis* that emerges from mid-May to June at the same location. Near the warmer side of the hybrid zone (at 2600–2850°D_F; e.g., Appalachian Mountains of West Virginia and the Poconos of Pennsylvania) (Fig. 6.1) the delayed flight of hybrid types appears temporally isolated from both flights of the bivoltine *P. glaucus* (Fig. 6.3). In addition to the naturally occurring LF populations, we have also generated similar segregation in postdiapause adult emergences (EF versus LF) in hand-paired backcrosses in our lab using segregating morphospecies in offspring of the same mother (e.g., *Pc* × *PgPc*; Scriber 1990).

Population Phenologies

The phenological flight times (which relate to mating time) for the various *Papilio* species and populations are summarized in Fig. 6.3. *Papilio canadensis* (univoltine, with obligate diapause at all photoperiods) has flight times between May and June from the hybrid zone to Alaska (hatched lines). It is the only genotype to occur at fewer than 2250°D_F (from the Great Lakes to Alaska).

The hybrid zone has historically been defined by 2300 to 2800°D_F and can support the mid-July delayed LF as well as the *P. canadensis* EF in May and June. The gray bar indicates flight times of the LF recombinant hybrid swarms on the cooler side of the hybrid zone (2300–2750°D_F). Black bars indicate flights of the *P. appalachiensis* at the warmer side of the hybrid zone (2800–3300°D_F) in the mountains of West Virginia, Maryland, Pennsylvania, and south to Georgia, South Carolina, North Carolina, and Tennessee (Fig. 6.1). *Papilio appalachiensis* is univoltine and occurs 2 to 4 weeks after the first flight of *P. glaucus*, and 3 to 4 weeks before the second true generation of *P. glaucus*. Some *P. appalachiensis* types (Pavulaan and Wright 2002) occur in the mountains of northern Georgia (and North Carolina, South Carolina, and Tennessee) where we have shown that the thermal landscapes have also historically totaled 2700 to 3300°D_F (Scriber and Ording 2005).

The developmental degree-days required for *P. canadensis* postdiapause adult emergences is only 320 to 430°D_F compared to *P. glaucus* with 430 to 520°D_F required on average. The LF individuals do not emerge until 680 to 780°D_F have accumulated (thus producing the "delay" seen in natural populations and some lab backcrosses).

Rapid Isolation and Divergence in the LF

The physiological mechanisms involved in postdiapause delays may involve the gene for obligate diapause (od^+) (Hagen and Scriber 1995) and other (possible mtDNA) interactions. It seems feasible that this physiological/developmental postdiapause delay phenomenon could provide virtually immediate temporal isolation of certain male and female backcross genotypes from parental gene flow. With immediate delays in postdiapause emergences of some recombinant backcross genotypes (LF), but not others (EF), rapid allochronic reproductive isolation could occur (i.e., an allochronic, but sympatric mechanism of reproductive isolation).

We hypothesize that strong selection against the X-linked facultative diapause (od^-) trait (and the closely linked *Ldh*-100 allele) in regions with fewer than 2700°D_F, combined with divergent selection against *Pgd*-125 and *Pgd*-80 (or a closely linked trait) where 2250 to 2650°D_F occur, may have resulted in the observed genetic divergence of the reproductively isolated LF populations. High recombination rates in these interspecific hybrids, combined with divergent selection on X chromosomes of the EF and LF genotypes, could lead to parapatric hybrid speciation.

There is no evidence at present that host-plant shifts or changes in sex pheromones have driven this process, in contrast to many other speciation events in the Lepidoptera (as in corn borers [Roelofs et al. 2002; Thomas et al. 2003] and larch budmoths [Emalianov et al. 2004]). While various local host preferences and host-specific larval adaptations in *P. glaucus* and *P. canadensis* have been described (Scriber et al. 1991), there is no evidence of host-race formation in these very polyphagous tiger swallowtail butterflies. The temporal mating isolation in *Papilio* species here is not related to host-plant races (with different phenology or nutritional quality) as has been seen in treehoppers (Wood and Keese 1990; Wood et al. 1999). In addition, there are almost certainly no host-affiliated host mating behaviors (Lederhouse 1995; Deering and Scriber 2002; Stump and Scriber 2006), as may be involved in other potential sympatric speciation examples (Wood and Keese 1990; Bush 1994; Carroll et al. 1997; Craig et al. 1997; Caillaud and Via 2000; Filchak et al. 2000; Via 2001; Schwarz et al. 2005). We emphasize that temporal reproductive isolation in our delayed postdiapause LF emergences of males and females in the Battenkill populations and *P. appalachiensis* populations provides a different but equally effective mechanism to avoid introgression or competition from parental types (Scriber and Ording 2005).

The Nuts and Bolts

In the remainder of this chapter, we present detailed evidence from laboratory and field studies that relates to specific aspects of our model of parapatric hybrid divergence in *Papilio*. We conclude with a discussion of current molecular work in progress and future directions.

Genetics of Diapause Regulation

We conducted reciprocal interspecific hybridization and controlled environment rearing with parental species and some backcrosses in the lab. We observed that obligate diapause (od^+) and facultative diapause (od^-) are both controlled largely by loci on the X chromosome (Table 6.2) (Rockey et al. 1987a,b). Under long-day conditions in the lab, most *P. glaucus (Pg)* males and females developed directly (i.e., nondiapausing), and most *P. canadensis (Pc)* diapaused. The male *P. canadensis* (od^+, od^+) and male *P. glaucus* (od^-, od^-) transfer their X chromosomes to their daughters (females are the heterogametic sex). A *Pc* (mother) × *Pg* (father) cross produced nondiapause hybrid daughters (od^-, Y), while basically all daughters of the reciprocal hybrid cross (*Pg* mother × *Pc* father) appear to have obligate pupal diapause (od^+, Y). Sex-linked diapause inheritance is also evident for various hybrids and backcrosses (Table 6.2).

These results might also reflect interspecific genetic introgression near the historical hybrid zone limits. For example, of the 15 nondiapause *P. canadensis* males, 13 are from the hybrid zone (Marquette, Green Lake, Juneau, and Jackson counties, Wisconsin; and Roscommon County, Michigan). Of the seven nondiapause *P. canadensis* females, six were from the hybrid zone (three in Wisconsin and three in Isabella County, Michigan). Similarly, most of the *P. glaucus* diapausing males (N = 265) and females (N = 279) (Table 6.2) are from the northern edge of the distribution near the hybrid zone in Wisconsin, Illinois, Indiana, and the mountains of New York, Pennsylvania, West Virginia, North Carolina, and South Carolina (Fig. 6.1).

Nonoverlapping Emergences of EF and LF Populations

In 2002 and 2003, we conducted field-rearing at ambient environmental conditions for large numbers of larvae from females of the May to June EF and also the delayed mid-July LF populations. Insects were enclosed using large sleeve cages over wild black cherry *(Prunus serotina)* tree limbs, and pupae were collected near the end of September. Initial numbers of mothers for EF and LF ranged from 18 to 120 over the two years. In October we moved pupae to winter simulation controlled conditions (at 3–4°C total darkness) and brought them out in April to May and randomly assigned them to postdiapause chambers at different temperatures (14, 18, 22, 26°C, etc.).

The results show a clear pattern of delayed emergence for both LF males and females at all temperatures, compared to emergence times of *Papilio canadensis* EF and *P. glaucus* (Table 6.3). For example, at 26°C, the mean emergence times for LF males were 25.4 and 23.8 days (2003 and 2004, respectively), and for LF females 29.0 and 27.9 days. In contrast, male *P. canadensis* EF only took 13.1 and 11.1 days on average (females at 13.9 and 12.3 days), and *P. glaucus* males took 15.6 and 13.2 days (females at 14.7 and 12.2 days). The delay at 22°C was an additional few days greater for LF males and females compared to EF and *P. glaucus* (Table 6.3). Data from 18°C and 14°C showed LF delays that were larger still (3–7 weeks; Ording and Scriber in preparation). In only one or two cases among about 1200 pupae did the latest *P. canadensis* or *P. glaucus* males emerge after the earliest-appearing LF males and females. Thus, even though it is possible that long-lived and late EF individuals might encounter the earliest-flying LF individuals, there was basically no overlap temporally in emergence dates between EF and LF flights from these field-reared pupae (G.J.O. and J.M.S., unpublished).

Adult sex ratios from field-reared pupae were not significantly different than unity for either of the Battenkill River populations (EF or LF), indicting no Haldane selection against females (the heterogametic sex in Lepidoptera). This is in contrast to the results of some interspecific backcrosses (Hagen and Scriber 1995). As with wild-collected individuals, adult offspring of these LF were *P. appalachiensis*–like with a significantly higher *Pgd*-100 frequency than the EF, but with absolutely no evidence of successful introgression of the *Ldh*-100 allele (in several hundred specimens) (G.J.O., J.M.S., and R.J.M., unpublished). Their offspring also had the ability to detoxify both Magnoliaceae and Salicaceae (species-specific traits for *P. canadensis* and *P. glaucus*, respectively).

Is the Postdiapause Delay for LF in the Hybrid Zone Heritable for Both Sexes?

When reared under long-day conditions (16 or 18 hour, non-diapause-inducing) hybrid crosses involving a *P. glaucus* mother and *P. canadensis* father produced only diapausing daughters, while most sons developed directly (Table 6.2). We observed that these primary (F_1) hybrid daughters were significantly delayed in their emergence the next year compared to their diapausing brothers. This delay was from 16 to 20 days at 26°C (Table 6.3; see also Scriber et al. 2002). Unlike these primary F_1 hybrids, the wild LF and lab-paired LF × LF families (families number 19036, 19072, and 19075) produced sons that were as delayed as the daughters (Table 6.3). Various hybrid backcrosses also showed delays in the postdiapause emergences of both males and females.

We have seen both sons and daughters of the diapausing hybrid swarm of recombinant genotypes in the LF synchronously emerge later (postdiapause) than either *P. canadensis* (EF) and first-generation *P. glaucus* genotypes. We have also observed such delays among recombinant hybrid offspring of backcrosses reared in the lab. In addition to extensive X-chromosome recombination in backcrosses, we also see differential segregation of other species-diagnostic traits in diapause versus nondiapause (direct-developing) individuals. Within

TABLE 6.3
Mean Postdiapause Emergence Times of *Papilio* Parental Species, Primary F_1 Hybrids, and Late-flight (LF) Hybrid Swarm (wild captures and lab-paired LF × LF; three families: 19036, 19072, 19075)

Year	*Genotype*[a]	*Days male (N)*	*Days female (N)*	*Temperature (°C)*
Parentals				
2003	*Pc*	16.5 ± 2.4 (28)	17.0 ± 3.0 (36)	22
2004	*Pc*	15.9 ± 1.3 (18)	18.2 ± 2.0 (20)	22
2005	*Pc*	14.7 ± 1.2 (12)	15.1 ± 1.3 (15)	22
2002	*Pg*	19.2 ± 4.3 (45)	19.6 ± 3.9 (36)	22
2003	*Pg*	21.6 ± 4.7 (69)	23.1 ± 4.9 (64)	22
2004	*Pg*	22.1 ± 4.6 (37)	20.4 ± 3.3 (35)	22
2005	*Pg*	22.0 ± 4.2 (13)	25.6 ± 2.9 (16)	22
Primary hybrids (gc)				
2005	20011[b]	14.7 ± 1.2 (3)	27.4 ± 8.7 (5)	22
2005	20011[b]	12.5 ± 0.9 (12)	23.9 ± 5.9 (21)	22
2005	20015[b]	14.7 ± 0.6 (3)	32.0 ± 1.4 (2)	22
2005	20015[b]	14.5 ± 0.7 (2)	32.5 ± 13.4 (2)	22
LF hybrid swarm				
2003	(LF × LF)	28.2 ± 4.8 (5)	32.5 ± 7.8 (2)	22
2003	(LF × LF)	23.8 ± 4.3 (6)	42.4 ± 14.2 (5)	22
2003	(LF × LF)	24.1 ± 0.9 (11)	32.7 ± 10.3 (6)	22
2003	Wild LF	30.9 ± 3.9 (15)	34.4 ± 4.9 (19)	22
2004	Wild LF	30.9 ± 4.4 (17)	33.4 ± 4.5 (18)	22
Parentals				
2003	*Pc*	13.1 ± 1.7 (31)	13.7 ± 2.4 (36)	26
2004	*Pc*	11.1 ± 0.8 (20)	12.3 ± 0.9 (18)	26
2005	*Pc*	10.5 ± 0.8 (12)	11.1 ± 0.9 (15)	26
2002	*Pg*	14.0 ± 2.7 (75)	14.3 ± 3.4 (61)	26
2003	*Pg*	15.6 ± 2.5 (27)	14.7 ± 1.7 (34)	26
2004	*Pg*	13.3 ± 2.8 (34)	12.7 ± 2.5 (41)	26
2005	*Pg*	15.4 ± 1.1 (12)	15.3 ± 1.8 (18)	26
Primary hybrids (gc)				
2001–2002	*Pg* × *Pc*	8.0 ± 2.2 (13)	28.6 ± 3.8 (9)	26
2003	*Pg* × *Pc*	8.0 ± 2.1 (10)	24.8 ± 3.5 (5)	26
LF hybrid swarm				
2003	Wild LF	25.4 ± 4.3 (21)	29.0 ± 2.6 (15)	26
2004	Wild LF	23.8 ± 3.8 (22)	27.9 ± 2.8 (16)	26

[a] *Pc* = *P. canadensis*; *Pg* = *P. glaucus*.
[b] Different control chambers.

single-backcross families, the hindwing band widths and forewing lengths (of both males and females) were more *P. glaucus*–like in direct-developers and more *P. canadensis*–like in diapausers.

Does the Larger Size of LF versus EF Explain Its Delayed Postdiapause Emergence?

Field-rearing of EF, LF, and *P. glaucus* (in 2002 and 2003) resulted in pupae and postdiapause adults of significantly different sizes. The early *P. canadensis* EF had mean forewing lengths that ranged over both years from 44.1 to 44.8 mm for males and 43.4 to 45.4 mm for females. The hybrid swarm LF males averages ranged from 47.1 to 48.1 mm (females 47.6 to 49.5 mm). *Papilio glaucus* were the largest of all (males from 48.6 to 54.0 mm; females 51.7 to 54.4 mm), even though these were from southeastern Pennsylvania, near the northern edge of their range, where they are smallest for this species (see size cline, Scriber 1994). The large LF individuals emerge later than the EF individuals. However, under a wide range of temperature regimes the LF individuals are delayed significantly in their postdiapause emergences compared to *P. glaucus*, which are even larger (J.M.S., unpublished data). Clearly, size alone does not explain the postdiapause delay in LF emergences.

Intrinsic Genetic Incompatibilities among Hybrids

The selection against *P. glaucus* and certain recombinant hybrid genotypes in the hybrid zone may be due to intrinsic genetic incompatibilities other than those between the X-linked diapause (od) locus and *Ldh* allozyme locus. Multiple transects, large field samples, and allozyme analyses will help clarify the thermal constraint hypothesis. However, lab hybrid backcross rearing studies to experimentally remove natural selection (with "unlimited summer days") have allowed us to evaluate the possibility that genetic lethal linkages might be responsible for the elimination of certain recombinant genotypes.

An informative backcross between a dark-morph mother and a hybrid male (from a *Pg* mother × a *Pc* father) produced 20 females that directly developed (no diapause) (19 of these had the *P. glaucus*–diagnostic *Ldh*-100 and od^-; one had the *P. canadensis Ldh*-40). This result for recombinant X chromosomes lends support for strong (close) linkage between *Ldh*-100 and od^- and argues against lethal linkage. Although preliminary, the survival of four diapausing (od^+) females (all dark morphs; s^-, b^+) with X-linked *Ldh*-100 also indicates that the recombination of *Ldh*-100 and od^+ is possible and not necessarily lethal. However, more replicates of such backcrosses are needed to firmly reject this particular lethal-linkage hypothesis and to evaluate other potential genetic incompatibilities.

Rejection of the "lethal recombinant" concept for the *Ldh*/od^+ genetic incompatibility would not rule out other endogenous selection against hybrids due to other deleterious chromosomal interactions (such as X and Y, X and autosome, Y and autosome, or mtDNA and heterospecific chromosome) (see Wirtz 1999; Jiggins et al. 2001; Presgraves et al. 2003; Coyne and Orr 2004; Gemmell et al. 2004). In fact, strong selection has been observed against the recombinant backcross daughters that possess one or both of the *P. canadensis*–type X-linked allozyme alleles. In backcross families there tends to be higher mortality of female pupae (the heterogametic sex) than males (Hagen and Scriber 1989, 1995). Males are apparently buffered from such mortality (perhaps due to their homogamy) (see Rockey et al. 1987a, 1987b; also Presgraves 2002). In seven backcross families, the offspring expressing the four recombinant backcross X-linked allozyme genotypes clearly showed an overall shortage of females. Female backcross hybrid recombinants with one or both *P. canadensis* allozyme profiles showed female to male ratios of 0.61 for *Ldh*, 0.35 for *Pgd*, and 0.03 for both, compared to a ratio of 1.09 where offspring had *P. glaucus*–type allozymes (Hagen and Scriber 1995). Similar results have been obtained with recent backcross families (# 18006; G.J.O. and J.M.S., unpublished).

These data suggest that a region of the *P. canadensis* X chromosome between *Ldh* and *Pgd* (or separate regions near each locus) may contain the genes that disrupt pupal development in combination with *P. glaucus* genes (Haldane effect; see Hagen and Scriber 1995). Alternatively, recombinant backcross female offspring *(Pg × PgPc* or *Pg × PcPg)* carrying both *P. canadensis*–type alleles (*Pgd*-125 or *Pgd*-80, and *Ldh*-80 or *Ldh*-40) may survive poorly in relation to males with the same paternal haplotype, while those carrying *P. glaucus* alleles (or fewer *P. canadensis* alleles) do better.

The *Ldh*-80 or *Ldh*-40 and *Pgd*-100 combinations are delayed in their diapause emergences (as in *P. appalachiensis* and LF), but the *Ldh*-80 or *Ldh*-40 and *Pgd*-125 or *Pgd*-80 combinations may enter "permanent diapause" and die (Hagen and Scriber 1995). This differential offspring mortality of backcross females (especially with *Pgd*-125) could explain the increased frequencies of *Pgd*-100 observed in the 2200 to 2600°D_F region of the hybrid zone (independent of *Ldh*-100) (Fig. 6.4). Different causes may exist regarding Haldane effects for sterility versus inviability (Presgraves 2002). Resolving these possibilities requires more molecular markers.

Although thought to be functionally linked (i.e., complete linkage disequilibrium due to maternal inheritance of both), we have discovered that the Y-linked dark morph (b^+) and mtDNA are not associated in *P. glaucus*, possibly due to paternal leakage (Andolfatto et al. 2003). Also, since the Haldane effect is observed in progeny from nonmimetic yellow females (carrying the Y-linked b^-) as well as from mimetic females (Scriber et al. 1990), the Y-linked b^+ allele cannot be the sole source of hybrid inviability (further discussion in Hagen and Scriber 1995).

Nonconcordant Steep Clines for Species-Diagnostic Allozymes

From our expanded allozyme analysis, including midwestern as well as new eastern populations ($N = 70$ populations, more than 2000 males), a steep *Pgd*-100 cline is evident across latitudes, and the *Ldh*-100 frequencies show a complex and discordant pattern across latitude (Hagen 1990; Scriber and Ording 2005). However, when plotted against degree-days on the thermal landscape (Fig. 6.4), the *Ldh*-100 cline is clearly evident and appears even steeper (and at different thermal degree-day locations) than the *Pgd* cline. Even in areas such as near the Battenkill River, Vermont, where populations show extensive influxes of *P. glaucus*–like traits, we see no introgression of *Ldh*-100, no dark females, nor any "true" second generations (Scriber and Ording 2005).

Divergence between two species at "neutral" loci linked to traits under selection will depend on their recombination distances and the strength of selection (Charlesworth et al. 1998). However, allozymes themselves (possibly *Ldh* and *Pgd* in *Papilio*) may be under local selection (Wu 2001). In other organisms, different forms of *Ldh* enzymes show strong latitudinal clines and have different kinetic properties influencing hatching times, developmental rates, performance, and mortality at different temperatures (Schulte et al. 1997; Johns and Somero 2004). Recently, we have detected a rare *Ldh*-20 allele (or "hybrizyme") at 5% frequency in *P. appalachiensis,* and at 50% frequency in the Battenkill LF. Hybrizymes are rare alleles that show up only in hybrid zones, presumably as a result of (1) increased mutation rates, (2) positive selection on the allozyme itself, or (3) "purifying selection" against especially unfit multilocus hybrid genotypes, resulting in the rare allele increasing in relative frequency (Woodruff 1989; Hoffmann and Brown 1995; Schilthuizen et al. 2004; Seehausen 2004). The possibility of direct selection on these species-diagnostic *Papilio* allozymes in the hybrid zone (Fig. 6.4) cannot be ruled out.

Survival and Movement across the Hybrid Zone

The lack of movement across the historical hybrid zone may at least be partially due to differential survival of the pupal stage. Cold winter temperatures may kill or weaken diapausing pupae of *P. appalachiensis* and *P. glaucus* at locations north of the hybrid zone. This might explain the absence of *P. glaucus* in the hybrid zone, and *P. appalachiensis* north of the hybrid zone. Pupae of *P. glaucus* are known have lower tolerance of cold temperatures (14% survival) than *P. canadensis* in both Alaska and Michigan (75 to 90% survival) (Kukal et al. 1991). Conversely, hot summer periods (e.g., 33–36°C) that can occur for several consecutive days in mid-July south of the hybrid zone (see Scriber et al. 2002) will kill or severely stress *P. appalachiensis*, LF, and *P. canadensis* pupae before diapause for the winter. This might explain the general lack of these groups south of the hybrid zone, where *P. glaucus* survives this heat.

Hybrids Are More Polyphagous

The virtual inability of all neonate larvae of *P. canadensis* to survive and grow on tulip tree (*Liriodendron tulipifera*, of the Magnoliaceae) is an autosomal trait (Scriber 1986) found extensively across North America from 1980 to 1997. It has historically served as a species-diagnostic trait in itself (Table 6.1) (Scriber 1982; Hagen 1990). Natural interspecific hybridization is occurring well north of the historical hybrid zone (Donovan and Scriber 2003). Recent extensive movement of autosomal tulip tree detoxification genes in Michigan and Wisconsin since 1998 (Scriber 2002b) represents extremely rapid gene flow, especially when the host-plant tree distribution has not noticeably shifted northward itself. During 1998 to 2004, 60 *Papilio* families from northern Michigan showed an average survival on tulip tree of 24% to 60% (and only two families showed no survival), whereas tulip tree survival in these populations was previously zero (Scriber 2002b). In the Battenkill populations 46 EF families showed average family larval survival on tulip tree of 35% (where it was previously less than 10% in the 1980s), and the LF averaged 75% survival (reflecting additional introgression from *P. glaucus*).

Conversely, the phenolic glycosides in Salicaceae species such as quaking aspen *(Populus tremuloides)* have been shown toxic for *Papilio glaucus* (Lindroth et al. 1988). However, interspecific hybrids (and apparently *P. appalachiensis;* Scriber and Ording 2005) are known to possess enzymatic detoxification capabilities for excellent survival and rapid growth on both Magnioliaceae and Salicaceae (tulip tree and quaking aspen) (Scriber et al. 1989, 1999; Scriber 2004). These capabilities are autosomally determined (Scriber 1986). The ability to use other hosts such as cherry (e.g., *Prunus serotina*) and ash *(Fraxinus americana)* is shared by *Papilio appalachiensis*, the LF, *P. canadensis,* and *P. glaucus*.

As with most *P. glaucus*, Georgia (Clarke County) populations near Athens show little ability to detoxify and survive the neonate stages on quaking aspen (less than 6%) but have excellent tulip tree survival (greater than 84%). However, the Rabun and Habersham County populations in northern Georgia (southernmost part of the Appalachian Mountains) (Fig. 6.1) have the ability to detoxify both aspen (greater than 50% mean neonate survival) and tulip tree (67.6 % survival). These Georgia mountain populations are very close to type localities for *P. appalachiensis* (Pavulaan and Wright 2002), and individuals survived almost as well as primary lab hybrids (aspen survival greater than 81% and tulip tree survival greater than 80% [N = 16 $Pc \times Pg$ families and N = 49 $Pg \times Pc$ families]). Similar results were obtained with Battenkill LF hybrid swarm families (85% aspen and 80% tulip tree survival [N = 6 families]) (Scriber and Ording 2005). Since these Georgia mountain populations also had the ability to detoxify quaking aspen in the early 1980s (Scriber et al. 1987), this suggests historical gene flow prior to recent eight years of climate warming (or perhaps ancient gene flow, pre- or post-Pleistocene).

Oviposition Preferences

Bioassays with individual females in three-choice arenas (tulip tree, black cherry, and quaking aspen) have shown rank order of ovipositional host-plant preference to be genetically based, with X-linkage (Scriber et al. 1991; Scriber, 1993, 1994). Although other factors influence the specificity (strength) of oviposition preference (Bossart and Scriber 1999; Frankfater and Scriber 1999; Mercader and Scriber 2005), the three-choice bioassays used with *P. canadensis* and *P. glaucus* provide very useful information about introgression involving the X chromosome. The Battenkill EF populations have been very *P. canadensis*–like, with 39 to 41% preference for aspen in 2003 to 2005 (total N = 72). In contrast, the Battenkill LF populations are basically repelled by aspen (only 8 to 11% oviposition over three years, total N = 16 females) compared to 67 to 78% egg deposition on tulip tree (Mercader and Scriber 2007). This pattern of oviposition by LF females is much more *P. glaucus*–like. *Papilio glaucus* females from southern Pennsylvania only put

2 to 5% of their total eggs on quaking aspen in 2003 and 2004. In the future, we hope to monitor these EF and LF populations to determine the degree of introgression of X-linked rank-order oviposition profiles relative to other X-linked traits (e.g., *Pgd*, *Ldh*, diapause, dark-morph enablers) and autosomal traits (hindwing bands, forewing lengths, hexokinase allozymes, detoxification abilities on tulip tree, etc.) and in relation to molecular markers.

Are Traits from Ancient or Recent Hybridization?

The time (pre- or post-Pleistocene) of interspecific gene flow and genetic divergence in the putative Appalachian mountain refugium is not clear. Examination of the DNA sequences on the X chromosome, that permit locus-by-locus as well as total genome analyses will provide a tool (e.g., coalescence theory [P. Andolfatto, 2004, personal communication]) by which we can determine the divergence times and infer the age of introgressive hybridization in such populations. High rates of recombination of X-chromosome traits have been documented (A. Putnam and P. Andolfatto, 2005, personal communication) and can be seen in the offspring of backcross lab families (Ording et al., in preparation). It is uncertain how protected from further recombination these potentially coadapted gene complexes may be. For example, some plant detoxification abilities may be ancient (Scriber 2005; Scriber et al. 2007).

Molecular Work

In collaboration with Dr. Jeffrey Tompkins, at the Clemson University Genomics Institute, we are beginning to develop a physical map of the X (= Z) chromosome in *Papilio*. A 10 × coverage BAC library of the *P. glaucus* genome has just been completed (with funding from the National Institutes of Health, National Human Genome Research Initiative). We are assisting in the development of this BAC library (under leadership of Peter Andolfatto) using autosomal and Z-linked DNA sequences from a cDNA library of *P. glaucus*. Additional Z-linked markers will be added to the BAC scaffold to create a fine-scale physical map of the Z chromosome in *P. glaucus* and *P. canadensis,* which will complement the linkage map. Our preliminary results (P. Andolfatto, J.M.S. and A. Putnam unpublished data) have yielded a cDNA library of 247 unique, nonmitochondrial cDNA sequences that are greater than 600 base pairs in length. From these cDNA sequences, primers for titin, triosephosphate isomerase *(Tpi),* lactate dehydrogenase *(Ldh),* and kettin were identified with another Z-linked gene in Lepidoptera ("period"). All five markers were sequenced for males and females of *P. glaucus* and *P. canadensis*, with results supporting Z-linkage in these *Papilio*. Of these five markers, four have fixed differences between *P. glaucus* and *P. canadensis* and are suitable as quantitative trait locus (QTL) markers (Putnam et al. 2007). Since QTL markers are biased for detecting regions of large effect (e.g., Laurie et al. 1997; Gadau et al. 2002), we will use additional Z-linked and autosomal markers that will allow detection of regions of small effect.

We will use a two-part approach to assess current and historical interactions of gene flow and selection. Our goal is to distinguish between differential selection of certain parts of the Z chromosome and the persistence of ancient (ancestral) polymorphisms. The first step will be analyses of diagnostic markers in hybrid zone transects to estimate the intensity of selection (against hybrids) relative to locus-by-locus dispersal rates (Porter et al. 1997; Payseur et al. 2004). Second, we will obtain a cross section of normal variation at several dozen Z-chromosome loci in nonhybrid populations of *P. glaucus* and *P. canadensis*. By employing coalescence theory, this will allow us to make inferences about locus-by-locus and genomewide levels of divergence and historical gene flow between swallowtail species (Wakely and Hey 1997; Machado et al. 2002; Hey and Neilsen 2004).

Preliminary results (P. Andolfatto, 2003, personal communication) assessing variability in the five Z-linked gene fragments and mtDNA (CoI and CoII) indicate that average levels of nucleotide variability in *P. glaucus* and *P. canadensis* are comparable to those in *Drosophila* (Andolfatto 2001). The high level of polymorphism (1.3%) and divergence (2.8%) will facilitate the discovery of single nucleotide polymorphisms. The large variance among Z-linked loci in the number of shared and fixed polymorphisms, and the 10-fold variation in divergence times suggest differential divergence times and migration among loci (Putnam et al. 2007). It was estimated that recombination rates relative to mutation rate in *Papilio* are about 25 recombination events per mutation, which is much higher than estimated recombination rates in *Drosophila* (Haddrill et al. 2005).

We hope to eventually identify the sex-linked genes responsible for dark-morph females in Batesian mimicry by *Papilio* (the Y-linked b^+) (Clark and Sheppard 1962; Andolfatto et al. 2003) and its X-linked suppressor/enabler (s^+ or s^-) (Scriber et al. 1996; ffrench-Constant and Koch 2003). Of course diapause regulation genes would be of great significance, since they may also be involved in the postdiapause emergence delays that characterize both the Battenkill LF populations and *P. appalachiensis* (Pavulaan and Wright 2002; Scriber and Ording 2005). Determining the particular recombinant sequences involved in the delayed postdiapause emergences of LF would perhaps unlock the mystery of genetic divergence, reproductive isolation, and speciation mechanisms in these *Papilio* (Orr et al. 2005).

Future Efforts

The fitness of genotypes outside of the geographic place of origin will continue to be a fascinating aspect of introgression. Several additional questions loom, unresolved, in this *Papilio* system. First, the postzygotic egg and larval fitness of hybrid and backcross genotypes, while excellent for most reciprocal *Papilio* pairing combinations (Scriber et al. 2003), may not extend to hybrid pupal or adult offspring mating fitness (Hagen and Scriber 1995; Davies et al. 1997). Sexually selected traits in adults are often X-linked (Sperling 1994; Prowell

Haddrill, P., K. Thornton, B. Charlesworth, and P. Andolfatto. 2005. Multilocus patterns of nucleotide variability and the demographic and selection history of *Dropsophila melanogaster* populations. Genome Res. 15: 790–799.

Hagen, R. H. 1990. Population structure and host use in hybridizing subspecies of *Papilio glaucus* Lepidoptera: Papilionidae). Evolution 44: 1914–1930.

Hagen, R. H., and R. C. Lederhouse. 1985. Polymodal emergence of the tiger swallowtail, *Papilio glaucus* (Lepidoptera: Papilionidae): source of a false second generation in central New York State. Ecol. Entomol. 10: 19–28.

Hagen, R. H., and J. M. Scriber. 1989. Sex linked diapause, color, and allozyme loci in *Papilio glaucus*: linkage analysis and significance in a hybrid zone. Heredity 80: 179–185.

Hagen, R. H., and J. M. Scriber. 1995. Sex chromosomes and speciation in the *Papilio glaucus* group, pp. 211–228. In J. M. Scriber, Y. Tsubaki, and R. C. Lederhouse (eds.), The swallowtail butterflies: their ecology and evolutionary biology. Scientific Publishers, Gainesville, FL.

Hagen, R. H., R. C. Lederhouse, J. L. Bossart, and J. M. Scriber. 1991. *Papilio canadensis* and *P. glaucus* (Papilionidae) are distinct species. J. Lepidopt. Soc. 45: 245–258.

Hellmann, J. J. 2002. Butterflies as model systems for understanding and predicting the biological effects of climate change, pp. 93–126. In S. H. Schneider and T. L. Root (eds.), Wildlife responses to climate change. Island Press, Washington, DC.

Hey, J., and R. Neilsen. 2004. Multilocus methods for estimating population sizes, migration rates, and divergence time, with applications to the divergence of *Drosophila pseudoobscura* and *D. persimilis*. Genetics 167: 747–760.

Hoffman S. M., and W. Brown. 1995. The molecular mechanism underlying the "rare allele phenomenon" in a subspecific hybrid zone of the California field mouse, *Peromyscus californicus*. J. Mol. Evol. 41: 1165–1169.

Hoffmann, A. A., and M. W. Blows. 1994. Species borders: ecological and evolutionary perspectives. Trends Ecol. Evol. 9: 223–227.

Hoffmann, A. A., and H. J. Hercus. 2000. Environmental stress as an evolutionary force. Bioscience 50: 217–226.

Hoffmann, A. A., C. M. Sgro, and A. R. Weeks. 2004. Chromosomal inversion polymorphisms and adaptation. Trends Ecol. Evol. 19: 482–488.

Holzapfel, C. M., and W. E. Bradshaw. 2002. Protandry the relationship beteen emergence time and male fitness in the pitcher-plant mosquito. Ecology 83: 607–611.

Jiggins, C. D., and J. R. Bridle. 2004. Speciation in the apple maggot fly: a blend of vintages? Trends Ecol. Evol. 19: 111–114.

Jiggins C. D., M. Linares, R. A. Naisbit, C. Salazar, Z. H. Yang, and J. Mallet. 2001. Sex-linked sterility in a butterfly. Evolution 55: 1631–1638.

Johns, G. C., and G. N. Somero. 2004. Evolutionary convergence in adaptation of proteins to temperature: A4-lactate dehydrogenases of Pacific damselfishes (*Chromis* spp.) Mol. Biol. Evol. 21: 314–320.

Kukal O., M. P. Ayres, and J. M. Scriber. 1991. Cold tolerance of pupae in relation to the distribution of tiger swallowtails. Can. J. Zool. 69: 3028–3037.

Laurie, C., J. True, J. Liu, and J. Mercer. 1997. An introgression analysis of quantitative trait loci that contribute to a morphological difference between *Drosophila simulans* and *D. mauritania*. Genetics 145: 339–348.

Lederhouse, R. C. 1995. Comparative mating behavior and sexual selection in North American swallowtail butterflies pp. 117–131. In J. M. Scriber, Y. Tsubaki, and R. C. Lederhouse (eds.), Swallowtail butterflies: their ecology and evolutionary biology. Scientific Publishers, Gainesville, FL.

Lederhouse, R. C, M. P. Ayres, and J. M. Scriber. 1995. Physiological and behavioral adaptations to variable thermal environments in North American swallowtail butterflies, pp. 71–81. In J. M. Scriber, Y. Tsubaki, and R. C. Lederhouse (eds.), Swallowtail butterflies: their ecology and evolutionary biology. Scientific Publishers, Gainesville, FL.

Lindroth, R. L., J. M. Scriber, and M. T. S. Hsia. 1988. Chemical ecology of the tiger swallowtail: mediation of host use by phenolic glycosides. Ecology 69: 814–822.

Logan, J. A., and J. A. Powell. 2001. Ghost forests, global warming, and the mountain pine beetle (Coleoptera: Scolytidae). Am. Entomol. 47: 160–173.

Luebke, H. J., J. M. Scriber, and B. S. Yandell. 1988. Use of multivariate discriminant analysis of male wing morphometrics to delineate the Wisconsin hybrid zone for *Papilio glaucus glaucus* and *P. g. canadensis*. Am. Midl. Nat. 119: 366–379.

Machado, C., R. Kilman, J. Markert, and J. Hey. 2002. Inferring the history of speciation from multilocus DNA sequence data. Mol. Biol. Evol. 19: 472–488.

Mallet, J. 2005. Hybridization as an invasion of the genome. Trends Ecol. Evol. 20: 229–237.

Martinsen, G. D., T. G. Whitham, R. T. Turek, and P. Keim. 2001. Hybrid populations selectively filter gene introgression between species. Evolution 55: 1325–1335.

McKinnon, J. S., S. Mori, B. K. Blackman, L. David, D. M. Kingsley, L. Jamieson, J. Chou, and D. Schluter. 2004. Evidence for ecology's role in speciation. Nature 429: 294–297.

Mercader, R., and J. M. Scriber. 2005. Phenotypic plasticity in polyphagous *Papilio*: preferences, performances, and potential enhancement by hybridization, pp. 25–57. In T. N. Ananthakrishnan (ed.), Insect phenotypic plasticity: diversity of responses. Science Publishers, Plymouth, UK.

Mercader, R. J. and J. M. Scriber. 2007. Diversification of host use in two polyphagous butterflies: Differences in oviposition specificity or host rank hierarchy? Entom. Exp. et Appl. (Published article online: 26-Jul-2007. doi: 10.1111/j.1570-7458.2007.00598.x)

Mousseau, T. A., and D. A. Roff. 1989. Adaptation to seasonality in a cricket: patterns of phenotypic and genotypic variation in body size and diapause expression along a cline in season length. Evolution 43: 1483–1496.

Nygren, G. H. 2005. Latitudinal patterns in butterfly life history and host plant choice. Ph.D. dissertation. University of Stockholm, Sweden.

Nylin, S. 1988. Host plant specialization and seasonality in a polyphagous butterfly, *Polygonia c-album* (Nymphalidae). Oikos 53: 381–386.

Nylin, S., and K. Gotthard. 1998. Plasticity in life history traits. Annu. Rev. Entomol. 43: 63–83.

Nylin, S., C. Wiklund, and P.-O. Wickman. 1993. Absence of trade-off between sexual size dimorphism and early male emergence in a butterfly. Ecology 74: 1414–1427.

Orr, H. A., J. P. Masly, and D. C. Presgraves. 2005. Speciation genes. Curr. Opin. Genet. Devel. 14: 675–679.

Panhuis, T. M., R. Butlin, M. Zuk, and T. Tregenza. 2001. Sexual selection and speciation. Trends Ecol. Evol. 16: 364–371.

Parmesan, C., and G. Yohe. 2003. A globally coherent fingerprint of climate change impacts across natural systems. Nature 421: 37–42.

Pavulaan, H., and D. M. Wright. 2002. *Pterourus appalachiensis* (Papilionidae: Papilioninae), a newswallowtail butterfly from the Appalachian region of the United States. Taxonom. Rep. 3: 1–20.

Payseur, B. A., J. G. Krenz, and M. W. Nachman. 2004. Differential patterns of introgression across the X-chromosome in a hybrid zone between two species of house mice. Evolution 58: 2064–2078.

Porter, A. H., R. Wagner, H. Geiger, A. Scholl, and A. M. Shapiro. 1997. The *Pontia daplidice-edusa* hybrid zone in northwestern Italy. Evolution 51: 1561–1573.

Presgraves, D. C. 2002. Patterns of postzygotic isolation in Lepidoptera. Evolution 56: 1168–1183.

Presgraves, D. C., L. Balagopalan, S. M. Abmayr, and H. A. Orr. 2003. Adaptive evolution drives divergence of hybrid inviability gene between two species of *Drosophila*. Nature 423: 715–719.

Prowell, D. P. 1998. Sex linkage and speciation in Lepidoptera, pp. 109–319. In D. J. Howard and S. H. Berlocher (eds.), Endless forms: species and speciation. Oxford University Press, New York.

Putnam, A. S., J. M. Scriber, and P. Andolfatto. 2007. Discordant divergence times among Z-chromosome regions between two ecologically distinct swallowtail butterfly species. Evolution 61: 912–927.

Reinbold, K. 1998. Sex linkage among sexually selected traits. Behav. Ecol. Sociobiol. 44: 1–7.

Rhymer, J. M., and D. Simberloff. 1996. Extinction by hybridization and introgression. Annu. Rev. Ecol. Syst. 27: 83–109.

Rieseberg, L. 2001. Chromosomal rearrangements and speciation. Trends Ecol. Evol. 16: 351–358.

Rockey, S. J., J. H. Hainze, and J. M. Scriber. 1987a. A latitudinal and obligatory diapause response in three subspecies of the eastern tiger swallowtail *Papilio glaucus* (Lepidoptera: Papilionidae). Am. Midl. Nat. 118: 162–168.

Rockey, S. J., J. H. Hainze, and J. M. Scriber. 1987b. Evidence of a sex-linked diapause response in *Papilio glaucus* subspecies and their hybrids. Physiol. Entomol. 12: 181–184.

Roelofs, W. L., W. Liu, G. Hao, H. Jiao, A. P. Rooney, and C. E. Linn Jr. 2002. Evolution of moth sex pheromones via ancestral genes. Proc. Nat. Acad. Sci. USA 99: 13621–13626.

Roy, D. B., and T. H. Sparks. 2000. Phenology of British butterflies and climate change. Global Change Biol. 6: 407–416.

Schemske, D. W. 2000. Understanding the origin of species. Evolution 54: 1069–1073.

Schilthuizen M., R. F. Hoekstra, and E. Gittenberger. 2004. Hybridization, rare alleles and adaptive radiation. Trends Ecol. Evol. 19: 404–405.

Schluter, D. 2001. Ecology and the origin of species. Trends Ecol. Evol. 16: 372–380.

Schulte, P. M., M. Gomez-Chiarri, and D. A. Powers. 1997. Structural and functional differences in the promoter and 5´ flanking region of *Ldh*-B within and between populations of the teleost *Fundulus heteroclitus*. Genetics 145: 759–761.

Schwarz, D., B. J. Matta, N. L. Shakir-Botteri, and B. A. McPheron. 2005. Host shift to an invasive plant triggers rapid animal hybrid speciation. Nature 436: 546–549.

Scriber, J. M. 1982. Foodplants and speciation in the *Papilio glaucus* group, pp. 307–314. In J. H. Visser, and A. K. Minks (eds.), Proceedings of the 5th International Symposium on Insect Plant Relationships. PUDOC, Wagningen, Netherlands.

Scriber, J. M. 1986. Allelochemicals and alimentary ecology: heterosis in a hybrid zone? pp. 43–71. In L. Brattsten and S. Ahmad (eds.), Molecular mechanisms in insect plant associations. Plenum Press, New York.

Scriber, J. M. 1990. Interaction of introgression from *Papilio glaucus canadensis* and diapause in producing "spring form" eastern tiger swallowtail butterflies, *P. glaucus*. Great Lakes Entomol. 23: 127–138.

Scriber, J. M. 1993. Absence of behavioral induction in oviposition preference of *Papilio glaucus* (Lepidoptera: Papilionidae). Great Lakes Entomol. 26: 81–95.

Scriber, J. M. 1994. Climatic legacies and sex chromosomes: latitudinal patterns of voltinism, diapause size and host-plant selection in two species of swallowtail butterflies at their hybrid zone, pp. 133–171. In H. V. Danks (ed.), Insect life-cycle polymorphism: theory, evolution and ecological consequences for seasonality and diapause control. Kluwer Academic Publishers, Dordrecht, Netherlands.

Scriber, J. M. 1996. A new cold pocket hypothesis to explain local host preference shifts in *Papilio canadensis*. Entomol. Exp. Appl. 80: 315–319.

Scriber, J. M. 1998. Inheritance of diagnostic larval traits for interspecific hybrids of *Papilio canadensis* and *P. glaucus* (Lepidoptera: Papilionidae). Great Lakes Entomol. 31: 113–123.

Scriber, J. M. 2002a. Latitudinal and local geographic mosaics in host plant preferences as shaped by thermal units and voltinism. Eur. J. Entomol. 99: 225–39.

Scriber, J. M. 2002b. The evolution of insect-plant relationships: chemical constraints, coadaptation and concordance of insect/plant traits. Entomol. Exp. Appl. 104: 217–235.

Scriber, J. M. 2004. Non-target impacts of forest defoliator management options: decision for no spraying may have worse impacts on non-target Lepidoptera than *Bacillus thuringiensis* insecticides. J. Insect Conserv. 8: 241–261.

Scriber, J. M. 2007. A mini-review of the "feeding specialization/physiological efficiency" hypothesis: 50 years of difficulties, and strong support from the North American Lauraceae-specialist, *Papiolio troilus* (Papilionidae: Lepidoptera). Trends in Entomol. 4: 1–42.

Scriber, J. M., and R. C. Lederhouse. 1992. The thermal environment as a resource dictating geographic patterns of feeding specialization of insect herbivores, pp. 429–466. In M. R. Hunter, T. Ohgushi, and P. W. Price (eds.), Effects of resource distribution onanimal-plant interactions. Academic Press, New York.

Scriber, J. M., and G. J. Ording. 2005. Ecological speciation without host plant specialization: possible origins of a recently described cryptic *Papilio* species (Lepidoptera: Papilionidae). Entomol. Exp. Appl. 115: 247–263.

Scriber, J. M., M. H. Evans, and D. Ritland. 1987. Hybridization as a cause of mixed color broods and unusual color morphs of female offspring in the eastern tiger swallowtail butterfly, *Papilio glaucus*, pp. 119–134. In M. Huettle (ed.), Evolutionary genetics of invertebrate behavior. University of Florida Press, Gainesville.

Scriber, J. M., R. L. Lindroth, and J. Nitao. 1989. Differential toxicity of a phenolic glycoside from quaking aspen leaves by *Papilio glaucus* subspecies, their hybrids, and backcrosses. Oecologia 81: 186–191.

Scriber, J. M., R. Dowell, R. C. Lederhouse, and R. H. Hagen. 1990. Female color and sex ratio in hybrids between *Papilio glaucus glaucus* and *P. eurymedon*, *P. rutulus* and *P. multicaudatus* (Papilionidae). J. Lepid. Soc. 44: 229–244.

Scriber, J. M., B. L. Giebink, and D. Snider. 1991. Reciprocal latitudinal clines in oviposition behavior of *Papilio glaucus* and *P. canadensis* across the Great Lakes hybrid zone: possible sex-linkage of oviposition preferences. Oecologia 87: 360–368.

Scriber, J.M., R.H. Hagen, and R.C. Lederhouse. 1996. Genetics of mimicry in the tiger swallowtail butterflies, *Papilio glaucus* and *P. canadensis* (Lepidoptera: Papilionidae). Evolution 50: 222–236.

Scriber, J.M., K. Weir, D. Parry, and J. Deering. 1999. Using hybrid and backcross larvae of *Papilio canadensis* and *P. glaucus* to detect induced chemical resistance in hybrid poplars experimentally defoliated by gypsy moths. Entomol. Exp. Appl. 91: 233–236.

Scriber, J.M., K. Keefover, and S. Nelson. 2002. Hot summer temperatures may stop genetic introgression of *Papilio canadensis* south of the hybrid zone in the North America Great Lakes region? Ecography 25: 184–192.

Scriber, J.M., A. Stump, and M. Deering. 2003. Hybrid zone ecology and tiger swallowtail traitclines in North America, pp. 367–391. In C.L. Boggs, W.B.Watt., and P.R. Ehrlich (eds.), Butterflies: ecology and evolution taking flight. University of Chicago Press, Chicago.

Scriber, J.M., M.L. Larsen, and M.P. Zalucki. 2007. *Papilio aegeus* host plant range evaluated experimentally on ancient angiosperms. Austral. J. Entom. 46: 65–74.

Seehausen, O. 2004. Hybridization and adaptive radiation. Trends Ecol. Evol. 19: 198–207.

Shaw K.L. and Y.M. Parsons. 2002. Divergent mate recognition behavior and its consequences for genetic architectures of speciation. Amer. Nat. 159: 561–575.

Sperling, F.A.H. 1990. Natural hybrids of *Papilio* (Insecta: Lepidoptera): poor taxonomy or interesting evolutionary problem? Can. J. Zool. 68: 1790–1799.

Sperling, F.A.H. 1994. Sex-linked genes and species differences in Lepidoptera. Can. Entomol. 165: 233–242.

Sperling, F.A.H. 2003. Butterfly molecular systematics: from species definitions to higher level phylogenies pp. 431–458. In C.L. Boggs, W.B.Watt., and P.R. Ehrlich (eds.), Butterflies: ecology and evolution taking flight. University of Chicago Press, Chicago.

Stump, A.D., and J.M. Scriber. 2006. Sperm precedence in experimental interspecific multiple matings of hybridizing North American tiger swallowtail butterflies (Lepidoptera: Papilionidae). J. Lepidopt. Soc. 60: 65–78.

Stump, A.D., A. Crim, F.A.H. Sperling, and J.M. Scriber. 2003. Gene flow between Great Lakes region populations of the Canadian tiger swallowtail butterfly, *Papilio canadensis,* near the hybrid zone with *P. glaucus* (Lepidoptera: Papilionidae). Great Lakes Entomol. 36: 41–53.

Tauber, M.J., C.A. Tauber, and S. Masaki. 1986. Seasonal adaptations of insects. Oxford University Press, New York.

Tesar, D., and J.M. Scriber. 2003. Growth season constraints in climatic cold pockets: tolerance of subfreezing temperatures and compensatory growth by tiger swallowtail butterfly larvae. Holarctic Lepidopt. 7: 39–44.

Thomas, C., D.A. Cameron, and A. Green. 2004. Extinction risk from climate change. Nature 427: 145–148.

Thomas, C.D., E. J. Bodsworth, R.J.Wilson, A.D. Simmons, Z.G. Davies, M. Musche, and L. Conradt. 2001. Ecological processes at expanding range margins. Nature 411: 577–581.

Thomas, M.B., and S. Blanford. 2003. Thermal biology in insect-parasite interactions. Trends Ecol. Evol. 18: 344–350.

Thomas, Y., M.T. Bethenod, L. Pelozuelo, B. Frerot, and M. Bourguet. 2003. Genetic isolation between two sympatric host-plant races of the European corn borer, *Ostrinia nubilalis* Hubner. I. Sex pheromone, moth emergence timing, and parasitism. Evolution 57: 261–273.

Thompson, J.N. 1988. Evolutionary ecology of the relationship between oviposition preference and performance of offspring in phytophagous insects. Entomol. Exp. Appl. 47: 3–14.

Via, S. 2001. Sympatric speciation in animals: the ugly duckling grows up. Trends Ecol. Evol. 16: 381–390.

Valella, P., and J.M. Scriber. 2005 . Latitudinal variation in photoperiodic induction of pupal diapause in the spicebush swallowtail butterfly, *Papilio troilus*. Holarctic Lepidopt. 10: 37–41.

Wakeley, J., and J. Hey. 1997. Estimating ancestral population parameters. Genetics 145: 1091–1106.

Wirtz, P. 1999. Mother species–father species: unidirectional hybridization in animals with female choice. Anim. Behav. 58: 1–12.

Wood, T.K., and M. Keese. 1990. Host plant induced assortative mating in *Enchenopa* treehoppers. Evolution 44: 619–628.

Wood, T.K., K.J. Tilmon, A.B. Shantz, C.K. Harris, and J. Pesek. 1999. The role of host plant fidelity in initiating insect race formation. Evol. Ecol. Res. 1: 317–332.

Woodruff, D.S. 1989. Genetic anomalies associated with Cerion hybrid zones: the origin and maintenance of new electromorphic variants called hybrizymes. Biol. J. Linn. Soc. 36: 281–294.

Wu, C.-I. 2001. The genic view of the process of speciation. J. Evol. Biol. 14: 851–865.

Zakharov, E., M.S. Caterino, and F.A.H. Sperling. 2004. Molecular phylogeny, historical biogeography, and divergence time estimates for swallowtail butterflies of the genus *Papilio* (Lepidoptera: Papilionidae). Syst. Biol. 53: 193–215.

Zonneveld, C. 1996. Being big or emerging early: protandry and the trade-off between size and emergence in male butterflies. Am. Nat. 147: 946–965.

SEVEN

Host Shifts, the Evolution of Communication, and Speciation in the *Enchenopa binotata* Species Complex of Treehoppers

REGINALD B. COCROFT, RAFAEL L. RODRÍGUEZ, AND RANDY E. HUNT

Speciation in animals is promoted by the evolution of behavioral differences that reduce attraction, mating, and fertilization between individuals in diverging populations (Mayr 1963; West-Eberhard 1983; Eberhard 1985, 1994, 1996; Coyne and Orr 2004). Behavioral traits involved in communication between the sexes often provide the most immediate contributions to reproductive isolation (Blair 1955; Claridge 1990; Ryan and Rand 1993; Bridle and Ritchie 2001; Gerhardt and Huber 2002; Kirkpatrick and Ravigné 2002). Consequently, identifying the evolutionary forces that lead to changes in sexual communication is necessary to understand the evolution of behavioral isolation and its contribution to divergence and speciation.

The relationship between sexual communication and speciation depends on the extent of interactions between individuals from the diverging populations. When geography and/or ecology prevent such interactions during the speciation process, differences in mating signals and preferences may become important upon secondary contact, at which point the differences may be enhanced by selection against hybridization (Howard 1993; Kelly and Noor 1996; Coyne and Orr 1997; Jiggins and Mallet 2000; Servedio and Noor 2003). In contrast, when the geography and ecology of speciation do lead to interactions between individuals from the diverging populations (i.e., when speciation occurs in sympatry), behavioral causes of reproductive isolation are important from the outset (Kondrashov et al. 1998; Kondrashov and Kondrashov 1999; Kirkpatrick and Ravigné 2002).

Sympatric speciation is implicated in the diversification of host-specific, plant-feeding insects (Mallet 2001; Berlocher and Feder 2002; Bush and Butlin 2004). In these insects, which comprise a large fraction of animal diversity (Price 2002; Bush and Butlin 2004), speciation is often associated with changes in host-plant use. Host shifts have widespread consequences for life-history traits, and aspects of insect adaptation to their host plants can lead to assortative mating as a pleiotropic effect (Wood and Keese 1990; Craig et al. 1993; Wood 1993; Feder 1998; Berlocher and Feder 2002; see also Jiggins et al. 2005). Colonization of a new host environment may also have a profound influence on the evolution of communication.

In this chapter we examine the role of communication systems in the diversification of the *Enchenopa binotata* species complex of treehoppers (Hemiptera: Membracidae), a clade of 11 sap-feeding species distributed across eastern North America. Based on the career-long series of studies by T. K. Wood and colleagues, this group provides one of the most widely cited examples of sympatric speciation in plant-feeding insects (Tauber and Tauber 1989; Wood 1993; Berlocher and Feder 2002; Coyne and Orr 2004). The *E. binotata* complex is especially promising for studies of the relationship between host shifts and behavioral isolation (Landolt and Phillips 1997; Etges 2002), because of the rich understanding of its natural history, comparative biology, and communication behavior.

We first discuss the role of ecological factors in promoting assortative mating among populations of *E. binotata* on ancestral and novel species of host plant. We then examine the role of sexual communication in behavioral isolation among *E. binotata* species in the present. We explore sources of divergent sexual and natural selection that could alter the evolutionary trajectory of mating signals and preferences after a host shift. We also consider ways in which developmental influences on sexual communication may affect gene flow between populations on ancestral and novel hosts (before any evolutionary change in communication systems has occurred), and how developmental plasticity may generate changes in sexual selection regimes. We will argue that for plant-feeding insects, sexual communication systems provide an important link between host use, assortative mating, and divergent selection and may be a key component of sympatric speciation through host shifts.

Ecological Isolation in the *E. binotata* Complex

Thomas K. Wood and colleagues developed the *E. binotata* complex as a model for evaluating the role of host shifts in promoting genetic divergence in sympatry (Wood and Guttman 1983; Wood 1993). (A complete list of Wood's publications is provided by Deitz and Bartlett [2004].) The central message of this research is that successful colonization of a novel host can lead to immediate assortative mating between populations on the original and novel hosts, facilitating a response to divergent selection imposed by differences between the host species.

The most immediate factor contributing to assortative mating after a host shift in *Enchenopa* is a change in life-history timing caused by differences in host-plant phenology (Wood and Guttman 1982; Wood and Keese 1990; Wood et al. 1990). Species in the *E. binotata* complex have one generation per year, with eggs deposited into host-plant tissue in late summer and fall, and developing in the spring (Wood 1993). Egg development is triggered by hydration from the flow of sap through plant stems, which happens on different schedules in different host species (Wood et al. 1990). As a consequence, egg hatch occurs at different times on different hosts. Depending on the rate of nymphal development on different hosts, the timing of adult eclosion and mating will also differ (Wood 1993). In an elegant experiment that isolated the effects of allochrony from other potential sources of reproductive isolation, Wood and Keese (1990) manipulated the phenology of *Celastrus scandens* plants containing egg masses of *E. binotata* from a single population to create a series of age classes. This experiment resulted in significant assortative mating by age, even though age classes were only five days apart (Wood and Keese 1990). Assortative mating apparently was a consequence of differences in the timing of female receptivity, as well as in male mortality schedules.

Assortative mating after a host shift is also promoted by host fidelity. In an experiment in which males and females of six species were placed in a common cage and allowed to choose mates and host plants, most matings occurred on the female's natal host (Wood 1980). A later study showed assortative mating by host among experimentally host-shifted populations, as a consequence of female and male host fidelity (Wood et al. 1999). Females also show high host fidelity in oviposition (Wood 1980).

Host shifts lead not only to assortative mating, but also to divergent natural selection. When females were forced to oviposit on nonnatal hosts, female fecundity and survivorship, as well as hatching rates and nymphal survivorship, were lower on nonnatal hosts (Wood and Guttman 1983). Experimental shifts to novel host-plant species also revealed the presence of the genetic variation necessary for successfully colonizing a new host (Tilmon et al. 1998). Thus, the hypothesis that local adaptation to a novel host can occur in sympatry with the original host is supported by the combination of host fidelity during oviposition and mating, assortative mating among allochronically shifted age classes, and host-related ecological divergent selection.

In *Enchenopa* treehoppers, allochrony and host fidelity reduce, but do not eliminate, interactions among individuals on ancestral and novel hosts. The potential for interactions depends on the degree of allochrony among host-shifted populations, the extent of host fidelity, and the dispersal distance between plants (Wood 1980; Wood et al. 1999). There can be substantial overlap in mating periods between species on different hosts. In a study in which populations of six *E. binotata* species were established on natal host plants in a common environment, the average date of mating differed by only one to two days between species on certain hosts (Wood and Guttman 1982); as a consequence, the three- to four-week mating periods overlapped almost completely.

Overlap of mating periods leads to interactions between individuals from different hosts, at least in experimental populations in close proximity. In a study by Wood (1980), over 40% of courtships were between males and females of different species. Strikingly, however, only 6% of the matings were between heterospecifics. The common occurrence of mixed-species courtship, coupled with the rarity of mismating, suggested that behavioral isolation resulting from courtship interactions may be even stronger than ecological isolation. Dispersal between host plants by mate-searching males occurs not only in cage experiments, but also in the field. For example, in one sample of individuals collected during the breeding season on *Cercis canadensis* in Missouri, 88% of the males were *E. binotata* 'Cercis,' while 12% of the males were from another host-associated species in the complex (R.B.C., unpublished data). (*Note:* The species in this complex have not yet been formally named, and here we refer to them using the name of their host plants.)

We suggest that behavioral isolation arising from differences in sexual communication systems may act synergistically with allochrony and host fidelity to promote isolation and facilitate divergence among host-shifted populations. That is, while the isolation that arises as a consequence of differences in host phenology is clearly important in initiating divergence, the evolution of behavioral isolation may be important for completing the process of speciation (Rundle and Nosil 2005).

In the following sections, we examine the relationships between host shifts, sexual communication, and assortative mating in the *E. binotata* complex. We suggest that sexual communication can play several important roles. First, in the early stages of a host shift, a number of aspects of female choice and male signaling behavior may contribute to assortative mating even before any divergence in signals or preferences has occurred. Second, host shifts are likely to promote the diversification of sexual communication systems, contributing to the speciation process and/or facilitate

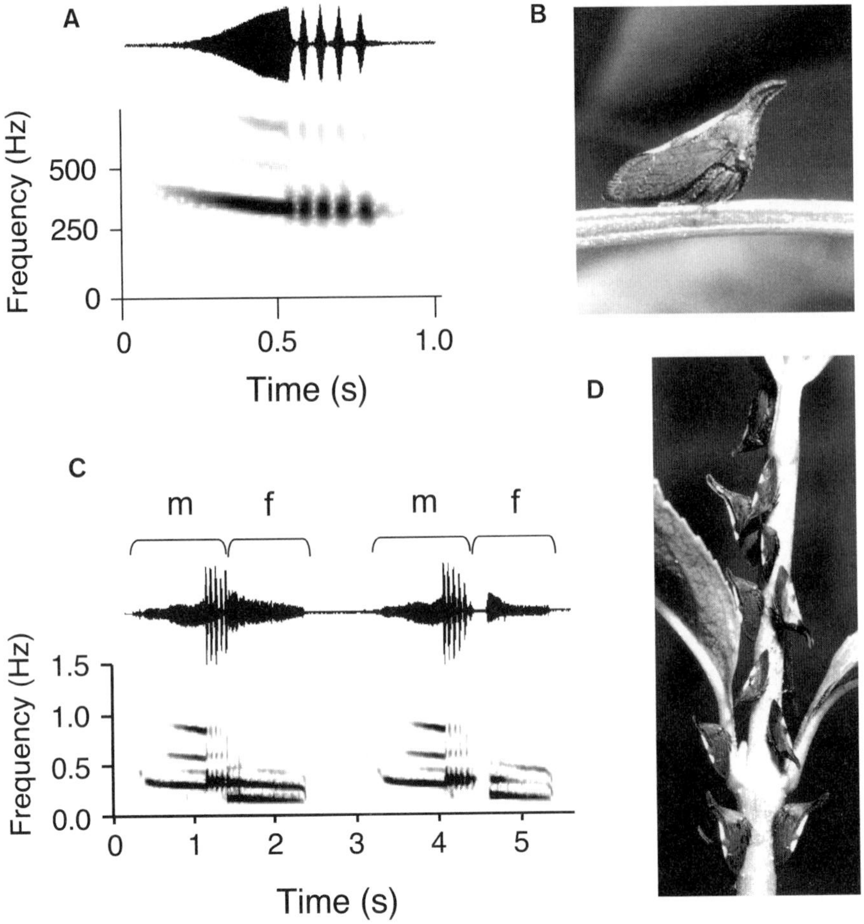

FIGURE 7.1. A. Signal of a male *Enchenopa binotata* 'Ptelea' (waveform *[above]* and spectrogram), showing the typical whine-and-pulses structure. B. Solitary mate-searching male of *E. binotata* 'Robinia.' C. Portion of a male-female duet in *E. binotata* 'Viburnum' (waveform and spectrogram). D. Aggregation of male and female *E. binotata* 'Viburnum'; grouped males often chorus for extended periods. Photo in part B by R. B. Cocroft, photo in part D by C. P. Lin, with permission.

species coexistence. Third, female preferences for male signal traits provide not only a cause of assortative mating, but also a source of divergent selection. Finally, female choice for condition-dependent male signal traits may accelerate the process of adaptation to novel host plants.

Behavioral Sources of Assortative Mating

Evidence for Behavioral Isolation

What is the evidence that sexual communication systems contribute to behavioral isolation among species in *E. binotata* complex? We first describe the nature of the communication that takes place between males and females. We then review the evidence that signals contribute to reproductive isolation in *Enchenopa* and describe patterns of variation in signal traits in relation to host use and geography.

The first evidence that behavioral interactions during courtship contribute to mate recognition came from the studies described above (Wood 1980; Wood and Guttman 1982), which showed that although males frequently courted heterospecific females when species from different hosts were placed in close proximity, mismatings were rare. Insight into the mechanisms underlying female mate choice was provided by Hunt (1994), whose study of *E. binotata* 'Cercis' revealed that males court females using substrate-borne vibrational signals that travel through the stems and petioles of their host plants.

For all species in the *E. binotata* complex, the male advertisement signal consists of a tone that drops in frequency (the "whine"), followed by a series of pulses (Fig. 7.1A). Males produce bouts of two or more signals in close succession. Signaling often occurs within a mate-searching strategy in which males move between a series of stems or plants, signaling on each one and waiting for a female response (Fig. 7.1B). When a female alternates her own response signals with those of the male (Fig. 7.1C), the male searches locally while continuing to duet. When the male finds the female, he climbs on her side; mating begins when the female allows the male to couple his genitalia, climb down from her side, and face away from her. At this point duetting ceases. As an alternative to active searching, males may remain in one location and chorus with other males, usually in the presence of one or more females (Fig. 7.1D). The lack of continued courtship during copulation, together with the striking similarity of male genitalia across the complex (Pratt and Wood 1993), suggests that premating sexual communication is the primary source of behavioral isolation in the *E. binotata* complex.

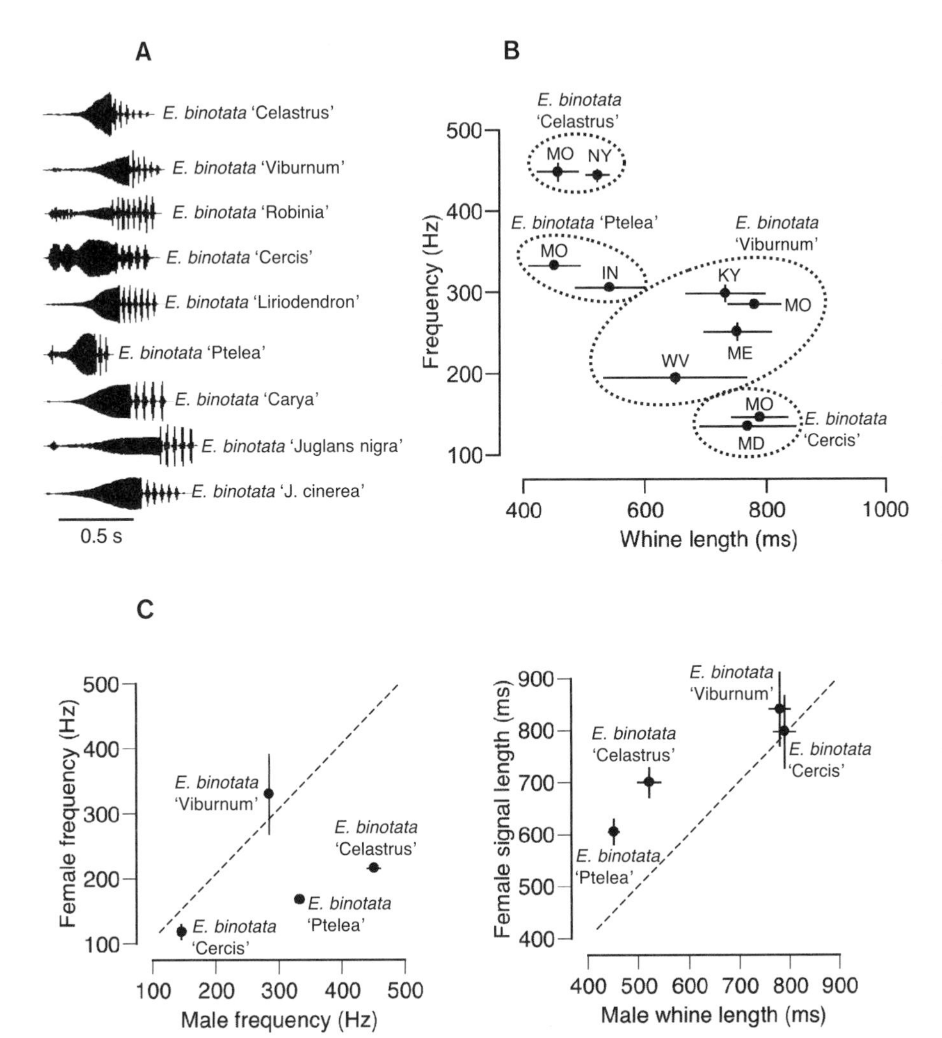

FIGURE 7.2. A. Waveforms of male signals for nine species in the *Enchenopa binotata* complex, showing similarity in signal structure. B. Variation in the frequency and whine length of male signals for four species in the *E. binotata* complex (U.S. state abbreviations indicate recording localities, which for each species span >600 km). C. Correspondence between the frequency and length of male and female signals for four species from Missouri. Dashed lines indicate a 1:1 relationship. Shown are mean signal values ±1 SE. Part C modified from Rodríguez and Cocroft (2006), with permission from Blackwell Publishing.

As a first step in evaluating the role of vibrational mating signals in behavioral isolation in the *E. binotata* complex, we presented receptive females of one species (*E. binotata* 'Viburnum') with playbacks of vibrational mating signals of males from their own population, as well as mating signals from sympatric populations of six other species. Females readily engaged in duets with male signals from their own population, but rarely with those from closely related species within the complex (Rodríguez et al. 2004). Female responses were also influenced by individual male identity within species, suggesting the presence of sexual selection. In a similar playback experiment, females of *E. binotata* 'Cercis' also responded to mating signals only of conspecific males (R.E.H., unpublished data). Females thus choose among potential mates on the basis of between-species and between-individual signal variation, and their decision to respond to a male will influence the likelihood of being found by him.

Although playback experiments with recorded natural signals revealed that signal differences can contribute to reproductive isolation among current-day species in the *E. binotata* complex, they provided only indirect evidence of the signal traits important for female choice. Furthermore, they provided no information on how female preferences differ among species, and whether female choice may exert divergent sexual selection on male signals. Our next steps, then, were to characterize variation in signal traits within and between species, and to characterize variation in female preferences for male signals among species.

Variation in Male and Female Signals

The evolutionary processes acting on communication systems in a group of closely related species result in a pattern of variation in signals and preferences among those species. Describing this variation can shed light on the processes that produced it; for example, marked differences in signals and preferences among closely related species, along with variation in the same traits among geographically separated populations within species, are "signatures" of the role of sexual selection in speciation (Panhuis et al. 2001). Describing geographic variation is less difficult for male signals than for female preferences and is the first step we have taken here.

Male advertisement signals are qualitatively similar among species in the *E. binotata* complex (Fig. 7.2A). However, signals differ quantitatively among species in multiple

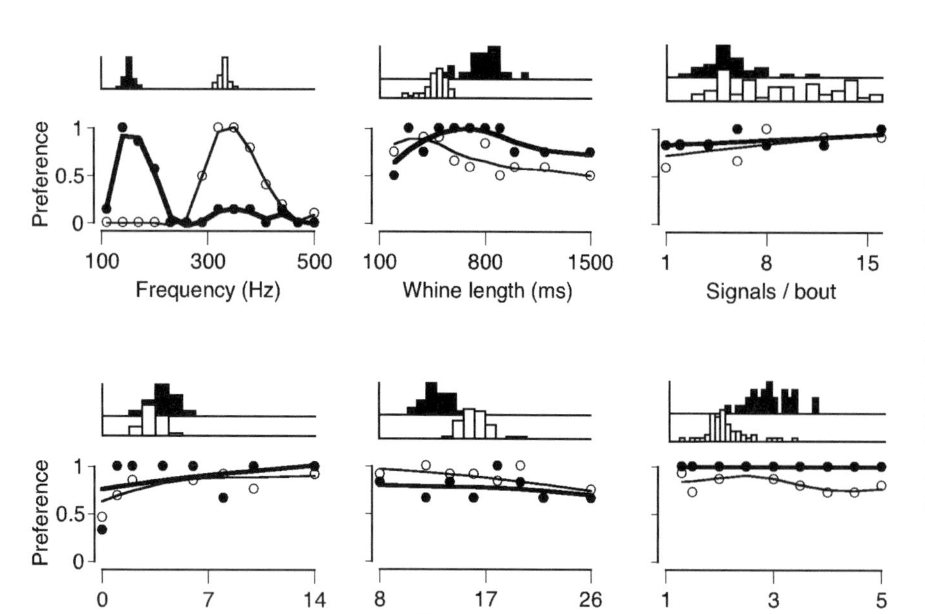

FIGURE 7.3 Variation in female preferences and the distribution of male signal traits in four species in the *Enchenopa binotata* complex. The curves and symbols indicate female preferences, and histograms indicate the distribution of male signal traits. Filled symbols and bars indicate *E. binotata* 'Cercis,' open symbols and bars indicate *E. binotata* 'Ptelea.' See Rodríguez et al. (2006) for a detailed analysis of preference-trait relationships.

characteristics, especially in frequency and length (Fig. 7.2B), but also in the number and rate of the pulses that follow the whine. In general, signal variation among species on different host plants is greater than geographic variation within species on a single host plant. As an example, here we show variation in signal frequency and length among populations of *E. binotata* 'Celastrus,' 'Cercis,' 'Ptelea,' and 'Viburnum' in eastern North America (Fig. 7.2B). Note that for three of the species, there are substantial differences in frequency between species, but little variation among geographically separated populations within a species. This pattern is typical of most other species in the complex (R.B.C., R.L.R., and R.E.H., unpublished), but there is one notable exception: for the species on *Viburnum,* geographic variation in mating signals is comparable to that between species (Fig. 7.2B). In general, there is a close association between speciation, host use, and signal divergence in the *E. binotata* complex. Given the consistent differences among signals of species on different hosts, the female preferences for the signals of conspecific males shown by Rodríguez et al. (2004) should lead to behavioral isolation among most species throughout their range.

In the duetting communication system of the *E. binotata* complex, species differences in female response signals may be as important as those in male signals. As with male advertisement signals, species differ in the frequency and length of female response signals (Fig. 7.2C) (Rodríguez and Cocroft 2006). Species also differ in the timing relationship between male and female signals during duetting interactions. Although we have not examined geographic variation in female signals, comparisons based on one population each from four species show that male and female signal traits covary (Fig. 7.2C). This pattern suggests that signal evolution in both sexes is correlated and/or influenced by shared or similar sources of selection. These sources of selection may be preferences for signal traits in both sexes, or selection on signal transmission exerted by the filtering properties of host plants (see "Sensory Drive and Signal Divergence," below). Alternatively, the signals of one sex may show a correlated response to selection on signals of the other.

Understanding signal evolution in the *E. binotata* complex will require characterizing these potential sources of selection on male and female signals. We begin by assessing the contribution of female preferences to male signal divergence, namely, the contribution of selection exerted by female choice.

Variation in Female Preferences

We have characterized female preferences for male signal traits in four *E. binotata* species (Rodríguez et al. 2006). For this study, we used female response signals as an assay of their preferences for computer-generated male signals. From the pattern of female responses we generated preference functions, which show the probability of response as a function of variation in the male trait. For some signal traits, female preference functions were closed (i.e., females preferred intermediate values), while for other traits preferences were open ended (females preferred higher or lower values). Between species, there were differences in the shape, preferred value, and strength (i.e., how strongly females discriminated against unattractive signals) of preferences for the same signal traits (Fig. 7.3).

Species differences in female preferences suggest that divergent sexual selection exerted by female choice may be responsible for differences in male signals across the *E. binotata* complex. We tested the predictions of this hypothesis by taking advantage of between-species variation in the shape and strength of female preferences for different signal traits (Rodríguez et al. 2006). Briefly, for closed preferences the most preferred value should predict the mean of the male signal trait, with the match between preferred value and mean signal trait being greater for stronger preferences. By contrast, for open preferences the male trait values should be shifted toward the preferred extreme of the range. In addition, there should be a greater variability in signal traits for open than for closed preferences. We tested these predictions in four species in the *E. binotata* complex and found support for all of them (Rodríguez et al. 2006). These results point to female mate choice as an important agent of signal evolution in this complex.

Comparison of female preferences and the distribution of male signal traits can be used not only to infer past selection on male signals, but also to formulate hypotheses about the nature of current selection on male signals exerted by female choice. These hypotheses should ideally be tested within the framework of a comprehensive understanding of *E. binotata* mating systems and determinants of mating success, to which we now turn.

Mating Systems

Enchenopa mating systems can be described as a cursorial polygyny (Shuster and Wade 2003), in which reproductive females form aggregations, and males move between groups of females. In general, *E. binotata* females mate once (only about 5% mate a second time), while males may mate multiply (Wood 1993).

Measures of the relative variance in mating success between males and females reveal a high opportunity for sexual selection (L. E. Sullivan and R.B.C., unpublished). The large variance in male mating success is generated in part by the relatively asynchronous schedule of female receptivity: in an outdoor enclosure experiment in which life-history timing was close to that in the field, typically only a few females mated each day. Most courtship and mating occurred during a relatively narrow time window during the morning (activity ceased during the heat of the day), and males rarely mated more than once in a day. As a result, certain males had the opportunity to mate with multiple females, at the expense of other males. Preliminary results indicate that the number of signals a male produces—a potential reflection of condition—is important in mate choice under natural conditions.

The contribution of sexual communication to reproductive isolation and divergence may be increased by interactions with other behavioral sources of assortative mating, and with effects of host plants on signal variation. We next consider the consequences of mate choice in the context of shifts to novel host plants.

Communication in a New Host Environment

A new host plant represents a novel environment with potentially far-reaching effects on the function and evolution of communication and mating systems (Funk et al. 2002). The idea that changes in host-plant use can promote differences in mating signals is supported by species using chemical cues in mate choice (Landolt and Phillips 1997), where differences may arise either as a pleiotropic effect of host-plant adaptation (Funk 1998) or as a direct consequence of developing on a novel host (Etges 1998). Communication systems using substrate-borne signals might be expected to be especially sensitive to changes in host-plant use, given that host-plant tissues constitute the transmission channel. Here we examine how the communication system may play a role in assortative mating in the early stages of a host shift, and how traits involved in communication may evolve in response to differences in sexual and natural selection between hosts.

Substrate-Related Signal Variation

In considering the effect of a host shift on the evolution of vibrational communication systems, the first issue that arises is how an individual's signals may differ when produced on different plant species. Our finding is that when the same males are recorded on a host and a nonhost, the spectral and fine-temporal characteristics of their vibrational signals are unchanged. We have confirmed this result in two independent experiments with *Enchenopa*, using various host-plant and non-host-plant species (Sattman and Cocroft 2003; Rodríguez et al., in press), and in another experiment with *Umbonia* treehoppers (Cocroft et al. 2006). Thus, signal differences between species on different hosts are not replicated by simply moving individuals between those hosts. The constancy of *Enchenopa* mating signals across substrates is in part a consequence of the use of relatively pure tones (Sattman and Cocroft 2003).

Although the signals of male *E. binotata* retain their basic structure across plant substrates, males do reduce their signaling effort on nonhosts. In an experiment in which male *E. binotata* 'Ptelea' were recorded both on their own host and on a nonhost, the males produced fewer signals in a bout on the nonhost, and the signals they did produce were shorter (Sattman and Cocroft 2003). We view these changes in signaling behavior as host fidelity in mate searching, and they may influence the probability of mating between individuals on different hosts, because both changes will reduce a male's attractiveness (Fig. 7.3) (Rodríguez et al. 2006). Female preferences for male signal traits may thus interact with host fidelity to generate assortative mating in the early stages of a host shift, whether or not any divergence in sexual communication systems has occurred.

Origins of Diversity in Sexual Communication

Insect populations colonizing a novel host plant might acquire differences in signal or preference traits through developmental plasticity (Etges 2002; West-Eberhard 2003,

2005). We evaluated the role of plasticity in signal divergence in the *E. binotata* complex by rearing full-sib families of one species, *E. binotata* 'Ptelea,' on two different host-plant species: their native host and the host of another species in the complex *(Robinia pseudoacacia)*. The second host was chosen because it provides a relatively benign rearing environment, increasing the sample size of individuals reaching adulthood. Preliminary analysis of this experiment (Rodríguez et al., in press) indicates that overall host-plant effects have a small role in generating signal variation. However, there is substantial genetic variation in, as well as diversity in the reaction norms of, most signal traits. Indeed, full-sib families often differed so much in their signal reaction norms that their attractiveness ranking is expected to change across plant species, even if female preferences do not show similar plasticity. Such reaction norm diversity has the potential to alter selection regimes (Lynch and Walsh 1998; Rodríguez and Greenfield 2003; Greenfield and Rodríguez 2004), fostering divergent selection. Phenotypic plasticity in communication systems thus has the potential to promote divergence following a host shift.

In addition to potential changes in the regimes of sexual selection arising from developmental plasticity, host shifts may change selection on the communication system as a consequence of differences in the communication environment, a possibility we examine in the next section.

Sensory Drive and Signal Divergence

Because each host-plant species provides a unique environment for vibrational communication, use of different hosts has the potential to impose divergent natural selection on signals. The sensory drive hypothesis (Endler and Basolo 1998) predicts that when populations occupy environments that differ in their effects on signal transmission and perception, they will experience divergent natural selection on long-range communication systems (see Boughman 2001; Leal and Fleishman 2004; Maan et al. 2006).

Plant stems and leaves impose a frequency filter on vibrational signals (Michelsen et al. 1982; Cokl and Virant-Doberlet 2003; Cocroft and Rodríguez 2005). This phenomenon is easily observed by introducing a broadband noise signal with equal energy across a range of frequencies into a plant stem: the same signal recorded at various distances from the source will contain very unequal energy in different frequencies, with frequencies that transmit efficiently having a substantially higher amplitude. Host plants of closely related species can differ greatly in physical structure and have the potential to impose very different filters on the vibrational signals transmitted through them. Such differences among plant species would provide a source of divergent natural selection on signal frequency, with potentially important consequences for signal evolution and assortative mating.

The sensory drive hypothesis for vibrational signals predicts that if plant species differ predictably in their transmission characteristics, insects will use signals that transmit efficiently through tissues of their host. Support for the matching of insect songs to plant transmission characteristics comes from studies of vibrational communication in green stinkbugs, *Nezara viridula,* which use a variety of different host plants. Cokl et al. (2005) found that the low-frequency songs of *N. viridula* are transmitted efficiently in some of their commonly used hosts. Addressing the hypothesis that sensory drive leads to signal divergence, however, requires comparison among closely related species using different hosts. Henry and Wells (2004) tested for signal-environment matching in two green lacewing species *(Chrysoperla),* one inhabiting conifers and the other inhabiting a variety of meadow plants. The authors examined degradation during signal propagation using a hemlock and a grass species, but found no evidence of differential signal transmission.

Enchenopa treehoppers provide a powerful test of the sensory drive hypothesis for vibrational signals, because most species in the complex use only a single species of host plant, unlike the species studied by Henry and Wells (2004) and Cokl et al. (2005), which use a range of different hosts. Furthermore, the most important signal feature for assortative mating between *E. binotata* species on different hosts is frequency (Rodríguez et al. 2006), which could be under strong selection if plants differ in the efficiency with which they transmit signals of different frequencies.

We have measured the signal transmission properties of two of the host-plant species used by members of the *E. binotata* species complex (G. D. McNett and R.B.C., unpublished). For *E. binotata* 'Cercis,' the signal frequency used by males transmits with less attenuation than other frequencies in the range of the species complex. Male *E. binotata* 'Ptelea' also use a frequency that transmits well in their host plant, although substrate filtering in 'Ptelea' is weaker. Filtering properties of host-plant stems may thus impose divergent selection on the frequency of male vibrational signals. Because signal frequency is so important in mate choice, it may be a "magic trait" (Gavrilets 2004): one that greatly facilitates speciation because it is involved in both adaptation to the novel environment and assortative mating. Filtering properties of additional hosts are currently being investigated.

Environmental differences that can influence signal evolution are not limited to transmission characteristics. Predators can directly influence the evolution of mating systems and possibly facilitate speciation in cases where they exploit sexual signals (Bradbury and Vehrencamp 1998; Endler and Basolo 1998; Zuk and Kolluru 1998; Bailey 2006). Different host plants may harbor different predators or parasitoids, with a concomitant change in natural selection on communication systems. Environmental sources of noise that can interfere with signal reception are also an important influence on the evolution of communication systems (Endler

Jiggins, C. D., and J. Mallet. 2000. Bimodal hybrid zones and speciation. Trends Ecol. Evol. 15: 250–255.

Jiggins, C. D., I. Emelianov, and J. Mallet. 2005. Assortative mating and speciation as pleiotropic effects of ecological adaptation: examples in moths and butterflies, pp. 451–473. In M. Fellowes, G. Holloway, and J. Rolff (eds.), Insect evolutionary ecology. Royal Entomological Society, London.

Kelly, J. K., and M. A. F. Noor. 1996. Speciation by reinforcement: a model derived from studies of *Drosophila*. Genetics 143: 1485–1497.

Kirkpatrick, M., and V. Ravigné. 2002. Speciation by natural and sexual selection: models and experiments. Am. Nat. 159: S23–S35.

Kokko, H., and J. Lindstrom. 1996. Evolution of female preference for old mates. Proc. R. Soc. Lond. B 263: 1533–1538.

Kondrashov, A. S., and F. A. Kondrashov. 1999. Interactions among quantitative traits in the course of sympatric speciation. Nature 400: 351–354.

Kondrashov, A. S., L. Y. Yampolsky, and S. A. Shabalina. 1998. On the sympatric origin of species by means of natural selection, pp. 90–98. In D. J. Howard and S. H. Berlocher (eds.), Endless forms: species and speciation. Oxford University Press, New York.

Landolt, P. J., and T. W. Phillips. 1997. Host plant influences on sex pheromone behavior of phytophagous insects. Annu. Rev. Entomol. 42: 371–391.

Leal, M., and L. J. Fleishman. 2004. Differences in visual signal design and detectability between allopatric populations of *Anolis* lizards. Am. Nat. 163: 26–39.

Lorch, P. D., S. Proulx, L. Rowe, and T. Day. 2003. Condition-dependent sexual selection can accelerate adaptation. Evol. Ecol. Res. 5: 867–881.

Lukhtanov, V. A., N. P. Kandul, J. B. Plotkin, A. V. Dantchenko, D. Haig, and N. E. Pierce. 2005. Reinforcement of pre-zygotic evolution in *Agrodiaetus* butterflies. Nature 436: 385–389.

Lynch, M., and B. Walsh. 1998. Genetics and analysis of quantitative traits. Sinauer, Sunderland, MA.

Maan, M. E., K. D. Hofker, J. J. M. van Alphen, and O. Seehausen. 2006. Sensory drive in cichlid speciation. Am. Nat. 167: 947–954.

Mallet, J. 2001. The speciation revolution. J. Evol. Biol. 14: 887–888.

Marshall, J. L., M. L. Arnold, and D. J. Howard. 2002. Reinforcement: the road not taken. Trends Ecol. Evol. 17: 558–563.

Mayr, E. 1963. Animal species and evolution. Belknap Press, Cambridge, MA.

Messina, F. J. 2004. Predictable modification of body size and competitive ability following a host shift by a seed beetle. Evolution 58: 2788–2797.

Michelsen, A., F. Fink, M. Gogala, and D. Traue. 1982. Plants as transmission channels for insect vibrational songs. Behav. Ecol. Sociobiol. 11: 269–281.

Miyatake, T., and T. Shimizu. 1999. Genetic correlations between life-history and behavioral traits can cause reproductive isolation. Evolution 53: 201–208.

Nosil, P., B. J. Crespi, and C. P. Sandoval. 2003. Reproductive isolation driven by the combined effects of ecological adaptation and reinforcement. Proc. R. Soc. Lond. B 270: 1911–1918.

Panhuis, T. M., R. Butlin, M. Zuk, and T. Tregenza. 2001. Sexual selection and speciation. Trends Ecol. Evol. 16: 364–371.

Pratt, G., and T. K. Wood. 1993. Genitalic analysis of males and females in the *Enchenopa binotata* (Say) complex (Membracidae: Homoptera). Proc. Entomol. Soc. Wash. 95: 574–582.

Price, P. W. 2002. Resource-driven terrestrial interaction webs. Ecol. Res. 17: 241–247.

Proulx, S. R. 1999. Mating systems and the evolution of niche breadth. Am. Nat. 154: 89–98.

Proulx, S. R. 2001. Female choice via indicator traits easily evolves in the face of recombination and migration. Evolution 55: 2401–2411.

Proulx, S. R., T. Day, and L. Rowe. 2002. Older males signal more reliably. Proc. R. Soc. Lond. B. 269: 2291–2299.

Rice, W. R., and E. E. Hostert. 1993. Laboratory experiments on speciation: what have we learned in 40 years? Evolution 27: 1673–1653.

Rodríguez, R. L., and R. B. Cocroft. 2006. Divergence in female duetting signals in the *Enchenopa binotata* species complex of treehoppers (Hemiptera: Membradicae). Ethol. 112: 1231–1238.

Rodríguez, R. L., and M. D. Greenfield. 2003. Genetic variance and phenotypic plasticity in a component of female mate choice in an ultrasonic moth. Evolution 57: 1304–1313.

Rodríguez, R. L., L. E. Sullivan, and R. B. Cocroft. 2004. Vibrational communication and reproductive isolation in the *Enchenopa binotata* species complex of treehoppers (Hemiptera: Membracidae). Evolution 58: 571–578.

Rodríguez, R. L., K. Ramaswamy, and R. B. Cocroft. 2006. Evidence that female preferences have shaped male signal evolution in a clade of specialized plant-feeding insects. Proc. R. Soc. B. 273: 2585–2593.

Rodríguez, R. L., L. M. Sullivan, R. L. Snyder, and R. B. Cocroft. In press. Host shifts and the beginning of signal divergence. Evolution.

Rundle, H. D., and P. Nosil. 2005. Ecological speciation. Ecol. Lett. 8: 336–352.

Ryan, M. J., and A. S. Rand. 1993. Species recognition and sexual selection as a unitary problem in animal communication. Evolution 47: 647–657.

Sattman, D. A., and R. B. Cocroft. 2003. Phenotypic plasticity and repeatability in the mating signals of *Enchenopa* treehoppers, with implications for reduced gene flow among host-shifted populations. Ethol. 109: 981–994.

Servedio, M. R., and M. A. F. Noor. 2003. The role of reinforcement in speciation: theory and data. Annu. Rev. Ecol. Syst. 34: 339–364.

Shuster, S. M., and M. J. Wade. 2003. Mating systems and strategies. Princeton University Press, Princeton, NJ.

Tauber, C. A., and M. J. Tauber. 1989. Sympatric speciation in insects: perception and perspective, pp. 307–344. In D. Otte and J. A. Endler (eds.), Speciation and its consequences. Sinauer, Sunderland, MA.

Tilmon, K. J., T. K. Wood, and J. D. Pesek. 1998. Genetic variation in performance traits and the potential for host shifts in *Enchenopa* treehoppers (Homoptera: Membracidae). Ann. Entomol. Soc. Am. 91: 397–403.

Tomkins, J. L., J. Radwan, J. S. Kotiaho, and T. Tregenza. 2004. Genic capture and resolving the lek paradox. Trends Ecol. Evol. 19: 324–328.

Tregenza, T., V. L. Pritchard, and R. K. Butlin. 2000. Patterns of trait divergence between populations of the meadow grasshopper, *Chorthippus parallelus*. Evolution 54: 574–585.

Via, S. 1990. Ecological genetics in herbivorous insects: the experimental study of evolution in natural and agricultural systems. Annu. Rev. Entomol. 35: 421–446.

West-Eberhard, M. J. 1983. Sexual selection, social competition, and speciation. Q. Rev. Biol. 58: 155–183.

West-Eberhard, M. J. 2003. Developmental plasticity and evolution. Oxford University Press, New York.

West-Eberhard, M. J. 2005. Developmental plasticity and the origin of species differences. Proc. Natl. Acad. Sci. USA 102: 6543–6549.

Wood, T. K. 1980. Divergence in the *Enchenopa binotata* Say complex (Homoptera: Membracidae) effected by host plant adaptation. Evolution 34: 147–160.

Wood, T. K. 1993. Speciation of the *Enchenopa binotata* complex (Insecta: Homoptera: Membracidae), pp. 299–317. In D. R. Lees and D. Edwards (eds.), Evolutionary patterns and processes. Academic Press, New York.

Wood, T. K., and S. I. Guttman. 1982. Ecological and behavioral basis for reproductive isolation in the sympatric *Enchenopa binotata* complex (Homoptera: Membracidae). Evolution 36: 233–242.

Wood, T. K., and S. I. Guttman. 1983. *Enchenopa binotata* complex: sympatric speciation? Science 220: 310–312.

Wood, T. K., and M. C. Keese. 1990. Host-plant-induced assortative mating in *Enchenopa* treehoppers. Evolution 44: 619–628.

Wood, T. K., K. L. Olmstead, and S. I. Guttman. 1990. Insect phenology mediated by host-plant water relations. Evolution 44: 629–636.

Wood, T. K., K. J. Tilmon, A. B. Shantz, C. K. Harris, and J. D. Pesek. 1999. The role of host-plant fidelity in initiating insect race formation. Evol. Ecol. Res. 1: 317–332.

Zuk, M., and G. R. Kolluru. 1998. Exploitation of sexual signals by predators and parasitoids. Q. Rev. Biol. 73: 415–438.

reached the sphere when hawthorn volatiles were present, and even fewer (~11%) when the source sphere was baited with the dogwood blend (Fig. 8.1A). Hawthorn and dogwood flies displayed similar patterns (Fig. 8.1B, 8.1C, 8.1D). A total of 73% of hawthorn flies reached the sphere when it was seeded with the hawthorn blend (N = 277), compared to 5% for apple-baited and 17% for dogwood-blend spheres. Likewise, around 76% of dogwood flies reached spheres baited with their natal dogwood blend (N = 189), while 12% of flies reached apple spheres and 20% hawthorn spheres. One interesting finding was that a fair proportion of both hawthorn and dogwood flies initially took flight to each other's fruit volatiles in the tunnel (Fig. 8.1B, 8.1C, 8.1D), possibly due to chemical similarities in the hawthorn and dogwood blends (Linn et al. 2005a). However, upwind flight was quickly arrested for hawthorn and flowering dogwood flies, with the majority of individuals not reaching nonnatal odor spheres (Fig. 8.1B, 8.1C, 8.1D).

The flight tunnel results were nearly identical for three pairs of apple- and hawthorn-fly populations tested from Grant, Michigan, Fennville, Michigan, and Urbana, Illinois, as well as for a laboratory colony of Geneva, New York, apple flies established from the wild in the 1970s (Linn et al. 2003). Thus, the host races showed a consistent preference for their natal blend regardless of geographic location. Moreover, Urbana, Illinois, hawthorn flies reared for two generations in the lab on apple fruit displayed the same behavioral responses as hawthorn flies reared directly from field-collected haws, discounting an effect of the larval host-fruit environment on adult fly behavior (Linn et al. 2003).

Some geographic variation was observed, however, for dogwood flies among three populations surveyed from Granger, Indiana, Raccoon Lake, Indiana, and Byron, Georgia (Linn et al. 2005a). The dogwood-fly populations from the two sites in Indiana showed similar levels of fruit-odor discrimination between natal and nonnatal blends as displayed by the apple and hawthorn host races (Fig. 8.1C). However, dogwood flies from the southern Byron, Georgia, population, while preferring their natal dogwood blend, appeared to be less discriminatory than the other flies (65% of Byron flies reached dogwood-blend spheres, ~40% hawthorn spheres, and ~25% apple spheres; N = 64) (Fig. 8.1D). The cause for the reduced specificity of flowering dogwood flies at the Byron, Georgia, site remains to be resolved but could involve increased gene flow from local hawthorn-fly populations.

Analysis of the orientation behavior of individual flies in flight tunnel tests to all three volatile blends presented in succession revealed population-level variation in response patterns (Linn et al. 2005a). As expected, apple, hawthorn, and dogwood flies showed maximal response to their respective natal fruit blend, with significant decreases in their responses to nonnatal blends. However, some flies of each taxon did respond to nonhost volatiles, and further analysis showed that in almost every case, these flies also responded to their natal blend. Indeed, within parental apple-, hawthorn-, and dogwood-fly populations an average of about 60% of flies reached natal-blend spheres only, 15 to 20% both natal- and nonnatal-blend spheres, 3% nonnatal spheres only, and 15 to 20% did not respond to any blend (see Fig. 8.2A, 8.2B for results for apple and hawthorn flies). Thus, there appears to be in each host population a proportion of flies with broad response specificity to multiple hosts. Linn et al. (2005a) proposed that these flies could be a source for host shifting, as they would not require a specific volatile mix to find a host on which to mate or in which to oviposit.

When tested at the higher 2000 μg dose, apple, hawthorn, and dogwood flies all displayed dramatically reduced levels of completed flight to their natal fruit blend (Linn et al. 2004). Although almost all flies tested took flight from the release stand, only 2 dogwood flies out of 46 tested from Granger, Indiana, reached the source sphere when it was baited with the dogwood blend at the higher dose. Moreover, no apple (N = 44) or hawthorn fly (N = 34) assayed from Grant, Michigan, reached the 2000 μg baited source sphere. The test results indicate that parental flies initially oriented to the 2000 μg dose of their natal fruit blend, but their upwind flight was arrested by high amounts of volatiles in the odor plume. It is possible that behavioral antagonism to high-blend dose could be a consequence of flies discriminating against overripe fruit, suboptimal for larval survivorship.

Field Trials of Fruit Blends Demonstrate Both Preference and Avoidance

Field trials conducted in the 2002 and 2003 field seasons confirmed the relevance of the flight tunnel results to nature. In 2002, three-way choice experiments involving triads of sticky sphere traps triangulated 1 m apart and baited with the apple blend or hawthorn blend, versus no odor were performed at paired apple orchards and hawthorn tree stands in Fennville, Michigan, and Geneva, New York. At both the Michigan and New York sites, resident apple and hawthorn flies were preferentially captured on the traps baited with their natal blend (Linn et al. 2003).

The 2002 studies were expanded upon in 2003 by testing resident flies for both their attraction to and avoidance of apple, hawthorn, and dogwood blends (Forbes et al. 2005). The basic experimental design in 2003 was a two-way choice experiment between red, sticky spheres baited with a given host-fruit blend versus a blank, odorless control sphere positioned 1 m away in the tree canopy. Tests were performed at the same apple orchard and hawthorn copse in Fennville, Michigan, at which the 2002 tests were conducted, as well as two flowering dogwood tree stands near Cassopolis, Michigan, and Granger, Indiana. Once again, resident apple, hawthorn, and dogwood flies were captured significantly more often on their natal blends versus blank

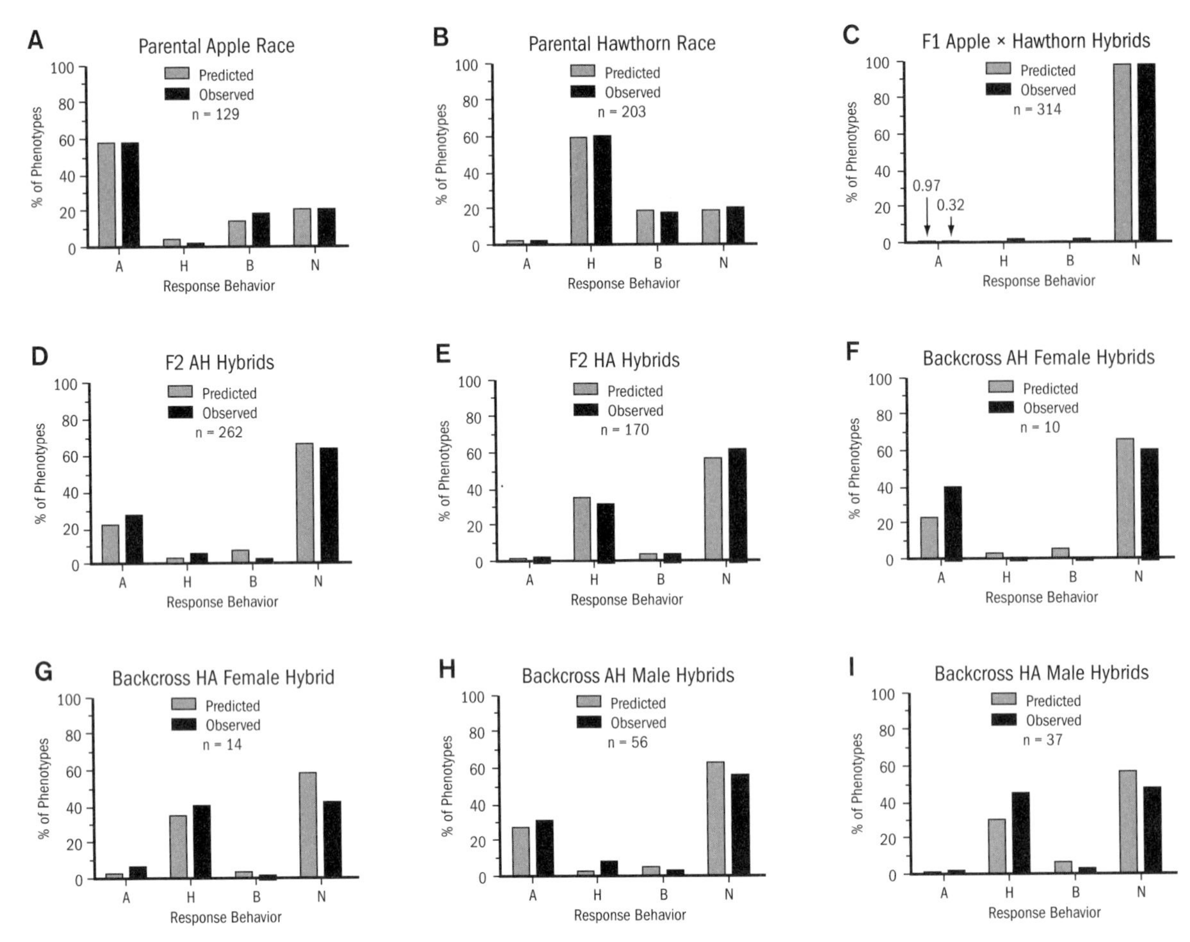

FIGURE 8.2. Predicted response patterns for the maximum likelihood (ML) solution of the four-component genetic model (gray bars) plotted against the observed percentages (dark bars) in 200 μg dose flight tunnel tests. Shown are the responses for parental apple and hawthorn flies, and F_1, F_2, and backcross hybrids reaching apple blend only (A), hawthorn blend only (H), both apple and hawthorn blend (B), and neither apple nor hawthorn blend (N) baited spheres. Maximum likelihood estimates obtained for allele frequencies and gametic disequilibrium values between the *P* preference and *A1* avoidance locus in the model for the apple race were $Pah^+ = 0.489$, $A1a^- = 0.761$, and $r = 0.339$ (excess of $Pah^+ / A1a^-$ and $Ph^+/A1h^-$ gametes), and for the hawthorn race $Pah^+ = 0.220$, $A1a^- = 0.343$, and $r = -0.371$ (excess of $Ph^+/A1a^-$ and $Pah^+/A1h^-$ gametes).

spheres (>60% on natal-blend spheres versus paired blank spheres; Forbes et al. 2005). In contrast, resident flies were captured significantly less often on nonnatal blends versus blank controls (from 25 to 39% capture on nonnatal versus blank control spheres). Thus, not only do flies preferentially orient to their natal fruit odor, but they also avoid nonnatal volatiles. It is important to note that values from the two-way trapping experiments likely underestimate the level of fruit-odor discrimination in the field. The red spheres used to capture flies are a visual attractant to flies. Thus, increased capture rates for natal blends represent the degree to which flies are attracted to host-fruit volatiles above and beyond the extent to which they are drawn to the visual stimulus of the red sphere. In contrast, capture rates for nonnatal blends represent the degree to which flies are antagonized by fruit volatiles from accepting an inherently positive visual stimulus.

Our two-way field studies also showed that preference and avoidance behaviors by *R. pomonella* flies extend to certain individual volatile compounds, as well as to entire fruit blends (Forbes et al. 2005). However, not every natal or nonnatal compound affected fly capture. For example, hawthorn flies did not appear to orient to or avoid butyl hexanoate, a minor component in the hawthorn blend and a major attractant to apple flies. Moreover, hawthorn flies were not preferentially trapped on 3-methylbutan-1-ol–baited spheres in the field. However, 3-methylbutan-1-ol was shown to be an essential component for enhancing the upwind flight in flight tunnel tests when included in the hawthorn blend (Nojima et al. 2003a).

It is tempting to speculate that the lack of response of hawthorn flies to certain key ingredients of the apple and dogwood blends may have predisposed *R. pomonella* to shift to these host fruits. Menken and Roessingh (1998) have suggested that such a mechanism may exist for host shifts wherein the presence of low levels of a particular compound in the natal host may facilitate a shift to a novel host that has the same component chemical.

However, in the case of butyl hexanoate, it is not clear whether the lack of response is the ancestral condition for hawthorn flies or is due to ongoing gene flow from sympatric apple fly populations. Also, it is not certain whether hawthorn flies shifted to flowering dogwood or vice versa. In addition, while hawthorn flies may not avoid certain individual compounds in the apple and dogwood blends, in our field tests they did tend to avoid the apple and dogwood blends in their entireties. Thus, the relevance of the hawthorn race's indifference to butyl hexanoate and 3-methylbutan-1-ol for facilitating host shifts remains to be determined. Further testing of individual compounds and subcomponent mixes, as well as determination of the physiological and genetic bases for fruit-odor preference and avoidance, are needed to address this issue. In this regard, tests of hawthorn-fly populations from the southern United States, outside of the range of overlap with apple flies, may help to resolve whether the indifference of hawthorn flies to butyl hexanoate observed in the Fennville, Michigan, field trials is consistent across the range of the hawthorn race and is the ancestral state for *R. pomonella*.

Combined mixtures of natal plus nonnatal blends also resulted in significantly reduced captures of flies (Forbes et al. 2005). This finding reiterates that apple, hawthorn, and dogwood flies do not merely fail to recognize the odor of alternative host fruit but also generally tend to avoid them. Moreover, avoidance of nonnatal blends appears to be stronger than the inherent attraction invoked by components of the natal blend. In five of six instances, the relative proportions of flies caught on the combined blend were similar to the reductions seen when nonnatal blends were tested alone. For the one exception (hawthorn + apple blend tested in the hawthorn copse), the proportion captured on the combined blend (compared to blank spheres) was intermediate between the apple and hawthorn blends tested alone, but still much lower than that observed for the natal hawthorn blend. Similar antagonistic responses to a mix of natal and nonnatal blends were found in flight tunnel assays by Linn et al. (2005b). The results for the combined blend tests in the field and laboratory underscore that chemosensory changes accompanying *R. pomonella* host shifts appear to involve not just the derivation of a new preference for a novel fruit that overcomes an ancestral aversion behavior, but also the rapid evolution of avoidance for ancestral fruit volatiles.

Genetic Analysis of Fruit-Odor Discrimination

Discerning the genetic architecture of host-fruit-odor discrimination differences among apple, hawthorn, and dogwood flies is important for addressing a number of questions concerning sympatric host-race formation and speciation. First, many models of sympatric speciation are based on the need for gametic disequilibrium to evolve between complementary habitat preference and performance alleles due to the "selection-recombination antagonism" of Felsenstein (1981) (see also Berlocher and Feder 2002; Fry 2003). For *Rhagoletis*, diapause traits related to host fruiting phenology are important for differential fly survivorship on apple, hawthorn, and dogwood (Filchak et al. 2000). This raises the question of whether genes controlling diapause in *R. pomonella* are physically linked to loci responsible for fruit-odor choice.

Second, expansion of two-locus preference and performance models to more realistic, quantitative genetic scenarios has shown that spreading the genetic basis for habitat choice among an increasing number of loci can lessen the likelihood for sympatric race formation and speciation (Fry 2003). Consequently, it is important to determine whether host-fruit-odor discrimination in *Rhagoletis* is highly polygenic or due to only a moderate number of genetic changes between host races and siblings species.

Third, little is known about the physiological basis for host-odor discrimination: are differences between newly formed races likely to involve changes in the peripheral olfactory system [e.g., odor receptors in olfactory receptor neurons (ORNs) of the antennae] or central nervous system processing (e.g., neural connections of ORNs to antennal lobes or modifications in direct or indirect glomerular innervation of mushroom bodies, central complex, lateral accessory lobes, and the lateral protocerebrum)?

Finally, genetic studies also bear on the role of larval conditioning in host choice. Many of the previous flight tunnel assays of apple, hawthorn, and dogwood flies were performed on field-collected larvae reared to adulthood in the laboratory (Linn et al. 2003, 2004, 2005a, 2005b). It is therefore conceivable that the observed fruit-odor fidelity among taxa could be partly explained by larval conditioning, rather than by heritable preference differences (Coyne and Orr 2004). Rearing of hawthorn flies for one generation in the laboratory on apple did not affect the preference of the hawthorn race for hawthorn volatiles (Linn et al. 2003), tending to discount the conditioning hypothesis. However, detailed analysis of hybrid crosses is needed to confirm the genetic basis of host-odor discrimination.

F_1 Hybrids Display Reduced Behavioral Responses to Fruit Volatiles

To investigate the genetic basis of fruit-odor discrimination, we constructed reciprocal crosses between all pairwise combinations of apple, hawthorn, and dogwood flies. An unexpected finding from these crosses was that all combinations of F_1 hybrids between fly taxa were insensitive to host-fruit volatiles in flight tunnel assays (see Fig. 8.2C for results for apple × hawthorn hybrids; Linn et al. 2004a). The vast majority of F_1 hybrids did not respond to apple, hawthorn, or dogwood fruit volatile blends in wind tunnel assays at doses of 200 μg that normally elicit maximal upwind flight of parental flies to source spheres (Linn et al.

2004a). A fraction of F_1 flies (30–57%) did respond when an order of magnitude higher blend dose (2000 μg) was used in the assays, a dose that results in arrestment of the upwind flight of parental flies to source spheres (Linn et al. 2004a). Almost every F_1 fly responding to 2000 μg doses of fruit volatiles (~97%) oriented to a 1:1 combination of their parent's blends, mixtures that also tend to antagonize the upwind flight of parent flies at low doses (Linn et al. 2004b). In addition, about 25% of F_1 hybrids also reached spheres baited with high doses of one or both of their parent's natal fruit blends (Linn et al. 2004a). Finally, F_1 apple × hawthorn hybrids responding at 2000 μg displayed a bias for orienting to their mother's natal blend (Linn et al. 2004a).

The reduced behavioral response pattern seen for F_1 flies implies that hybrids have an altered chemosensory system and are hampered in their ability to detect and orient to fruit odor. As a result, hybrids may suffer a fitness disadvantage finding host fruit in nature, a hypothesis that we are currently testing through mark-release-recapture experiments in the field. If true, then host-specific mating would play a dual role in sympatric race formation and speciation, serving as an important postzygotic, as well as premating, reproductive isolating barrier (Linn et al. 2004a). The cause for the reduction in hybrid response remains to be definitively determined (although see below) but could be due alone or in combination to F_1 flies being behaviorally compromised by possessing different alleles for avoidance to the volatiles of both parental fruit or to genetic incompatibilities between fly taxa disrupting normal development of the chemosensory system.

Genetic Analysis of F_2 Hybrids

Analysis of the odor response patterns of F_2 hybrids in flight tunnel assays suggest that only a modest number of loci underlie the differences in fruit volatile discrimination among apple, hawthorn, and dogwood flies (Dambroski et al. 2005). In contrast to F_1 hybrids, many F_2 hybrids (30–65%; N = 819 total flies tested for all crosses) reached 200 μg baited source spheres in the flight tunnel (see Fig 8.2D, 8.2E for apple × hawthorn fly results; Dambroski et al. 2005). A subset of the low-dose-responding F_2 flies were also tested at 2000 μg (N = 176 flies). Most (63–100%) of these 200 μg responders failed to reach the high-dose source sphere, implying that they were antagonized by the elevated volatile level. A fair proportion of F_2 hybrids therefore had behavioral response patterns that mirrored those seen for parental apple, hawthorn, and dogwood flies, implying that fruit-odor discrimination variation among taxa is not overly polygenic.

F_2 flies also showed a grandmaternal effect suggestive of cytonuclear gene interactions (Dambroski et al. 2005). The vast majority of F_2 flies responding to 200 μg doses of fruit volatiles reached spheres baited with the natal blend of their respective maternal grandmothers (Fig. 8.2D, 8.2E). The maternal grandmother effect was consistent across all of the different types and directions of F_2 crosses, as well as collecting sites (Dambroski et al. 2005). The only exception was one of the apple × dogwood crosses, where equal proportions of flies responded to the apple and to the dogwood blend tested alone. Within a given cross type, responding F_2 males and females in 200 μg assays showed similar preferences for the maternal grandmother blend, discounting sex-linkage as a major factor affecting host-fruit-odor discrimination (Dambroski et al. 2005).

Genetic Analysis of Backcross Hybrids

Flight tunnel results for backcross progeny mirrored those for F_2 hybrids (Dambroski et al. 2005). Sample sizes for certain backcrosses were small and flies were tested only to the 200 μg dose. Nevertheless, 33 to 66% of apple × hawthorn and hawthorn × dogwood backcross progeny reached spheres baited with parental fruit blends (Fig. 8.2F, 8.2G). Moreover, like F_2 hybrids, the paternal genotype in the backcrosses had little effect on the behavioral pattern of progeny (Fig. 8.2F, 8.2G). Responding backcross hybrids primarily oriented to the natal fruit blend of their mother (male F_1 backcrosses) or maternal grandmother (female F_1 backcrosses). The response patterns were very similar for responding males and females within a given backcross type, further discounting sex-linkage for major host-discrimination loci hybrids (Dambroski et al. 2005).

Physiological Basis for Fruit-Odor Discrimination

Olfactory discrimination in *Rhagoletis* could evolve via changes in the central processing centers of the brain and/or through variation in peripheral chemoreception. Based on single-cell sensillum electrophysiology of fly antennae, Olsson et al. (2006a) showed that apple, hawthorn, and dogwood flies, as well as *R. mendax* (the blueberry maggot), possess similar classes of ORNs responding to host and nonhost volatiles. None of the taxa possessed significantly more ORNs tuned to one set of compounds than another, and topographical mapping indicated that ORN locations did not differ morphologically between apple, hawthorn, dogwood, and blueberry flies (Olsson et al. 2006a). Therefore, differences in host-plant preference among fly populations do not appear to be a function of altering receptor neuron specificity to host or nonhost volatiles. Apple, hawthorn, dogwood, and blueberry flies all have broad sensory palettes that can detect all the volatiles from the various different fruit (Olsson et al. 2006a). Differences among fly populations were found, however, with respect to ORN response characteristics, including both sensitivity to specific volatiles and temporal firing patterns (Olsson et al. 2006b). These peripheral sensory differences could contribute to fruit-odor discrimination variation among apple, hawthorn, and dogwood flies.

Single-cell recordings were also made for F_1 hybrids between apple, hawthorn, and dogwood host populations, flies that displayed a reduced behavioral response to fruit volatiles in flight tunnel assays. These recordings revealed

TABLE 8.1

Four-Component Genetic Model for Fruit-Odor Discrimination in the Apple and Hawthorn Host Races of *R. pomonella*

Locus	*Hawthorn-Race Allele Frequency*		*Apple-Race Allele Frequency*		*Function/Effect*
Preference (*P*)	h^+	0.780	h^+	0.511	Orient to hawthorn blend
	ah^+	0.220	ah^+	0.489	Orient to apple and hawthorn blend
Avoidance (*A1*)	h^-	0.657	h^-	0.239	Avoid high-dose hawthorn blend
	a^-	0.343	a^-	0.761	Avoid high-dose apple blend
Avoidance (*A2*)	a^-	1.000	h^-	1.000	Avoid low-dose nonnatal blend
mtDNA haplotype	*h*	1.000	*a*	1.000	Cytonuclear interactions

NOTE: Shown are hypothesized genetic loci, autosomal alleles, and mtDNA haplotypes segregating in apple and hawthorn populations, along with maximum likelihood estimates of their frequencies, and the function (effect) of the genes on the behavioral responses of flies to host-fruit volatiles in the flight tunnel. See text, appendix, and Table 8.2 for additional details concerning the model, including predicted phenotypes for multilocus genotypes.

distinct and diverse hybrid response profiles for certain ORNs not seen in parent populations (S. B. Olsson, et al. 2006), consistent with compromised peripheral chemoreception in F_1 flies. Abnormalities in the peripheral system of F_1 hybrids imply some form of genetic and physiological breakdown in the development and function of the ORNs, possibly due to the misexpression and/or presence of novel combinations of receptors in hybrid neurons (S.B. Olsson, et al. 2006). However, analysis of ORNs for fruit-odor-responding and fruit-odor-nonresponding F_2 hybrid flies revealed that both behavioral classes of flies shared similar diverse sets of electroantennal response profiles as F_1 hybrids (S. B. Olsson, 2006, pers. comm.). Hence, the observed differences in ORN response patterns in hybrid flies may not be the root cause for olfactory dysfunction and reduced chemosensory orientation in F_1 and F_2 progeny.

A Genetic Model for Fruit-Odor Discrimination

To clarify the nature of fruit-odor discrimination among *Rhagoletis* flies, we explored a series of genetic models attempting to describe the phenotypic pattern of variation observed for parental, F_1, and second-generation hybrid flies in the flight tunnel assays. We discovered that a four-component genetic model closely matched the observed fruit-odor responses for flies (Fig. 8.2). The four elements of the model are (1) one autosomal locus affecting host-fruit-odor preference (designated *P*) and two autosomal loci affecting fruit-odor avoidance (designated *A1* and *A2*); (2) chemosensory incompatibilities generated by heterozygous combinations of avoidance genes; (3) mitochondrial and/or cytoplasmic effects on nuclear gene expression; and (4) epistatic interactions among preference and avoidance loci. Details concerning the nature of the loci, allelic variation, gene interactions underlying the model, and phenotypes associated with specific multilocus genotypes are discussed in the appendix and Tables 8.1 and 8.2.

The four-component model highlights several important points. First, the model underscores that although the phenotypic pattern of fruit-odor discrimination is complex in parental taxa and hybrids, the pattern does not necessitate an undue number of genetic changes to explain.

Second, the model shows that shifts in host discrimination to a novel fruit do not require the de novo creation of new chemosensory pathways. For example, an allelic substitution at the *P* preference gene causing receptors for apple volatiles to be underexpressed in neurons targeting antagonistic pathways and overexpressed in neurons targeting agonist pathways, could facilitate a shift from hawthorn to apple. Thus, while future genetic mapping studies may change the specific details of the model, the general principles of the model are likely to remain valid and provide insight into rapid sympatric differentiation.

Third, the model raises several issues concerning the evolutionary dynamics of host-fruit-odor discrimination: given that nonresponding flies suffer at least some detrimental fitness effects due to a reduced ability to locate prime host fruit, how did the $A2h^-$ allele arise and fix in the derived apple race if an initial state of underdominance existed with all carriers of the new mutation being less fit heterozygotes? One explanation centers on epistatic interactions in which the detrimental effects of $A2h^-/A2a^-$ heterozygotes in a Pah^+/Pah$^+$, $A1h^+/A1h^+$, hawthorn mtDNA genetic background are not expressed, and these flies orient to apple odor. This would create a ridge in the adaptive fitness surface enabling avoidance of hawthorn fruit odor to evolve. In this regard, the sole F1 apple × hawthorn hybrid out of 314 flies tested in the flight tunnel that responded to parental fruit volatiles at 200 μg reached an apple-blend-baited sphere (Linn et al. 2004a). It is also possible that the fixation of $A2h^-$ in the apple race occurred via a two-step process involving an intermediate *A2* null allele (since lost) that did not cause avoidance to either apple or hawthorn fruit volatiles.

TABLE 8.2

Four-Component Genetic Model with Cytonuclear Gene Interactions Describing *R. pomonella* Behavioral Responses to Fruit Volatile Blends in Flight Tunnel Assays

mtDNA (cytoplasm)		*Hawthorn mtDNA Haplotype*			*Apple mtDNA Haplotype*		
Nuclear Loci		*A2* Locus			*A2* Locus		
P LOCUS	*A1* LOCUS	A^-/A^-	A^-/H^-	H^-/H^-	A^-/A^-	A^-/H^-	$\mathbf{H^-/H^-}$
ah^+/ah^+	h^-/h^-	**A/B**	A	A	A	A	**A**
	h^-/a^-	**N**	N	A	A	N	**A**
	a^-/h^-	**N**	N	N	A	N	**A**
	a^-/a^-	**H**	N	N	B	N	**A**
ah^+/h^+	h^-/h^-	**B**	N	B	A	N	**A**
	h^-/a^-	**B**	N	B	A	N	**A**
	a^-/h^-	**H**	N	H	A	N	**A**
	a^-/a^-	**H**	N	H	B	N	**A**
h^+/ah^+	h^-/h^-	**H**	N	H	A	N	**A**
	h^-/a^-	**H**	N	B	A	N	**A**
	a^-/h^-	**H**	N	H	A	N	**B**
	a^-/a^-	**H**	N	H	B	N	**B**
h^+/h^+	h^-/h^-	**N**	N	N	N	N	**N**
	h^-/a^-	**H**	N	H	N	N	**N**
	a^-/h^-	**H**	N	H	H	N	**N**
	a^-/a^-	**H**	N	H	H	N	**H/N**

NOTE: Shown are behavioral response phenotypes designated under the model for multilocus genotypes segregating for preference (*P*) and avoidance loci *A1* and *A2* in different mtDNA (cytoplasmic) genetic backgrounds. Behavioral responses: A, fly reaches apple blend baited sphere only; H, fly responds to hawthorn blend; B, fly responds to both blends; N, no response to either blend. Maternally inherited allele is listed first in genotype. For a given mtDNA background, the bold-text column represents the expected phenotypes under the model for parental hawthorn- or apple-race flies, the middle column for F_1 hybrids, all three columns for F_2 hybrids (ratio 1:2:1), and the middle and one of the outer columns the backcross hybrids, depending upon the direction of the mating (ratio 1:1).

Theoretical Significance of the Fruit-Odor Discrimination Studies

The finding that apple and hawthorn flies prefer their natal fruit blends is consistent with general verbal and mathematical models of sympatric speciation. Mayr (1947, 1963) articulated a two-locus genetic model for sympatric host-race formation involving habitat preference and performance genes, which was later fully developed by Bush (1975). The "Bush model" (Fry 2003) envisions two alleles segregating at the host-preference locus, one allele resulting in preferential choice for a novel habitat and the other for the ancestral habitat. A similar polymorphism for differential survivorship on hosts segregates at the habitat performance locus.

The Bush model is not without its difficulties, however. One problem concerns the behavior of heterozygotes for the preference locus. If there is not complete dominance, then heterozygotes may often be indiscriminate in their host choice, resulting in extensive interhost gene flow. A second problem concerns the need for linkage disequilibrium to develop between complimentary habitat preference and performance alleles, or the "selection-recombination antagonism" of Felsenstein (1981). However, under an assumption of dominance for host-preference genes and fairly strong disruptive selection between habitats, significant progress toward sympatric speciation can be made under the two-locus model (Fry 2003). Expansion of the Bush model to more realistic, quantitative genetic scenarios has shown that additional performance loci can enhance the potential for sympatric divergence (Fry 2003). But this is not the case for habitat choice; spreading the genetic basis for habitat preference among an increasing number of loci lessens the likelihood for race formation and speciation (Fry 2003). However, it is still possible for a series of performance and preference alleles to sequentially arise and differentially fix in populations, leading to completion of the speciation process (Kawecki 1996; Fry 2003).

Our studies of fruit-odor discrimination imply that current models of habitat-specific mating may not accurately depict how phytophagous insects choose their hosts, short-changing the potential for sympatric divergence. Building upon the Bush scenario, most models assume that host choice is determined by the aggregate effect of alleles imparting positive preferences for different host-plant species (i.e., individuals differ in their rank order preferences for different potential hosts, with an insect most often choosing the plant with the highest ranking based on the relative number of preference alleles it possesses favoring the plant over others [Fry 2003]). But our results imply that insects also actively avoid certain hosts. Negative-effect genes for disregarding alternative plants may have as much to do with host use as preference alleles.

Inclusion of negative-response genes in models of population divergence has important implications for sympatric speciation. Avoidance alleles could adversely affect the host-seeking behavior of heterozygous (hybrid) individuals, a problem associated with the two-locus Bush model. Unlike preference alleles, insects heterozygous for alternative habitat avoidance genes may have compromised chemosensory systems, as observed for F_1 *Rhagoletis* hybrids. Consequently, these individuals may have difficulty locating suitable host plants (fruit) and suffer severe fitness consequences in nature. Habitat-specific mating would therefore serve a dual role as an important postmating, as well as prezygotic, reproductive barrier to gene flow during sympatric race formation. As with sexual selection, hybrids would have difficulty securing mates because they possess abnormal phenotypes that preclude them from fully participating in the mating leks formed on alternative host trees (see review by Servedio and Noor 2003). Selection for nonnatal-host avoidance would therefore not only generate positive assortative mating, but also directly reduce the inclusive fitness of disassortatively mating individuals. When hybrids are unfit, reinforcement is also possible (Servedio and Noor 2003), providing additional avenues for sympatric divergence to proceed toward completion through selection for increased mate discrimination based on traits not affecting habitat performance.

In contrast to host preference, the addition of more habitat-avoidance loci would also enhance, rather than weaken, the possibility of sympatric speciation. Increasing the number of antagonism genes that an insect possess for a certain host can increase its likelihood of avoiding the host, while not necessarily diminishing the effects of alternative avoidance genes for disdaining other hosts. Consequently, more avoidance loci can increase the problems multilocus heterozygotes and individuals of mixed ancestry have in efficiently locating suitable host plants (i.e., the greater the number of avoidance genes that exist, the harder it would be to segregate and assort genotypes for avoidance to just one host in backcross and later-generation hybrids). At the same time, more avoidance loci would increase the opportunities for newly arising avoidance mutations to be linked to host-performance loci within host populations, facilitating sympatric race formation.

Generality of Results to Other Phytophagous Insects

An important question is whether our finding of chemosensory incompatibility and grandmaternal effects for *Rhagoletis* hybrids is common in other phytophagous insects. With respect to a general role for chemical-mediated avoidance in host choice, it is clear that nonhost compounds often act as feeding deterrents to phytophagous insects (Bernays and Chapman 1987; Bernays et al. 2000; Bernays 2001). However, even when recognized, host antagonism is still generally perceived through the same lens as preference, in which the cumulative additive effects of avoidance alleles for alternative plants determines acceptance behavior (Bernays 2001). The potential adverse incompatibilities generated by alternative avoidance alleles are not considered.

With respect to the general issue of genomic incompatibility, hybrid dysfunction is a key feature of speciation theory and is a ubiquitous phenomenon in nature. Inviability and sterility due to negative epistatic interactions occurring in the mixed genomes of hybrids are central to the Bateson-Dobzhansky-Muller model for reproductive isolation (Lynch and Force 2000). However, these incompatibilities are generally viewed as being genomewide in their scope (Wu and Polopoli 1994; Wu and Hollocher 1998) and having allopatric (geographic) origins (Mayr 1963). In contrast, the developmental/physiological incompatibility that we hypothesize as compromising the chemosensory system of *R. pomonella* hybrids is more restricted in scope, likely involving just a few loci altering expression patterns of receptors in ORNs, disrupting the output signals of heterozygous ORNs, or affecting central nervous system processing or interpretation of ORN input signals.

Although much progress has been made in recent years, the genetics of host-plant choice for phytophagous insects is still in its relative infancy (reviewed in Berlocher and Feder 2002; Drès and Mallet 2002). Most studies of host choice in insects have reported that F1 hybrids are either intermediate in their acceptance behaviors between parental taxa or show a dominance preference for one over the other parental plant (Jaenike 1987; R'Kha et al. 1991; Lu and Logan 1995; Sheck and Gould 1995; Messina and Slade 1997; Sezer and Butlin 1998a, 1998b; Possidente et al. 1999; Via et al. 2000; Craig et al. 2001; Campan et al. 2002; Emelianov et al. 2003). The only other apparent case of impaired F1 response is an EAG study of hybrids from another pair of *Rhagoletis* species, *R. pomonella* × *R. mendax* (Frey and Bush 1996). In this instance, hybrids showed reduced EAG amplitudes to certain volatile chemicals.

Studies of F_2 phytophagous insect hybrids have generally yielded similar results as F_1 experiments (Jaenike 1987; Lu and Logan 1995; Sezer and Butlin 1998a, 1998b; Possidente et al. 1999; Hawthorn and Via 2001) and, in certain instances, have implied that only a limited number of genes may be involved in host discrimination (Sezer and Butlin 1998a, 1998b; Via et al. 2000; Craig et al. 2001). Thompson (1988) discovered a paternal effect in hybrids of two *Papilio* swallowtail butterfly species. The effect is likely due to sex-linkage of oviposition preference genes on the X chromosome combined with female heterogamy in lepidoterans. Huettel and Bush (1972) reported that F_1 hybrid females between different species of *Procecidochares* gall-forming tephritid flies displayed ovipositional preference for their mother's host plant. They interpreted this result as evidence of host conditioning, although it is more likely indicative of a maternal effect.

At face value, the empirical data for phytophagous insects provides little evidence for host-choice incompatibility. However, it is possible that the general lack of support could be related to experimental methodology. Many tests of host preference for insects are conducted in confined choice arenas in which it may be difficult for an insect to avoid a plant in an environment in which it is receiving mixed cues. In the case of our flight tunnel testing for *Rhagoletis*, a key factor is the development of biologically active synthetic fruit blends. These blends allow for a testing procedure requiring a positive, directed response by flies to a specific, isolated chemosensory cue to indicate host acceptance. More refined experimental designs may reveal additional incompatibilities for other phytophagous hybrids.

Future Directions and Conclusions

Discerning the genetic and physiological bases and evolutionary origins for host-related chemosensory specialization in insects is critical for understanding the genesis of biodiversity and adaptive radiation on a grand scale, since there are more phytophagous host specialists than any other life form. Our studies have built upon the trailblazing legacies of Tom Wood and Ron Prokopy, providing a window into the physiology and genetics of fruit-odor discrimination, a key trait involved in host-specific mating and sympatric speciation for *R. pomonella*. Our data imply that fruit-odor discrimination has an intricate genetics, involving altered (reduced) olfaction response in hybrids, cytoplasmic-nuclear gene interactions (maternal effects), and possibly epistasis. These phenomena raise a number of fascinating questions and possibilities concerning the evolutionary trajectory and mechanistic basis for chemosensory-related behavioral changes in *R. pomonella* that have facilitated host shifts and adaptive radiation onto novel plants. In particular, we have discovered an important and overlooked aspect of habitat-specific mating bearing directly on sympatric host-race formation and speciation for phytophagous insects; both preference for novel hosts and avoidance responses to ancestral hosts evolve rapidly during host shifts.

Many questions remain to be answered, however, concerning host-avoidance behavior in *Rhagoletis*. Although the four-component model provides a good fit to the empirical data, further genetic analysis is required to confirm the hypothesis. In this regard, mapping studies involving F_2 and backcross hybrids are currently being conducted using cDNA and microsatellite markers developed for *Rhagoletis* (Roethele et al. 1997, 2001; Velez et al. 2006) to discern quantitative trait loci possessing the preference and avoidance characteristics predicted by the model. In addition, more detailed studies of the fly's chemosensory system (especially the central nervous system) are needed to identify the physiological basis for fruit-odor incompatibility in hybrids. Also, computer simulations are being conducted modeling the population genetic consequences of avoidance loci and host-finding incompatibility in hybrids. Preliminary results imply that these considerations can facilitate sympatric divergence, providing a means to enhance the partial reproductive isolation of host races and propel them toward species status. This has historically been an unresolved question of great theoretical interest with respect to sympatric speciation models (Berlocher and Feder 2002).

Whatever the final outcome of these studies, it is clear that what once seemed to be a relatively straightforward understanding of the basis for host choice (preference) may have greatly oversimplified the biology of the process. The resulting plot twist would have enthralled the rebellious scientific streaks in Tom Wood and Ron Prokopy: an underappreciation of the potential for habitat-specific mating to serve as a reproductive barrier to gene flow during ecological speciation, regardless of geographic context.

Acknowledgments

The authors wish to thank the following individuals for their assistance, moral support, and conversational input: Stewart Berlocher, Guy Bush, Hattie Dambroski, Ken Filchak, Rick Harrison, Charles Linn Jr., Kirsten Pelz, Thomas Powell, Wendell Roelofs, Jim Smith, Lukasz Stelinski, Uwe Stolz, Mike Taylor, Sebastian Velez, Frank Wang, Frank Wang Jr., John Wise, Xianfa (Frank) Xie, the USDA/APHIS/PPQ facility at Niles, Michigan, and the Trevor Nichols Research Station of Michigan State University at Fennville, Michigan. This research was supported, in part, by grants from NSF, an NRI grant from the USDA, and support from the Twenty-first Century Fund of the state of Indiana to J.L.F. The authors dedicate this work to the memory of Tom Wood, Ron Prokopy, and Sridhar Polavarapu, friends and giants in the field of sympatric speciation and *Rhagoletis* research.

Appendix: The Four-Component Genetic Model for Fruit-Odor Discrimination

Five major features of the flight tunnel data require explanation. First, parent apple, hawthorn, and dogwood flies reached odor source spheres in low-dose 200 μg assays, but not in high-dose 2000 μg trials (Linn et al. 2003, 2004). Second, Linn et al. (2005a) found that parent flies within host populations displayed variation in their 200 μg response pattern: about 60% reached natal-blend spheres only, 15% to 20% both natal and nonnatal-blend spheres, 3% nonnatal spheres only, and 15% to 20% did not respond to any blend (Fig. 8.2A, 8.2B). Third, F_1 hybrids did not reach source spheres in 200 μg trials (Fig. 8.2C) (Linn et al. 2004). Fourth, an average of almost 50% of F_2 and backcross hybrids reached 200 μg source spheres (Fig. 8.2D, 8.2E, 8.2F, 8.2G, 8.2H, 8.2I). Fifth, responding second-generation hybrids showed a strong bias for orienting to the natal blend of their maternal grandmother, with a few flies responding to both grandmaternal and paternal blends, and a few to just the maternal grandpaternal blend (Fig. 8.2D, 8.2E, 8.2F, 8.2G, 8.2H, 8.2I).

To account for the 200 μg data, we hypothesize a four-component genetic model (see Tables 8.1 and 8.2). The first element consists an autosomal fruit-odor preference gene (*P*) whose alleles differentially are involved in the attraction of flies to host-specific volatiles, and two autosomal loci (*A1* and *A2*) whose alleles are involved in the avoidance of flies to specific volatile blends. The *P* locus segregates for two alleles within the apple and hawthorn host races. One preference allele (Ph^+) results in flies differentially orienting to hawthorn volatiles (the ancestral state), while the derived Pah^+ allele results in flies having a broad response capacity to orient to both hawthorn and apple fruit odor. The *Pah* allele accounts for the observation that of the 15 to 25% of apple and hawthorn flies that reached nonnatal fruit-odor spheres, almost all also responded to their natal fruit blend (Linn et al. 2005a).

We hypothesize that the *A1* and *A2* avoidance loci perform different biological functions in flies. The *A1* locus normally acts in a threshold manner. When flies are exposed to high concentrations of volatiles, adaptation of peripheral receptor neurons occurs, resulting in arrestment of upwind flight. The upper concentration threshold could be an inherent property of the fly's nervous system physiology, or it could potentially represent an adaptive response for flies to avoid overripe and suboptimal fruit that will soon drop (abscise) from trees. (*Note: Rhagoletis* flies oviposit only into unabscised fruit "on the vine"). The threshold effect of the *A1* locus would account for the observation that parent apple and hawthorn flies essentially all take flight, but fail to reach source spheres, baited with high doses (2000 μg) of their natal fruit blend (Linn et al. 2004). We postulate that the *A1* avoidance gene segregates for two alleles. One allele ($A1h^-$) results in hawthorn flies being deterred by high concentrations of the hawthorn blend (the ancestral state), while the alternate allele ($A1a^-$) is the derived state in the apple race and results in apple flies avoiding high concentrations of apple volatiles.

The *A2* gene primarily controls the avoidance behavior of flies to nonnatal fruit volatiles. Incompatibilities generated by the *A2* locus in flies of mixed ancestry—the second component of the four-component model—are largely responsible for nonresponse phenotypes in F_1 and F_2 hybrids. In contrast to the *P* and *A1* loci, we contend that the host races are essentially fixed for different alleles at the *A2* avoidance locus. The hawthorn race possesses the ancestral gene $A2a^-$ that results in hawthorn flies avoiding apple volatiles at concentrations found in ripe apple fruit, while the derived $A2h^-$ allele in the apple race causes apple flies to avoid hawthorn fruit volatiles. Most importantly, $A2a^-/A2h^-$ heterozygotes have compromised chemosensory systems that render them incapable of orienting to either the apple or hawthorn fruit blend in 200 μg dose trials. Heterozygotes fail to respond because they possess functional neural pathways for antagonist inputs from both apple and hawthorn fruit or they are hampered by a developmentally altered chemosensory system with a reduced sensitivity to fruit volatiles. As the avoidance alleles $A2h^-$ and $A2a^-$ are fixed in alternate states between apple and hawthorn flies, F_1 hybrids between the races are all heterozygotes and do not respond to 200 μg doses of either the apple or hawthorn fruit blend. However, segregation of apple and hawthorn avoidance genes at the *A2* locus in second-generation F_2 and backcross hybrids results in a significant proportion of these flies (30–65%) being homozygous and displaying parental apple- or hawthorn-fly response phenotypes.

The third component of the model involves cytonuclear gene interactions that influence the penetrance of autosomal preference and avoidance genes. The cytonuclear gene interactions are largely responsible for the observed bias of F_2 and backcross flies in responding to the natal blend of their maternal grandmother. For example, in an apple-fly mtDNA cytoplasm, we hypothesize that the threshold level for the $A1h^-$ allele is reduced, resulting in flies avoiding lower concentrations of hawthorn volatiles typical of newly ripening hawthorn fruit. Similarly, when maternally inherited in a hawthorn-fly mtDNA background, the threshold level for the $A1a^-$ allele can also be reduced, causing flies to avoid the apple blend at lower concentrations.

The fourth element of the model is epistatic interactions among nuclear preference and avoidance genes. These epistatic interactions help explain the response variation to natal and nonnatal blends observed within the host races. They are also significant because they provide a potential ridge of high fitness in the adaptive surface for host discrimination that may have been responsible for the evolution of increased apple fruit-odor preference and hawthorn avoidance in apple flies despite most $A2a^-/A2h^-$ heterozygotes being nonresponsive to fruit volatiles.

To estimate relevant model parameters (i.e., allele frequencies and gametic disequilibrium values [*r*] for preference and avoidance genes in apple and hawthorn populations), we performed a maximum likelihood (ML) analysis based on a multinomial probability distribution, using expected phenotype frequencies generated by the model to calculate the likelihood of obtaining the observed data for parental, F_1, F_2, and backcross flies. Expected phenotype frequencies were derived by assuming that parental flies tested as adults in the flight tunnel and used in genetic crosses represented a random sample of individuals drawn from the two host populations. Under this assumption, gamete frequencies could be estimated based on Mendelian principles for any suite of model allele frequency and disequilibrium values, and gametes randomly combined to generate expected genotype distributions for parent, F_1, F_2, and backcross flies. Behavioral response phenotypes for genotypes were then assigned based on the hypothesized relationships given in Table 8.2. The overall likelihood for a particular set of model allele frequencies and disequilibrium values ($-\ln L$) was calculated as the negative sum of the individual log likelihoods for obtaining the observed data for parents, F_1, F_2, and backcross progeny. To find an approximate ML solution, calculations were performed allowing allele frequencies to vary independently in the races at the preference and *A1* avoidance loci between 0 and 1 in increments of 0.001, with fixed differences assumed between apple and hawthorn flies for the *A2* locus and mtDNA haplotypes. Gametic disequilibrium values (standardized correlation coefficient *r*) between alleles segregating at the preference and *A1* locus were also simultaneously varied between –1 and 1 in increments of 0.001 for each set of allele frequencies analyzed within the races.

References Cited

Becerra, J.X. 1997. Insects on plants: macroevolutionary chemical trends in host use. Science 276: 253–256.

Berlocher, S.H. 1999. Host race or species? Allozyme characterization of the "flowering dogwood fly," a member of the *Rhagoletis pomonella* complex. Heredity 83: 652–662.

Berlocher, S.H. 2000. Radiation and divergence in the *Rhagoletis pomonella* species group: inferences from allozymes. Evolution 54: 543–557.

Berlocher, S.H., and J.L. Feder. 2002. Sympatric speciation in phytophagous insects: moving beyond controversy? Annu. Rev. Entomol. 47: 773–815.

Berlocher, S.H., B.A. McPheron, J.L. Feder, and G.L. Bush. 1993. A revised phylogeny of the *Rhagoletis pomonella* (Diptera: Tephritidae) sibling species group. Ann. Entomol. Soc. Am. 86: 716–727.

Bernays, E.A. 2001. Neural limitations in phytophagous insects: implications for diet breadth and evolution of host affiliation. Annu. Rev. Entomol. 46: 703–727.

Bernays, E.A., and R.F. Chapman. 1987. The evolution of deterrent responses in plant feeding insects, pp. 159–173. In R.F. Chapman, E.A. Bernays, J.G. Stoffolano (eds.), Perspectives in chemoreceptors and behavior. Springer-Verlag, New York.

Bernays, E. A., S. Oppenheim, R.F. Chapman, H. Kwon, and F. Gould. 2000. Taste sensitivity of insect herbivores to deterrents is greater in specialists than in generalists: a behavioral test of the hypothesis with two closely related caterpillars. J. Chem. Ecol. 26: 547–563.

Boller, E.F., and R.J. Prokopy. 1976. Bionomics and management of *Rhagoletis*. Annu. Rev. Entomol. 21: 223–246.

Bush, G.L. 1966. The taxonomy, cytology, and evolution of the genus *Rhagoletis* in North America (Diptera: Tephritidae). Bull. Mus. Comp. Zool. 134: 431–562.

Bush, G. L. 1969a. Sympatric host race formation and speciation in frugivorous flies of the genus *Rhagoletis* (Diptera: Tephritidae). Evolution 23: 237–251.

Bush, G.L. 1969b. Mating behavior, host specificity, and the ecological significance of sibling species in frugivorous flies of the genus *Rhagoletis* (Diptera: Tephritidae). Am. Nat. 103: 669–672.

Bush, G.L. 1975. Sympatric speciation in phytophagous parasites, pp. 187–206. In P. W. Price (ed.), Evolutionary strategies of parasitic insects and mites. Plenum, New York.

Bush, G.L. 1992. Host race formation and sympatric speciation in *Rhagoletis* fruit flies (Diptera: Tephritidae). Psyche 99: 335–357.

Campan, E., A. Couty, Y. Carton, M. H. Pham-Delègue, and L. Kaiser. 2002. Variability and genetic components of innate fruit odour recognition in a parasitoid of *Drosophila*. Physiol. Entomol. 27: 243–250.

Coyne, J.A., and H.A. Orr. 2004. Speciation. Sinauer, Sunderland, MA.

Craig, T.P., J.D. Horner, and J.K. Itami. 2001. Genetics, experience, and host-plant preference in *Eurosta solidaginis*: implications for host shifts and speciation. Evolution 55: 773–782.

Dambroski, H.R., C. Linn Jr., S.H. Berlocher, A.A. Forbes, W. Roelofs, and J.L. Feder. 2005. The genetic basis for fruit odor discrimination in *Rhagoletis* flies and its significance for sympatric host shifts. Evolution 59: 1953–1964.

Diehl, S.R., and G.L. Bush. 1989. The role of habitat preference in adaptation and speciation, pp. 527–553. In D. Otte and J.A. Endler (eds.), Speciation and its consequences. Sinauer, Sunderland, MA.

Dobzhansky, T. 1937. Genetic nature of species differences. Am. Nat. 71: 404–420.

Dobzhansky, T. 1951. Genetics and the origins of species. Columbia University Press, New York.

Drès, M., and J. Mallet. 2002. Host races in plant-feeding insects and their importance in sympatric speciation. Phil. Trans. R. Soc. Lond. B 357: 471–492.

Emelianov, I., F. Simpson, P. Narang, and J. Mallet. 2003. Host choice promotes reproductive isolation between host races of the larch budmoth *Zeiraphera diniana*. J. Evol. Biol. 16: 208–218.

Farrell, B.D. 1998. Inordinate fondness explained: why are there so many beetles? Science 281: 539–559.

Feder, J.L., and G.L. Bush. 1989. Gene frequency clines for host races of *Rhagoletis pomonella* (Diptera: Tephritidae) in the midwestern United States. Heredity 63: 245–266.

Feder, J.L., and K.E. Filchak. 1999. It's about time: the evidence for host plant-mediated selection in the apple maggot fly, *Rhagoletis pomonella*, and its implications for fitness trade-offs in phytophagous insects. Entomol. Exp. Appl. 91: 211–225.

Feder, J.L., S.B. Opp, B. Wlazlo, K. Reynolds, W. Go, and S. Spizak. 1994. Host fidelity as an effective premating barrier between sympatric races of the apple maggot fly. Proc. Natl. Acad. Sci. USA 91: 7990–7994.

Feder, J.L., S.H. Berlocher, and S.B. Opp. 1998. Sympatric host race formation and speciation in *Rhagoletis* (Diptera: Tephritidae): a tale of two species for Charles D, pp. 408–441. In S. Mopper and S. Strauss (eds.), Genetic structure in natural insect populations: effects of host plants and life history. Chapman and Hall, New York.

Fein, B.L., W.H. Reissig, and W.L. Roelofs. 1982. Identification of apple volatiles attractive to the apple maggot. J. Chem. Ecol. 8: 1473–1487.

Felsenstein, J. 1981. Homage to Santa Rosalia, or why are there so few kinds of animals? Evolution 35: 124–138.

Filchak, K.E., J.B. Roethele, and J.L. Feder. 2000. Natural selection and sympatric divergence in the apple maggot, *Rhagoletis pomonella*. Nature 407: 739–742.

Forbes, A.A., and J.L. Feder. 2006. Divergent preferences of *Rhagoletis* host races for mock host fruits. Entomol. Exp. Appl. 119: 121–127.

Forbes, A.A., J. Fisher, and J.L. Feder. 2005. Habitat avoidance: overlooking an important aspect of host specific mating and sympatric speciation? Evolution 59: 1552–1559.

Frey, J.E., and G.L. Bush. 1996. Impaired host odor perception in hybrids between the sibling species *Rhagoletis pomonella* and *R. mendax*. Entomol. Exp. Appl. 80: 163–165.

Fry, J. D. 2003. Multilocus models of sympatric speciation: Bush versus Rice versus Felsenstein. Evolution 57: 1735–1746.

Futuyma, D. 1986. Evolutionary biology, 2nd ed. Sinauer, Sunderland, MA.

Hawthorn, D. J., and S. Via. 2001. Genetic linkages of ecological specialization and reproductive isolation in pea aphids. Nature 412: 904–907.

Huettel, M. D., and G. L. Bush. 1972. The genetics of host selection and its bearing on sympatric speciation in Procecidochares (Diptera: Tephritidae). Entomol. Exp. Appl. 15: 465–480.

Jaenike, J. 1987. Genetics of oviposition-site preference in *Drosophila tripunctata*. Heredity 59: 363–369.

Johnson, P. A., and U. Gullberg. 1998. Theory and models of sympatric speciation, pp. 79–89. In D. J. Howard and S. H. Berlocher (eds.), Endless forms: species and speciation. Oxford University Press, New York.

Kawecki, T. J. 1996. Sympatric speciation driven by beneficial mutations. Proc. R. Soc. Lond. B 263: 1515–1520.

Kondrashov, A. S, L. Y. Yampolsky, and S. A. Shabalina. 1998. On the sympatric origin of species by means of natural selection, pp. 90–98. In D. J. Howard and S. H. Berlocher (eds.), Endless forms: species and speciation. Oxford University Press, New York.

Linn, C. Jr., J. L. Feder, S. Nojima, H. Dambroski, S. H. Berlocher, and W. Roelofs. 2003. Fruit odor discrimination and sympatric race formation in *Rhagoletis*. Proc. Natl. Acad. Sci. USA 100: 11490–11493.

Linn, C. Jr., H. R. Dambroski, J. L. Feder, S. H. Berlocher, S. Nojima, and W. Roelofs. 2004. Postzygotic isolating factor in sympatric speciation in *Rhagoletis* flies: reduced response of hybrids to parental host-fruit odors. Proc. Natl. Acad. Sci. USA 101: 17753–17758.

Linn, C. Jr., S. Nojima, H. R. Dambroski, J. L. Feder, S. H. Berlocher, and W. Roelofs. 2005a. Variability in response specificity of apple, hawthorn, and flowering dogwood-infesting *Rhagoletis* flies to host fruit volatile blends: implications for sympatric host shifts. Entomol. Exp. Appl. 116: 55–64.

Linn, C. Jr., S. Nojima, and W. Roelofs. 2005b. Antagonist effects of non-host fruit volatiles on chemically-mediated discrimination of host fruit by *Rhagoletis pomonella* flies infesting apple, hawthorn (*Crataegus* spp.), and flowering dogwood (*Cornus florida*). Entomol. Exp. Appl. 114: 97–105.

Lu, W., and P. Logan. 1995. Inheritance of host-related feeding and ovipositional behaviors in *Leptinotarsa decemlineata* (Coleoptera: Chrysomelidae). Environ. Entomol. 24: 278–287.

Lynch, M., and A. G. Force. 2000. The origin of interspecific genomic incompatibility via gene duplication. Am. Nat. 156: 590–605.

Maynard-Smith, J. 1966. Sympatric speciation. Am. Nat. 100: 637–650.

Mayr, E. 1947. Ecological factors in speciation. Evolution 1: 263–288.

Mayr, E. 1963. Animal species and evolution. Harvard University Press, Cambridge, MA.

Menken, S. B. J., and P. Roessingh. 1998. Evolution of insect-plant associations: sensory perception and receptor modifications direct food specialization and host shifts in phytophagous insects, pp. 145–156. In D. J. Howard and S. H. Berlocher (eds.), Endless forms: species and speciation. Oxford University Press, Oxford, UK.

Messina, F. J., and A. F. Slade. 1997. Inheritance of host-plant choice in the seed beetle *Callosobruchus maculates* (Coleoptera: Bruchidae). Ann. Entomol. Soc. Am. 90: 848–855.

Mitter, C., B. Farrell, and B. Wiegmann. 1988. The phylogenetic study of adaptive zones: has phytophagy promoted insect diversification? Am. Nat. 132: 107–128.

Nojima, S., C. Linn Jr., B. Morris, A. Zhang, and W. Roelofs. 2003a. Identification of host fruit volatiles from hawthorn (*Crataegus* spp.) attractive to hawthorn-origin *Rhagoletis pomonella* flies. J. Chem. Ecol. 29: 321–336.

Nojima, S., C. Linn Jr., and W. Roelofs. 2003b. Identification of host fruit volatiles from flowering dogwood (*Cornus florida*) attractive to dogwood-origin *Rhagoletis pomonella* flies. J. Chem. Ecol. 29: 2347–2357.

Olsson, S. B., C. E. Linn Jr., and W. L. Roelofs. 2006a. The chemosensory basis for behavioral divergence involved in sympatric host shifts. I. Characterizing olfactory receptor neuron classes responding to key host volatiles. J. Comp. Physiol. A. 192: 279–288

Olsson S. B., C. E. Linn Jr., and W. L. Roelofs. 2006b. The chemosensory basis for behavioral divergence involved in sympatric host shifts. II. Olfactory receptor neuron sensitivity and temporal firing pattern to individual key host volatiles. J. Comp. Physiol. A. 192: 289–300.

Olsson S. B., C. E. Linn, Jr, A. Michel, H. R. Dambrowski, S. H. Berlocher, J. L. Feder, and W. L. Roelofs. 2006c. Receptor expression and sympatric speciation: unique olfactory receptor neuron responses in F1 hybrid Rhagoletis populations. J. Exp. Biol. 209: 3729–3741.

Possidente, B., M. Mustafa, and L. Collins. 1999. Quantitative genetic variation for oviposition preference with respect to phenylthiocarbamide in *Drosophila melanogaster*. Behav. Genet. 29: 193–198.

Prokopy, R. J. 1968. Visual responses of apple maggot flies, *Rhagoletis pomonella* (Diptera: Tephritidae): orchard studies. Entomol. Exp. Appl. 11: 403–422.

Prokopy, R. J., and B. D. Roitberg. 1984. Foraging behavior of true fruit flies. Am. Sci. 72: 41–49.

Prokopy R. J , E. W. Bennett, and G. L. Bush. 1971. Mating behavior in *Rhagoletis pomonella* (Diptera: Tephritidae). I. Site of assembly. Can. Entomol. 103: 1405–1409.

Prokopy, R. J., E. W. Bennett, and G. L. Bush. 1972. Mating behavior in *Rhagoletis pomonella*. II. Temporal organization. Can. Entomol. 104: 97–104.

Prokopy, R. J., V. Moericke, and G. L. Bush. 1973. Attraction of apple maggot flies to odor of apples. Environ. Entomol. 2: 743–749.

Prokopy, R. J., M. Aluja, and T. A. Green. 1987. Dynamics of host odor and visual stimulus interaction in host finding behavior of apple maggot flies, pp. 161–166. In V. Labeyrie, G. Fabres, and D. Lachaise (eds.), Insects–plants. Junk, Dordrecht, Netherlands.

Rice, W. R., and E. E. Hostert 1993. Laboratory experiments on speciation: what have we learned in forty years. Evolution 47: 1637–1653.

R'Kha, S., P. Capy, and J. R. David. 1991. Host-plant specialization in the *Drosophila melanogaster* species complex: a physiological, behavioral and genetical analysis. Proc. Natl. Acad. Sci. USA 88: 1835–1839.

Roethele, J. B., J. L. Feder, S. H. Berlocher, M. E. Kreitman, and D. Lashkari. 1997. Towards a molecular genetic linkage map for the apple maggot fly, *Rhagoletis pomonella* (Diptera:Tephritidae): a comparison of alternative strategies. Ann. Entomol. Soc. Am. 90: 470–479.

Roethele, J.B., J. Romero-Severson, and J. L. Feder. 2001. Evidence for broad-scale conservation of linkage map relationships between *Rhagoletis pomonella* and *Drosophila melanogaster*. Ann. Entomol. Soc. Am. 94: 936–947.

Schluter, D. 2001. Ecology and the origin of species. Trends Ecol. Evol. 16: 372–380.

Servedio, M. R., and M. A. F. Noor. 2003. The role of reinforcement in speciation: theory and data. Annu. Rev. Ecol. Evol. Syst. 34: 339–364.

Sezer, M., and R. K. Butlin. 1998a. The genetic basis of oviposition preference differences between sympatric host races of the brown planthopper *(Nigaparvata lugens)*. Proc. R. Soc. Lond. B. 265: 2399–2405.

Sezer, M., and R. K. Butlin. 1998b. The genetic basis of host plant adaptation in the brown planthopper *(Nilaparvata lugens)*. Heredity 80: 499–508.

Sheck, A. L., and F. Gould. 1995. Genetic analysis of differences in oviposition preferences of *Heliothis virescens* and *H. subflexa* (Lepidoptera: Noctuidae). Environ. Entomol. 24: 341–347.

Simpson, G. G. 1953. The major features of evolution. Columbia University Press, New York.

Smith, J. J., and G. L. Bush. 1997. Phylogeny of the genus *Rhagoletis* (Diptera: Tephritidae) inferred from DNA sequences of mitochondrial cytochrome oxidase II. Mol. Phylogenet. Evol. 7: 33–43.

Thompson, J. N. 1988. Evolutionary genetics of oviposition preference in swallowtail butterflies. Evolution 42: 1223–1234.

Tilmon, K. J., T. K. Wood, and J. D. Pesek. 1998. Genetic variation in performance traits and the potential for host shifts in *Enchenopa* treehoppers (Homoptera: Membracidae): Ann. Entomol. Soc. Am. 91: 397–403.

Velez, S., M. S. Taylor, M. A. F. Noor, N. F. Lobo, and J. L. Feder. 2006. Isolation and characterization of microsatellite loci from the apple maggot fly, *Rhagoletis pomonella* (Diptera: Tephritidae). Mol. Ecol. Notes 6: 90–92.

Via, S. 2001. Sympatric speciation in animals: the ugly duckling grows up. Trends Ecol. Evol. 16: 381–390.

Via, S., A. C. Bouck, and S. Skillman. 2000. Reproductive isolation between divergent races of pea aphids on two hosts. II. Selection against migrants and hybrids in the parental environments. Evolution 54: 1626–1637.

Walsh, B. D. 1864. On phytophagic varieties and phytophagic species. Proc. Entomol. Soc. Phil. 3: 403–430.

Walsh, B. D. 1867. The apple-worm and the apple maggot. J. Hort. 2: 338–343.

Weis, A. R., and M. Berenbaum. 1989. Herbivorous insects and green plants, pp. 123–162. In W. G. Abrahamson (ed.), Plant animal interactions. McGraw Hill, New York.

Wood, T. K. 1980. Divergence in the *Enchenopa binotata* complex (Homoptera: Membracidae) effected by host plant adaptation. Evolution 34: 147–160.

Wood, T. K., and S. Guttman. 1982. Ecological and behavioral basis for reproductive isolation in the sympatric *Enchenopa binotata* complex (Homoptera: Membracidae). Evolution 36: 233–242.

Wood, T. K., and M. Keese. 1990. Host-plant-induced assortative mating in *Enchenopa* treehoppers. Evolution 44: 619–628.

Wood ,T. K., K. J. Tilmon, A. B. Shantz, and C. K. Harris. 1999. The role of host-plant fidelity in initiating insect race formation. Evol. Ecol. Res. 1: 317–332.

Wu, C. I., and H. Hollocher. 1998. Subtle is nature: the genetics of species differentiation and speciation, pp. 339–351. In D. J. Howard and S. H. Berlocher (eds.), Endless forms: species and speciation. Oxford University Press, Oxford, UK.

Wu, C. I., and M. F. Palopoli. 1994. Genetics of postmating reproductive isolation in animals. Annu. Rev. Genet. 28: 283–308.

Zhang, A., C. Linn Jr., S. Wright, R. Prokopy, W. H. Reissig, and W. L. Roelofs. 1999. Identification of a new blend of apple volatiles attractive to the apple maggot, *Rhagoletis pomonella*. J. Chem. Ecol. 25: 1221–1232.

NINE

Comparative Analyses of Ecological Speciation

DANIEL J. FUNK AND PATRIK NOSIL

For much of the twentieth century, the study of speciation had two major emphases. One was evaluating the geographic circumstances under which speciation occurred, and specifically whether geographic isolation (allopatry) was required (Mayr 1942, 1947; Bush 1969; Futuyma and Mayer 1980; Coyne and Orr 2004). The other was deciphering the genetic architecture of speciation, that is, the roles played by chromosomal translocations, and the kinds, numbers, linkage and epistatic relationships, and phenotypic effects of the genes involved in reproductive isolation (Noor et al. 2001; Orr 2001; Rieseberg 2001; Ortiz-Barrientos et al. 2002). Only recently have the contributions of natural selection to speciation been added as a major empirical focus of speciation studies. A few notable early articles notwithstanding (e.g., Dodd 1989; reviewed by Rice and Hostert 1993), the contributions of adaptive ecological divergence to reproductive isolation have primarily been evaluated—with increasing frequency—over the last decade or two (see reviews by Schluter 2000, 2001; Berlocher and Feder 2002; Drès and Mallet 2002; Funk et al. 2002; Coyne and Orr 2004; Rundle and Nosil 2005). And only over the last several years have approaches been developed to rigorously isolate the contributions of natural selection from other potential causes of the evolution of reproductive isolation (Schluter and Nagel 1995; Funk 1996, 1998; Fitzpatrick 2002; Funk et al. 2002, 2006). These comparative approaches take advantage of the recent availability of molecular genetic data, phylogenetic frameworks, and meta-analytic approaches. Such comparative approaches offer opportunities to tease out generalities about the complex and heterogeneous process of speciation that cannot be otherwise obtained. This chapter demonstrates how particular comparative approaches can be used to identify and quantify ecological contributions to reproductive isolation and speciation. Because this volume is about herbivorous insects, our chapter illustrates these points with reference to insect systems.

This chapter has six parts. First, we discuss the nature and study of ecological speciation. Second, we present some advantages of evaluating herbivorous insect systems for such investigations. Third, we describe recently developed comparative approaches that have provided insights on this topic. Fourth, we analyze available herbivore data sets to illustrate these methods and the evidence they provide on ecology's role in herbivore speciation. Fifth, we highlight important issues pertaining to the conduct and interpretation of these comparative studies. Sixth, we suggest future directions for advancing the multidimensional study of ecological speciation through the creative application of comparative approaches.

Ecological Speciation

Although the contribution of natural selection to speciation was not an explicit empirical focus of most of the twentieth century, it was often assumed to be an important factor in the evolutionary diversification of lineages. Indeed, such ideas characterized elements of the evolutionary synthesis. At a macroevolutionary scale, an association between ecology and diversification seemed evident in the oft-observed union of ecological diversity/novelty with species richness in particular taxa (Lack 1947; Simpson 1953; Ehrlich and Raven 1964; Van Valen 1965; Stanley 1979; Mitter et al. 1988; Bernatchez and Wilson 1998; Schluter 2000). Simpson (1953) explained such patterns as "adaptive radiations" driven by the evolution of "key innovations." Such key innovations were traits that provided a lineage with access to a previously underexploited ecological niche and associated resources, namely, a new "adaptive zone." Subsequent partitioning of these new resources, it was argued, would promote the formation and coexistence of a large number of ecologically varied species (e.g., Ehrlich and Raven 1964; Mitter et al. 1988; Hodges and Arnold 1995).

At a microevolutionary scale, midcentury verbal models described how the individual speciation events that cumulatively constitute an adaptive radiation might be causally connected to ecology (Muller 1942; Mayr 1947, 1963). Consider two populations that occupy different habitats/environments. These populations are expected to adapt to their respective habitats and genetically diverge as a consequence. Thus, at loci involved in this local adaptation, alternative alleles will be expected to rise to high frequency or fixation in the two populations. This genetically based adaptive divergence might be expected to incidentally promote the evolution of reproductive isolation between these populations for two reasons. First, divergent selection may have incidental (pleiotropic) effects on reproductive isolation. For example, consider divergent selection on body size in populations occupying alternative habitats. If ecologically adaptive divergence in body size occurs between populations of a species that exhibit size-assortative mating preferences, sexual isolation may be a pleiotropic consequence (Nagel and Schluter 1998; McKinnon et al. 2004). Second, incidental reproductive isolation might evolve if traits that directly affect reproductive isolation are genetically correlated with traits actually under selection for local adaptation. This association between divergent adaptation and reproductive isolation might reflect physical linkage of the underlying loci on a chromosome, so that recombinational separation of their associated alleles is unlikely and reproductive isolation effectively evolves as a consequence of genetic "hitchhiking" (Barton 1995; Hawthorne and Via 2001). Alternatively, this association can owe to other causes of linkage disequilibrium between alleles at adaptive and reproductive loci (Rundle and Nosil 2005).

In recognition of the increasing interest in testing the above models, a term—ecological speciation—was coined to define and delimit the phenomenon under study. To wit, Schluter (2001, p. 372) stated that "the ecological hypothesis of speciation is that reproductive isolation evolves ultimately as a consequence of divergent selection on traits between environments." Because models of sympatric speciation are often inherently ecological and because much study of ecological factors in speciation has been conducted by workers investigating sympatric speciation, ecological speciation and sympatric speciation are commonly, and mistakenly, viewed as interchangeable concepts. By contrast, the original verbal models described above were discussed in terms of allopatric populations. Divergent sources of natural selection can certainly promote ecological divergence and associated reproductive isolation between allopatric populations, and this may be expected to occur readily, given the absence of homogenizing gene flow (Felsenstein 1976; Slatkin 1987; Hendry et al. 2001). Thus, the study of ecological speciation is properly viewed as focusing on the role of ecology, not geography, in the speciation process.

The study of ecological speciation involves exploring the various mechanisms by which divergent selection on ecological traits can cause reproductive isolation between populations. These mechanisms are intuitive when the ecological trait that is diverging under natural selection is the very same trait that yields reproductive isolation, as may occur in certain instances of pleiotropy. Such cases reflect the fact that particular reproductive barriers are inherently ecological. Consider the following scenario: two populations diverge under selection in response to their different habitats. As a result, they evolve a tendency to remain within their local habitat and to survive best in that habitat. This situation will be expected to result in the evolution of at least three inherently ecological reproductive barriers between these populations. These populations will experience habitat isolation because a reluctance to leave the natal habitat restricts opportunities for interbreeding between populations and thus contributes to assortative mating (Rice and Salt 1990; Feder et al. 1994; Johnson et al. 1996; Via 1999). They will experience immigrant inviability (Mallet 1989; Mallet and Barton 1989; Funk 1998; Via et al. 2000; Nosil 2004; Nosil et al. 2005) because those individuals that do migrate to the other habitat will experience higher mortality there and thus are less likely to survive long enough to mate with the resident populations or to produce their full complement of offspring if they do mate. They may also experience extrinsic hybrid inviability if those hybrid offspring that are produced are poorly adapted to either parental habitat and thus experience high mortality in both (Craig et al. 1997; Via et al. 2000; Rundle and Whitlock 2001; Rundle 2002). These examples illustrate an a priori expectation that ecological divergence could commonly contribute to speciation in habitat-specific taxa.

A more challenging task is identifying the potential contributions of ecological divergence to reproductive barriers that are not inherently ecological (Rundle and Nosil 2005). These barriers include sexual isolation (based on behavioral mating preferences), cryptic postmating isolation (due to lower rates of fertilization or zygote development following hybrid matings), and intrinsic hybrid inviability (due to fundamental genetic and developmental incompatibilities between genomes). For such barriers, the contributions of ecology are not obvious and would occur through the invisible effects of pleiotropy and genetic correlations on nonecological traits. Yet the incidental effects on such barriers could represent a major element of ecology's contributions to speciation. Efficiently identifying such contributions requires controlled comparative analyses (see below).

Herbivorous Insect Exemplars

A variety of taxa have contributed to our understanding of ecological speciation. Freshwater fishes have played a central role in these studies (Schluter 1996, 2000; Lu and Bernatchez 1999; Hendry et al. 2000), with various bird (Morton 1975; Slabbekoorn and Smith 2002; Patten et al. 2004; Bearhop et al. 2005), lizard (Odgen and Thorpe 2002), plant (Bradshaw and Schemske 2003; Ramsey et al. 2003), and other taxa (see table 1 in Rundle and Nosil 2005) also providing critical insights. With due respect to the broad and often seminal contributions provided by studies of these taxa, we would argue that herbivorous insect systems are especially likely to offer characteristics that provide copious and general insights (Funk et al. 2002). A few of these characteristics are discussed below.

Most fundamentally, herbivorous insects account for about one-quarter of all multicellular species (Strong et al. 1984). The phylogenetically diverse insect lineages that have evolved the herbivorous habit thus provide many speciation events to account for and many opportunities to study this phenomenon. Indeed, insect herbivores are commonly invoked to illustrate adaptive radiation. This is best exemplified by Ehrlich and Raven's classic paper (1964), in which both insect and plant diversification is described as the consequence of an iterative series of reciprocal adaptive zone invasions. From the insect perspective, this model suggests that a given insect lineage that evolves the capacity to overcome the chemical defenses of a particular plant lineage may thereafter partition and speciate/radiate onto the new host species offered by that plant lineage.

From a more mechanistic perspective, the strong and intimate patterns of ecological specificity that characterize many herbivorous insect lineages may increase opportunities for the individual ecological speciation events that drive such adaptive radiations. Many insect herbivores use but a single plant family, genus, or even species as hosts (Strong et al. 1984; Bernays and Chapman 1994). The great majority of herbivores develop to maturity in direct association with the host plant, which effectively represents their entire habitat. In many groups, this association persists into the adult stage so that virtually all life activities occur in association with the host. However, phylogenetic analyses show that insect lineages do change host associations over evolutionary time (Farrell et al. 1992; Janz and Nylin 1998; Janz et al. 2001; Nosil 2002; Nosil and Mooers 2005). Possible evidence for such incipient "host shifts" are provided by accumulating cases of geographic variation in the particular hosts used by populations of a particular herbivore species (Futuyma and Peterson 1985; Singer et al. 1992; Funk 1998; Nosil et al. 2006a, 2006b; Thompson 2005). Given the intimacy of their host associations, such herbivore populations may commonly be exposed to strongly divergent ecological selection pressures (Bierbaum and Bush 1990; Filchak et al. 2000; Sandoval and Nosil 2005) and their associated incidental effects on reproductive isolation. This is especially true given the preference of many insect herbivores to mate on their native host plant, a habit that will automatically confer a degree of habitat isolation and immigrant inviability (see above) between divergently host-adapted insect populations.

Given these considerations, it makes sense that some of the earliest thought on the relationships between ecology, reproductive isolation, and species status derive from a student of herbivorous insects. Just after Darwin wrote his "big book," Benjamin Walsh wrote about what he called "phytophagic varieties" and "phytophagic species" (Walsh 1864, 1867). Walsh studied apparently related insect populations that were associated with different host plants and noted gradations in the degree to which these populations differed from one another in subtle anatomical traits. Walsh thus speculated that being associated with different plants reduced opportunities for interbreeding and allowed for differentiation and the formation of new species. Later authors, still of the premolecular era, invoked host-plant association as a means of identifying the possible existence of cryptic (i.e., sibling) species, providing early discussions of this phenomenon (Brown 1958, 1959). Among the study organisms that inspired Walsh were *Rhagoletis* fruit flies, the insects that Guy Bush later famously posited to be undergoing sympatric speciation (Bush 1969). Over the last 15 or more years, Bush's student Jeffrey Feder and colleagues have conducted a great diversity of empirical studies that have established the hawthorn and apple "host races" of *Rhagoletis* as perhaps the best-supported case of ongoing sympatric speciation (Feder et al. 1988, 1994, 1997; Feder 1998; Filchak et al. 2000).

In so doing, the *Rhagoletis* work has also provided some of the most compelling examples of inherently ecological reproductive barriers. First, Feder et al. (1994) elegantly demonstrated that each host race exhibits strong "host fidelity" via a rare field study of this phenomenon. This was accomplished by observing that apple and hawthorn flies released away from host plants returned to their native hosts the great majority of the time, a tendency that seems likely to confer strong habitat isolation between the host races. Second, these host races have ecologically diverged in response to the different phenologies of their host plants (apples produce the fruits used by *Rhagoletis* larvae earlier than do hawthorns) in a manner that influences the timing of pupal diapause initiation and adult eclosion. Such timing is critical as, for example, individuals emerging in winter will not survive (Feder et al. 1997; Filchak et al. 2000). Hybrids between these host races will emerge at the wrong time and suffer elevated mortality, yielding postmating isolation through ecologically based hybrid inviability.

Another influential pioneer in the de facto study of herbivore ecological speciation was Tom Wood, to whom this book is dedicated. Wood and colleagues studied habitat isolation among the various host-specific taxa of the *Enchenopa binotata* complex of treehoppers (Wood and Guttman 1982; Wood et al. 1999). Most famously, they ingeniously evaluated the effects of host association on temporal isolation, another ecological reproductive barrier (Wood and Keese 1990; Wood et al. 1990). These treehoppers lay eggs in the stems of their host plant. These eggs are induced to hatch by spring sap flow, but sap flow begins at different times in different host taxa. The result is that different host-associated treehopper taxa begin and complete development at different times. Because *Enchenopa* mating activity occurs over a fairly narrow time window, an observed result of these host-imposed differences in developmental schedule was restricted temporal overlap in mating activity between certain host-associated taxa. Wood's work thus provided some of the earliest detailed empirical evidence for the effect of ecological divergence on reproductive isolation in natural systems. Several other herbivore systems have importantly contributed to our understanding of the ecology of reproductive isolation (Berlocher and Feder 2002; Funk et al. 2002 for review).

Comparative Approaches and Ecological Speciation

When evaluating an evolutionary mechanism, one can ask three different types of questions: Can this mechanism possibly occur? Does this mechanism actually occur? Is this mechanism generally important? The first question can sometimes be answered entirely by theory and an understanding of basic biology. An example is the verbal models of ecological speciation provided during the evolutionary synthesis. These models seem intuitively plausible, based on our fundamental understandings of natural selection and genetic architecture. The second question requires empirical evidence on a given study system. An example is the host-associated habitat isolation documented between *Rhagoletis* host races. This observation provides a degree of corroboration for the hypothesis that ecological factors *can* indeed promote the evolution of reproductive isolation. The third question requires a comparative approach in which the mechanism is simultaneously evaluated in multiple study systems that have been selected at random with respect to past evidence on the occurrence of the mechanism of interest. This strategy allows the rigorous evaluation of how regularly the mechanism occurs in nature. The present chapter describes some comparative approaches and how they may be used to evaluate the hypothesis that ecological speciation is a generally important evolutionary phenomenon in herbivorous insects.

Although our focus here is on evaluating experimentally investigated patterns of reproductive isolation, many readers will first think of phylogenetic analyses when the term "comparative approach" is used, as this is an arena in which such methods are routinely applied (Harvey and Pagel 1991). For example, consider a clever comparative phylogenetic approach that rigorously addressed the question of whether herbivory generally promotes insect diversification (Mitter et al. 1988). This seminal study employed the "sister-group approach," where sister groups are defined as monophyletic lineages (clades) that share their most recent common ancestor. For this study, the authors consulted the literature to find as many documented sister groups as possible where one lineage was primitively herbivorous and its sister lineage was nonherbivorous. They then evaluated the frequency with which the herbivorous lineage in a comparison included more species than its sister group. Because sister groups are, by definition, of equal age, and because all available study comparisons were evaluated (thus avoiding possible bias), this approach allowed the authors to rigorously investigate whether there was a tendency for herbivore lineages to exhibit accelerated species-level diversification rates. Statistical support for this hypothesis indicated that indeed they did. Sister group approaches have since been applied in further studies of insect diversification (e.g., Wiegmann et al. 1993; Farrell 1998).

Mitter et al.'s macroevolutionary approach influenced one of us (D.J.F.) to develop a microevolutionary analog. This approach sought to isolate the contributions of host-plant-associated sources of natural selection (i.e., divergent ecological adaptation) to reproductive isolation and thereby provide a rigorous test of the ecological speciation hypothesis (Funk 1996, 1998). This approach compared levels of reproductive isolation between pairs of allopatric insect populations of two types: those associated with the same host-plant taxon in nature (same-host population pairs) versus those associated with different host-plant taxa (different-host population pairs). This study predicted that if host-associated selection was promoting reproductive isolation, then different-host pairs should tend to be more reproductively isolated than same-host pairs, other things being equal. This is because different-host pairs should have experienced all the factors (genetic drift, sexual selection, etc.) promoting the evolution of reproductive isolation between same-host populations, but may have additionally experienced divergent host-associated adaptation. The initial application of this approach, to *Neochlamisus bebbianae* leaf beetles, yielded results that were in accord with the predictions of the ecological speciation hypothesis (Funk 1996, 1998). Each population in different-host pairs most readily rested/fed on and grew/survived best on its native hosts, while same-host pairs exhibited no population differentiation in these traits. These patterns were consistent with the notion that ecological divergence between different-host populations was promoting two inherently ecological reproductive barriers: habitat isolation and immigrant inviability, respectively. More intriguingly, the different-host pairs also exhibited considerable sexual isolation, while the same-host pairs mated randomly. This same approach was advantageously applied to a much larger number of population comparisons in *Timema* stick insects in a study that demonstrated a statistical tendency for different-host populations to be more sexually isolated than same-host populations (Nosil et al. 2002).

The sexual isolation findings illustrate the utility of comparative approaches for identifying incidental ecological contributions to nonecological reproductive barriers. Simply demonstrating sexual isolation between different-host pairs would not have been sufficient to invoke a role for natural selection, as this isolation could have evolved entirely by other mechanisms, such as genetic drift. It is only by comparing different-host to same-host pairs that the additional contributions of ecology to reproductive isolation can be identified. Yet this is an approach that has rarely been applied. Thus, while evidence on divergence in host preference or host-associated mating times has been increasingly interpreted in the context of ecological speciation (Wood and Keese 1990; Feder et al. 1994; Funk 1998; Via 1999; Rundle and Nosil 2005 for review), little progress has been made in evaluating the incidental evolutionary contributions of ecology to critical reproductive barriers such as sexual isolation and intrinsic hybrid inviability in insect herbivores. (For a nonherbivore example, see informative applications of same-habitat versus different-habitat population comparisons to sexual isolation in stickleback fishes [Rundle et al. 2000; Boughman et al. 2005; Vines and Schluter 2006]).

The same-host versus different-host approach might be viewed as a special case of a broader comparative strategy that asks the following question: to what *degree* does ecological divergence promote the evolution of reproductive isolation? In the special case of the same-host versus different-host approach, the potential continuum of ecological divergence is reduced to two categorical states. On the broader view, the ecological speciation hypothesis yields the prediction that a positive relationship should exist between degree of ecological divergence and degree of reproductive isolation, the strength of which can be quantified. This approach requires the collection of data on ecological variables for the compared populations that can be used to calculate quantitative indices of ecological divergence between them. It also requires sampling population pairs all along the "speciation continuum," that is, at varying stages of evolutionary divergence and levels of reproductive isolation. This approach thus facilitates the investigation of speciation as an extended process.

The rigor and biological relevance of this more general comparative approach is further improved by incorporating a third variable in addition to ecological divergence (ED) and reproductive isolation (RI). Recall the above argument that comparing different-host versus same-host RI could provide insights on the role for ecology, "other things being equal." This critical "other thing" is time. Consider the following scenario: a study happens to evaluate ecologically divergent populations that have been separated and evolutionarily diverging for long periods of time and ecologically similar populations that have been diverging for only short time periods. In this situation, the predicted positive relationship between ED and RI will be observed. However, under these circumstances this relationship (hereafter referred to as the ecology-isolation association) may exist independent of any causal association between the variables of interest. After all, both ED and RI are expected to generally increase with time due only to mutation and random factors (Coyne and Orr 2004). Thus, our hypothetical study's finding might simply reflect the greater time and opportunity for overall differentiation under genetic drift that was experienced by the different-host populations, rather than any contributions from natural selection. Fortunately, the contributions of time can be accounted for in comparative analyses by incorporating a third aspect of evolutionary differentiation: genetic distance (GD). Neutral genetic markers should provide an estimate of GD that is proportional to time under the assumption of a molecular clock. Thus, GD is commonly used as a surrogate for time in comparative analyses (e.g., Coyne and Orr 1989). In a statistical analysis of pairs of populations for which data on GD, ED, and RI are available, the effects of time can be statistically removed by calculating residuals of RI on GD and using these residuals in regression analyses of their association with ED. This approach thus allows rigorous analyses of the time-independent ecology-isolation association per se (Funk et al. 2002, 2006).

This strategy is related to the approach described by Coyne and Orr in the 1989 article cited above. In that study, these authors plotted and evaluated the relationship between GD (time) and RI across pairs of *Drosophila* species. They used this approach (hereafter, 'the Coyne and Orr approach') to evaluate the form and statistical significance of the relationship between time and RI in order to make inferences on the time course of speciation. The present method effectively extends the Coyne and Orr approach into a third, ecological, dimension of evolutionary differentiation (Funk et al. 2002). Thus, it analogously allows one to plot residual (i.e., time-corrected) RI as a function of ED across pairs of populations. From such a plot, one can make observations on the 'ecological course' of speciation by observing the shape of the plot; one can obtain a parameter estimate of the ecology-isolation association by quantifying its direction (positive or negative) and strength (e.g., via partial correlations); and one can evaluate whether this relationship is statistically significant. All of these provide insights into the nature of ecological contributions to speciation within a given higher taxon, such as *Drosophila*.

This new method was recently applied for the first time, thus providing information on the ecology-isolation association in *Drosophila* by adding information on ED to the classic Coyne and Orr data sets (Funk et al. 2006). However, this new study was not restricted to *Drosophila*. Instead this method was applied separately to each of the data sets that had previously been evaluated by the Coyne and Orr approach and cited in Coyne and Orr's recent (2004) book, *Speciation*. Eight disparate taxa were thereby included: angiosperms, *Drosophila*, Lepidoptera, darters (percid fishes of the genus *Etheostoma*), other fishes, frogs, birds, and pigeons/doves (see citations in Funk et al. 2006). Thus, this paper included eight simultaneous applications of this new approach. It required the consultation of over 200 literature sources to collect the required ecological data from over 500 species. This intensive approach reflects the major goal of Funk et al. (2006): to evaluate the degree to which a positive ecology-isolation association is a general feature of biologically disparate taxa. In other words, this study evaluated whether available data suggest that speciation is commonly an inherently ecological process. Given this singular goal, the details of these relationships within each of the eight study taxa were not a focus of interest. Rather, each of the eight taxa was specifically used to obtain parameter estimates of the ecology-isolation association that could be used in cross-taxon analyses. This study revealed ecology-isolation associations to be positive for seven of eight study taxa (and only slightly negative for the eighth) ($p < 0.05$; binomial test). As a group, the mean ecology-isolation association across taxa proved to be significantly greater than zero ($p = 0.004$; *t*-test). This positive association was also observed when data were separately analyzed with respect to different components of RI and different ecological traits, and was robust to a variety of approaches to data treatment and analysis. This study thus provided highly consistent support for the hypothesis that ecology plays an evolutionarily general role in speciation.

Herbivore Analyses and Insights

Funk et al. (2006) provided compelling results. However, the focus of this paper on the cross-taxon ecology-isolation association precluded a consideration of the additional insights that can be gleaned from a fuller exploitation of this comparative approach. Here, we provide a more detailed inspection of the few herbivore taxa for which appropriate data (i.e., on GD, ED, and RI across multiple population pairs) are available for comparative analyses of the sorts described above. We analyze these data to illustrate these methods and associated issues, and to evaluate initial patterns in the ecology-isolation associations of herbivorous insects. Given the theme of this chapter, the only reproductive barriers evaluated here are nonecological barriers, that is, those for which a comparative approach proves especially useful in identifying the incidental evolutionary contributions of ecological divergence to the evolution of reproductive isolation.

Study Taxa

We treat four data sets, conveniently representing taxa from four divergent phylogenetic lineages of insects: Diptera, Lepidoptera, Coleoptera, and Orthopteroidea. The first three are holometabolous, undergoing a metamorphosis between morphologically divergent larval and adult stages via a quiescent pupal stage. The last is hemimetabolous, requiring no metamorphosis, with immatures that differ from adults primarily in size and the absence of wings.

DROSOPHILA

Drosophila (Diptera: Drosophilidae) are atypical "herbivores" that are probably best described as saprophytic. Although many taxa use angiosperm hosts as sites of oviposition, larval development, and adult feeding, both larvae and adults often actually feed on yeasts and other microorganisms associated with decaying plant tissues. Females lay their eggs on or close to the larval substrate, and larvae complete development on a single plant individual. The degree of host specificity varies greatly, with some species using a single host plant and others using hosts from many different plant families. Mating occurs on host substrates in some species and away from hosts in others.

LEPIDOPTERA

Lepidoptera are moths and butterflies. These insects typically lay eggs on host plants. These eggs hatch into caterpillar larvae that feed and develop on host foliage. In some species, caterpillars complete development on the natal plant, while in others individual caterpillars move among and feed on different plant individuals or even on different host taxa. The degree of host specificity exhibited by larvae and ovipositing females varies considerably among species but is typically lower than in *N. bebbianae* and *T. cristinae*. Pupation may occur on or off of the host. Typically, adults are generalist nectarivores and mating is not closely tied to host plants.

NEOCHLAMISUS BEBBIANAE

Neochlamisus bebbianae (Karren 1972) is an eastern North American species of leaf beetle (Coleoptera: Chrysomelidae) that is closely associated with its host plants during all life-history stages (Brown 1943, 1946, 1952, 1961; Karren 1972; Funk 1998; Brown and Funk 2005). Eggs are laid on host foliage, and larvae feed and develop entirely on the natal plant, which is also the usual site of pupation. Adults feed and mate on the host, on which they are active for many weeks both preceding and following an adult winter diapause. *Neochlamisus bebbianae* populations are associated with particular species from six taxonomically disparate tree genera: red maple (*Acer rubrum*, Aceraceae), certain alders (*Alnus rugosa/incana*, *A. serrulata*, Betulaceae), river birch (*Betula nigra*, Betulaceae), American hazel (*Corylus americana*, Corylaceae), certain oaks (*Quercus* spp., Fagaceae), and Bebbs willow (*Salix bebbiana*, Salicaceae). In contrast to this pattern of apparent oligophagy, other *Neochlamisus* species tend to specialize on a single host-plant genus or species. Recent work indicates that populations of *N. bebbianae* associated with different host plants are divergent in host preference and performance traits, with populations most readily using the host with which they are associated in nature (Funk 1998; Funk et al. 2002; Egan and Funk 2006; D.J.F., unpublished data). Based on this and other evidence, *N. bebbianae* is considered to represent a complex of six partially ecologically and evolutionarily differentiated "host forms," each comprising those populations associated with a particular host plant (i.e., the alder host form, the maple host form, etc.).

TIMEMA CRISTINAE

Timema cristinae walking sticks (Phasmatodea: Timemidae) are wingless insects that inhabit the chaparral of a restricted area of southern California (Vickery 1993; Crespi and Sandoval 2000). Each of the 21 described species in this genus uses between one and a few host-plant taxa, and *T. cristinae* uses two hosts: *Ceanothus spinosus* (Rhamnaceae) and *Adenostoma fasciculatum* (Rosaceae) (Sandoval 1994a, 1994b). Most life activities, including mating, occur on host plants, although oviposition is often near rather than on the host. Experimental work has demonstrated divergent and specialized host preferences in populations associated with different host species (Nosil et al. 2006a, 2006b), but no evidence for divergence in the capacity of nymphs to develop on a given host (Sandoval and Nosil 2005). Survivorship of immatures is nonetheless host specific because host-associated populations exhibit morphological divergence in the traits promoting crypsis on one or the other of these structurally divergent plants, leading to high predation rates for insects on the "wrong" host (Sandoval 1994b; Nosil 2004). Hosts usually occur in homogeneous patches of modest size, and rates of migration between adjacent patches by *T. cristinae* are low (as in Nosil et al. 2002, 2003; Nosil and Crespi 2004).

Data Sets

DROSOPHILA

GD and RI estimates derive from the seminal studies of Coyne and Orr (1989, 1997). GD is estimated as Nei's *D* (1972) from allozyme data. Three components of RI were available. Sexual isolation estimates were derived from laboratory mate-choice trials. Postmating isolation represents an index that combines hybrid inviability and hybrid sterility. Total isolation represents the combined contributions of these pre- and postmating components. ED was calculated as the proportion of host-plant genera that are not shared by the two species, that is:

ED = 1 − [number of host genera shared by species 1 and 2/ (number of genera used by species 1 only + number of genera used by species 2 only + number of genera shared by species 1 and 2)]

The data treated here are those analyzed in Funk et al. (2006).

LEPIDOPTERA

Lepidoptera: GD and RI estimates derive from Presgraves (2002). GD was based on estimates of Nei's *D* that were derived from allozyme and DNA sequence data. Three aspects of RI were available. For sympatric species, data on premating isolation was presented in the form of the presence or absence of reported observations of hybridization between species in nature. Postmating isolation is represented by hybrid inviability data. Total postmating isolation represents a combination of hybrid inviabilitiy and hybrid sterility. (Hybrid sterility data were not presented in Presgraves' [2002] article and so are not separately evaluated here.) ED was calculated as for *Drosophila* except that the proportion of shared host-plant families rather than genera was evaluated. The postmating data treated here are those analyzed in Funk et al. (2006), whereas the premating data have not been previously analyzed in an ecological context.

NEOCHLAMISUS BEBBIANAE

We evaluate each of the eight unique pairwise host-form comparisons for which appropriate data are available. Two of these comparisons derive from Funk (1998): one is a comparison between birch and maple host forms and the other is a comparison between geographically divergent maple-host-form populations from Georgia and New York (representing a same-host comparison). The remaining host-form comparisons derive from a recent study (D.J.F. and S.P.E., unpublished): alder versus hazel, alder versus maple, alder versus willow, hazel versus maple, hazel versus willow, and maple versus willow host forms. GD is estimated as mean pairwise divergences in ca. 1 kb DNA sequences of the cytochrome oxidase I (COI) mitochondrial gene, corrected for multiple hits using the Kimura two-parameter model (see Funk 1999 for details). For RI, sexual isolation was quantified on a scale of −1 to +1 via transformations of the Levene's isolation index, as in Funk 1998 (0 = random mating, 1 = completely assortative mating = complete sexual isolation) (Ehrman and Petit 1968; Spieth and Ringo 1983). ED derives from female host-fidelity data collected during pairwise choice trials in Petri dishes. In these trials, individual beetles of a given host form were simultaneously presented with foliage of their native host and with foliage from the native host-plant of the host form with which they were being compared. Beetles were observed at regular intervals, at which times the plant on which they were resting was scored (Funk 1998; D.J.F. and S.P.E., unpublished). From these data, we calculated ED for a given host form comparison as the absolute value of the difference in the mean proportions of observations in a trial in which beetles from each host form were resting on one of the two test plants (the reference plant). For example, for the maple and willow host forms:

ED = |% observations of maple host form beetles on maple foliage − % observations of willow host form beetles on maple foliage|

Note that the test plant chosen as the reference has no effect on results because absolute values are evaluated. For the maple_{GA} versus maple_{NY} comparison, ED was based on responses to maple versus birch foliage because these populations were tested during a year of studies in which they were also compared to the birch host form (Funk 1998). This approach yields ED values that can range from 0.0 (indicating identical relative test-plant usage by the two host forms, and thus no ED) to 1.0 (indicating that each host form rested only on its native host, indicating complete ED).

TIMEMA CRISTINAE

GD is estimated as mean pairwise divergence in 467 bp sequences of COI and 409 bp sequences of an internal transcribed spacer (ITS-2) of nuclear ribosomal RNA genes. Sequence divergences were corrected for multiple hits using the Kimura two-parameter model. For RI, sexual isolation was quantified on a scale from −1 to +1 as for *N. bebbianae*, using the I_{PSI} index (Rolán-Alvarez and Caballero 2000), as in the original Nosil et al. (2002) study. Postmating, prezygotic isolation occurs by several mechanisms sometimes referred to as cryptic RI (Price et al. 2001; Coyne and Orr 2004; Nosil and Crespi 2006). Cryptic isolation for a given population pair was estimated as the difference in mean fitness (fecundity) of females participating in within-population versus between-population matings:

cryptic isolation = 1 − mean lifetime oviposition of females used in between-population matings/mean lifetime oviposition of females used in within-population matings.

Intrinsic postzygotic isolation was analogously evaluated by estimating:

intrinsic F_1 hybrid egg inviability = 1 − % of eggs hatched in broods derived from between-population crosses/% eggs hatched in broods derived from within-population crosses

(data from Nosil et al. 2006c). ED for all *T. cristinae* analyses was categorically scored according to the same-host versus different-host status of population pairs.

Data Independence

Statistical significance testing requires that data points present independent estimates of the parameter of interest. With respect to the present study, this issue is relevant in two contexts. First, the importance of historical independence reflects the fact that species sharing a common ancestry will share evolutionarily inherited commonalities. This degree of nonindependence between related species may bias significance tests by inflating degrees of freedom (Felsenstein 1985; Harvey and Pagel 1991). For this reason, methods have been introduced that use phylogenetic relationships to identify sets of species comparisons that are less affected by such historical nonindependence. Applying such methods yields phylogenetically adjusted data sets that include fewer comparisons but can be more confidently used for significance testing. Here, significance testing for *Drosophila* and Lepidoptera employs data sets that were phylogenetically adjusted following the methods used by the authors of the original Coyne and Orr data sets. The ability to make such phylogenetic adjustments, however, often breaks down at the intraspecific level, where treelike phylogenetic relationships among populations may not be recoverable or even exist due to such factors as gene flow and incomplete lineage sorting (Funk and Omland 2003). Moreover, at such low levels of evolutionary divergence the length of shared history may be anticipated to be sufficiently small that it represents a minimal source of bias, and that including genetic distance in analyses of the association between ED and RI effectively accounts for shared ancestry (Thorpe 1996; Rundle et al. 2000). Hence, we do not address the issue of historical independence for the *N. bebbianae* and *T. cristinae* data sets.

Second, statistical independence requires that a given study population not be used in more than one pairwise comparison. This is because a set of comparisons that involve the same population do not represent true replicate tests of a hypothesis. Any particular tendencies of that population may have similar effects on the results of all comparisons in which it participates, biasing cross-comparison results and effectively inflating degrees of freedom. This issue is potentially relevant for population comparisons of the *N. bebbianae* and *T. cristinae* data sets. For *T. cristinae*, population comparisons for postmating analyses are statistically independent, as no study population was used in more than one comparison. However, this is not the case for the sexual isolation analyses of *N. bebbianae* or *T. cristinae*, in which particular populations are used multiple times. Such statistical nonindependence can be accounted for in significance testing using randomization approaches such as Mantel tests (Manly 1997). Unfortunately, the minimum number of comparisons necessary to obtain significant results with this method (10) was greater than the number of *N. bebbianae* comparisons evaluated here. Thus, although we present a p value for the *N. bebbianae* ecology-isolation association to allow comparisons across taxa, this value should be interpreted with caution. For *T. cristinae*, however, such Mantel tests (using 10,000 randomizations) were informatively applied (to $n = 28$ pairwise population comparisons).

An important distinction must be made here: while nonindependence should be removed or accounted for in significance testing, it does not clearly compromise parameter estimation (Sokal and Rohlf 1995). That is, when estimating a parameter such as the partial r value for the ecology-isolation association, the analysis of all comparisons, independent or not, may be justified. Including all the data provides the most information for parameter estimation, and simulations have shown explicitly that estimating the evolutionary correlation between continuous traits without prior phylogenetic adjustment does not lead to biased estimates of the correlation (Martins and Garland 1991). Thus, in the current study we conduct parameter estimation using all the data, while for significance testing we account for historical and statistical nonindependence as described above.

Data Analysis

For the first three data sets, our analyses adopted the regression-based approach briefly described above (Funk et al. 2002, 2006). These analyses were based on quantifying the association of ED with RI, independent of time. Values for this association were obtained from parametric regression analyses of residuals. These analyses treated GD (as a surrogate for divergence time) and ED as independent variables, and RI as the dependent variable. Specifically, residuals of RI on GD were calculated prior to determining the association of these residuals with ED using regression analysis. Before conducting these analyses, we first evaluated whether the ED-GD and RI-GD relationships of the data were linear, as assumed by linear regression. Deviations from linearity were detected in only a few cases, and in these instances we found that a quadratic regression provided a proper fit to the data. We thus applied linear and/or quadratic regression as appropriate to derive the residuals used in our analyses (see Funk et al. [2006] for a detailed description of this approach). We conducted separate analyses for each reproductive barrier for each data set. Because the RI variable for the "hybridization" data of the Lepidoptera data set was binomial (yes or no), the residuals of RI on GD were estimated using logistic rather than linear/quadratic regression for subsequent analysis.

It was not possible to apply the regression approach to the *T. cristinae* data for three reasons. First, ED was a categorical rather than continuous variable. Second, as described above, comparisons used in sexual isolation analyses were not statistically independent, and multivariate Mantel tests–which would be required to control for GD using such non-independent data–often perform poorly in these circumstances (Raufaste and Rousset 2001). Third, for the postmating forms of RI, genetic data were not available for the specific population pairs that were evaluated. Thus, analyses of the ecology-isolation association from this same-host versus different-host approach involved (1) univariate Mantel tests for the sexual isolation data, and (2) standard *t*-tests for the independent comparisons offered by the postmating data.

Nonetheless, strong, but indirect, evidence suggests that the associations between ED and RI observed for the *T. cristinae* analyses (see below) are independent of divergence time. Two points support this argument. First, there was no association between GD and RI among *T. cristinae* populations (e.g., for mtDNA and sexual isolation data from 66 population pairs [Nosil et al. 2003]: $r = -0.11$, $p > 0.50$, Mantel test; similar results were obtained from nuclear DNA). This indicates that variation in divergence time does not explain variation in RI. Given this observation it seems unlikely that time would nonetheless confound observed associations between ED and RI. Second, different-host population pairs show no mean tendency to be more genetically divergent than same-host pairs (e.g., for mtDNA and sexual isolation data from 36 different-host pairs and 30 same-host pairs [Nosil et al. 2003]: Mantel test $t = 0.97$; $p = 0.83$; similar results were obtained from nuclear DNA). These results provide more specific evidence that observed differences are not biased by time. This is the same reasoning offered by Coyne and Orr (1989) to argue that time did not affect the overall differences they observed between components of RI and between allopatric versus sympatric population pairs.

As in Funk et al. (2006), we also evaluated whether the ecology-isolation association is consistently positive when evaluated across the variety of taxa treated here. To accomplish this, we first separately calculated the mean parametric association (partial *r*) value for each taxon-specific data set by averaging the values provided by analyses of different reproductive barriers for that taxon. We then used a one-tailed, one-sample *t*-test to evaluate whether the mean of these four data set means was significantly greater than zero, as per the a priori prediction of our hypothesis. By contrast, under the null hypothesis of no taxonomically general relationship between these two aspects of evolutionary divergence, means of individual data sets should randomly vary around zero, and the mean across data sets should not differ significantly from it.

Results and Interpretations

The findings we report here only overlap with Funk et al. (2006) in including parameter values for the ecology-isolation association in *Drosophila* and Lepidoptera (Table 9.1). Following the approach of that work, however, the first results we discuss here concern patterns of taxonomic generality in these association values. The most macroevolutionarily important observation from these analyses is that in four of four taxa, mean ecology-isolation association values were positive (*Drosophila* = 0.07, Lepidoptera = 0.09, *N. bebbianae* = 0.83, *T. cristinae* = 0.58). The random probability of all four of these values being in the predicted direction is $p = 0.0625$ (one-tailed binomial test). Interestingly, a more powerful parametric analysis yielded a comparable result ($p = 0.0635$, one-sample *t*-test), reflecting the high variance and small number of these associations. The consistency of these positive associations is particularly striking given the somewhat crude nature of our ED estimates. These cross-taxon results are thus consistent with the hypothesis that ED contributes to the evolution of RI across phylogenetically and biologically disparate herbivorous insect lineages.

At the level of the individual analyses of each reproductive barrier, consistency is also evident. In 8 of 10 of these analyses, association values are positive (Table 9.1). In 4 of 10 analyses the association is significantly positive at $p \leq 0.026$ (and $p = 0.075$ in a fifth analysis). These five analyses are distributed across all four taxa and represent both premating and postmating data. With this in mind, it is worth reiterating that all analyzed reproductive barriers are non-ecological in the sense that there is no a priori biological reason to assume that ED (here, in host-associated traits) should generally confer RI. Instead, these findings are most readily interpreted as supporting the verbal models in which RI evolves as an incidental consequence of divergent adaptation due to genetic pleiotropy or linkage.

TABLE 9.1

Comparative Analyses of the Association Between Ecological Divergence and Aspects of Reproductive Isolation that are Not Essentially Ecological in Nature

	Ecological Divergence	*Reproductive Isolation*		*Parameter Estimation*			*Significance Testing*	
Taxon	Based On:	Pre- or Postmating	Barriers Evaluated	*N*	Parameter	Strength of ED-RI Association	*N*	*p*
Drosophila	% shared hosts	Pre-	Sexual isolation	30	*r*	**0.18**	14	0.464
Drosophila	% shared hosts	Post-	Hybrid inviability + infertility	35	*r*	−0.27	19	0.210
Drosophila	% shared hosts	Both	Total isolation	50	*r*	**0.31**	13	0.075~
Lepidoptera	% shared hosts	Pre-	Natural hybridization?	51	*r*	**0.23**	32	0.015*
Lepidoptera	% shared hosts	Post-	Hybrid inviability	66	*r*	**0.17**	46	0.131
Lepidoptera	% shared hosts	Post-	Total postmating isolation	48	*r*	−0.14	33	0.468
N. bebbianae	Host preferences	Pre-	Sexual isolation	8	*r*	**0.83**	8	0.011*
T. cristinae	Same- vs. different-host	Pre-	Sexual isolation	28	$Y_1 - Y_2 / (Y_1 + Y_2)$	**0.50**	28	0.006*
T. cristinae	Same- vs. different-host	Post-	Cryptic isolation (fecundity)	6	$Y_1 - Y_2 / (Y_1 + Y_2)$	**0.81**	6	0.026*
T. cristinae	Same- vs. different-host	Post-	Hybrid inviability	6	$Y_1 - Y_2 / (Y_1 + Y_2)$	**0.44**	6	0.340

NOTE: N = number of species or population pairs analyzed. For the regression-based analyses, r values indicate the strength of the association between ecological divergence and reproductive isolation after time (genetic distance) has been statistically removed. Parameter estimates are based on the unadjusted data sets while significance testing is based on the phylogenetically adjusted data sets. For Lepidoptera, time-independent "natural hybridization" (yes/no) was analyzed using the residuals from a logistic regression. The p value for the *N. bebbianae* analyses must be interpreted with caution as individual test populations were used in multiple comparisons, which are thus not statistically independent. For the same-host versus different-host analyses, effects of ecological divergence on reproductive isolation are quantified by subtracting the mean same-host reproductive isolation from the mean different-host reproductive isolation and dividing by the absolute value of the sum of these values. Parameter estimates indicating the predicted positive ecology-isolation association are presented in bold. For the Lepidoptera, "total postmating isolation" represented a combined index of hybrid inviability and hybrid sterility; we do not present a separate analysis of sterility as these data were not provided in the original article. Abbreviations: pre-, premating reproductive barriers; post-, postmating reproductive barriers * = $p < 0.05$, ~ = $0.05 < p < 0.10$.

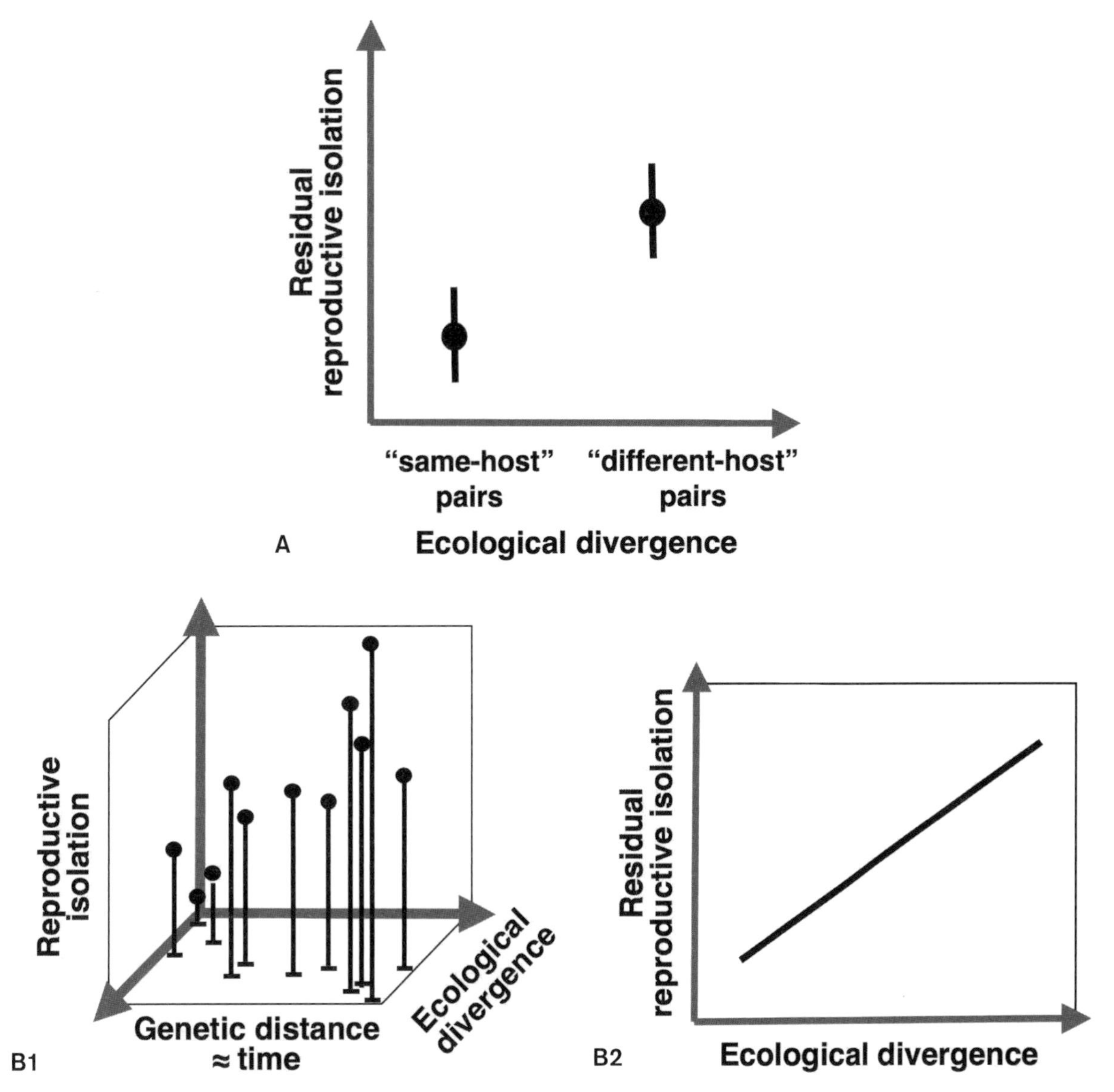

FIGURE 9.1. Comparative analyses of ecological contributions to herbivore speciation. See text for further methodological details and Fig. 9.2 and Table 9.1 for results of the application of these methods. A. The same-host versus different-host approach (Funk 1996, 1998). This approach compares reproductive isolation between pairs of populations expected to have experienced a lesser degree of ecologically divergent selection (same-host population pairs) with those expected to have experienced a greater degree of divergent selection (different-host pairs). A finding that different-host pairs tend to be more reproductively isolated than same-host pairs thus supports the hypothesis that ecological divergence promotes speciation. The plot illustrates predicted patterns. B. The regression-based approach (Funk et al. 2002, 2006). This approach simultaneously evaluates quantitative estimates of genetic distance (a surrogate for time), ecological divergence, and reproductive isolation. It statistically removes the contributions of time and thereby explicitly isolates and quantifies the association between ecological divergence and time-independent residual reproductive isolation. B1. A hypothetical data set, illustrating the distribution of paired population or species comparisons (data points) with respect to the three analyzed aspects (axes) of evolutionary differentiation. B2. A hypothetical best-fit line illustrating the predicted positive association between degrees of ecological divergence and of residual (i.e., time-adjusted) reproductive isolation. (B1 and B2 are modified from Funk et al. 2006; copyright 2006, National Academy of Sciences, U.S.A.).

At a yet finer level, plotting results for individual pairwise population comparisons for each taxon and reproductive barrier allows an inspection of the shape of the 'ecological course' of speciation in these taxa (Fig. 9.1). This is analogous to Coyne and Orr's (1989) visual inspection of their plots of the time course of speciation in *Drosophila*. In this case, doing so reveals few obvious patterns that are not already summarized by the ecology-isolation association values. For *Drosophila* and Lepidoptera, perhaps the most general observation represents a counterpoint to one of the more famous inferences from Coyne and Orr's original article. In plotting their data, Coyne and Orr observed that high levels of RI were not observed at low levels of genetic distance. Together with other patterns, this observation indicated that RI does not evolve very soon after initial divergence in *Drosophila* and that speciation in these flies is thus a gradual process. In contrast, plots for both *Drosophila* and Lepidoptera (Fig. 9.2A, 9.2B, 9.2C, 9.2D, 9.2E, 9.2F) reveal considerable variation in the amount of RI observed at the lowest levels of ED, including instances of very high RI. This suggests that considerable

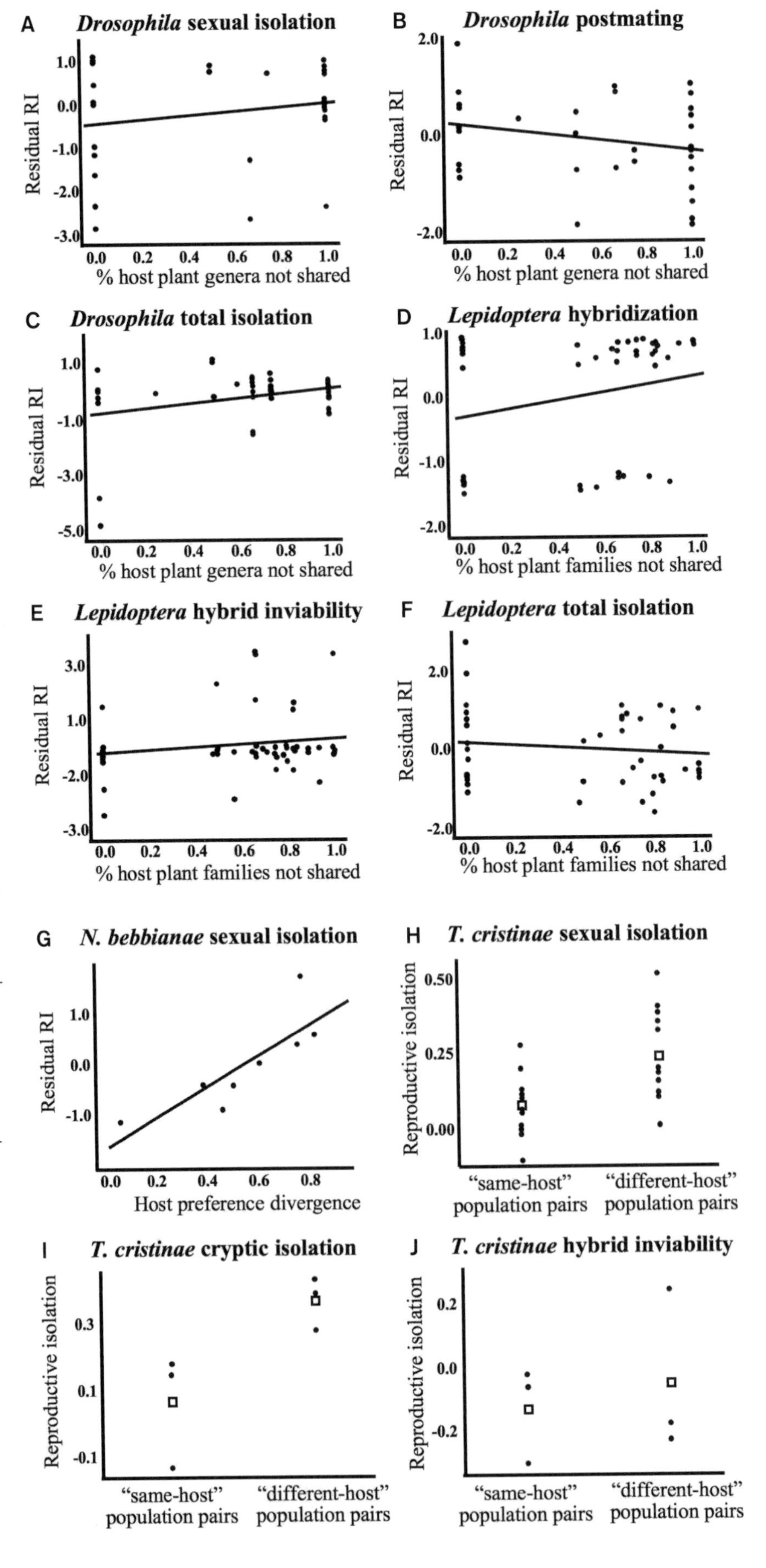

FIGURE 9.2. Raw results for the analyzed herbivore data sets. The x axes indicate aspects of ecological divergence evaluated, with degree of ecological divergence increasing from left to right. The y axes indicate amount of reproductive isolation. The reproductive barrier evaluated is indicated at the top of each plot, as is the study taxon analyzed. Each point represents the comparison of a particular pair of populations or species. Plots H through J illustrate the same-host versus different-host approach. Treatment means are represented by open boxes. Although time of divergence is not presented here, it does not affect conclusions on the effects of ecological divergence in these taxa, as described in the text. Note that in J the predicted positive ecology-isolation association is observed but is unusual in that reproductive isolation is negative for both types of population pairs, but less negative for different-host pairs. Plots A through E and plot G illustrate the regression-based approach, with best-fit lines depicting the ecology-isolation association. Plot F represents a modification of this approach that employs logistic regression to evaluate a qualitative estimate of reproductive isolation (whether or not hybridization between species pairs has been observed in nature). These plots show the association between ecological divergence and nonecological components of reproductive isolation to be positive in 8 of 10 analyses.

ED (at least as estimated here) may not be a strict requirement of speciation in these taxa and that the speciation process is not necessarily one of ecological gradualism. Intriguingly, quite different patterns are revealed in the plots of *N. bebbianae* and *T. cristinae* (Fig. 9.2G, 9.2H, 9.2I, 9.2J). For these taxa, the highest levels of RI are indeed only reached in population comparisons of high ED. Furthermore, in *N. bebbianae*, the increase in RI with increasing ED is strikingly linear and gradual (D.J.F. unpublished), consistent with the idea that the evolution of sexual isolation among the host forms of these beetles is very tightly associated with their evolutionary divergence in host preferences.

The difference among study taxa in these plotted patterns could be a consequence of variation in data quality. For example, perhaps *Drosophila* and Lepidoptera would exhibit patterns more similar to *N. bebbianae* and *T. cristinae* if their host associations were more adequately documented, resulting in more accurate estimates of ED for their species comparisons. Or perhaps more accurate ecology-isolation association values can be obtained from the analysis of conspecific populations than by evaluation of longer-diverged species. Alternatively, these patterns might reflect real biological differences and their implications for ecological speciation. Based on the basic host-associated biology of these taxa (described earlier), one might readily guess that *Drosophila* and Lepidoptera would typically be subjected to less intensive host-associated ecological selection pressures than *N. bebbianae* and *T. cristinae*, due to a trend toward less intimate host-insect associations in the former pair of study taxa. Higher levels of host specialization and a greater number of tightly host-associated life-history stages in *N. bebbianae* and *T. cristinae* support this claim. Further, one could argue that *Drosophila* should be least subjected to divergent ecological selection because it feeds on microorganisms in addition to the host itself. One could similarly argue that *N. bebbianae* may be subjected to the most intense host-related selection because, unlike *T. cristinae*, its host forms are physiologically adapted to their particular hosts, to the extent that some host forms are completely incapable of development on each other's hosts (Funk 1998). So, if reproductive isolation evolves in proportion to the intensity of suspected host-associated selection pressures, one might predict the same linear order of association values observed here: *Drosophila* < Lepidoptera < *T. cristinae* < *N. bebbianae*. This post hoc thought experiment is clearly somewhat subjective and in no way statistically defensible (see also the comments on sampling bias below). However, it highlights the biological heterogeneity that underlies the consistently positive ecology-isolation associations observed here and their relevance for comparative analyses of ecological speciation.

Comparative Caveats

One goal of this chapter is to advocate and describe the use of comparative approaches for studying ecological speciation. In the preceding sections, a number of issues have been raised that are important to consider when adopting such approaches. The present section highlights a few such topics that deserve introduction or further elaboration.

Model System and Comparative Approaches

Approaches to studying the mechanisms responsible for biological phenomena vary from focused investigations of individual study systems to broad comparisons across disparate taxa. The "model system" approach can provide a detailed and integrated understanding of a biological system. Such intensive study may provide considerable confidence in inferences about proximate mechanisms, especially when these can be directly observed. This approach is most useful for evaluating whether a hypothesized mechanism can and does occur, and how it does so. By contrast, the confidence placed in the results of comparative hypothesis testing is based on indirect statistical inference and the capacity to rule out alternative explanations for observed patterns across taxa. The comparative approach is, however, necessary to evaluate how general and biologically/evolutionarily important a mechanism or phenomenon of interest appears to be. Comparative approaches additionally provide means of testing hypotheses that are not generally accessible using the model system approach.

Advances in biology depend on both approaches. In the context of ecological speciation in herbivorous insects, detailed and multifaceted studies of model systems such as *Rhagoletis* fruit flies (Feder et al. 1988, 1994, 1997; Filchak et al. 2000) and *Enchenopa* treehoppers (Wood and Guttman 1982; Wood and Keese 1990; Wood et al. 1990, 1999; see Funk et al. 2002 for a summary of additional systems) have provided a detailed understanding of various mechanisms by which ecological divergence causes reproductive isolation. Meanwhile, by showing a cross-taxa statistical association between the evolution of herbivory and increased rates of species-level diversification, the comparative phylogenetic approach adopted by Mitter et al. (1988) provided an evolutionarily general finding that helped motivate the development of ecological speciation as a field of study. Findings from the comparative analyses presented in this chapter, although somewhat preliminary, furthermore support the hypothesis that ecological divergence consistently promotes the evolution of reproductive isolation and thus may be an evolutionarily general contributor to insect speciation. Moreover, this approach documented positive ecology-isolation associations for reproductive barriers that are not intrinsically ecological in nature. By contrast, directly documenting the proximate genetic and physiological mechanisms that explain, say, the statistical association between ecological divergence and the host-independent sexual isolation observed for a single taxon such as *N. bebbianae* might represent an enormous enterprise requiring intensive experimental and molecular analysis.

Correlation, Causation, and Replication

While both approaches have advantages, they also have limitations. One limitation that is commonly attributed to comparative approaches is the well-known scientific aphorism: "correlation does not equal causation." This important truism reflects the fact that when two variables (A and B) are statistically associated this may not be because A causes B, as the investigator imagines. It could also be because B causes A, or because A and B are associated through one or more common causes (C, D, etc.) that have not been evaluated. This issue is relevant for all correlational analyses, but it is not, as is sometimes supposed, specific to comparative analyses. Rather it applies to all statistical hypothesis testing. A statistical result derived from an ANOVA-based study of two populations is no more immune to alternative interpretations than is one derived from regression analyses of data from many taxa. Indeed, both types of analyses are special cases of general linear models (Sokal and Rohlf 1995). Regardless of the approach employed, findings may be explainable by alternative hypotheses, which should be explored. Yet the investigation of any question has to begin somewhere and statistically supported comparative findings are not deserving of extra skepticism.

The criticism of comparative inferences as "only correlational" can be seen to be somewhat ironic to the degree that results from single-system studies are interpreted as evidence on the existence of general biological phenomena. Consider a hypothetical study of a pair of populations. The investigators examine diverse aspects of the biology of these two populations in great detail and demonstrate that they are highly ecologically divergent and also highly reproductively isolated through the contributions of various reproductive barriers. Moreover, the genetic basis and biochemical pathways underlying many of the traits of interest are worked out. Clearly, much has been learned about the biology of these populations and many biological conclusions have been made with a high degree of confidence. However, with respect to the inherently correlational/comparative hypothesis that ED exhibits an evolutionary tendency to promote RI (i.e., that ecological speciation is an important evolutionary phenomenon), these intensive studies represent a sample size of $N = 1$. This observation is closely tied to the problem of pseudoreplication (Hurlbert 1984), whereby inferences are made at levels of generality higher than is merited by the level at which replicated observations are made. Only through comparative studies (for which N = many) can inferences on the general evolutionary importance of biological phenomena be legitimately made.

Bias in Comparative Analyses

As with any other sort of statistical inference, making accurate inferences from comparative analyses does assume that units of study have been selected in a manner that is random with respect to the issue being studied. That is, they require the avoidance of potential sampling bias. In the Mitter et al. (1988) study, the authors avoided such bias by intensively searching the literature to identify all phylogenetically supported sister groups in which one lineage was herbivorous and the other was not. Doing so avoided the possibility of unconsciously biasing results in the direction of their favored hypothesis. Such bias could otherwise have been introduced if, for example, the authors had evaluated only those sister groups with which they were already familiar. In that case, it would not be surprising if the data set were biased, given that more scientists may be more prone to remember cases they find to be especially interesting and supportive of their hypotheses. The issue of potential bias is one reason why the work by Funk et al. (2006) was necessary and (hopefully) compelling. Although investigations of several specific study systems had already provided support for associations between ED and RI in particular cases, some of these systems may have been selected for study because of an a priori suspicion that they would exhibit the predicted relationship and thus be suitable for studying it in detail. It is only problematic if general inferences were based on the collective finding of such studies, in which case such bias would yield an overestimate of the natural frequency and importance of ecological speciation. The potential for such sampling bias was avoided by Funk et al. (2006) due to the fortuitous publication of Coyne and Orr's (2004) list of previously published Coyne and Orr studies. Because these original studies were not conducted with ecological factors in mind, there is no reason to suspect them to represent a biased sample with respect to the hypothesis actually under test in Funk et al. (2006). However, the cross-taxon analyses of herbivore taxa in the present chapter may indeed suffer somewhat from this bias. This is because the continued development of *N. bebbianae* and *T. cristinae* as study systems has indeed been encouraged by initial findings that were consistent with a positive ecology-isolation association.

Commensurable Comparisons

One challenge of comparative studies is that the heterogeneous biologies of disparate taxa and the limited availability of appropriate data make it difficult to compare apples with apples rather than with oranges. For such reasons, the components and quantification of ED and RI in our analyses were somewhat different for each of our four focal taxa. Such variation does not compromise our general interpretations about the observed consistency of positive ecology-isolation associations. (Indeed, the fact that positive associations were observed despite this heterogeneity offers a further argument for their generality.) However, it does limit the variety of rigorous comparisons that can be made and the confidence in the associated inferences. For example, the earlier rankings of partial correlations for the ecology-isolation association among our study taxa would be more compelling and easily interpreted if these values were derived from a more homogeneous set of ED and RI variables. Using the same variables across taxa thus allow for greater biological control in such comparisons.

Choosing Ecological Traits and Indices

When quantifying ED between pairs of populations or species for comparative studies of ecological speciation, any of a variety of ecological traits might be chosen for this purpose. If the goal is to document whether a positive ecology-isolation association exists, one should choose traits suspected to have played an important role in the adaptation of one's study organisms to their respective environments. Those traits that have experienced the greatest amount of divergent selection pressures between populations should be expected to have ED indices that exhibit the strongest relationship with RI. In the work of Funk et al. (2006), ED indices were separately calculated using data from the literature on each of three traits: habitat, diet, and size. For the sake of simplicity, a single estimate of ED was used for the analyses in this chapter, in each case based on host-preference and host-association data for the study taxa. In most insect herbivores, the host plant is a clear locus of ecological adaptation, and thus host-associated insect traits presumably offer good sources for evaluating the ecology-isolation association. However, even for a particular ecological trait, ED indices can be calculated in various ways. And, too, some ways of calculating ED for that trait may better capture the specific aspect of the ecological trait on which RI-promoting selection is acting. For example, in using insect-host associations to quantify ED, one must select the taxonomic level at which host-plant sharing is evaluated. Here, we somewhat arbitrarily chose the genus level for *Drosophila* and the family level for Lepidoptera, based on initial inspection of the degree of variation in host association among insect study species at various taxonomic levels. However, it is quite possible that species-level host associations would have provided more informative estimates of ED. This might be true, for example, if host-associated adaptive divergence is largely driven by the presence of particular plant chemicals that are diagnostic of specific plant species, but not of higher plant taxa. In this case, an insect's association with particular plant genera and families would be largely incidental, while the species-level associations would be highly indicative of the selective factors involved in the evolutionary divergence of these insects' host ranges (Yotoko et al. 2005). There are yet other ways that divergence in host association might be quantified. For example, one might assign different hierarchical levels of plant taxonomy different numerical values and then quantify ED as the lowest taxonomic level (numerical value) at which at least one host is shared. Or one might evaluate divergence in the degree of host specificity (absolute number of hosts used), and so forth. In sum, choosing how to quantify ED is a critical element of the comparative regression-based analyses of ecological speciation discussed here.

Opportunities and Directions

The comparative approaches described here offer a wealth of opportunities for future studies of the generalities and particularities of ecological speciation. The profusion of herbivorous-insect lineages, species, life histories, and host associations provide a panoply of systems to choose from in order to address specific questions of interest. The experimental tractability of these study systems further argues for their suitability for intensive cross-taxa investigations. Thus, the most basic requirement for further advancing the comparative study of ecological speciation in herbivorous insects, or any other group, is the detailed study of additional taxa. Specifically, information on GD, RI, and ED is required for a sufficient number of population/species pairs per study taxon to allow for accurate parameter estimates and powerful tests of statistical significance. By sampling populations at various stages of evolutionary divergence, the resulting comparative studies will furthermore allow the investigation of speciation as a process (e.g., Mallet et al. 1998; Drès and Mallet 2002; Etges 2002). By statistically comparing results across analyses of disparate taxa, general evolutionary patterns and principles might be inferred (Funk et al. 2006; this chapter).

To date, comparative studies of speciation have generally been post hoc in the sense that they analyze GD and RI (and now ED) estimates that have been cobbled together using data from diverse literature sources and by workers using heterogeneous methods. Alternatively, comparative studies could be planned a priori and conducted by workers or (in view of the scale of these endeavors) research groups using consistent approaches. This strategy would increase commensurability, reduce statistical noise, and presumably increase the quality of data and results. The *N. bebbianae* and *T. cristinae* studies described here illustrate this strategy. Perhaps not coincidentally, these taxa yielded the strongest ecology-isolation associations. Two non-herbivore examples of this strategy deserve mention. Mendelson (2003) chose phylogenetically independent pairs of darter species for study and consistently evaluated GD and both premating and postmating RI (but not ED) between them. A very recent article on centrarchid fishes takes a similar approach but also evaluates size to estimate ED and evaluate the ecology-isolation association (Bolnick et al. 2006).

Besides addressing broad questions of evolutionary generality, these approaches can evaluate more specific yet important questions about speciation. For example, we presently know very little about the relative importance of various reproductive barriers for speciation because multiple barriers have been studied in very few taxa (but see McMillan et al. 1997; Funk 1998; Lu and Bernatchez 1998; Mendelson 2003; Ramsey et al. 2003; Moyle et al. 2004; Nosil et al. 2005). Thus, we call for comparative regression-based studies that evaluate diverse reproductive barriers for population/species pairs at various stages of evolutionary divergence. Such data would allow the investigation of various outstanding issues, including the following: Do prezygotic or postzygotic barriers contribute more to speciation, and which barriers evolve earliest in the speciation process? Does divergent selection drive ecological speciation more through its effects on ecological or on non-ecological reproductive barriers? More specifically, to what degree is postzygotic isolation a function of intrinsic (genetic-developmentally based) versus extrinsic (ecology-based) factors? Do different barriers play important roles at different stages of the speciation process?

Just as evaluating the degree to which ED promotes different reproductive barriers is important to understanding ecological speciation, so too is evaluating the degree to which different ecological traits are associated with reproductive isolation. These latter comparisons might prove informative of the particular sources of selection that drive speciation and how and why these vary among taxa. Such insights might, for example, help explain why some groups of organisms are more prone to adaptive radiation than others. This chapter only touches the surface of the deep well of questions that can be investigated via the creative application of comparative approaches to the study of ecological speciation generally, and to the amazing diversity of study systems offered by herbivorous insects in particular. We hope that researchers will further invest in the collection of data necessary for such informative studies in the future.

Acknowledgments

I (D.J.F.) first met Tom Wood as a graduate student attending my second professional meeting (the annual meeting of the Entomological Society of America, in Baltimore). After being introduced to Tom by my lab-mate and Tom's former master's student, Mark Keese, I felt myself fortunate to be invited to tag along with this gruff and charismatic man. At that meeting I also had the pleasure of meeting Kelley Tilmon for the first time, when she was still a student of Tom's. My last interaction with Tom was when he e-mailed me some *Enchenopa* information I needed for a review article I was writing. He had generously put together quite a bundle of helpful materials, yet still apologized for having taken a (very) little while to send it along, briefly mentioning that he'd not been feeling his best lately. In between those interactions, Tom Wood's articles played a critical role in inspiring my own studies of insect speciation. A few weeks after his e-mail, Tom passed away. Our thanks go out to both Tom and Kelley for their passion and their critical contributions to advancing the evolutionary study of insect herbivores. We also thank Bill Etges for helpful comments on an earlier version of this manuscript.

References Cited

Barton, N.H. 1995. Linkage and the limits to natural selection. Genetics 140: 821–841.

Bearhop, S., W. Fiedler, R.W. Furness, S.C. Votier, S. Waldron, J. Newton, G.J. Bowen, P. Berthold, and K. Farnsworth. 2005. Assortative mating as a mechanism for rapid evolution of a migratory divide. Science 310: 502–504.

Berlocher, S.H., and J.L. Feder. 2002. Sympatric speciation in phytophagous insects: moving beyond controversy? Annu. Rev. Entomol. 47: 773–815.

Bernatchez, L., and C.C. Wilson. 1998. Comparative phylogeography of nearctic and palearctic fishes. Mol. Ecol. 7: 431–452.

Bernays, E.A., and R.F. Chapman. 1994. Host-plant selection by phytophagous insects. Chapman and Hall, London.

Bierbaum, T.J., and G.L. Bush. 1990. Genetic differentiation in the viability of sibling species of *Rhagoletis* fruit flies on host plants, and the influence of reduced hybrid viability on reproductive isolation. Entomol. Exp. Appl. 55: 105–118.

Bolnick, D.I., T.J. Near, and P.C. Wainwright. 2006. Body size divergence promotes post-zygotic reproductive isolation in centrarchids. Evol. Ecol. Res. 8: 903–913.

Boughman J.W., H.D. Rundle, and D. Schluter. 2005. Parallel evolution of sexual isolation in sticklebacks. Evolution 59: 361–373.

Bradshaw, H.D., and D.W. Schemske. 2003. Allele substitution at a flower colour locus produces a pollinator shift in monkeyflowers. Nature 426: 176–178.

Brown, C.G., and D.J. Funk. 2005. Aspects of the natural history of *Neochlamisus* (Coleoptera: Chrysomelidae): fecal case-associated life history and behavior, with a method for studying insect constructions. Ann. Entomol. Soc. Am. 98: 711–725.

Brown, W.J. 1943. The Canadian species of *Exema* and Arthrochlamys (Coleoptera: Chrysomelidae). Can. Entomol. 75: 119–131.

Brown, W.J. 1946. Some new Chrysomelidae, with notes on other species (Coleoptera). Can. Entomol. 78: 47–54.

Brown, W.J. 1952. Some species of Phytophaga (Coleoptera). Can. Entomol. 84: 335–342.

Brown, W.J. 1958. Sibling species in the Chrysomelidae. Proc. Tenth Intl. Congr. Entomol. 1: 103–109.

Brown, W.J. 1959. Taxonomic problems with closely related species. Annu. Rev. Entomol. 4: 77–98.

Brown, W.J. 1961. Notes on North American Chrysomelidae (Coleoptera). Can. Entomol. 93: 966–977.

Bush, G.L. 1969. Sympatric host-race formation and speciation in frugivorous flies of the genus *Rhagoletis* (Diptera, Tephritidae). Evolution 23: 237–251.

Coyne, J.A., and H.A. Orr. 1989. Patterns of speciation in *Drosophila*. Evolution 43: 362–381.

Coyne, J.A., and H.A. Orr. 1997. "Patterns of speciation in *Drosophila*" revisited. Evolution 51: 295–303.

Coyne, J.A., and H.A. Orr. 2004. Speciation. Sinauer, Sunderland, MA.

Craig, T.P., J.D. Horner, and J.K. Itami. 1997. Hybridization studies on the host races of *Eurosta solidaginis*: implications for sympatric speciation. Evolution 51: 1552–1560.

Crespi, B.J., and C.P. Sandoval. 2000. Phylogenetic evidence for the evolution of ecological specialization in *Timema* walking-sticks. J. Evol. Biol. 13: 249–262.

Dodd, D.M.B. 1989. Reproductive isolation as a consequence of adaptive divergence in *Drosophila pseudoobscura*. Evolution 43: 1308–1311.

Drès, M., and J. Mallet. 2002. Host races in plant-feeding insects and their importance in sympatric speciation. Phil. Trans. R. Soc. Lond. B 357: 471–492.

Egan, S.P., and D.J. Funk. 2006. Individual advantages to ecological specialization: insights on cognitive constraints from three conspecific taxa. Proc. R. Soc. Lond. B 273: 843–848.

Ehrlich, P.R., and P.H. Raven. 1964. Butterflies and plants: a study in coevolution. Evolution 18: 586–608.

Ehrman, L., and C. Petit. 1968. Genotype frequency and mating success in the *willistoni* species group of *Drosophila*. Evolution 22: 649–658.

Etges, W.J. 2002. Divergence in mate choice systems: does evolution play by rules? Genetica 116: 151–166.

Farrell, B.D. 1998. "Inordinate fondness" explained: why are there so many beetles? Science 281: 555–559.

Farrell, B.D., C. Mitter, and D.J. Futuyma. 1992. Diversification at the insect-plant interface: insights from phylogenetics. Bioscience 42: 34–42.

Feder, J.L. 1998. The apple maggot fly, *Rhagoletis pomonella*: flies in the face of conventional wisdom about speciation? pp. 130–144. In D.J. Howard, and S.H. Berlocher (eds.). Endless forms: species and speciation. Oxford University Press, Oxford, UK.

Feder, J.L., C.A. Chilcote, and G.L. Bush. 1988. Genetic differentiation between sympatric host races of the apple maggot fly *Rhagoletis pomonella*. Nature 336: 61–64.

Feder, J.L., S.B. Opp, B. Wlazlo, K. Reynolds, W. Go, and S. Spsisak. 1994. Host fidelity is an effective premating barrier between sympatric races of the apple maggot fly. Proc. Natl. Acad. Sci. USA 91: 7990–7994.

Feder, J.L., J.B. Roethele, B. Wlazlo, and S. H. Berlocher. 1997. Selective maintenance of allozyme differences among sympatric host races of the apple maggot fly. Proc. Natl. Acad. Sci. USA 94: 11417–11421.

Felsenstein, J. 1976. The theoretical population genetics of variable selection and migration. Annu. Rev. Genetics 10: 253–280.

Felsenstein, J. 1985. Phylogenies and the comparative method. Am. Nat. 125: 1–15.

Filchak, K.E., J.B. Roethele, and J.L. Feder. 2000. Natural selection and sympatric divergence in the apple maggot *Rhagoletis pomonella*. Nature 407: 739–742.

Fitzpatrick, B. M.2002. Molecular correlates of reproductive isolation. Evolution 56: 191–198.

Funk, D.J. 1996. The evolution of reproductive isolation in *Neochlamisus* leaf beetles: a role for selection. Ph.D. dissertation, State University of New York, Stony Brook.

Funk, D.J. 1998. Isolating a role for natural selection in speciation: host adaptation and sexual isolation in *Neochlamisus bebbianae* leaf beetles. Evolution 52: 1744–1759.

Funk, D.J. 1999. Molecular systematics of cytochrome oxidase I and 16S from *Neochlamisus* leaf beetles and the importance of sampling. Mol. Biol. Evol. 16: 67–82.

Funk, D.J., and K.E. Omland. 2003. Species-level paraphyly and polyphyly: frequency, causes, and consequences, with insights from animal mitochondrial DNA. Annu. Rev. Ecol. Syst. 34: 397–423.

Funk, D.J., K.E. Filchak, and J.L. Feder. 2002. Herbivorous insects: model systems for the comparative study of speciation ecology. Genetica 116: 251–267.

Funk, D.J., P. Nosil, and W.J. Etges. 2006. Ecological divergence exhibits consistently positive associations with reproductive isolation across disparate taxa. Proc. Natl. Acad. Sci. USA 103: 3209–3213.

Futuyma, D.G., and G.C. Mayer. 1980. Non-allopatric speciation in animals. Syst. Zool. 29: 254–271.

Futuyma, D.J., and S.C. Peterson. 1985. Genetic variation in the use of resources by insects. Annu. Rev. Entomol. 30: 217–238.

Harvey, P.H., and M.D. Pagel. 1991. The Comparative method in evolutionary biology. Oxford University Press, Oxford, UK.

Hawthorne, D.J., and S. Via. 2001. Genetic linkage of ecological specialization and reproductive isolation in pea aphids. Nature 412: 904–907.

Hendry, A.P., J.K. Wenburg, P. Bentzen, E.C. Volk, and T.P. Quinn. 2000. Rapid evolution of reproductive isolation in the wild: evidence from introduced salmon. Science 290: 516–518.

Hendry, A.P., T. Day, and E.B. Taylor. 2001. Population mixing and the adaptive divergence of quantitative traits in discrete populations: a theoretical framework for empirical tests. Evolution 55: 459–466.

Hodges, S.A., and M.L. Arnold, M.L. 1995. Spurring plant diversification: are floral nectar spurs a key innovation? Proc. R. Soc. Lond. B 262: 343–348.

Hurlbert, S.H. 1984. Pseudoreplication and the design of ecological field experiments. Ecol. Monog. 54: 187–211.

Janz, N., and S. Nylin. 1998. Butterflies and plants: a phylogenetic study. Evolution 52: 486–502.

Janz, N., K. Nyblom, and S. Nylin. 2001. Evolutionary dynamics of host-plant specialization: a case study of the tribe Nymphalini. Evolution 53: 783–796.

Johnson, P.A., F.C. Hoppensteadt, J.J. Smith, and G.L. Bush. 1996. Conditions for sympatric speciation: a diploid model incorporating habitat fidelity and non-habitat assortative mating. Evol. Ecol. 10: 187–205.

Karren, J.B. 1972. A revision of the subfamily Chlamisinae north of Mexico (Coleoptera: Chrysomelidae). Univ. Kansas Sci. Bull. 49: 875–988.

Lack, D. 1947. Darwin's finches. Cambridge University Press, Cambridge, UK.

Lu, G., and Bernatchez, L. 1998. Experimental evidence for reduced hybrid viability between dwarf and normal ecotypes of lake whitefish (*Coregonus clupeaformis* Mitchell). Proc. R. Soc. Lond. B 265: 1025–1030.

Lu, G., and L. Bernatchez. 1999. Correlated trophic specialization and genetic divergence in sympatric lake whitefish ecotypes (*Coregonus clupeaformis*): support for the ecological speciation hypothesis. Evolution 53: 1491–1505.

Mallet, J. 1989. The genetics of warning colour in Peruvian hybrid zones of *Heliconius erato* and *H. melpomene*. Proc. R. Soc. Lond. B 236: 163–185.

Mallet, J., and N.H. Barton. 1989. Strong natural selection in a warning-color hybrid zone. Evolution 43: 421–431.

Mallet, J., C.D. Jiggins, and W.O. McMillan. 1998. Mimicry and warning colour at the boundary between races and species, pp. 390–403. In D.J. Howard and S.H. Berlocher (eds.). Endless forms: species and speciation. Oxford University Press, Oxford, UK.

Manly, B.F.J. 1997. Randomization, bootstrap and Monte Carlo methods in biology. Chapman and Hall, London.

Martins, E.P., and T. Garland Jr. 1991. Phylogenetic analysis of the correlated evolution of continuous characters. Evolution 45: 534–557.

Mayr, E. 1942. Systematics and the origin of species. Columbia University Press, New York.

Mayr, E. 1947. Ecological factors in speciation. Evolution 1: 263–288.

Mayr, E. 1963. Animal species and evolution. Harvard University Press, Cambridge, MA.

McKinnon, J.S., S. Mori, B.K. Blackman, D.L. Kingsley, L. Jamieson, J. Chou, and D. Schluter. 2004. Evidence for ecology's role in speciation. Nature 429: 294–298.

McMillan, O., C. Jiggins, and J. Mallet. 1997. What initiates speciation in passion-vine butterflies? Proc. Natl. Acad. Sci. USA 94: 8628–8633.

Mendelson, T.C. 2003. Sexual isolation evolves faster than hybrid inviability in a diverse and sexually dimorphic genus of fish (Percidae: *Etheostoma*). Evolution 57: 317–327.

Mitter, C., B. Farrell, and B. Wiegmann. 1988. The phylogenetic study of adaptive zones: has phytophagy promoted insect diversification? Am. Nat. 132: 107–128.

Morton, E.S. 1975. Ecological sources of selection on avian sounds. Am. Nat. 109: 17–34.

Moyle, L.C., Olson, M.S., and Tiffin, P. 2004. Patterns of reproductive isolation in three angiosperm genera. Evolution 58: 1195–1208.

Muller, H.J. 1942. Isolating mechanisms, evolution and temperature. Biol. Symp. 6: 71–125.

Nagel, L., and D. Schluter. 1998. Body size, natural selection, and speciation in sticklebacks. Evolution 52: 209–218.

Nei, M. 1972. Genetic distance between populations. Am. Nat. 106: 283–292.

Noor, M.A.F., K.L. Grams, L.A. Bertucci, and J. Reiland. 2001. Chromosomal inversions and the reproductive isolation of species. Proc. Natl. Acad. Sci. USA 98: 12084–12088.

Nosil, P. 2002. Transition rates between specialization and generalization in phytophagous insects. Evolution 56: 1701–1706.

Nosil, P. 2004. Reproductive isolation caused by visual predation on migrants between divergent environments. Proc. R. Soc. Lond. B. 271: 1521–1528.

Nosil, P., and B.J. Crespi. 2004. Does gene flow constrain adaptive divergence or vice versa? A test using ecomorphology and sexual isolation in *Timema cristinae* walking-sticks. Evolution 58: 102–112.

Nosil, P. and B.J. Crespi. 2006. Cryptic reproductive isolation driven by ecological divergence. Proc. R. Soc. Lond. B. 273: 991–997.

Nosil, P., and A. Mooers. 2005. Testing hypotheses about ecological specialization using phylogenetic trees. Evolution 59: 2256–2263.

Nosil, P., B.J. Crespi, and C.P. Sandoval. 2002. Host-plant adaptation drives the parallel evolution of reproductive isolation. Nature 417: 441–443.

Nosil, P., B.J. Crespi, and C.P. Sandoval. 2003. Reproductive isolation driven by the combined effects of ecological adaptation and reinforcement. Proc. R. Soc. Lond. B 270: 1911–1918.

Nosil, P., T.H. Vines, and D.J. Funk. 2005. Perspective: Reproductive isolation caused by natural selection against immigrants from divergent habitats. Evolution 59: 705–719.

Nosil, P., B. J. Crespi, and C. P. Sandoval. 2006a. The evolution of host preference in allopatric versus parapatric populations of *T. cristinae* walking-sticks. J. Evol. Biol. 19: 929–942

Nosil, P., B.J. Crespi, C.P. Sandoval, and M. Kirkpatrick. 2006b. Migration and the genetic covariance between habitat preference and performance. Am. Nat. 167: E66–E78.

Nosil, P., B.J. Crespi, R. Gries, and G. Gries. 2006c. Natural selection and divergence in mate preference during speciation. Genetica 129: 309–327.

Ogden, R., and R.S. Thorpe. 2002. Molecular evidence for ecological speciation in tropical habitats. Proc. Natl. Acad. Sci. USA 99: 13612–13615.

Orr, H. A. 2001. The genetics of species differences. Trends Ecol. Evol. 16: 343–350.

Ortiz-Barrientos, D., J. Reiland, J. Hey, and M.A. F. Noor. 2002. Recombination and the divergence of hybridizing species. Genetica 116: 167–178.

Patten, M.A., J.T. Rotenberry, and M. Zuk. 2004. Habitat selection, acoustic adaptation, and the evolution of reproductive isolation. Evolution 58: 2144–2155.

Presgraves, D.C. 2002. Patterns of postzygotic isolation in Lepidoptera. Evolution 56: 1168–1183.

Price, C. S.C., C.H. Kim, C.J. Gronlund, and J.A. Coyne. 2001. Cryptic reproductive isolation in the *Drosophila simulans* species complex. Evolution 55: 81–92.

Ramsey, J., H.D. Bradshaw Jr., and D. W. Schemske. 2003. Components of reproductive isolation between the monkeyflowers *Mimulus lewissii* and *M. cardinalis* (Phrymaceae). Evolution 57: 1520–1534.

Raufaste N, and F. Rousset. 2001. Are partial mantel tests adequate? Evolution 55: 1703–1705.

Rice, W.R., and E.E. Hostert. 1993. Laboratory experiments on speciation: what have we learned in 40 years. Evolution 47: 1637–1653.

Rice, W.R., and G.W. Salt. 1990. The evolution of reproductive isolation as a correlated character under sympatric conditions: experimental evidence. Evolution 44: 1140–1152.

Rieseberg, L.H. 2001. Chromosomal rearrangements and speciation. Trends Ecol. Evol. 16: 351–358.

Rolán-Alvarez, E., and M. Caballero. 2000. Estimating sexual selection and sexual isolation effects from mating frequencies. Evolution 54: 30–36.

Rundle, H.D. 2002. A test of ecologically dependent postmating isolation between sympatric sticklebacks. Evolution 56: 322–329.

Rundle, H.D., and P. Nosil. 2005. Ecological speciation. Ecol. Lett. 8: 336–352.

Rundle, H.D., and M. Whitlock. 2001. A genetic interpretation of ecologically dependent isolation. Evolution 55: 198–201.

Rundle, H.D., L. Nagel, J.W. Boughman, and D. Schluter. 2000. Natural selection and parallel speciation in sympatric sticklebacks. Science 287: 306–308.

Sandoval, C.P. 1994a. The effects of relative geographic scales of gene flow and selection on morph frequencies in the walking stick *Timema cristinae*. Evolution 48: 1866–1879.

Sandoval, C.P. 1994b. Differential visual predation on morphs of *Timema cristinae* (Phasmatodeae: Timemidae) and its consequences for host range. Biol. J. Linn. Soc. 52: 341–356.

Sandoval, C.P., and P. Nosil. 2005. Counteracting selective regimes and host preference evolution in ecotypes of two species of walking-sticks. Evolution 59: 2405–2413.

Schluter, D. 1996. Ecological speciation in postglacial fishes. Phil. Trans. R. Soc. Lond. B 351: 807–814.

Schluter, D. 2000. The ecology of adaptive radiation. Oxford University Press, Oxford, UK.

Schluter, D. 2001. Ecology and the origin of species. Trends Ecol. Evol. 16: 372–380.

Schluter D., and L.M. Nagel. 1995. Parallel speciation by natural selection. Am. Nat. 146: 292–301.

Simpson, G.G. 1953. The major features of evolution. Columbia University Press, New York.

Singer, M.C., D. Ng., D. Vasco, and C.D. Thomas. 1992. Rapidly evolving associations among oviposition preferences fail to constrain evolution of insect diet. Am. Nat. 139: 9–20.

Slabbekoorn, H., and T.B. Smith. 2002. Habitat-dependent song divergence in the little greenbul: an analysis of selection pressures on acoustic signals. Evolution 56: 1849–1858.

Slatkin, M. 1987. Gene flow and the geographic structure of natural populations. Science 236: 787–792.

Sokal, R.R., and F.J. Rohlf. 1995. Biometry, 3rd ed. W.H. Freeman and Company, New York.

Spieth, H.T., and J.M. Ringo. 1983. Mating behavior and sexual isolation in *Drosophila* pp. 223–284. In M. Ashburner, H.L. Carson, and J.N. Thompson Jr. (eds.). The genetics and biology of *Drosophila*, vol. 3c. Academic Press, New York.

Stanley, S.M. 1979. Macroevolution: pattern and process. W.H. Freeman, San Francisco.

Strong, D.R., J.H. Lawton, and R. Southwood. 1984. Insects on plants: community patterns and mechanisms. Blackwell Scientific, London.

Thompson, J.N. 2005. The geographic mosaic of coevolution. University of Chicago Press, Chicago.

Thorpe, R.S. 1996. The use of DNA divergence to help determine the correlates of evolution of morphological characters. Evolution 53: 1446–1457.

Van Valen, L. 1965. Morphological variation and width of the ecological niche. Am. Nat. 99: 377–390.

Via, S. 1999. Reproductive isolation between sympatric races of pea aphids. I. Gene flow restriction and habitat choice. Evolution 53: 1446–1457.

Via, S., A.C. Bouck, and S. Skillman. 2000. Reproductive isolation between divergent races of pea aphids on two hosts. II. Selection against migrants and hybrids in the parental environments. Evolution 54: 1626–1637.

Vickery, V.R. 1993. Revision of *Timema* Scudder (Phasmatoptera: Timematodea) including three new species. Can. Entomol. 125:657–692.

Vines, T.H., and D. Schluter. 2006. Strong assortative mating between allopatric sticklebacks as a by-product of adaptation to different environments. Proc. R. Soc. Lond. B 273: 911–916.

Walsh, B.J. 1864. On phytophagous varieties and phytophagous species. Proc. Entomol. Soc. Philadelphia 3: 403–430.

Walsh, B.J. 1867. The apple-worm and the apple maggot. J. Hort. 2: 338–343.

Wiegmann, B.M., C. Mitter, and B. Farrell. 1993. Diversification of carnivorous parasitic insects: extraordinary radiation or specialized dead-end? Am. Nat. 142: 737–754.

Wood, T.K., and S.I. Guttman. 1982. Ecological and behavioral basis for reproductive isolation in the sympatric *Enchenopa binotata* complex (Homoptera, Membracidae). Evolution 36: 233–242.

Wood, T.K., and M.C. Keese. 1990. Host-plant-induced assortative mating in *Enchenopa* treehoppers. Evolution 44: 619–628.

Wood, T.K., K.L. Olmstead, and Guttman, S.I. 1990. Insect phenology mediated by host-plant water relations. Evolution 44: 629–636.

Wood, T.K., K.J. Tilmon, A.B. Shantz, C.K. Harris, and J. Pesek. 1999. The role of host-plant fidelity in initiating insect race formation. Evol. Ecol. Res. 1: 317–332.

Yotoko, K.S. C., P.I. Prado, C.A.M. Russo, and V.N. Solferini. 2005. Testing the trend towards specialization in herbivore-host plant associations using a molecular phylogeny of *Tomoplagia* (Diptera: Tephritidae). Mol. Phylo. Evol. 35: 701–711.

TEN

Sympatric Speciation: Norm or Exception?

DOUGLAS J. FUTUYMA

Host-specific herbivorous insects have inspired speculation about sympatric speciation at least since the 1860s, when Walsh (1864) described the now famous host races of the apple maggot (Berlocher and Feder 2002). Even Ernst Mayr, who lamented that "sympatric speciation is like the Lernaean Hydra which grew two new heads whenever one of its old heads was cut off," admitted that "host races [of phytophagous insects] are a challenging biological phenomenon, and constitute the only known case indicating the possible occurrence of incipient sympatric speciation" (Mayr 1963, p. 460). He immediately added, though, that "even in this case a process of sympatric speciation is neither established nor even probable."

Mayr's (1963) thorough analysis of the theoretical and empirical weaknesses of sympatric speciation in animals did not slay the Hydra. Review articles and books (e.g., Schilthuizen 2001; Via 2001; Dieckmann et al. 2004) continue to provide enthusiastic affirmation, and one could easily form an impression from reading the contemporary literature that it is as reasonable to assume a sympatric as an allopatric or parapatric origin of sister species. My aim in this chapter is to suggest that at least as it applies to herbivorous insects, the current enthusiasm for sympatric speciation exceeds both theoretical and empirical justification.

My attention to this topic stems from an interest in speciation that predates even my formal education in biology, and from having adopted host-specific insects as research material in the 1970s. The problem of the evolution of host specificity leads quickly to the issue of speciation. In reviewing the still sparse literature on this subject, Gregory Mayer and I concluded (Futuyma and Mayer 1980) that the evidence did not support the growing acceptance of sympatric speciation. In particular, we argued that the evidence from the best and most widely cited case—Guy Bush's study of the apple maggot *Rhagoletis pomonella* (Bush 1969)—was unconvincing, in part because there was no conclusive evidence of a genetic basis for the morphological and phenological differences that the supposed host races display. Soon afterward, Felsenstein (1981) provided a typically incisive theoretical analysis of a common verbal model, in which assortative mating evolves between two subpopulations adapted to different resources because it prevents production of poorly adapted hybrid offspring. The key issue, Felsenstein noted, is that recombination breaks down the association among loci that govern ecological adaptation and mate preference. The plausibility of sympatric speciation then depends on the recombination rate, the number of loci, and the strength of divergent selection. Felsenstein concluded that sympatric speciation "would be nearly impossible" if mating were based on a "two-allele" model of mate preference, and thus explained why sympatric speciation does not produce "a different species on every bush" (p. 124).

Bush and his student Scott Diehl quite appropriately responded that Felsenstein's objection did not apply to insects that mate only on their host plant, and in which reproductive isolation is a by-product of divergent host preference (Diehl and Bush 1989). Natural selection in this case *favors* linkage disequilibrium between loci that affect "performance" on and "preference" for two different host species. In Diehl and Bush's simulations, distinct host-specific species quickly formed, exhibiting complete linkage disequilibrium and complete genetic divergence. At about the same time, Bush's student Jeffrey Feder, together with other colleagues, confirmed genetic differences between the host races of *R. pomonella* (Feder et al. 1988; Feder and Bush 1989), initiating a long-term program of study that has made this the most convincing instance of incipient sympatric speciation. Since then, host races have been claimed for several insects (reviewed by Drès and Mallet 2002). These cases, as well as pairs or groups of host-divergent sister species, are often interpreted as the result of sympatric divergence and incipient speciation. Among these, of course, are the *Enchenopa* treehoppers to which Tom Wood dedicated his extraordinary energy and inspiring quest for understanding. I think Tom may have viewed my skepticism as a measure of the empirical difficulties that any attempt to demonstrate sympatric speciation must face. I immensely enjoyed our interactions, and I like to think he felt the same way.

Theory

Gavrilets (2004) has provided a detailed, clear exposition of the many mathematical and simulation models of speciation, and Coyne and Orr (2004) summarize and analyze empirical studies within a theoretical framework. Several of their general points bear repeating. First, any hypothesis of speciation by natural selection must consider how divergently selected (e.g., ecologically adaptive) traits are genetically coupled to (and therefore result in) reproductive isolation. Coupling might be based on pleiotropy (whereby ecological adaptation and reproductive isolation are controlled by the same genes and might even be considered the same trait) or by linkage disequilibrium (in which alleles conferring reproductive isolation are nonrandomly associated with alleles for divergently selected traits).

Second, speciation in allopatry is almost inevitable (Coyne and Orr 2004, p. 85), given enough time, because all the possible causes of speciation can drive divergence of separated populations, unimpeded by gene flow. Third, allopatric speciation caused by divergent ecological or sexual selection can be as rapid as sympatric or parapatric speciation (Gavrilets 2004, p. 405). Fourth, some of the models that have been most enthusiastically received may have a limited realm of application. For example, Dieckmann and Doebeli (1999), like several previous authors, model a single trait that is both the object of competition-induced disruptive ecological selection and the basis for assortative mating (a model of pleiotropic association). Although examples of such "magic traits" (Gavrilets 2004) have been described, sexual selection models (involving a signal emitted by one sex and a response by the other, both characters being different from the ecologically selected trait) probably apply to many more species of animals. Dieckmann and Doebeli's (1999) model of speciation has been criticized on theoretical grounds as well (Gavrilets 2004, p. 393; Gavrilets 2005; Polechová and Barton 2005). For example, it assumes that the population has maximal genetic variation for the trait from the outset, that variation is maintained both by strong frequency-dependent selection and by an unrealistically high mutation rate, and that mate choice is not costly (Gavrilets 2004). "Starting with low levels of genetic variation, realistically low mutation rates, increasing the number of loci, introducing costs of being choosy, or increasing the population size will all act against sympatric speciation" in this model (Gavrilets 2005, p. 698).

In the remainder of this chapter, I will consider mostly Bush's (1975) scenario of sympatric speciation by host shift in host-specific phytophagous insects that mate on the host plant. In this scenario, evolution of exclusive preference for different host-plant species, by adults of both sexes, is tantamount to assortative mating (hence, prezygotic isolation). (Importantly, the degree to which this supposition is realized by actual insects has not been explicitly demonstrated in any species. In *Rhagoletis*, for instance, there is opportunity for off-host mating during a period of dispersal between adult eclosion in the soil and later return to the host plant.) In Bush's model, the evolution of divergent exclusive preferences for ancestral and novel host plants is driven by a trade-off in fitness on the two plants, owing to physiological, phenological, or morphological characters that affect performance or "host-specific fitness." Note that although there is a pleiotropic relationship between host preference and reproductive isolation, divergence to the point of two distinct, specialized host preferences depends on linkage disequilibrium between preference and performance alleles at different loci.

Fry (2003a) has provided the most complete exploration of Bush's verbal model. In Fry's simulations, two traits, "host preference" and "host-specific fitness" (of an individual occupying one host species or the other), are polygenic (controlled by varying numbers of loci). Speciation has been completed when only two genotypes exist at the preference loci, with only "+" alleles fixed in one species and only "−" alleles fixed in the other species. Since these two genotypes exclusively use different hosts, they will not encounter each other when mating and so will be reproductively isolated. Each of the fitness loci exhibits a trade-off, enhancing fitness on one host at the expense of fitness on the other. A critical feature of Fry's simulations (and of many models of sympatric speciation) is that the population initially occupies both hosts and is fixed for the "+" allele at half of the preference and fitness loci, and for the "−" allele at the other half of these two classes of loci. Thus, the model "bypasses the question of how the initial host expansion occurred" (p. 1737). It entails strong frequency-dependent (soft) selection, because of the implicit competition for resources. Fry's results indicate that speciation can occur under a fairly broad range of conditions, but the selection coefficient (s) must be quite large at each fitness locus. Because of recombination, both among loci of the same class (e.g., preference loci) and between the preference and fitness loci, stronger selection is required for speciation as the number of preference loci is increased. (For example, the minimal per-locus s required for speciation is a whopping 0.35 if there are 10 preference loci.) Thus, even in a model that entails some favorable assumptions (which bear examination; see below), speciation is "by no means automatic," as Fry notes; it requires strong selection and a small number of preference loci.

Broader population genetic models of niche evolution also bear on the likely occurrence of sympatric speciation by host shift. Holt and Gaines (1992; see also Holt 1996; Kirkpatrick and Barton 1997; Tadeusz et al. 2002) modeled a single population that occupies both a primary habitat (e.g., an ancestral host) in which population growth is positive (a "source" habitat) and a secondary habitat (e.g., a novel host) in which population growth is negative (a "sink" habitat). An allele that enhances fitness in the sink habitat cannot increase if that habitat does not contribute descendants to the gene pool. In general, moreover, alleles that enhance fitness in the habitat that contributes more to the growth of the entire population increase in frequency, by sheer force of numbers. Thus, even if both habitats provide positive population growth, but the primary habitat contributes much more growth, alleles that enhance fitness in the primary habitat increase, as will the difference in mean fitness between the two habitats (see also Servedio and Kirkpatrick 1997; Servedio and Noor 2003). As adaptation to the majority habitat increases, relative to the minority habitat, the intensity of stabilizing selection on habitat preference also increases (Fry 1996; Whitlock 1996). Evolution of the niche (e.g., host association) is therefore likely to be "conservative" if a population presented with a novel or rare secondary habitat retains access to an abundant primary habitat to which it is well adapted. Note that these models concern the step that Fry's (2003a) simulation bypassed: initial expansion into a new habitat or resource. These models thus indicate that a sympatric expansion or shift of host use is unlikely if there is an initially large disparity in mean fitness between the hosts. By contrast, this poses less of a constraint on speciation for an allopatric population that has access mostly or exclusively to a novel host.

Biological Considerations

To judge how commonly speciation is likely to proceed according to the Bush model, it would be useful to know how often host-specific insects meet its assumptions. Below, I consider the evidence on the genetic basis of host utilization, trade-offs, and colonization of novel host plants and show that it is meager. I do not attempt an exhaustive review of these topics.

Genetics of Host Utilization

Host preference, by the feeding stage (e.g., lepidopteran larvae) or by ovipositing females, is often characterized as the proportion of choices made for one plant type over another by insects presented with both plants simultaneously. The possible inadequacies of such choice tests, compared with tests of acceptability of each plant presented separately, include the observations that many insects do not have the opportunity for comparison shopping, and that such tests do not tell us if individuals have a heightened acceptance of the favored plant, a heightened aversion to the disfavored plant, or both. As I will note below, it may be important to recognize the separate components of host preference and whether or not they are genetically independent.

The only formal quantitative trait locus (QTL) analysis of host preference that I know of has been performed in crosses between alfalfa- and clover-preferring genotypes of pea aphids *(Acyrthosiphon pisum)* (Hawthorne and Via 2001; Via and Hawthorne 2002). The authors detected seven QTLs that affected acceptance of one or the other plant in no-choice tests. This is a minimal estimate because each QTL is a chromosome region that might contain more than one gene that affects the trait, and because other loci would not have been detected if they have small effects on the character. Estimates of gene number in other species are based on a variety of less precise, classical quantitative genetic approaches. For example, data from between-species crosses were interpreted to indicate that oviposition preference is polygenic in the noctuid moths *Heliothis virescens/H. subflexa* (Sheck and Gould 1996), sex-linked and probably determined by few loci in the butterflies *Papilio glaucus/P. canadensis* (Scriber et al. 1991), and polygenically determined, with both sex-linked and autosomal genes, in *P. oregonius/P. zelicaon* (Thompson 1988b). Feeding preference of two "host races" (probably sibling species) of the rice brown planthopper *(Nilaparvata lugens)* for different grasses is apparently based on allelic difference at a single locus (Sezer and Butlin 1998a). Different food preferences of populations or strains of *Drosophila tripunctata* (Jaenike 1987) and of the milkweed bug *Oncopeltus fasciatus* (Leslie and Dingle 1983) were reported to have a polygenic basis.

Data on the number of loci affecting fitness on different host plants seem even sparser. At least five loci underlie the greater resistance of *D. sechellia* than of *D. simulans* to the toxin (octanoic acid) in the host of *D. sechellia* (Jones 1998). Resistance to a toxin in a novel host plant has different genetic bases, each involving one or a few genes of large effect, in two populations of a flea beetle *(Phyllotreta nemorum)* (Nielsen 1997; de Jong et al. 2000). Differential survival of the host races of rice brown planthopper on each other's host was estimated to be based on one to three allelic differences (Sezer and Butlin 1998b). At least four QTLs affect fecundity of pea aphids on alfalfa and/or clover; most of these are associated with (i.e., are identical or closely linked to) QTLs that affect host acceptance (Hawthorne and Via 2001; Via and Hawthorne 2002).

From these few studies, it appears that the number of loci underlying the evolution of host-use traits ranges from few (a favorable condition for sympatric divergence) to a considerable number (less favorable). The data suggest that host-preference traits are most often polygenic.

An interesting point emerges from Hawthorne and Via's QTL analysis of pea aphids: fecundity (a fitness component) and acceptance (a preference component) were genetically associated and perhaps based on the same loci. This association was not surprising, since Caillaud and Via (2000) had earlier reported that each host race hardly fed at all on the host of the other, and attributed the ecological specialization

(i. e., differential survival) of these forms to their behavior, since each will simply starve if confined on the other form's host plant. In this instance, the difference in fitness appears not to be a distinct physiological characteristic from host preference (Coyne and Orr 2004, p. 163). Recall that our assessment of the conditions for sympatric speciation (by the Bush model) depends on our assessment of the conditions for evolution of distinct host preferences. In the Bush model, disruptive selection of host preference exists only because of an underlying trade-off with respect to another character that reduces fitness on one host plant even as it enhances fitness on the other. This trade-off might be based on any of many features that affect the insect's ability to contend with the plants' morphology, toxins, or phenology, but it must be a different feature than preference if it is to drive the divergent evolution of preference, that is, to cause evolution of reproductive isolation. Without such a trade-off, there is no selection against genotypes that accept both hosts, and therefore no basis for the development of *distinct* host-associated populations.

Most tests of the survival or fecundity of insects on plants do not determine if these fitness components are reduced by a physiological, morphological, or phenological character of this kind, or simply by failure to consume the plant. If failure to feed is the cause, and if this is a manifestation of host preference, then such studies merely document divergent host preference but shed little or no light on the possible cause of its evolution. (A different issue arises if different life-history stages have different preferences; then divergent evolution of, say, larval preference could be driven by divergent adult preferences in oviposition.)

Trade-offs

Trade-offs in fitness on different host plants have been fundamental to most thinking on both the evolution of host specificity and sympatric speciation (Futuyma and Moreno 1988; Jaenike 1990; Coyne and Orr 2004). However, although almost all models of sympatric speciation assume trade-offs, not all models of host specificity do. For example, a population may evolve specialized preference for one of several plants simply because mean fitness is highest on that plant. A shift to a certain plant species, and possibly evolution of exclusive preference for that plant, might be favored if the plant provides enemy-free space (Jeffries and Lawton 1984; for examples, see Berdegue et al. 1996; Gratton and Welter 1999; Thomas et al. 2003). Likewise, an oligophagous population could evolve increased preference for one of its hosts because of differences in the mean fitness of individuals that develop on the different plants. Fry (1996; see also Whitlock 1996; Kawecki et al. 1997) has modeled a population in which the fitness effect of mutations at some loci is somewhat different on two different hosts. The initial population has both a majority host (on which more than half the population develops) and a minority host. Because of the demographic basis of selection mentioned earlier (Holt and Gaines 1992), mutations that increase fitness primarily on the majority host have a greater chance of fixation, and are fixed faster, than those that enhance fitness on the minority host. Moreover, deleterious mutations can rise to higher frequencies (due to drift) if they affect fitness mostly on the minority host. Hence the difference in mean population fitness between the majority and minority hosts increases, and preference for the majority host can evolve in the absence of trade-offs between the fitness effects of any loci on the two hosts.

The Bush model of sympatric speciation assumes a genetic trade-off in fitness between different hosts. It is this fitness trade-off that imposes selection for divergent, specialized host preferences by different genotypes (and against generalist genotypes). The trade-off must exist among genotypes within a population, such that alleles of at least one gene (or states of at least one genetically determined character) that enhance fitness on one host diminish fitness on another host. Detection and demonstration of trade-offs can present substantial difficulties (Fry 1993). It is not sufficient to show that each host-specific population has higher mean fitness on its normal host than on the other population's host, because this is what we would expect from any model of the evolution of divergent host affiliation, such as Fry's (1996) model, in allopatric populations. That is, apparent trade-offs may result from adaptation to different hosts after the evolution of host specificity, rather than having caused the evolution of host specificity (Futuyma 1983).

Researchers often look for trade-offs in fitness among genotypes within populations by estimating genetic correlations in fitness-related characters across host plants and by estimating correlations between fitness on different hosts in selection experiments. Genetic correlation studies dominate the literature. Failure to find the negative genetic correlation that would signal a trade-off can be explained away by supposing that it exists at some loci but is masked by positively correlated effects at other loci (often referred to as variation in "general vigor"). Service and Rose (1985; see also Joshi and Thompson 1995) postulated that variation in general vigor may be especially confounding if the insects are reared and tested in a novel environment such as a laboratory. (In a study of the moth *Alsophila pometaria*, Futuyma and Philippi [1987] found mostly positive genetic correlations in performance on several hosts in both field and laboratory.) Fry (1993) has argued that there is no reliable statistical method for discerning a negative genetic correlation that is masked by variation in general vigor. He instead advocates selection experiments, in which adaptation to one host species should be accompanied by reduced fitness on other species if trade-offs indeed exist (Fry 2003b). However, it is important that experimenters guard against decreases in population fitness caused by inbreeding or by hitchhiking of deleterious alleles with selected trait loci. Such effects can result from intense selection in small populations and are commonly observed in selection experiments.

Evidence for trade-offs from genetic correlations has been reported for *Aphis fabae* (one of three correlations); (Mackenzie 1996) and *Enchenopa* from *Viburnum* (with respect to one other plant) (Tilmon et al. 1998). Diminished fitness on a normal host as a correlated response to selection on a novel host has been reported in the spider mite *Tetranychus urticae* (Gould 1979; Fry 1990). Berlocher and Feder (2002) cite examples in which reduced fitness in F_1 hybrids between host-specific populations is interpreted as evidence of a trade-off, but this need not be so. Suppose, following Fry (1996), that loci A and B affect fitness only on hosts 1 and 2 respectively. In this case, drift in each population can fix an allele at the unexpressed locus that reduces fitness in the heterozygote, in the absence of host-associated fitness effects.

On the other hand, even granting the confounding effects of general vigor and possible laboratory artifacts, the paucity of evidence for genetic trade-offs in insects' physiological capacity for growth and survival on different plant species is breathtaking. Among estimates of genetic correlations, nonsignificant or positive genetic correlations (the opposite of a trade-off) have been reported in (inter alia) the aphids *Myzus persicae* (1000 clones tested) (Weber 1985), *A. fabae* (two of three correlations among three hosts) (Mackenzie 1996), and *Acyrthosiphon pisum* (among clones from either clover or alfalfa) (Via 1991); *Enchenopa* treehoppers from *Viburnum* (with respect to two other plants) (Tilmon et al. 1998); the chrysomelid beetles *Leptinotarsa decemlineata* (Hare and Kennedy 1986), *Oreina elongata* (Ballabeni et al. 2003), *Ophraella notulata* and *O. slobodkini* (Keese 1998), *Deloyala guttata* (Rausher 1984), and *Stator limbatus* (Fox et al. 1994); the coccinellid beetles *Epilachna vigintioctomaculata* and *E. pustulosa* (Ueno et al. 1999, 2003); the flies *Liriomyza sativae* (Via 1984) and *Drosophila tripunctata* (Jaenike 1989); and the lepidopterans *Colias philodice* (Karowe 1990), *Euphydryas editha* (Ng 1988), *Papilio oregonius* (Thompson 1996), and *Alsophila pometaria* (Futuyma and Philippi 1987). These include tests both on sets of normal hosts of the insect population and normal/novel host pairs (Joshi and Thompson 1995). Selection of the chrysomelid beetles *Callosobruchus maculatus* and *Leptinotarsa decemlineata* for adaptation to novel hosts did not diminish fitness on the normal hosts (Fox 1993; Lu et al. 1997).

I am not aware of any biochemical or physiological evidence for (or a priori reasons to expect) trade-offs in adaptation of an insect to different plants because of differences in their secondary compounds. Several investigators (e.g., Neal 1987; Appel and Martin 1992; Singer 2001) have been unable to detect an energetic cost of detoxification, and many insects and other animals have a large complement of genes for cytochrome P450s and other detoxification enzymes that may provide genetically independent capacities to detoxify diverse compounds (see Berenbaum 2002). Comparisons of growth and assimilation efficiencies of related specialist and generalist insects have provided little evidence for a physiological cost of generalization (Futuyma and Wasserman 1981; Moran 1986).

The studies cited above test for trade-offs in growth and survival that might stem from physiological effects of plant chemistry (or perhaps morphology), which has been widely considered an important basis for the evolution of host specificity (Futuyma and Moreno 1988; Jaenike 1990; Singer 2001). However, trade-offs could exist in other aspects of host use and impose selection for specialized host preference. For example, different host-associated genotypes may each be better at escaping predation when on their favored host plant, as is the case with cryptic forms of some color-polymorphic stick insects (see Sandoval and Nosil 2005). Trade-offs may also occur if a plant species that provides enemy-free space is inferior with respect to growth and survival for physiological reasons (Gratton and Welter 1999). Specialists may process information more effectively (Bernays and Wcislo 1994), and there is evidence that specialists find, choose, and use hosts more rapidly than generalists (Janz and Nylin 1997; Bernays and Funk 1999). Almost certainly, matching between plant phenology and insect life-history events creates trade-offs, when potential host species differ in temporal availability of the insect's key resource. As exemplified by *Rhagoletis* (Feder and Filchak 1999) and *Enchenopa* (Wood 1993), this may well be the most common basis for fitness trade-offs in phytophagous insects. As Berlocher and Feder (2002) note, an insect can eclose only once, so a life history synchronized with one host may readily incur a fitness disadvantage on a phenologically different plant.

Colonization

The conditions for establishment of a population on a novel plant species, and their implications for sympatric divergence, bear more examination than they have received. Experimental data show that independent of density, insects placed on novel plants that represent actual or potential host shifts often display great reduction in fitness components such as larval growth rate, survival, and fecundity (e.g., Wood 1993; Funk 1998; Keese 1998; Nosil et al. 2005). That is, hard selection may be substantial. Berlocher and Feder (2002) note, by contrast, that most sympatric models assume soft, density-dependent and frequency-dependent selection, whereby a genotype that accepts a novel plant increases in density and enjoys an immediate advantage because of escape from competition or predation on the normal host plant.

How do these factors affect the likelihood of sympatric divergence? If selection is mostly soft, so that genetic change is not required for establishment on the new host but instead follows it, how strong is selection likely to be for alleles that improve fitness? Intuitively, it seems that if the insect is physiologically preadapted to the novel host, the scope for soft selection is unlikely to be as great as for hard selection. Therefore, the strength of soft selection might not be nearly as great as models suggest it must be to drive the evolution of reproductive isolation (as in Fry 1996). In contrast, the potential for strong selection is clearly high if fitness on the novel host is greatly reduced because of physiological inability to survive. But in this case, hard selection is likely, and establishment on the host is more difficult and may be possible only by a genetic minority of the population.

The prospect of population establishment by this genetic minority may depend critically on the genetic basis of host-related traits: if fitness on a novel host depends on a complex of genetically independent polygenic traits, the necessary genotype may form very rarely by recombination and will almost certainly be destroyed by recombination if the ancestral genotype is abundant. With some exceptions (e.g., Singer at al. 1988; Bossart 2003), insects' behavioral responses (acceptance of or preference for the novel plant) are genetically largely uncorrelated with postingestive physiological features (performance) that affect growth and survival (Thompson 1988a; Sezer and Butlin 1998a; Forister 2005; Gassmann et al. 2006). In tests of very closely related species on each other's host plant, it is often the case that very few individuals (or none) feed or survive (Futuyma et al. 1995; Funk 1998), suggesting that survival may require rare mutations of large effect or rare multilocus combinations of alleles.

The terms "preference" and "performance" both summarize many underlying characters, each of which could prevent successful host utilization (Bernays and Chapman 1994; Schoonhoven et al. 1998). In lepidopteran larvae, for instance, feeding may be interrupted after contact with the leaf surface, after exploratory biting, or after ingestion, each of which may be affected by distinct positive and negative stimuli (Miller and Strickler 1984). Closely related insects can differ in their responses to different plant compounds that may be perceived as deterrents or as stimulants (van Drongelen 1979; Menken 1996). A host shift might therefore entail loss of a deterrent response to some compounds in a new host and acquisition of a positive response to other compounds, perhaps based on changes in binding affinity of a number of receptor proteins (Roessingh et al. 1999). A host shift may further entail a loss of responsiveness to feeding or oviposition stimulants in the ancestral host. All these elements of host acceptance and rejection might have separate genetic bases. Furthermore, we do not know whether or not feeding by immature stages, feeding by adults (in those insects that do so), and oviposition are genetically independent behaviors. That larval responses to different plant species can differ from adult responses is well known (e.g., Wiklund 1981; Thompson et al. 1990) and suggests that these traits may be somewhat genetically independent.

The biology of insects' interactions with plants therefore suggests that preference and performance are not likely to be simple traits affected by anonymous, interchangeable polygenes but are, rather, complexes of poorly understood component traits. As such, preference for and performance on a novel plant may each require a combination of specific alleles at multiple specific loci. Thus, acceptance of a novel plant may require not merely that the insect have at least four "plus" alleles at any of six loci, but that it carry the specific alleles A_1, B_2, C_2, and D_3, each of which affects a critical component trait. If so, recombination will be all the more effective in preventing growth of a genetically distinctive subpopulation, since a smaller fraction of possible gene combinations yield adequate fitness. To the extent that preference and performance conform to the genetic architecture I postulate, the high initial level of gene flow and recombination inherent in sympatric divergence may make establishment on the novel host unlikely. In contrast, recombination poses no problem if there is directional selection for adaptation to a novel plant in an allopatric population.

Evidence on Sympatric Speciation

Theory indicates that sympatric speciation is possible, but it also says that the conditions under which it is likely are considerably more stringent than the conditions for allopatric speciation. Coyne and Orr (2004) note that critical parameters in sympatric models, such as selection coefficients, or the cost of mate (or host-plant) choosiness, are seldom measured; Waxman and Gavrilets (2005) conclude that claims of wide conditions favoring sympatric speciation are "usually based on models incorporating unrealistic assumptions or using unreasonable initial conditions and numerical values of parameters" (p. 1151). For these reasons, it seems prudent to consider sympatric speciation an onerous hypothesis that should not be invoked without considerable evidence, and to assume that speciation in any given case has entailed low gene flow between spatially separated populations (i.e., allopatric or parapatric speciation), unless there is substantial evidence to the contrary. I agree with Coyne and Orr (2004) that this null hypothesis can be rejected, and the hypothesis that two species or host races arose sympatrically may be accepted, if evidence on their biogeographic and evolutionary history makes an allopatric phase very unlikely. In agreement with Berlocher and Feder (2002), a combination of several kinds of indirect evidence would more strongly favor a sympatric interpretation: key biological features (e.g., mating almost exclusively on the host plant), favorable genetic architecture (known in very few cases), strong selection, and evidence of trade-offs in fitness on the two hosts.

As noted above, the observation that the mean performance of each species or population is higher on its normal host than on the other host is not adequate evidence for a trade-off that drove the evolution of exclusive, specialized host preference (and consequent assortative mating), because we expect to see this pattern under any model of adaptation and speciation. Furthermore, the evolution of specialized preferences cannot have been driven by specialized preferences: failure to grow or survive is not evidence of a trade-off if it is caused by failure to feed. Except when evidence from biogeography is exceptionally favorable, it is very difficult to rule out a history of allopatry for fully reproductively isolated species (Chesser and Zink 1994; Losos and Glor 2003), especially in view of repeated, extensive Pleistocene changes in the distribution of plant species (and, presumably, of their associated host-specific insects). Thus, I consider sister species to provide far less informative evidence than sympatric host races, which, as Drès and Mallet (2002) note, are partially reproductively isolated and thus show the interplay between selection and gene flow that is inherent in sympatric differentiation. Below, I briefly summarize my assessment of those herbivorous insects that have been most studied with respect to possible sympatric speciation. Drès and Mallet (2002), Berlocher and Feder (2002), and Coyne and Orr (2004) also discuss these and other examples, and their conclusion may differ from mine in some cases.

Collections of *Acyrthosiphon pisum*, the pea aphid, from alfalfa and red clover in eastern United States display pronounced allozyme differences and feed hardly at all when confined on each other's host plant (Caillaud and Via 2000). The low survival of each on the other's host plant may well be a consequence of failure to feed and thus be a manifestation of the same trait (Coyne and Orr 2004). They may well be strongly isolated species, rather than host races. Pea aphids and these host plants have been introduced into North America from Europe, where alfalfa, clover, and other legumes likewise harbor genetically differentiated pea aphids; in fact, allozymes suggest that the American "races" may have originated in Europe (Simon et al. 2003). A QTL analysis of pea aphids has provided some of the best evidence of the genetic architecture of host-related features (Hawthorne and Via 2001), but little can be said about the geography of pea aphid speciation.

Samples of the aphid *Cryptomyzus galeopsidis* from two primary hosts *(Ribes rubrum and R. nigrum)* differ in frequency at one allozyme locus. Females display continuous variation in host preference, while males from one host show strong host preference (Guldemond 1990; Guldemond et al. 1994). Evidence on fitness trade-offs is sparse, and the relative contributions of feeding preference and postingestive effects are unknown. Drès and Mallet (2002) consider these probable host races, but the present evidence seems too sparse to judge their status, much less their bearing on sympatric speciation.

Craig, Itami, and colleagues (e.g., Itami et al. 1998; Craig et al. 2001) have extensively characterized the host races of *Eurosta solidaginis* (Diptera: Tephritidae), which forms stem galls on the goldenrods *Solidago altissima* and *S. gigantea*. The several components of reproductive isolation between them include host preference, phenology, sexual isolation, lowered mean survival on each other's host plants, and postzygotic isolation (greatly reduced preadult survival of most hybrid genotypes). In at least some geographic regions, the *S. gigantea* form is fixed for an mtDNA haplotype or for certain allozymes, suggesting that gene exchange is rare or absent. Whether or not gene flow between the races exceeds a token level is unclear, and it is not clear why these forms should not be considered distinct species. That several other insect taxa display parallel host-associated differentiation suggests a history of vicariance, that is, allopatric divergence (Berlocher and Feder 2002).

The so-called host races or biotypes of the rice brown planthopper, *Nilaparvata lugens* (Sezer and Butlin 1998a, 1998b), likewise appear to be distinct species that show no evidence of hybridization (Berlocher and Feder 2002). The same is true of the nine sibling species of the *Enchenopa binotata* complex of treehoppers that Tom Wood discovered and studied with such admirable insight and depth. Tom made a strong case for the plausibility of sympatric speciation, especially by showing that phenological differences in sap flow among host-plant species directly affect egg hatch, and thus create allochronic separation in mating time that is quite substantial between at least two pairs of sister species (Wood and Keese 1990; Wood 1993). Extrinsically imposed temporal "vicariance" can presumably be as effective as spatial vicariance in reducing gene flow and allowing divergent adaptation, once a phenologically divergent plant has been colonized. Thus, allochronic isolation is theoretically quite different from the Bush model of divergence in host preference and suffers less from the problem of recombination. The *Enchenopa* species represent a plausible case of sympatric speciation, but demonstrating this is quite another issue. I maintain that a sympatric origin of fully distinct species cannot be inferred from the fact of their sympatry.

To my surprise and amusement, Berlocher and Feder (2002; see their online supplementary material) say, "Ironically, in light of the famous critique of sympatric speciation by Futuyma and Mayer [64], Futuyma's primary research organisms, species of the chrysomelid beetle genus *Ophraella*, are candidates for sympatric speciation"—since they are host specific, mate on the host plant, and are sympatric in some cases. Although I am grateful for the citation, I feel compelled to state that it had not occurred to me to view *Ophraella* species in this light. For the record, except for sibling species that cannot be identified as museum specimens, the geographic ranges of most of the species are fairly well known (LeSage1986; Futuyma 1990), and most of them do not suggest sympatric origins. Five recent speciation events in *Ophraella* are implied by mtDNA sequence differences and phylogeny (Funk et al. 1995). *Ophraella communa* and *O. bilineata* diverged perhaps one to two million years ago. The geographic range of *O. bilineata* is in the Great Plains, north of *O. communa*, and I have failed to find sympatric or even fairly close populations of these species despite extensive effort. *Ophraella arctica* is most closely related to, and may well be conspecific with, *O. bilineata*; the sites at which *O. arctica* has been collected lie far to the north of the known range of *O. bilineata*. *Ophraella conferta*, found in northeastern United States, is replaced to the south by *O. sexvittata*, which has been distinguished from it on the basis of minor morphological differences. They differ in mtDNA sequence no more than do conspecific populations of other species, overlap partially in diet (both have been collected on several closely related species of *Solidago*), and are probably conspecific. The closest relative of *O. nuda*, a morphologically distinctive species that has been collected only in southeastern Alberta, appears to be *O. artemisiae*, which I described and have collected in Arizona, Colorado, and Minnesota. The host plant of *O. artemisiae* does not occur in sites or habitats inhabited by *O. nuda*, and there is no evidence that these species are anywhere sympatric, although the possibility cannot be ruled out. *Ophraella notulata* and *O. slobodkini* are sister (and sibling) species that can produce F_1 hybrids (Keese 1996). Nevertheless, the substantial differences between them in mtDNA sequence and allozymes (Futuyma and McCafferty 1990; Funk et al. 1995) imply a rather old speciation event, perhaps about seven million years ago. The known distribution of *O. slobodkini* lies in and near Florida, but its host plant (*Ambrosia artemisiifolia*, common ragweed) is a very broadly distributed annual weed, and a few specimens from interior North America might represent this species. *Ophraella notulata*, in contrast, is a habitat specialist, feeding on a shrub that grows in Atlantic and Gulf coastal salt marshes. Although the distributions of these species abut, there is considerable opportunity for spatial separation. Overall, then, the distributions of *Ophraella* species are at least as compatible with an allopatric as with a sympatric origin.

The larch budmoth, *Zeiraphera diniana*, includes populations on larch and pine that differ in several phenotypic traits, including female sex pheromone and male pheromone response. That they differ strongly in genetic markers on some, but not all, chromosomes provides evidence for ongoing gene flow, countered by divergent selection in some chromosome regions (Emelianov et al. 2004). This appears to be an instance of well-marked host races that should perhaps be considered a single species. Whether the genetic differences developed sympatrically is not known, but the maintenance of divergence despite gene flow makes this a promising case for further study.

It is a tribute to Guy Bush's, Jeff Feder's, and Stewart Berlocher's insight and industry that the one fairly convincing example of sympatric origin of host races, in my opinion, is the renowned apple maggot, *Rhagoletis pomonella* (Bush 1969; Filchak et al. 2000; Feder et al. 2003a). The ancestral hawthorn-feeding and the recently arisen apple-feeding races differ genetically in host preference and phenology, the latter giving rise to a substantial fitness trade-off that cannot be plausibly ascribed to divergence after establishment of reproductive isolation; the apple race enjoys reduced parasitism even as it suffers reduced viability; genetic differentiation persists despite clearly evident ongoing gene flow; and historical documentation adds support to the sympatric scenario. As Coyne and Orr (2004) note, it is still possible for skeptics to suppose that the apple race has had a long history of allopatric association with an undiscovered host plant (although perhaps the skeptics should bear the burden of demonstrating this). Feder et al. (2003a, 2003b) have found evidence that the apple-specific diapause alleles lie in chromosome inversions that arose in Mexico. This suggests that the formation of adaptive complexes of linked genes may have occurred in allopatry. If this is so, and if the consequent gene complexes were essential for the formation of host races, divergence in host association may have had both an allopatric component and a sympatric component, namely, the origin of and establishment of a correlation between phenology and preference for apple.

At this time, therefore, there appears to be one good example *(Rhagoletis)* of sympatrically arisen host races—although even this may have an important allopatric component—and a few suggestive cases that may yet yield such evidence.

The Allopatric Alternative

Like other animals (and plants!), phytophagous insects display geographic variation, often with a demonstrably genetic basis, in features that may confer reproductive isolation, specifically, in their host associations (Fox and Morrow 1981; Claridge 1993). Examples include the chrysomelid beetles *Leptinotarsa decemlineata* (Hsaio 1978; Lu and Logan 1993), *Oreina elongata* (Balabeni et al. 2003), *Neochlamisus bebbianae* (Funk 1998), *Phratora vitellinae* (Rowell-Rahier 1984), and *Stator limbatus* (Fox et al. 1994); the lepidopterans *Papilio polyxenes* (Blau and Feeny 1983), *Euphydryas editha* (Radtkey and Singer 1995), and *Depressaria pastinacella* (Zangerl and Berenbaum 2003); and the flies *Drosophila mojavensis* (Etges et al. 1999) and *Tephritis conura* (Römstock and Arnold 1987).

An allopatric host shift has two likely components. First, a population evolves a positive feeding and oviposition response to a new plant species, perhaps because it provides predator-free space, or simply because it is much more abundant than the insect's normal host. Preference for a specific resource can evolve if its great abundance more than compensates for lower quality (e.g., Templeton and Rothman 1981). Indeed, scarcity of the normal host should provide strong selection for use of an alternative plant. If the insect is not physiologically preadapted to the novel plant, survival on the novel host may be initially very low and may require a specific multifactorial genotype as postulated above. In that case, adaptation to the novel host may be facilitated if the insect population is sustained in low numbers on its (uncommon) ancestral host, or if a low rate of gene flow into the colony from populations on the ancestral host supplies genetic variation (Tadeusz et al. 2002). The second component of a host shift is loss of the ancestral host from the diet. This is facilitated by absence of strong competition (and thus no or little frequency-dependent selection maintaining polymorphism in host use). Use of the ancestral host may be lost due to selection if there is a functional trade-off, to fixation by random genetic drift of alleles that reduce fitness on the minority (ancestral) host (Kawecki 1994; Fry 1996), and/or to selection for exclusive preference for the majority (novel) host as the disparity in fitness on the two hosts grows (Fry 1996). Experimental evidence shows that a population confined to a single host species may evolve preference for that host; selection for preference does not require that the population be faced with a choice (Wasserman and Futuyma 1981).

The same factors that might impel evolution of a new specialized host preference—and thereby possible reproductive isolation—in sympatry can operate in allopatry, without the opposing effects of gene flow and recombination. Which, then, absent compelling evidence, is the more parsimonious hypothesis?

Conclusions

Until quite recently, there was exceedingly little evidence for "the origin of species by means of natural selection." Even though Mayr (1963) argued that drift-induced genetic changes in small populations might initiate selection toward new equilibria, based on gene interactions rather than extrinsic factors, he and most other students of speciation believed that ecological selection was largely responsible for the genetic changes that incidentally cause reproductive isolation. But evidence of a common genetic foundation for the ecological differences between species and reproductive isolation was almost completely lacking. This is still the case, by and large, but the problem has now been recognized, and both direct and indirect evidence for a role of ecological selection in speciation has mounted (see Funk et al. 2006 and summaries by Schluter 2000; Coyne and Orr 2004). Indeed, "ecological speciation" (Schluter 2000; Rundle and Nosil 2005) or "adaptive speciation" (Dieckmann et al. 2004) is a prominent theme in contemporary speciation studies.

Rundle and Nosil (2005) define ecological speciation as "the process by which barriers to gene flow evolve between populations as a result of ecologically based divergent selection" (p. 336). Almost all models of sympatric speciation involve exactly this process. But ecological speciation need not be sympatric; as Rundle and Nosil (2005) say, ecological speciation includes "the special case in which selection favours opposite, usually extreme, phenotypes within a single population" (p. 336), but it is not limited to this special case.

Focused study of ecological speciation is so recent that our understanding of it is rudimentary. We know little about the ecological factors that may impose divergent selection on characters that confer prezygotic or (especially) postzygotic reproductive isolation, or about the functional genetic relationship between adaptation and reproductive isolation. Almost certainly, speciation in host-specific phytophagous insects is frequently a consequence of adaptation to different host plants. In many clades, closely related species are specialized on different hosts. Reproductive isolation in many such groups may be a consequence of mating on the host plant, and even when this is not the case, developing on the wrong host often results in low fitness and therefore extrinsic postzygotic isolation (Nosil et al. 2003, 2005). Adaptation to different plants may also pleiotropically affect sexual isolation (Funk 1998) and might underlie intrinsic genetic incompatibilities that result in postzygotic isolation; these possibilities have been little investigated.

Speciation as an effect of divergent adaptation to host plants, like ecological speciation generally, need not be sympatric (Funk et al. 2002). I suspect that host specialization and sympatric speciation have come to be associated (and to some, perhaps, nearly synonymous) by historical accident. Before the biological species concept was articulated (Mayr 1942), entomologists noted and were puzzled by the revelation, in many insects, of morphologically indistinguishable host-associated "biological races" of what they took to be single species (Thorpe 1930). Many of these, such as *Rhagoletis* flies on blueberry and apple, proved to be complexes of reproductively isolated sibling species (see Mayr 1963). Their supposition that these were incipient species, diverging in sympatry, was reinforced by evidence (much since questioned) that new host preferences might be "induced" by the experience of accidentally developing on a new plant. The contentious role of phytophagous insects as exemplars of sympatric speciation has thereby overshadowed their similarity to other kinds of animals. Sibling species, ecological niche differences between sister species, and multiple-niche polymorphism are not peculiar to phytophagous insects or parasites; they are well documented in birds, fishes, and other animals. And geographic variation in ecological niche (host use) and genetic compatibility is as characteristic of phytophagous insects as it is of taxa in which sympatric speciation is seldom postulated.

I agree with other authors (e.g., Coyne and Orr 2004; Gavrilets 2004) that sympatric speciation is theoretically possible, under particular circumstances. There is fairly compelling prima facie evidence of sympatric speciation in some fishes, even if its mechanism is obscure (Coyne and Orr 2004; Barluenga et al. 2006). I agree that biological characteristics of many insects, such as *Enchenopa*, make it likely that they would meet some of the conditions for sympatric speciation, and I think evidence from some host races, especially of *Rhagoletis pomonella*, goes a long way toward showing a sympatric origin of reproductive isolation. All in all, it seems quite likely that some species of host-specific insects that mate on their host plants have originated sympatrically. But plausibility does not amount to evidence. Theory tells us that in general, sympatric speciation is an "onerous concept that should be used only where it is really necessary," as Williams (1966, p. 4) said of adaptation. Speciation by divergence of spatially separated populations is theoretically likely under very broad conditions, has been abundantly documented (Coyne and Orr 2004), and is generally a more parsimonious hypothesis than sympatric speciation.

If we aspire to rigor in evolutionary biology, we will conservatively require evidence, not just plausibility, and I believe that in the case of sympatric speciation that means more research. At the risk of seeming hopelessly antiquated, I still agree with Maynard Smith (1976) that "it is in the nature of science that once a position becomes orthodox it should be subjected to criticism. . . . It does not follow that, because a position is orthodox, it is wrong."

Acknowledgments

Tom Wood was a dedicated, rigorous biologist who hoped to demonstrate sympatric speciation, unquestionably added to its respectability, and still recognized that there remained work to be done. Indeed, he left unfinished what might have been the most definitive step in his research program. I am grateful to Kelley Tilmon for inviting me to contribute to this volume in his memory. I thank Jerry Coyne, Matt Forister, Dan Funk, and an anonymous reviewer for helpful comments on the manuscript, without implying that they all necessarily agree with everything they were kind enough to read.

References Cited

Appel, H.M., and M.M. Martin. 1992. Significance of metabolic load in the evolution of host specificity of *Manduca sexta*. Ecology 73: 216–228.

Ballabeni, P., K. Gottbard, A. Kayumba, and M. Rahier. 2003. Local adaptation and ecological genetics of host-plant specialization in a leaf beetle. Oikos 101: 70–78.

Barluenga, M., K.N. Stölting, W. Salzburger, M. Muschick, and A. Meyer. 2006. Sympatric speciation in Nicaraguan crater lake cichlid fish. Nature 439: 719–723.

Berdegue, M., J.T. Trumble, J.D. Hare, and R. A. Redak. 1996. Is it enemy-free space? The evidence for terrestrial insects and freshwater arthropods. Ecol. Entomol. 21: 203–217.

Berenbaum, M.L. 2002. Postgenomic chemical ecology: from genetic code to ecological interactions. J. Chem. Ecol. 28: 873–896.

Berlocher, S.H., and J.L. Feder. 2002. Sympatric speciation in phytophagous insects: moving beyond controversy? Annu. Rev. Entomol. 47: 773–815.

Bernays, E.A., and R.F. Chapman. 1994. Host-plant selection by phytophagous insects. Chapman and Hall, London.

Bernays, E.A., and D.J. Funk. 1999. Specialists make faster decisions than generalists: experiments with aphids. Proc. R. Soc. Lond. B. 266: 151–156.

Bernays, E.A., and W. Wcislo. 1994. Sensory capabilities, information processing and resource specialization. Q. Rev. Biol. 69: 187–204.

Blau, W.S., and P. Feeny. 1983. Divergence in larval responses to food plants between temperate and tropical populations of the black swallowtail butterfly. Ecol. Entomol. 8: 249–257.

Bossart, J.L. 2003. Covariance of preference and performance on normal and novel hosts in a locally polyphagous butterfly population. Oecologia 135: 477–486.

Bush, G.L. 1969. Sympatric host race formation and speciation in frugivorous flies of the genus *Rhagoletis* in North America (Diptera, Tephritidae). Evolution 23: 237–251.

Bush, G.L. 1975. Sympatric speciation in phytophagous parasitic insects, pp. 187–207. In P.W. Price (ed.), Evolutionary strategies in parasitic insects and mites. Plenum, New York.

Caillaud, M.C., and S. Via. 2000. Specialized feeding behavior influences both ecological specialization and assortative mating in sympatric host races of pea aphids. Am. Nat. 156: 606–621.

Chesser, R.T., and R.M. Zink. 1994. Modes of speciation in birds: a test of Lynch's method. Evolution 48: 490–497.

Claridge, M.F. 1993. Speciation in insect herbivores: the role of acoustic signals in leafhoppers and planthoppers, pp. 285–297. In D.R. Lees and D. Edwards (eds.), Evolutionary patterns and processes. Academic Press, London.

Coyne, J., and H. A. Orr. 2004. Speciation. Sinauer, Sunderland, MA.

Craig, T.P.,J.D. Horner, and J.K. Itami. 2001. Genetics, experience, and host-plant preference in *Eurosta solidaginis*: implications for sympatric speciation. Evolution 51: 1552–1560.

de Jong, P.W.,H.O. Frandsen, L. Rasmussen, and J.K. Nielsen. 2000. Genetics of resistance against defences of the host plant *Barbaraea vulgaris* in a Danish flea beetle population. Proc. R. Soc. Lond. B. 267: 1663–1670.

Dieckmann, U., and M. Doebeli. 1999. On the origin of species by sympatric speciation. Nature 400: 354–357.

Dieckmann, U., M. Doebeli, J. A.J. Metz, and D. Tautz (eds.). 2004. Adaptive speciation. Cambridge University Press, Cambridge, UK.

Diehl, S.D., and G.L. Bush. 1989. The role of habitat preference in adaptation and speciation, pp. 345–365. In D. Otte and J. A. Endler (eds.), Speciation and its consequences. Sinauer, Sunderland, MA.

Drès, M., and J. Mallet. 2002. Host races in plant-feeding insects and their importance in sympatric speciation. Phil. Trans. R. Soc. Lond. B. 357: 471–492.

Emelianov, I., F. Marec, and J. Mallet. 2004. Genomic evidence for divergence with gene flow in host races of the larch budmoth. Proc. R. Soc. Lond. B. 271: 97–105.

Etges, W.J., W.R.Johnson, G. A. Duncan, G. Huckins, and W. B. Heed. 1999. Ecological genetics of cactophilic *Drosophila*, pp. 164–214. In R. Robichaux (ed.), Ecology of Sonoran desert plants and plant communities. University of Arizona Press, Tucson.

Feder, J.L., and G.L. Bush. 1989. A field test of differential host plant usage between two sibling species of *Rhagoletis pomonella* fruit flies (Diptera: Tephritidae) and its consequences for sympatric models of speciation. Evolution 43: 1813–1819.

Feder, J.L., and K.E. Filchak. 1999. It's about time: the evidence for host plant–mediated selection in the apple maggot fly, *Rhagoletis pomonella*, and its implications for fitness trade-offs in phytophagous insects. Entomol. Exp. Appl. 91: 211–225.

Feder, J.L., C.A. Chilcote, and G.L. Bush. 1988. Genetic differentiation between sympatric host races of the apple maggot fly *Rhagoletis pomonella*. Nature 336: 61–64.

Feder, J.L., S.H. Berlocher, J. B. Roethele, H. Dambroski, J.J. Smith, W.L. Perry. 2003a. Allopatric genetic origins for sympatric host-plant shifts and race formation in *Rhagoletis*. Proc. Nat. Acad. Sci. U.S.A. 100: 10314–10319.

Feder, J.L., J.B. Roethele, K. Filchak, J. Niedbalski, and J. Romero-Severson. 2003b. Evidence for inversion polymorphism related to sympatric host race formation in the apple maggot fly, *Rhagoletis pomonella*. Genetics 163: 939–953.

Felsenstein, J. 1981. Skepticism towards Santa Rosalia, or why are there so few kinds of animals? Evolution 35: 124–138.

Filchak, K.E., J.B. Roethele, and J.L. Feder. 2000. Natural selection and sympatric divergence in the apple maggot *Rhagoletis pomonella*. Nature 407: 739–742.

Forister, M.L. 2005. Independent inheritance of preference and performance in hybrids between host races of *Mitoura* butterflies (Lepidoptera: Lycaenidae). Evolution 59: 1149–1155.

Fox, C.W. 1993. A quantitative genetic analysis of oviposition preference and larval performance on two hosts in the bruchid beetle *Callosobruchus maculatus*. Evolution 47: 166–175.

Fox, C.W., K.J. Waddell, and T. A. Mousseau. 1994. Host-associated fitness variation in a seed beetle (Coleoptera: Bruchidae): evidence for local adaptation to a poor quality host. Oecologia 99: 329–336.

Fox, L.R., and P. A. Morrow. 1981. Specialization: species property or local phenomenon? Science 211: 887–893.

Fry, J.D. 1990. Tradeoffs in fitness on different hosts: evidence from a selection experiment with a phytophagous mite. Am. Nat. 136: 569–580.

Fry, J.D. 1993. The "general vigor" problem: can antagonistic pleiotropy be detected when genetic covariances are positive? Evolution 47: 327–333.

Fry, J.D. 1996. The evolution of host specialization: are trade-offs overrated? Am. Nat. 148: S84–S107.

Fry, J.D. 2003a. Multilocus models of sympatric speciation: Bush versus Rice versus Felsenstein. Evolution 57: 1735–1746.

Fry, J.D. 2003b. Detecting ecological trade-offs using selection experiments. Ecology 84: 1672–1678.

Funk, D.J. 1998. Isolating a role for natural selection in speciation: host adaptation and sexual isolation in *Neochlamisus bebbianae* leaf beetles. Evolution 52: 1744–1759.

Funk, D.J., D.J. Futuyma, G. Ortí, and A. Meyer. 1995. A history of host associations and evolutionary diversification for *Ophraella* (Coleoptera: Chrysomelidae): new evidence from mitochondrial DNA. Evolution 49: 1008–1017.

Funk, D.J., K.E. Filchak, and J.L. Feder. 2002. Herbivorous insects: model systems for the comparative study of speciation ecology. Genetica 116: 251–267.

Funk, D.J., P. Nosil, and W.J. Etges. 2006. Ecological divergence exhibits consistently positive associations with reproductive isolation across disparate taxa. Proc. Natl. Acad. Sci. USA 103: 3209–3213.

Futuyma, D.J. 1983. Evolutionary interactions among herbivorous insects and plants, pp. 207–221. In D.J. Futuyma and M. Slatkin (eds.), Coevolution. Sinauer, Sunderland, MA.

Futuyma, D.J. 1990. Observations on the taxonomy and natural history of *Ophraella* Wilcox (Coleoptera: Chrysomelidae), with a description of a new species. J. NY Entomol. Soc. 98: 163–186.

Futuyma, D.J., and G.C. Mayer. 1980. Non-allopatric speciation in animals. Syst. Zool. 29: 254–271.

Futuyma, D.J., and S. S. McCafferty. 1990. Phylogeny and the evolution of host plant associations in the leaf beetle genus *Ophraella* (Coleoptera, Chrysomelidae). Evolution 44: 1885–1913.

Futuyma, D.J., and G. Moreno. 1988. The evolution of ecological specialization. Annu. Rev. Ecol. Syst. 19: 207–233.

Futuyma, D.J., and T.E. Philippi. 1987. Genetic variation and covariation in responses to host plants by *Alsophila pometaria* (Lepidoptera: Geometridae). Evolution 41: 269–279.

Futuyma, D.J., and S. S. Wasserman. 1981. Foodplant specialization and feeding efficiency in the tent caterpillars *Malacosoma disstria* and *M. americanum*. Entomol. Exp. Appl. 30: 106–110.

Futuyma, D.J., M.C. Keese, and D.J. Funk. 1995. Genetic constraints on macroevolution: the evolution of host affiliation in the leaf beetle genus *Ophraella*. Evolution 49: 797–809.

Gassmann, A.J., A. Levy, T. Tran, and D.J. Futuyma. 2006. Adaptations of an insect to a novel host plant: a phylogenetic approach. Funct. Ecol. 20: 478–485.

Gavrilets, S. 2004. Fitness landscapes and the origin of species. Princeton University Press, Princeton, NJ.

Gavrilets, S. 2005. "Adaptive speciation"—it is not that easy: a reply to Doebeli et al. Evolution 59: 696–699.

Gould, F. 1979. Rapid host range evolution in a population of the phytophagous mite *Tetranychus urticae* Koch. Evolution 33: 791–802.

Gratton, C., and S.C. Welter. 1999. Does "enemy-free space" exist? Experimental host shifts of an herbivorous fly. Ecology 80: 773–785.

Guldemond, J.A. 1990. Choice of host plant as a factor in reproductive isolation of the aphid genus *Cryptomyzus* (Homoptera, Aphididae). Ecol. Entomol. 15: 43–51.

Guldemond, J.A., A.F.G. Dixon, and W. T. Tigges. 1994. Mate recognition in *Cryptomyzus* aphids: copulation and insemination. Entomol. Exp. Appl. 73: 67–75.

Hare, J.D., and G.G. Kennedy. 1986. Genetic variation in plant-insect associations: survival of *Leptinotarsa decemlineata* populations on *Solanum carolinense*. Evolution 40: 1031–1043.

Hawthorne, D.J., and S. Via. 2001. Genetic linkage of ecological specialization and reproductive isolation in pea aphids. Nature 412: 904–907.

Holt, R.D. 1996. Demographic constraints in evolution: towards unifying the evolutionary theories of senescence and niche conservatism. Evol. Ecol. 10: 1–11.

Holt, R.D., and M.S. Gaines. 1992. Analysis of adaptation in heterogeneous landscapes: implications for the evolution of fundamental niches. Evol. Ecol. 6: 433–447.

Hsiao, T.H. 1978. Host plant adaptation among geographic populations of the Colorado potato beetle. Entomol. Exp. Appl. 24: 437–447.

Itami, J.K., T.P Craig, and J.D. Horner. 1998. Factors affecting gene flow between the host races of *Eurosta solidaginis*, pp. 375–407. In S. Mopper and S. Y. Strauss (eds.), Genetic structure and local adaptation in natural insect populations: effects of ecology, life history, and behavior. Chapman and Hall, New York.

Jaenike, J. 1987. Genetics of oviposition-site preference in *Drosophila tripunctata*. Heredity 59: 363–369.

Jaenike, J. 1989. Genetic population structure of *Drosophila tripunctata:* patterns of variation and covariation of traits affecting resource use. Evolution 43: 1467–1482.

Jaenike, J. 1990. Host specialization in phytophagous insects. Annu. Rev. Ecol. Syst. 21: 243–273.

Janz, N., and S. Nylin. 1997. The role of female search behaviour in determining host plant range in plant feeding insects, a test of the information processing hypothesis. Proc. R. Soc. Lond. B. 264: 701–707.

Jeffries, M.J., and J. H. Lawton. 1984. Enemy-free space and the structure of biological communities. Biol. J. Linn. Soc. 23: 269–286.

Jones, C.D. 1998. The genetic basis of *Drosophila sechellia*'s resistance to a host plant toxin. Genetics 149: 1899–1908.

Joshi, A., and J.N. Thompson. 1995. Trade-offs and the evolution of host specialization. Evol. Ecol. 9: 82–92.

Karowe, D.N. 1990. Predicting host range evolution: colonization of *Coronilla varia* by *Colias philodice* (Lepidoptera: Pieridae). Evolution 44: 1637–1647.

Kawecki, T.J. 1994. Accumulation of deleterious mutations and the long-term cost of being a generalist. Am. Nat. 144: 833–838.

Kawecki, T.J., J.D. Fry, and N.H. Barton. 1997. Mutational collapse of fitness in marginal habitats and the evolution of ecological specialization. J. Evol. Biol. 10: 407–429.

Keese, M.C. 1996. Feeding responses of hybrids and the inheritance of host-use traits in leaf-feeding beetles (Coleoptera: Chrysomelidae). Heredity 76: 36–42.

Keese, M.C. 1998. Performance of two monophagous leaf feeding beetles (Coleoptera: Chrysomelidae) on each other's host plant: do intrinsic factors determine host plant specialization? J. Evol. Biol. 11: 403–419.

Kirkpatrick, M., and N.H. Barton. 1997. Evolution of a species' range. Am. Nat. 150: 1–23.

LeSage, L. 1986. A taxonomic monograph of the Nearctic galerucine genus *Ophraella* Wilcox (Coleoptera: Chrysomelidae). Mem. Entomol. Soc. Canada 133: 1–75. Ottawa, ON.

Leslie, J.F., and H. Dingle. 1983. A genetic basis of oviposition preference in the large milkweed bug, *Oncopeltus fasciatus*. Entomol. Exp. Appl. 34: 215–220.

Losos, J.B., and R.E. Glor. 2003. Phylogenetic comparative methods and the geography of speciation. Trends Ecol. Evol. 18: 220–227.

Lu, W., and P. Logan. 1993. Induction of feeding on potato in Mexican *Leptinotarsa decemlineata* (Coleoptera: Chrysomelidae). Environ. Entomol. 22: 759–765.

Lu, W.H., G.G. Kennedy, and F. Gould. 1997. Genetic variation in larval survival and growth and response to selection by Colorado potato beetle (Coleoptera: Chrysomelidae) on tomato. Environ. Entomol. 26: 67–75.

Mackenzie, A. 1996. A trade-off for host plant utilization in the black bean aphid, *Aphis fabae*. Evolution 50: 155–162.

Maynard Smith, J. 1976. Group selection. Quart. Rev. Biol. 51: 277–283.

Mayr, E. 1942. Systematics and the origin of species. Columbia University Press, New York.

Mayr, E. 1963. Animal species and evolution. Harvard University Press, Cambridge, MA.

Menken, S.B.J. 1996. Pattern and process in the evolution of insect-plant associations: *Yponomeuta* as an example. Entomol. Exp. Appl. 80: 297–305.

Miller, J.R., and K. L. Strickler. 1984. Finding and accepting host plants, pp. 127–157. In W.J. Bell and R. Cardé (eds.), Chemical ecology of insects. Sinauer, Sunderland, MA.

Moran, N. 1986. Benefits of host plant specificity in *Uroleucon* (Homoptera: Aphididae). Ecology 67: 108–115.

Neal, J.J. 1987. Metabolic costs of mixed-function oxidase induction in *Heliothis zea*. Entomol. Exp. Appl. 43: 175–179.

Ng, D. 1988. A novel level of interactions in plant-insect systems. Nature 334: 611–612.

Nielsen, J.K. 1997. Genetics of the ability of *Phyllotreta nemorum* larvae to survive on an atypical host plant, *Barbarea vulgaris* ssp. *arcuata*. Entomol. Exp. Appl. 82: 37–44.

Nosil, P., B.J. Crespi, and C. Sandoval. 2003. Reproductive isolation driven by the combined effects of ecological adaptation and reinforcement. Proc. R. Soc. Lond. B. 270: 1911–1918.

Nosil, P., T.H. Vines, and D.J. Funk. 2005. Perspective: reproductive isolation caused by natural selection against immigrants from divergent habitats. Evolution 59: 705–719.

Polechová, J., and N.H. Barton. 2005. Speciation through competition: a critical review. Evolution 59: 1194–1210.

Radtkey, R., and M.C. Singer. 1995. Repeated reversals of host preference evolution in a specialist insect herbivore. Evolution 49: 351–359.

Rausher, M.D. 1984. Tradeoffs in performance on different hosts: evidence from within- and between-sire variation in the beetle *Deloyala guttata*. Evolution 38: 582–595.

Roessingh, P., K.H. Hora, J.J. A. van Loon. 1999. Evolution of gustatory sensitivity in *Yponomeuta* caterpillars: sensitivity to the stereoisomers dulcitol and sorbitol is localized in a single sensory cell. J. Comp. Physiol. A. 184: 119–126.

Römstock, M., and H. Arnold. 1987. Populationsökologie und Wirtswahl bei *Tephritis conura* Loew-Biotypen (Dipt.: Tephritidae). Zool. Anz. 219: 83–102.

Rowell-Rahier, M. 1984. The food plant preferences of *Phratora vitellinae* (Coleoptera: Chrysomelidae): a laboratory comparison of geographically isolated populations and experiments on conditioning. Oecologia 64: 375.

Rundle, H.D., and P. Nosil. 2005. Ecological speciation. Ecol. Lett. 8: 336–352.

Sandoval, C.P., and P. Nosil. 2005. Counteracting selective regimes and host preference evolution in ecotypes of two species of walking-sticks. Evolution 59: 2405–2413.

Schilthuizen, M. 2001. Frogs, flies, and dandelions: speciation—the evolution of new species. Oxford University Press, New York.

Schluter, D. 2000. The ecology of adaptive radiation. Oxford University Press, Oxford, UK.

Schoonhoven, L.M., T. Jermy, and J.J. A. van Loon. 1998. Insect-plant biology: from physiology to evolution. Chapman and Hall, London.

Scriber, J.M., B.L. Giebink, and D. Snider. 1991. Reciprocal latitudinal clines in oviposition behaviour of *Papilio glaucus* and *P. canadensis* across the Great Lakes hybrid zone: possible sex-linkage of oviposition preferences. Oecologia 87: 360–368.

Servedio, M.R., and M. Kirkpatrick. 1997. The effects of gene flow on reinforcement. Evolution 51: 1764–1772.

Servedio, M.R., and M. A. F. Noor. 2003. The role of reinforcement in speciation: theory and data. Annu. Rev. Ecol. Evol. Syst. 34: 339–364.

Service, P.M., and M.R. Rose. 1985. Genetic covariation among life history components: the effect of novel environments. Evolution 39: 943–945.

Sezer, M., and R.K. Butlin. 1998a. The genetic basis of oviposition preference differences between sympatric host races of the brown planthopper *(Nilaparvata lugens)*. Proc. R. Soc. Lond. B. 265: 2399–2405.

Sezer, M., and R. K. Butlin. 1998b. The genetic basis of host plant adaptation in the brown planthopper *(Nilaparvata lugens)*. Heredity 80: 499–508.

Sheck, A.L., and F. Gould. 1996. The genetic basis of differences in growth and behavior of specialist and generalist herbivore species: selection in hybrids of *Heliothis virescens* and *Heliothis subflexa* (Lepidoptera). Evolution 50: 831–841.

Simon, J.C., S. Carre, M. Boutin, N. Priunier-Leterme, B. Sabater-Muñoz, A. Latorre, and R. Bournoville. 2003. Host-based divergence in populations of the pea aphid: insights from nuclear markers and the prevalence of facultative symbiosis. Proc. R. Soc. Lond. B. 270: 1703–1712.

Singer, M.C., D. Ng, and C.D. Thomas. 1988. Heritability of oviposition preference and its relationship to offspring performance within a single insect population. Evolution 42: 977–985.

Singer, M.S. 2001. Determinants of polyphagy by a woolly bear caterpillar: a test of the physiological efficiency hypothesis. Oikos 93: 194.

Tadeusz, J., T.K. Kawecki, D. Robert, and R.H. Holt. 2002. Evolutionary consequences of asymmetric dispersal rates. Am. Nat. 160 333–347.

Templeton, A.R., and E.D. Rothman. 1981. Evolution in fine-grained environments. II. Habitat selection as a homeostatic mechanism. Theor. Popul. Biol. 19: 326–340.

Thomas, Y., M.-T. Bethenod, L. Pelozuelo, B. Frérot, and D. Bourguet. 2003. Genetic isolation between two sympatric host races of the European corn borer moth, *Ostrinia nubilalis* Hübner. I. Sex pheromone, moth emergence timing, and parasitism. Evolution 57: 261–273.

Thompson, J.N. 1988a. Evolutionary ecology of the relationship between oviposition preferences and performance of offspring in phytophagous insects. Entomol. Exp. Appl. 47: 3–14.

Thompson, J.N. 1988b. Evolutionary genetics of oviposition preference in swallowtail butterflies. Evolution 42: 1223–1234.

Thompson, J.N. 1996. Trade-offs in larval performance on normal and novel hosts. Entomol. Exp. Appl. 80: 133–139.

Thompson, J.N., W. Wehling, and R. Podolsky. 1990. Evolutionary genetics of host use in swallowtail butterflies. Nature 344: 148–153.

Thorpe, W.H. 1930. Biological races in insects and allied groups. Biol. Rev. 5: 177–212.

Tilmon, K.J., T.K. Wood, and J.D. Pesek. 1998. Genetic variation in performance traits and the potential for host shifts in *Enchenopa* treehoppers (Homoptera: Membracidae). Ann. Entomol. Soc. Am. 91: 397–403.

Ueno, H.,N. Fujiyama, K. Irie, Y. Sato, and H. Katakura. 1999. Genetic basis for established and novel host plant use in a herbivorous ladybird beetle, *Epilachna vigintioctomaculata*. Entomol. Exp. Appl. 72: 1–6.

Ueno, H.,N. Fujiyama, I. Yao, Y. Sato, and H. Katakura. 2003. Genetic architecture for normal and novel host-plant use in two local populations of the herbivorous ladybird beetle, *Epilachna pustulosa*. J. Evol. Biol. 16: 883–895.

van Drongelen, W. 1979. Contact chemoreception of host plant specific chemicals in larvae of various *Yponomeuta* species (Lepidoptera). J. Comp. Physiol. 134: 265–279.

Via, S. 1984. The quantitative genetics of polyphagy in an insect herbivore. II. Genetic correlations in larval performance within and among host plants. Evolution 38: 896–905.

Via, S. 1991. The genetic structure of host plant adaptation in a spatial patchwork: demographic variability among reciprocally transplanted pea aphid clones. Evolution 45: 827–852.

Via, S. 2001. Sympatric speciation in animals: the ugly duckling grows up. Trends Ecol. Evol. 16: 381–390.

Via, S., and D.J. Hawthorne. 2002. The genetic architecture of ecological specialization: correlated gene effects on host use and habitat choice in pea aphids. Am. Nat. 159: S76-S88.

Walsh, B.D. 1864. On phytophagic varieties and phytophagic species. Proc. Entomol. Soc. Phila. 3: 403–430.

Wasserman, S.S., and D.J. Futuyma. 1981. Evolution of host plant utilization in laboratory populations of the southern cowpea weevil, *Callosobruchus maculatus* Fabricius (Coleoptera: Bruchidae). Evolution 35: 605–617.

Waxman, D., and S. Gavrilets. 2005. 20 questions on adaptive dynamics. J. Evol. Biol. 18: 1139–1154.

Weber, G. 1985. Genetic variability in host plant adaptation of the green peach aphid, *Myzus persicae*. Entomol. Exp. Appl. 38: 49–56.

Whitlock, M.C. 1996. The Red Queen beats the jack-of-all-trades: the limitations on the evolution of phenotypic plasticity and niche breadth. Am. Nat. 148: S65-S77.

Wiklund, C. 1981. Generalist vs. specialist oviposition behaviour in *Papilio machaon* (Lepidoptera) and functional aspects on the hierarchy of oviposition preferences. Oikos 36: 163–170.

Williams, G.C. 1966. Adaptation and natural selection: a critique of some current evolutionary thought. Princeton University Press, Princeton, NJ.

Wood, T.K. 1993. Speciation of the *Enchenopa binotata* complex (Insecta: Homoptera: Membracidae), pp. 299–317. In D. R. Lees and D. Edwards (eds.), Evolutionary patterns and processes. Academic Press, London.

Wood, T.K., and M. C. Keese. 1990. Host-plant-induced assortative mating in *Enchenopa* treehoppers. Evolution 44: 619–628.

Zangerl, A.R., and M. R. Berenbaum. 2003. Phenotype matching in wild parsnip and parsnip webworms: causes and consequences. Evolution 57: 806–815.

PART II

CO- AND MACROEVOLUTIONARY RADIATION

ELEVEN

Host-Plant Use, Diversification, and Coevolution: Insights from Remote Oceanic Islands

GEORGE K. RODERICK AND DIANA M. PERCY

Insects and flowering plants are among the most diverse macroorganisms on earth, and their mutual interactions provide little doubt that each group is in part responsible for the other's diversity (Hairston et al. 1960; Ehrlich and Raven 1964; Strong et al. 1984; Novotny et al. 2006). However, exactly how diversification of flowering plants has affected the diversity of insects, and vice versa, is not well understood for the vast majority of plant and insect groups. Fossil evidence suggests that flowering plants were not associated with the initial modern diversification of insects that began 245 million years ago (Labandeira and Sepkoski 1993). Indeed, the basic diversity of the trophic machinery of insects was present in the fossil record 100 million years before the first fossil angiosperms. However, angiosperms do appear to be associated with a more recent diversification of insects, that is, diversification at the genus and species levels. One illustration of this association is Farrell's (1998) study of herbivorous beetles in which sister groups of beetles that feed on angiosperms were found to be consistently more diverse than their relatives feeding on gymnosperms and ferns. The approach used in Farrell's study makes some important assumptions, for example, that present feeding associations are indicative of past associations and that differences in numbers of species are not influenced unduly by other factors, such as asymmetrical extinction. The study demonstrates that the switch to angiosperms was associated with an increase in beetle diversity. But, exactly how does such an accelerated rate of diversification get started and to what extent do plants determine these processes?

For some insect and plant groups, we have a fairly good understanding of the factors associated with insect diversification. For example, studies of plant chemistry and diversification of Lepidoptera have provided evidence of the evolutionary arms races predicted by Ehrlich and Raven (1964) in their original hypothesis of insect-plant coevolution (Berenbaum and Feeny, this volume). Studies of pollination systems also illustrate how insects and plants are tightly tied to each other's diversity (e.g., yuccas and yucca moths, [Pellmyr and Thompson 1992]; figs and figwasps [Silvieus et al., this volume]). These systems illustrate complex patterns of coevolution that can themselves be a locus for diversification, as, for example, when such systems are exploited by complex layers of insect interlopers (Kjellberg et al. 2005). For each of these systems, progress in understanding has come through the close examination of the insect/plant part of the system. In this chapter, we use remote islands to focus on the evolutionary patterns of insect-plant interactions. We examine studies that have used the features of remote islands to tease apart factors associated with recent host-associated diversification in an attempt to understand how host plants contribute to the process of insect speciation. We also point to some areas where future work is likely to be especially fruitful.

Islands as a Model System

Islands have been attractive systems for the study of evolutionary and ecological phenomena, particularly as it relates to adaptation and diversification (Carlquist 1980; Simon 1987; Wagner and Funk 1995; Gillespie and Roderick 2002; Percy 2003b; Emerson et al. 2006). Important features of islands include the following: (1) Islands offer discrete geographic entities with defined boundaries. (2) Gene flow between islands within an archipelago is often limited, and gene flow between islands and a continental source pool may—depending on the level of isolation—be greatly reduced or nonexistent. The extent of isolation will also depend upon the dispersal abilities of the organisms and, in particular, their ability to disperse over water. A reduction of gene flow allows island populations to evolve in isolation from continental populations through processes such as genetic drift and local adaptation. (3) Islands are usually relatively small units so that the species can be more completely mapped and cataloged than is possible for continental regions. (4) Islands that are part of an archipelago offer multiple geographic replicates for studies of evolutionary and ecological processes. While islands can take many forms, here we limit our discussion to remote oceanic islands. As a group, remote oceanic islands are large enough to have a diversity of habitats. Further, because most are volcanic they can offer a diversity of geological landscapes, often with known ages and histories (e.g., Carson and Clague 1995). We examine how studies of island systems have contributed to an understanding of insect-plant interactions, specifically in the areas of host-associated specialization and speciation.

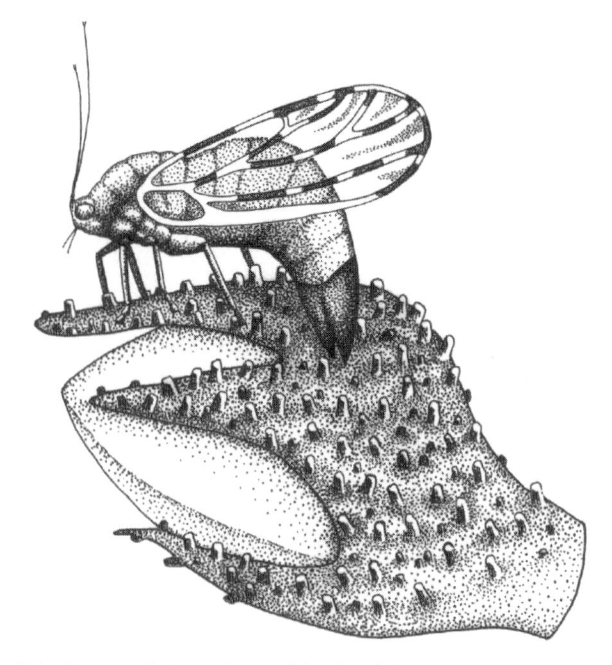

FIGURE 11.1. Endemic Canary Island psyllid ovipositing on an endemic broom. Drawing by D. Percy.

Lessons from Island Systems

Are Island Herbivorous Insects Specialists?

Theory suggests that species with more generalized habitat and feeding requirements should be expected to be more successful as colonists (Baker 1965; Lewontin 1965). Upon colonization such colonists may further expand their habitat requirement as a result of ecological release and expansion of range; specialist feeding might be expected to evolve only over evolutionary time (Gillespie and Roderick 2002). To this end, a number of studies have shown some insect colonists on remote islands to be generalists, feeding across a spectrum of plants. Some well-known recent invaders of islands have maintained extreme polyphagy, which may help explain their invasiveness; examples include the glassy-winged sharpshooter in French Polynesia (Grandgirard et al. 2006) and the two-spotted leafhopper in the Hawaiian Islands (Jones et al. 2000). Likewise, for a recent colonization of sub-Antarctic islands by aphids it has been suggested that polyphagy, along with parthenogenesis, has been important for success (Hulle et al. 2003). A large number of generalists have also been found among pollination systems in the Azores and Mauritius that include both native and introduced species (Olesen et al. 2002), and nonnative bees have been hypothesized to interrupt native pollinators in New Caledonia (Kato and Kawakita 2004). In a survey of the Laurisilva canopy of the Azores, Ribeiro et al. (2005) found that of 129 herbivore species, only 4 were clearly specialists and argued that the high proportion of generalists could be explained by adaptation to scleromorphic plants and a lack of predators, among other factors. However, the study grouped indigenous and nonindigenous taxa for both insects and plants, thus making it more difficult to tease apart historical from more recent insect plant associations. Moreover, the majority of insect occurrences in this study were concentrated on one plant species, which may suggest some specialization in itself.

Native herbivorous insects on remote islands show high levels of local endemism; for example, rates of endemism have been estimated at 98% for Hawaiian arthropods (Miller and Eldredge 1996) and 50% for Canary Island invertebrates (Juan et al. 2000). Such native herbivores are specialized in host use and some, highly so. In a study of Canary Island psyllids (Fig. 11.1), Percy (2002, 2003a, 2003b) documented that most were monophagous (70%) and a few oligophagous (30%), feeding on two to three closely related brooms (Fabaceae: Genisteae). This pattern of monophagy and oligophagy generally reflects the patterns of host associations for psyllids elsewhere and suggests that historical constraints, such as ancestral host use, may play a large role in determining patterns of host use for insects on islands. Similar constraints may explain host-use patterns in Canary Island leaf beetles (Garin et al. 1999), although one instance of island colonization in this system involved a switch to a novel plant family.

Monophagy and oligophagy on related hosts is also the pattern found in the sap-feeding planthopper genus *Nesosydne* (Hemiptera: Delphacidae) that has diversified within the Hawaiian Islands (Wilson et al. 1994; Roderick 1997). Here, 88% of the 81 species feed only on species within a single plant family, and 77% of species feed on a single plant species. This pattern of extreme specialization at the species level is also found for Delphacidae on other Pacific Islands, including the Marquesas (Wilson et al. 1994). Since most of these host records did not include the necessary observations of feeding, oviposition, and nymphal development required to establish host use, some of these records are likely spurious, and actual host specificity in these groups could be even higher. As with the Canary Island psyllids, patterns of specificity in island planthoppers reflect the patterns found in their most closely related continental relatives.

While it may be obvious that sap-feeding insects, such as psyllids and planthoppers, have a close association with their hosts that plays an important role in diversification, the means by which other types of host-plant interactions promote diversification is less clear. For example, a group of *Liparthrum* bark beetles in Macaronesia, which all share a similar habitat of dead or dying wood, have diversified, but with only a limited degree of host specialization (Jordal et al. 2004).

The variation in levels of host use across systems suggests that patterns of host use on islands may be largely dependent on the time period (ecological or evolutionary) over which the insect and host have interacted in an environment. A recent study of herb and herbivore colonization across a grade of island ages in a boreal archipelago off the coast of Sweden found that host-plant defenses increased over time in concert with the accumulation of specialist herbivores (Stenberg et al. 2006). Among groups of native insects that have diversified within islands, such as in the Canary psyllids and Hawaiian planthoppers, there is evidence of phylogenetic constraint in the patterns of host use and in the maintenance of broad patterns of monophagy (or oligophagy on related hosts). These constraints are evident despite switches to new hosts on islands. In island systems that display a signature of adaptive radiation in which one lineages diversifies to exploit a diversity of ecological habitats (Wagner and Funk 1995; Roderick and Gillespie 1998; Gillespie et al. 2001), evolutionary changes and shifts can be tracked and compared between different lineages. For this reason, we emphasize in the remainder of the chapter case studies of closely related groups of insects.

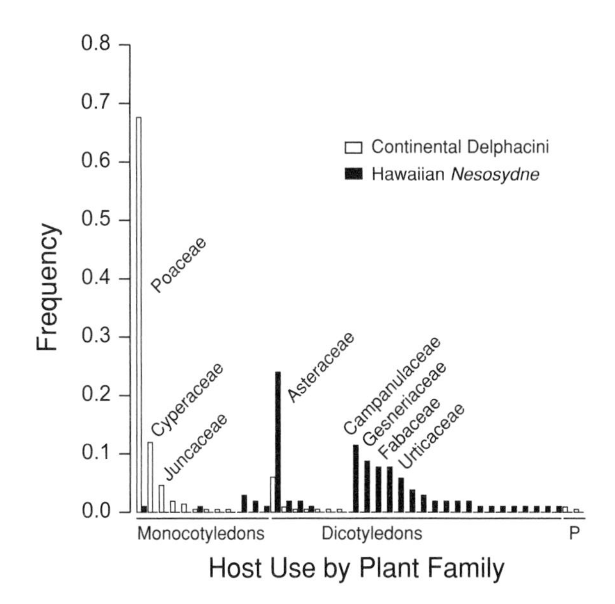

FIGURE 11.2. Diversity of host use by *Nesosydne* planthoppers (Hemiptera: Delphacidae) in the Hawaii Archipelago (after Roderick 1997). While continental Delphacini planthoppers (open bars) feed primarily on monocotyledons, grasses, and sedges, Hawaiian species (closed bars) feed on a wide diversity of plant families, mostly dicotyledons (see text). For both groups, a vast majority of species are monophagous. P denotes pteridophytes, ferns.

Host-Associated Diversification

The remarkable diversity of host plants associated with some adaptive radiations of island herbivorous insects might suggest that host switching is tied to insect diversification itself. Perhaps the most remarkable example of host shifts associated with adaptive radiation of herbivorous insects is that illustrated by the *Nesosydne* planthoppers noted above. In continental regions, the majority of species in the Delphacidae are restricted to monocots, especially grasses and sedges, with more than 85% of species feeding on three plant families (Denno and Roderick 1990; Cook and Denno 1994; Wilson et al. 1994; Roderick 1997). By contrast, the *Nesosydne* species in the Hawaiian Islands, all together, have been recorded on a remarkable total of 28 plant families, mostly dicots (Fig. 11.2). The plant family with the most recorded *Nesosydne* species is the Asteraceae with 25 planthopper species, while 12 plant families support only 1 *Nesosydne* species each, the remaining 15 families hosting 2 to 12 species.

To account for the dramatic host shifts that are associated with the island-dwelling delphacid planthoppers, a number of hypotheses have been proposed, including ecological release following initial colonization as a result of escape from competitors or predators (Wilson et al. 1994). Alternatively, multiple introductions, with each new colonist specializing on a different group of plants, may contribute to the diversity of hosts used by these planthopper species. Indeed, morphological studies by Asche (1997) suggest several colonization events, and perhaps as many as five, account for the 81 species of *Nesosydne*. Thus, colonization by a specialist species followed by subsequent diversification onto a number of unrelated hosts appears to have occurred several times in this group.

Observations of other plant feeding insect groups in the Hawaiian Islands likewise suggest that specialized species may have colonized remote islands and subsequently greatly expanded their host range. The genus of "true bugs" *Sarona* (Hemiptera: Miridae), comprises approximately 40 species in the Hawaiian Archipelago (Nishida 2002). The presumed ancestor was a specialist feeder restricted to a limited number of species in the Asteraceae (Asquith 1995, 1997). In the Hawaiian Islands, species of *Sarona* are still specialized, but as a whole the group feeds on species in 17 plant genera and 14 plant families. Asquith has hypothesized that species divergence is associated with host shifts following colonization. In another mirid genus, *Nesiomiris*, Gagne (1997) hypothesized that the precursor to the Hawaiian taxa was a group adapted to a genus in the Araliaceae. In the Hawaiian Islands host shifts have resulted in species feeding on different plant genera within the Araliaceae, but also plant species in the Urticaceae and Aquifoliaceae. A similar pattern of host shifts leading to an overall expansion of host usage has been suggested for species in another bug genus, *Cyrtolepis* (Gagne 1969; Asquith 1997). In groups that have looser, or less constrained, associations with their hosts, species diversification may be associated with habitat shifts rather than host switching (see also below). For example, species of *Nysius* bugs (Lygaeidae) and relatives in the Hawaiian Islands may have undergone habitat shifts along with their hosts (Usinger 1941). In the Hawaiian *Prognathogryllus* crickets (Orthoptera: Gryllidae), changes in host associations that are also coupled with geographical isolation may underlie species formation (Shaw 1995).

In each of these systems insects have diversified across different host species, and some of these shifts likely also involve geographic isolation. But, it is not known at what point, or how frequently, during diversification host switching occurred in these systems. It could be that broad expansions in host preference happen early relative to colonization of the islands, with more recent host shifts confined to closely related taxa. Indeed, there is evidence that host-associated diversification can happen rapidly. For example, moths in the *blackburni* group of *Omiodes* (*Hedylepta*; Pyralidae) in the Hawaiian Islands consist of seven species, two feeding on palms, one occasionally straying to banana, and five known only from banana (Zimmerman 1960). It appears that all of the banana species originated through host switching from species already present in the islands on native *Pritchardia* palms when the Polynesians introduced banana around 1000 years ago. Clearly, these systems will benefit from a modern reassessment employing molecular methods to pinpoint the timing of divergences associated with host shifts relative to the timing of the initial colonization of the islands by a lineage.

Relative Roles of Host Switching and Geographic Allopatry

Two studies of adaptive radiation of herbivorous insects in the Canary Islands are sufficiently complete to allow assessment of the role of host switches in speciation events. Both studies make use of phylogenetic hypotheses based on multiple data sets to identify sister groups that can be used in comparisons. In the first, Percy (2002, 2003a, 2003b) examined the patterns of host use in the arytainine psyllid (Hemiptera) radiation on broom (Fabaceae: Genisteae). Psyllids have colonized the Macaronesian islands several times, but only one colonization event has resulted in a significant radiation on broom. Percy used a phylogeny based on both DNA sequence data and morphology to examine conditions associated with the divergence of sister taxon pairs, and in particular to ask whether speciation was associated with allopatry or sympatry, and/or host shifts. In 10 of 11 comparisons, diversification was associated with geographic allopatry. Of these 10 allopatric events, 6 involved a host shift to a related host. In the other 4 allopatric events, the psyllids used the same host (i.e., a host shift was not involved in diversification). There were no cases of a switch in allopatry to an unrelated host. In areas where several psyllids shared the same host, resources appeared to be partitioned by ecological specialization within the host and temporal differences in psyllid phenology. In this system, Percy argued that host use by the psyllids is constrained by a narrow host preference to a group of plants within the legumes (i.e., Genisteae), and within the Canary Islands postcolonization divergence is closely related to the distribution and extent of legume diversity.

A second Canary Island system also emphasizes the role of allopatric isolation in divergence. Jordal and Hewitt (2004) examined a group of bark beetles in the genus *Aphanarthrum* (Curculionidae: Scolytinae) that feed and breed in woody *Euphorbia* spurges. As in Percy's study, Jordal and Hewitt used sister taxon pairs, identified by the phylogeny, to examine patterns of geographical and host use associated with putative speciation. Of 24 cladogenic (splitting) events, only 6—significantly lower than expected by chance—were associated with major changes in host use. The remaining cladogenic events involved taxa on closely related hosts or the same host. Jordal and Hewitt argued that the unexpectedly low frequency of host switching associated with cladogenic events suggests that geographical factors are more important than host-related factors in the diversification of this group. At the same time, they pointed out that a novel resource may provide the opportunity for diversification by providing new sets of niches. For example, the historical switch to the succulent growth form of *Euphorbia* likely provided the lineage with niches sufficient for 10 species. Accordingly, diversification in this system appears to be a consequence both of the availability of new resources (e.g., niche partitioning within a host species) and the factors that promote genetic divergence, such as genetic drift in allopatry.

Direct comparison of the results of these two studies is difficult as the psyllid-legume study distinguished between the same and closely related hosts while the beetle-*Euphorbia* study pooled the same and closely related hosts. Also, the psyllid study examined only species and subspecies pairs, while the beetle study also included higher-level cladogenesis. Nevertheless, it is clear that allopatric isolation is an important factor associated with cladogenesis in these systems, and although host shifts in sympatry are also observed in both systems (see below) it may be that sympatric host shifts are more likely subsequent to cladogenesis in allopatry (Table 11.1).

Sympatric Speciation

The studies reviewed above highlight the importance of allopatric isolation in the diversification of insects on islands, even in systems that utilize a diverse set of hosts. These studies also suggest that divergence in host-plant use may be important in allowing sibling species to co-occur in sympatry. Other studies of arthropods on islands have also emphasized the role of sympatry in ecological diversification (Kambysellis and Craddock 1997; Joy and Conn 2001; Shaw 2002; Gillespie 2004). Island systems may provide fertile conditions for sympatric speciation because of the large

TABLE 11.1
Putative Speciation Modes Associated with Cladogenetic Events for Continental Macaronesian Legume-Feeding Psyllids (Percy 2003a) and *Euphorbia*-Feeding Beetles (Jordal and Hewitt 2004) Demonstrating the Relative Roles of Geological Separation and Host Shifts

	Geography			
	Allopatry		Sympatry	
Host Group Taxon	SAME/SIMILAR HOST	SHIFT	SAME/SIMILAR HOST	SHIFT
Psyllids	4	6	0	2
Beetles	11	1	8	4

NOTE: For the psyllids the category "same/similar host" denotes the same host species, while shifts in allopatry are to hosts in the same or closely related genera, and shifts in sympatry to distantly related hosts. For the beetles, same/similar includes closely related species, and shifts indicate broader differences among host groups (see text).

numbers of species confined to a limited area and the potential for rapid changes in host use on a local scale (Gillespie and Roderick 2002). For example, on the island of Rapa in the Austral Islands of French Polynesia, 67 species of weevils have been described in the genus *Miocalles* (Coleoptera: Curculionidae: Cryptorhynchinae) (Paulay 1985). Collectively, these small flightless beetles feed on most of the 24 genera of native plants on this island, with varying degrees of host specificity. While there has been much interest in whether sympatric speciation has occurred in this system (see discussion in Coyne and Orr 2004), *Miocalles* has not been studied using modern molecular approaches. Ongoing research is focused on a detailed examination of this system to resolve whether the diversification happened in situ or represents the input from different lineages, perhaps from older, now nonexistent islands (Elin Claridge, 2007, personal communication).

Other sympatric species pairs have been studied more thoroughly (Table 11.1). Using a comparison of the Canary Island psyllids and their continental relatives, Percy (2003b) showed that one continental sympatric sister pair *(Arytainilla cytisi* and *A. telonicola)* occurs on distantly related hosts (Percy 2003b), while in the Canary Islands a set of subspecies of *A. modica* appears to be forming putative host races in sympatry, also on distantly related hosts. From these results, Percy suggested that switches to unrelated hosts may be a necessary condition for sympatric speciation in these insects. Reconstructing historical distributions is, of course, critical in these scenarios, and this emphasizes another benefit of using isolated and recently formed islands. In the case of the continental psyllid taxa, the divergence may have occurred in allopatry with subsequent post-divergence sympatry. However, the putative host races in the Canary Islands are restricted to the youngest islands (La Palma and El Hierro), and there is evidence that the host-race formation has occurred independently on each island.

In the *Euphorbia* beetles of the Canary Islands discussed above (Jordal and Hewitt 2004), two ecologically similar yet reproductively isolated species, *Aphanarthrum subglabrum* and *A. glabrum* ssp. *nudum*, were found to breed inside dead twigs of the same plant species, *E. lamarckii*, on the island of La Palma. A possible sympatric origin for these co-occurring species was studied in more detail by Jordal et al. (2006), and a phylogenetic analyses of the combined nuclear DNA data strongly supported a sister relationship between the two species. However, a network analysis of subdivided nonrecombinant segments at one locus suggested that *A. glabrum* had a closer relationship with another allopatric subspecies of *A. glabrum* and not the sympatric *A. subglabrum*. The authors used this result, coupled with a bimodal distribution of mtDNA haplotypes on La Palma, to argue that there had been multiple colonizations of this island by *A. glabrum*. Thus, the two La Palma species are thought to have diverged in allopatry before parallel colonization of this island. Jordal et al. suggested that the subsequent sympatric completion of divergence was due either to initial genetic incompatibility, or to morphological character displacement in male genitalia, or a combination of these factors.

Cospeciation and Coevolution

The data presented above show that host switches appear to be an important feature in the evolutionary history of host use by herbivorous insects on remote islands. A second mechanism that has been proposed for acquiring new hosts in a lineage is that of host tracking, with insect species divergence occurring after the diversification of its host. New molecular and analytical tools have made the study of such systems possible, increasing the likelihood that such patterns can be uncovered despite the potentially confounding effects of large amounts of host switching. With limited colonists and thus limited lineages, remote islands should be ideal laboratories in which to study host tracking.

When an insect phylogeny and a plant phylogeny are viewed in concert, strict host tracking may appear as a cospeciation event. Here, we distinguish between cospeciation, which we define in this context as cocladogenesis, and coevolution in the sense of Ehrlich and Raven (1964), which we use to mean reciprocal adaptive selection and speciation. Of course, coevolution may be a process behind a pattern of cospeciation, but it need not be. For example, the general occurrence of organisms on an island hot-spot archipelago as they colonize the next-youngest island when it becomes available (Wagner and Funk 1995; Roderick and Gillespie 1998) can lead to a pattern in which all such taxa show a similar phylogenetic history, simply because they all reflect a common underlying geological foundation. Thus, it is important to understand the nature of the interaction between lineages, including whether the cospeciation events are contemporaneous.

There are demonstrations of cospeciation in continental systems (e.g., milkweeds and beetles [Farrell 2001]), but these are few. The paucity of demonstrations of cospeciation suggests either the phenomenon is rare or it is difficult to detect. On islands, the most rigorous study to date of potential cospeciation is that of Percy et al. (2004), who examined the parallel diversification of psyllids and their legume (Fabaceae: Genisteae) hosts in lineages that include taxa on the islands of Macaronesia (reviewed above). High levels of parallel cladogenesis were found between the psyllids and legumes. However, when the molecular phylogenies were calibrated based on known geological ages of the Macaronesian islands, it appeared that the insect diversification was not contemporaneous with that of the host plant, but differed by several millions of years—the main plant radiation occurring approximately eight million years ago, while the insects only three million years ago (Fig. 11.3). The misinterpretation of cospeciation events in this case could be explained by a predominant pattern of host shifts between related hosts: Percy et al. estimated that more than 60% of the psyllid cladogenesis was associated with host switching between related hosts. Evidence for partially contemporaneous cospeciation was confined to a much more recent and localized radiation of legumes and psyllids in the Canary Islands. Although in one psyllid-legume association, colonization of the Canary Islands by both legumes and psyllids appears nearly contemporaneous, rapid cladogenesis after colonization, resulting in reduced phylogenetic resolution over these periods, makes the identification of specific putative cospeciation nodes difficult.

In the *Nesosydne* planthoppers of the Hawaiian Islands, parallel cladogenesis has also been documented between a group of planthoppers and their hosts in the silversword alliance (Asteraceae) on Maui and Hawaii, the two youngest islands of the Hawaiian chain (Roderick 1997). However, a broader analysis of *Nesosydne* planthoppers on other plant groups implies that host switching is a more common pattern of host-associated divergence in this group (Hasty 2005), and like in the psyllid/legume system, parallel cladogenesis may be restricted to relatively recent speciation events. Whether parallel cladogenesis is contemporaneous

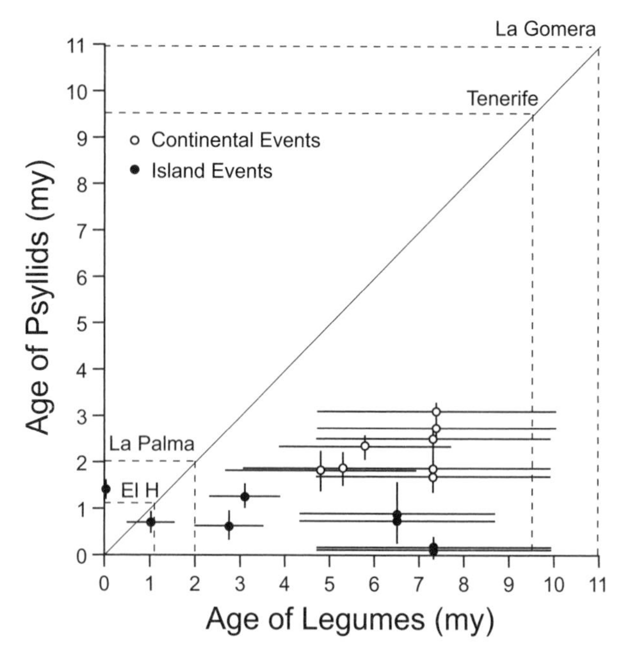

FIGURE 11.3. Comparative ages of 16 putative cospeciation events between continental and Macaronesian island psyllids and their legume hosts (after Percy et al. 2004). The legume nodes are much older than the psyllid nodes (compare to diagonal solid line), and all but one of the events is temporally implausible, suggesting that the diversification of psyllids followed that of their hosts. The majority of psyllid host switches were to related hosts. Continental cospeciation events are noted as open circles; island events as closed circles. Ages of the islands are shown with dashed lines: La Gomera 10–12 my, Tenerife 7.5–11 my, La Palma 2 my, El Hierro 1.1 my. Copyright 2004, from citation above. Reproduced by permission of Taylor and Francis Group, LCC., www.taylorandfrancis.com.

in this system awaits further calibration of the insect and plant phylogenies.

Parallel cladogenesis has also been documented in another system that includes some island species. Kawakita el al. (2004) studied a species-specific obligate mutualism between 18 species of *Glochidion* trees (Euphorbiaceae) and a corresponding 18 species of *Epicephala* moths (Gracillariidae). This system is pollination based, analogous to those of figs and figwasps (Silvieus et al., this volume) and yuccas and yucca moths (Pellmyr and Thompson 1992). Collections of *Epicephala* moths and corresponding *Glochidion* trees were made from Asia (Japan, Myanmar, Malaysia, Laos), Australia, Fiji, and New Caledonia. The study documented that parallel cladogenesis occurred more frequently than expected by chance. However, the evolutionary association was not perfect, and host shifts prompted questions as to how the one-to-one relationship between pollinator and hosts originated. Further studies of this system on more Pacific Islands are in progress (David Hembry, 2007, personal communication).

One assumption of the host-associated diversification of insects via host tracking is that insect populations can track changes in the distribution of their hosts over space and time. In this scenario, the insects are present on the host when it undergoes speciation. There is evidence from studies of populations within species that insects can respond remarkably quickly to changes in host distribution. The curculionid beetle *Brachyderes rugatus rugatus* is locally endemic on the island of La Palma in the Canary Islands, where it is reliant upon its pine host, *Pinus canariensis*. It is thought that *B. r. rugatus* colonized La Palma from Tenerife early in the emergence of the island's subarial terrains (Emerson et al. 2000). On the island of La Palma a network analysis shows that there were three expansions of *B. r. rugatus* populations from what appears to be a common point of origin. Emerson et al. suggested the species has tracked the expansion of its host, *P. canariensis*, perhaps in concert with other species associated with this plant. Definitive evidence for cospeciation will require a more rigorous assessment of the phylogeographic history of the pine and demonstration that the expansions were contemporaneous. At present, this study indicates that host-associated population expansion may be rapid and more complex than suggested solely from insect-plant distribution data.

In summary, it is clear that host tracking and cospeciation have occurred on islands. However, as Percy et al. (2004) noted, even in island systems where we have fairly good knowledge of extant herbivores and their hosts, the ability to document events of cospeciation is hindered by uncertainty within both insect and plant phylogenies, where the patterns can be obscured by the rapid rates of species radiations.

Rates of Host-Associated Diversification

Island systems have been used to illustrate some of the fastest rates of adaptive evolution leading to rapid speciation (Emerson and Kolm 2005; Shaw 2005). Ecological opportunity (Zimmerman 1960; Gillespie 2004), rapid recovery of genetic variation following a bottleneck (Carson 1990), and sexual selection (Shaw 2005) have all been implicated in contributing to evolutionary acceleration on remote islands. The extent to which each of these may contribute to rates of host-associated diversification in herbivorous insects is largely unknown, and a number of questions must first be answered before it can be addressed. For example, what was the diversity of hosts available to insects when they colonized an archipelago or single island? How large were the initial insect populations that colonized new hosts, how much genetic diversity did they contain, and how rapidly did they expand? It may be that new analytical approaches that tease apart simultaneously genetic signals attributable to historical population size, time since divergenece, and gene flow (e.g., the approach implemented program IM [Nielsen and Wakeley 2001; Hey and Nielsen 2004]) can be applied to herbivores across hosts on islands. With such methods it may be possible to reconstruct the historical demography associated with host shifts.

Sexual selection may play an underappreciated role in accelerating the rate at which specialized insect herbivores adapt to plants on islands. For example, the sap-feeding insects noted above, including psyllids (Percy et al. 2006) and Hawaiian planthoppers (O'Connell 1992), produce substrate-borne acoustic signals that are used in species recognition and courtship (Claridge and De Vrijer 1994; Percy et al. 2006). If the evolution of these signals is differentially affected by association with different plant species, a runaway selection process may facilitate rapid establishment of host associations and perhaps also contribute to host fidelity (Berlocher and Feder 2002).

Another factor that may contribute to enhanced rates of diversification on islands concerns the fact that many organisms on islands show a reduced tendency to disperse (see discussion in Gillespie and Roderick 2002). For example, many insects on islands are flightless, show wing reduction, or otherwise show reduced capacity for dispersal. While it has been noted that the number of flightless organisms on islands may be no less then on continents, the island species are all descendents of highly dispersive taxa. For example, in the Hawaiian *Nesosydne* planthoppers, only 3 of 81 species are fully winged at all times, and all are associated with tall trees (Denno 1994; Denno et al. 2001). The other species are either short winged (resulting in loss of flight ability) or polymorphic in wing length. Reduction in dispersal on islands appears to contribute to the effective level of isolation among allopatric populations of some arthropods both among and within islands, including *Drosophila* (Carson and Johnson 1975; Carson et al. 1990) and *Tetragnatha* spiders (Vandergast et al. 2004), but the effect of this is as yet untested among host-associated populations of insects. Even in dispersive species, genetic differences in traits such as oviposition can arise over relatively short distances (e.g., butterflies in habitat patches in Finland [Kuussaari et al. 2000]) so that any reduction in dispersal should amplify these effects.

Symbionts Associated with Island Herbivores

Some insect-plant interactions are obvious symbioses (e.g., plant-pollinator interactions). Less conspicuous are the other symbiotic organisms associated with insects, many of which may play a role in the evolution of host selection and host establishment in herbivorous insects. It is now well documented that likely all sap-feeding hemipteran insects have mutually obligate internal symbionts. In aphids, for example, these symbionts are bacteria that have evolved to produce essential nutrients for the aphids while undergoing a genome reduction affecting other cell functions (Moran and Wernegreen 2000). The bacteria are housed in specialized organs within the insect. Genomic sequencing is now providing a wealth of information about the interactive role of the symbiosis of these bacteria and their host insects. In psyllids, for example, the associated bacteria have an extremely reduced genome but still contain the machinery to produce essential amino acids (Nakabachi et al. 2006). Psyllids, like aphids, feed on phloem sap, which is nutritionally poor in essential amino acids, and the bacteria assist their hosts by supplementing a nutritionally poor food. Planthoppers also have symbiotic microorganisms, but they are yeastlike, rather than bacterial (Noda and Koizumi 2003). Like the aphids, the yeastlike symbionts of planthoppers are transmitted vertically and housed in special organs. Unfortunately, the yeastlike symbionts are less amenable to laboratory manipulations than the aphid bacteria, and their genomes are much larger—as a result their function is less well understood (Lu et al. 2004). In neither the Canary Island psyllid system nor the Hawaiian planthoppers have the roles of these symbionts been examined. It is tempting to speculate that the symbionts may in some way be associated with the ability of these insects to colonize new hosts or partition host resources. For instance, one study using psyllids suggested that the presence or absence of secondary endosymbionts is influenced by whether the insect host is gall-making or free living (Spaulding and von Dohlen 2001). A monophyletic group of circa 20 endemic Hawaiian psyllids may have repeatedly evolved the gall-forming habit on the host plant *Metrosideros* (D.M.P., personal observation). An analysis of the symbiont fauna in this psyllid group would make an interesting case study of the role of symbionts in host-resource use within an island system.

The role of other symbionts associated with island herbivores is less clear. For example, Hawaiian nitidulid beetles (Ewing and Cline 2005) that typically feed on plant sap and decaying plant matter vector a diversity of yeasts, previously unknown and also endemic to the Hawaiian Islands (Lachance et al. 2005). The nature of the interaction between the yeasts and the beetles is currently not understood. *Wolbachia* is another common symbiont associated with insects and has been documented in some island herbivores where it can have large effects on mating structure and other aspects of population biology (Charlat et al. 2005). The direct or indirect role of *Wolbachia* influencing host use is also not known. Because symbionts of various kinds affect the ability of organisms to invade or colonize new habitats (Richardson et al. 2000), it is reasonable to expect that symbionts will prove to be important in the evolution of host associations on islands.

Role of Plant Hybridization

Hybridization is a common phenomenon within the plant kingdom and has been implicated in the generation of an estimated 30 to 80% of all extant plant species (Wendel et al. 1991). When plants hybridize, the resulting taxa will have novel genotypes and may acquire, or lose, traits and adaptations from both parental species. The resulting effects on herbivorous insects can vary depending on the system. Hybrids can be more, or less, attractive to the herbivore; they may be the centers of unexpected herbivore or community diversity; and they may provide a pathway by which insects can colonize new hosts (Dungey et al. 2000; Czesak et al., this volume). Despite the prevalence of plant hybrids generally, including on islands, the role of hybrid plants on herbivorous insect diversity on islands has been examined in only a few studies.

In the Hawaiian Islands, Drew and Roderick (2005) examined insect communities across six separate hybrid zones formed by the hybridization of different members of the Hawaiian silversword alliance (Asteraceae). These plants all originated from one colonization event to the Hawaiian Islands some four to six million years ago (Baldwin 1997; Baldwin and Sanderson 1998) and thus share a similar genetic background. Hybrids form readily between members of the silversword alliance both in the field and in the laboratory. Drew and Roderick found that insect family-level richness was as high or higher on hybrid plants in four of the six hybrid zones, and results were generally consistent with hypotheses that predict either a breakdown in defenses against insects in hybrid plants or an increase in host-recognition cues in hybrids. However, the pattern was not universal, and a hypothesis that predicts an increase in hybrid resistance to herbivory was supported in two hybrid zones. Insect communities were found to be more similar across hybrid zones when the host plants were more closely related phylogenetically than when the host plants were more genetically dissimilar. This last finding suggests that hybrid zones that differ in the relatedness of the parental types may provide a mechanism for gauging the extent to which herbivorous taxa can switch to genetically novel hosts (Roderick and Metz 1997). Further work in this system is examining the specific responses of delphacid planthoppers to different hybrid systems. In the Canary Island psyllid system, the legume-feeding psyllids that are more oligophagous (feeding on several hosts) include among their hosts both parental types and hybrids (Percy and Cronk 2002; Percy 2003b). Like the planthoppers in the Hawaiian Islands, it is not clear yet whether expansion onto alternative hosts via hybrids is due to phenotypic or genotypic similarities to the original hosts, or whether the hybrids are in some way promoting a host shift and/or expansions in host preference.

Conclusions

The isolation and microcosmal nature of remote islands has provided a useful natural laboratory with which to examine the nature of host-associated diversification in insects. Unlike recently introduced island colonizers and invaders, which are largely host generalists, the majority of native host-associated insects on islands are endemic, and most of these show host specialization. Studies to date emphasize the combined roles of allopatric isolation and host switching in insect diversification. Rates of divergence can be rapid and likely involve a number of factors that may increase effective isolation among populations or accelerate host adaptation. Less well understood are the demographic conditions present at the time of host-associated divergence, and the roles of symbionts and plant hybrids in the diversification of insects on islands.

Acknowledgments

We thank Tom Wood for numerous animated and inspirational discussions probing all aspects of insect plant interactions, particularly those involving sap-feeders. Rosemary Gillespie, Quentin Cronk, and Kelley Tilmon gave helpful comments and suggestions. G.R. was supported by the U.S. National Science Foundation (9981513, 0342279, 0416268, 0451971), the University of California, and Institut National de la Recherche Agronomique (INRA), France. D.P. was supported by the Natural Environment Research Council (U.K.), with additional funding for fieldwork provided by the Carnegie Trust for the Universities of Scotland and the Louise Hiom Award (University of Glasgow); research on psyllid acoustics was supported by the Leverhulme Trust (U.K.).

References Cited

Asche, M. 1997. A review of the systematics of Hawaiian planthoppers (Hemiptera: Fulgoroidea). Pacific Sci. 51: 366–376.

Asquith, A. 1995. Evolution of *Sarona* (Heteroptera, Miridae), pp. 90–120. In W.L. Wagner and V.A. Funk (eds.), Hawaiian biogeography. Smithsonian Institution Press, Washington, DC.

Asquith, A. 1997. Hawaiian Miridae (Hemiptera: Heteroptera): the evolution of bugs and thought. Pacific Sci. 51: 356–365.

Baker, H.G. 1965. Characters and modes of origin of weeds, pp. 147–172. In H.G. Baker and G.L. Stebbins (eds.), The genetics of colonizing species. Academic Press, New York.

Baldwin, B.G. 1997. Adaptive radiation of the Hawaiian silversword alliance: congruence and conflict of phylogenetic evidence from molecular and non-molecular investigations. In T. Givnish and K. J. Sytsma (eds.), Molecular evolution and adaptive radiation. Cambridge University Press, Cambridge, UK.

Baldwin, B.G., and M.J. Sanderson. 1998. Age and rate of diversification of the Hawaiian silversword alliance (Compositae). Proc. Natl. Acad. Sci. USA 9: 9402–9406.

Berlocher, S.H., and J.L. Feder. 2002. Sympatric speciation in phytophagous insects: moving beyond controversy? Annu. Rev. Entomol. 29: 403–433.

Carlquist, S. 1980. Hawaii: a natural history. Pacific Tropical Botanical Garden, Lawai, Hawaii.

Carson, H.L. 1990. Increased genetic variance after a population bottleneck. Trends Ecol. Evol. 5: 228–230.

Carson, H.L., and D.A. Clague. 1995. Geology and biogeography of the Hawaiian Islands, pp. 14–29. In W.L. Wagner and V.A. Funk (eds.), Hawaiian biogeography evolution on a hot spot archipelago. Smithsonian Institution Press, Washington, DC.

Carson, H.L., and W.E. Johnson. 1975. Genetic variation in Hawaiian *Drosophila*. I. Chromosome and allozyme polymorphism in *Drosophila setosimentum* and *Drosophila ochrobasis* from the island of Hawaii. Evolution 29: 11–23.

Carson, H.L., J. P. Lockwood, and E.M. Craddock. 1990. Extinction and recolonization of local populations on a growing shield volcano. Proc. Natl. Acad. Sci. USA 87: 7055–7057.

Charlat, S., E.A. Homett, E.A. Dyson, P. P. Y. Ho, N. Thi Loc, M. Schilthuizen, N. Davies, G.K. Roderick, and G.D.D. Hurst. 2005. Prevalence and penetrance variation of male-killing *Wolbachia* across Indo-Pacific populations of the butterfly *Hypolimnas bolina*. Mol. Ecol. 14: 3525–3530.

Claridge, M.F., and P. De Vrijer. 1994. Reproductive behaviour: the role of acoustic signals in species recognition and speciation. In R. F. Denno and T.J. Perfect (eds.), Planthoppers: their ecology and management. Chapman and Hall, New York.

Cook, A.G., and R.F. Denno. 1994. Planthopper/plant interactions: feeding behavior, plant nutrition, plant defense, and host plant specialization, pp. 114–139. In R. F. Denno and T.J. Perfect (eds.), Planthoppers: their ecology and management. Chapman and Hall, New York.

Coyne, J.A., and H.A. Orr. 2004. Speciation. Sinauer, Sunderland, MA.

Denno, R.F. 1994. Life history variation in planthoppers, pp. 163–215. In R.F. Denno and T.J. Perfect (eds.), Planthoppers: their ecology and management. Chapman and Hall, New York.

Denno, R.F., and G.K. Roderick. 1990. Population biology of planthoppers. Annu. Rev. Entomol. 35: 489–520.

Denno, R.F., D.J. Hawthorne, B.L. Thorne, and C. Gratton. 2001. Reduced flight capability in British Virgin Island populations of a wing-dimorphic insect: the role of habitat isolation, persistence, and structure. Ecol. Entomol. 26: 25–36.

Drew, A.E., and G.K. Roderick. 2005. Insect biodiversity on plant hybrids within the Hawaiian silversword alliance (Asteraceae: Heliantheae-Madiinae). Environ. Entomol. 34: 1095–1108.

Dungey, H.S., B.M. Potts, T.G. Whitham, and H.F. Li. 2000. Plant genetics affects arthropod community richness and composition: evidence from a synthetic eucalypt hybrid population. Evolution 54: 1938–1946.

Ehrlich, P.R., and P.H. Raven. 1964. Butterflies and plants: a study in coevolution. Evolution 18: 586–608.

Emerson, B.C., and N. Kolm. 2005. Species diversity can drive speciation. Nature 434: 1015–1017.

Emerson, B.C., P. Oromi, and G.M. Hewitt. 2000. Colonization and diversification of the species *Brachyderes rugatus* (Coleoptera) on the Canary Islands: evidence from mitochondrial DNA COII gene sequences. Evolution 54: 911–923.

Emerson, B.C., S. Forgie, S. Goodacre, and P. Oromi. 2006. Testing phylogeographic predictions on an active volcanic island: *Brachyderes rugatus* (Coleoptera : Curculionidae) on La Palma (Canary Islands). Mol. Ecol. 15: 449–458.

Ewing, C.P., and A.R. Cline. 2005. Key to adventive sap beetles (Coleoptera: Nitidulidae) in Hawaii, with notes on records and habits. Coleopt. Bull. 59: 167–183.

Farrell, B.D. 1998. "Inordinate fondness" explained: why are there so many beetles? Science 281: 555–559.

Farrell, B.D. 2001. Evolutionary assembly of the milkweed fauna: cytochrome oxidase I and the age of *Tetraopes* beetles. Mol. Phylogen. Evol. 18: 467–478.

Gagne, W.C. 1969. New species and a revised key to the Hawaiian *Cyrtopeltis* Fieb. with notes on *Cyrtopeltis (Engytatus) hawaiiensis* Kirkaldy (Heteroptera: Miridae). Proc. Hawaiian Entomol. Soc. 20: 35–44.

Gagne, W.C. 1997. Insular evolution, speciation, and revision of the Hawaiian genus *Nesiomiris* (Hemiptera: Miridae). Bishop Museum Press, Honolulu.

Garin, C.F., C. Juan, and E. Petitpierre. 1999. Mitochondrial DNA plylogeny and the evolution of host-plant use in Palearctic *Chrysolina* (Coleoptera, Chrysomelidae) leaf beetles. J. Mol. Evol. 48: 435–444.

Gillespie, R.G. 2004. Community assembly through adaptive radiation in Hawaiian spiders. Science 303: 356–359.

Gillespie, R.G., and G.K. Roderick. 2002. Arthropods on islands: colonization, speciation, and conservation. Annu. Rev. Entomol. 47: 595–632.

Gillespie, R.G., F.G. Howarth, and G.K. Roderick. 2001. Adaptive Radiation, pp. 25–44. In S.A. Levin (ed.), Encyclopedia of biodiversity. Academic Press, New York.

Grandgirard, J., M.S. Hoddle, G.K. Roderick, J.N. Petit, D. Percy, R. Putoa, C. Garnier, and N. Davies. 2006. Invasion of French Polynesia by the glassy-winged sharpshooter, *Homalodisca coagulata* (Hemiptera: Cicadellidae): a new threat to the South Pacific. Pacific Sci. 60: 429–438.

Hairston, N.G., F.E. Smith, and L.B. Slobodkin. 1960. Community structure, population control, and competition. Am. Nat. 94: 421–425.

Hasty, G.L. 2005. Evolution and host use of Hawaiian Nesosydne planthoppers (Hemiptera: Delphacidae). Ph.D. dissertation, University of California, Berkeley.

Hey, J., and R. Nielsen. 2004. Multilocus methods for estimating population sizes, migration rates and divergence time, with applications to the divergence of *Drosophila pseudoobscura* and *D. persimilis*. Genetics 167: 747–760.

Hulle, M., D. Pannetier, J. C. Simon, P. Vernon, and Y. Frenot. 2003. Aphids of sub-Antarctic Iles Crozet and Kerguelen: species diversity, host range and spatial distribution. Antarct. Sci. 15: 203–209.

Jones, V.P., P. Anderson-Wong, P.A. Follett, P.J. Yang, D.M. Westcot, J.S. Hu, and D.E. Ullman. 2000. Feeding damage of the introduced leafhopper *Sophonia rufofascia* (Homoptera: Cicadellidae) to plants in forests and watersheds of the Hawaiian islands. Environ. Entomol. 29: 171–180.

Jordal, B.H., and G.M. Hewitt. 2004. The origin and radiation of Macaronesian beetles breeding in *Euphorbia*: the relative importance of multiple data partitions and population sampling. Syst. Biol. 53: 711–734.

Jordal, B.H., L.R. Kirkendall, and K. Harkestad. 2004. Phylogeny of a Macaronesian radiation: host-plant use and possible cryptic speciation in *Liparthrum* bark beetles. Mol. Phylogen. Evol. 31: 554–571.

Jordal, B.H., B.C. Emerson, and G.M. Hewitt. 2006. Apparent "sympatric" speciation in ecologically similar herbivorous beetles facilitated by multiple colonizations of an island. Mol. Ecol. 15: 2935–2947.

Joy, D.A., and J.E. Conn. 2001. Molecular and morphological phylogenetic analysis of an insular radiation in Pacific black flies *(Simulium)*. Syst. Biol. 52: 89–109.

Juan, C., K. Ibrahim, P. Oromi, and G.M. Hewitt. 2000. Colonization and diversification: towards a phylogeographic synthesis for the Canary Islands. Trends Ecol. Evol. 15: 104–109.

Kambysellis, M.P., and E.M. Craddock. 1997. Ecological and reproductive shifts in the diversification of the endemic Hawaiian *Drosophila*, pp. 475–509. In T.J. Givnish and K.J. Sytsma (eds.), Molecular evolution and adaptive radiation. Cambridge University Press, Cambridge, UK.

Kato, M., and A. Kawakita. 2004. Plant-pollinator interactions in New Caledonia influenced by introduced honey bees. Am. J. Bot. 91: 1814–1827.

Kawakita, A., A. Takimura, T. Terachi, T. Sota, and M. Kato. 2004. Cospeciation analysis of an obligate pollination mutualism: have Glochidion trees (Euphorbiaceae) and pollinating *Epicephala* moths (Gracillariidae) diversified in parallel? Evolution 58: 2201–2214.

Kjellberg, F., E. Jousselin, M. Hossaert-McKey, and J.Y. Rasplus. 2005. Biology, ecology and evolution of fig-pollinating wasps (Chalcidoidea, Agaonidae), pp. 539–572. In A. Raman, C.W. Schaefer, and T.M. Withers (eds.), Biology, ecology, and evolution of gall-inducing arthropods. Sciences Publishers, Enfield, NH.

Kuussaari, M., M. Singer, and I. Hanski. 2000. Local specialization and landscape-level influence on host use in an herbivorous insect. Ecology 81: 2177–2187.

Labandeira, C.C., and J.J. Sepkoski Jr. 1993. Insect diversity in the fossil record. Science 261: 310–315.

Lachance, M.A., C.P. Ewing, J.M. Bowles, and W.T. Starmer. 2005. *Metschnikowia hamakuensis* sp. nov., *Metschnikowia kamakouana* sp. nov., and *Metschnikowia mauinuiana* sp. nov., three endemic yeasts from Hawaiian nitidulid beetles. Int. J. Syst. Evol. Microbiol. 55: 1369–1377.

Lewontin, R.C. 1965. Selection for colonizing ability, pp. 77–94. In H.G. Baker and G.L. Stebbins (eds.), Genetics of colonizing species. Academic Press, New York.

Lu, Z.X., X.P. Yu, J.M. Chen, X.S. Zheng, H.X. Xu, J.F. Zhang, and L.Z. Chen. 2004. Dynamics of yeast-like symbiote and its relationship with the virulence of brown planthopper, *Nilaparvata lugens* Stal, to resistant rice varieties. J. Asia-Pacific Entomol. 7: 1–7.

Miller, S.E., and L.G. Eldredge. 1996. Numbers of Hawaiian species: supplement 1. Bishop Mus. Occ. Papers 45: 8–17.

Moran, N.A., and J.J. Wernegreen. 2000. Lifestyle evolution in symbiotic bacteria: insights from genomics. Trends Ecol. Evol. 15: 321–326.

Nakabachi, A., A. Yamashita, H. Toh, H. Ishikawa, H.E. Dunbar, N.A. Moran, and M. Hattori. 2006. The 160-kilobase genome of the bacterial endosymbiont *Carsonella*. Science 314: 267.

Nielsen, R., and J. Wakeley. 2001. Distinguishing migration from isolation: a Markov chain Monte Carlo approach. Genetics 158: 885–896.

Nishida, G. (ed.) 2002. Hawaiian terrestrial arthropod checklist. Bishop Museum Press, Honolulu.

Noda, H., and Y. Koizumi. 2003. Sterol biosynthesis by symbiotes: cytochrome P450 sterol C-22 desaturase genes from yeastlike symbiotes of rice planthoppers and anobiid beetles. Insect Biochem. Mol. Biol. 33: 649–658.

Novotny, V., P. Drozd, S.E. Miller, M. Kulfan, M. Janda, Y. Basset, and G.D. Weiblen. 2006. Why are there so many species of herbivorous insects in tropical rainforests? Science 213: 1115–1118.

O'Connell, C. 1992. Accoustic communication in Hawaiian planthoppers. Universisty of Hawaii, Manoa.

Olesen, J.M., I. Eskildsen, and S. Venkatasamy. 2002. Invasion of pollination networks on oceanic islands: importance of invader complexes and endemic super generalist. Divers. Distrib. 8: 181–192.

Paulay, G. 1985. Adaptive radiation on an isolated oceanic island: the Cryptorhynchinae (Curculionidae) of Rapa revisited. Biol. J. Linn. Soc. 26: 95–187.

Pellmyr, O., and J.N. Thompson. 1992. Multiple occurrence of mutualism in the yucca moth lineage. Proc. Natl. Acad. Sci. USA 89: 2927–2929.

Percy, D.M. 2002. Distribution patterns and taxonomy of some legume-feeding psyllids (Hemiptera: Psylloidea) and their hosts from the Iberian Penninsula, Morocco, and Macaronesia. J. Insect Syst. Evol. 33: 291–310.

Percy, D.M. 2003a. Legume-feeding psyllids (Hemiptera, Psylloidea) of the Canary Islands and Madeira. J. Nat. Hist. 37: 397–461.

Percy, D.M. 2003b. Radiation, diversity, and host-plant interactions among island and continental legume-feeding psyllids. Evolution 57: 2540–2556.

Percy, D.M., and Q.C.B. Cronk. 2002. Different fates of island brooms: contrasting evolution in *Adenocarpus*, *Genista* and *Teline* (Genisteae, Leguminosae) in the Canary Islands and Madeira. Am. J. Bot. 89: 854–864.

Percy, D.M., R.D.M. Page, and Q.C.B. Cronk. 2004. Plant-insect interactions: double-dating associated insect and plant lineages reveals asynchronous radiations. Syst. Biol. 53: 120–127.

Percy, D.M., G.S. Taylor, and M. Kennedy. 2006. Psyllid communication: acoustic diversity, mate recognition and phylogenetic signal. Invert. Syst. 20: 431–445.

Ribeiro, S.P., P.A.V. Borges, C. Gaspar, C. Melo, A.R.M. Serrano, J. Amaral, C. Aguiar, G. Andre, and J.A. Quartau. 2005. Canopy insect herbivores in the Azorean Laurisilva forests: key host plant species in a highly generalist insect community. Ecography 28: 315–330.

Richardson, D.M., N. Allsopp, C.M. D'Antonio, S. J. Milton, and M. Rejmanek. 2000. Plant invasions: the role of mutualisms. Biol. Rev. 75: 65–93.

Roderick, G.K. 1997. Herbivorous insects and the Hawaiian silversword alliance: coevolution or cospeciation? Pacific Sci. 51: 440–449.

Roderick, G.K., and E.C. Metz. 1997. Biodiversity of planthoppers (Hemiptera: Delphacidae) on the Hawaiian silversword alliance: effects of host plant history and hybridization. Mem. Mus. Victoria 56: 393–399.

Roderick, G.K., and R.G. Gillespie. 1998. Speciation and phylogeography of Hawaiian terrestrial arthropods. Mol. Ecol. 7: 519–531.

Shaw, K.L. 1995. Biogeographic patterns of two independent Hawaiian cricket radiations *(Laupala* and *Prognathogryllus)*, pp. 39–56. In W.L. Wagner and V. Funk (eds.), Hawaiian biogeography evolution on a hot spot archipelago. Smithsonian Institution Press, Washington, DC.

Shaw, K.L. 2002. Conflict between nuclear and mitochondrial DNA phylogenies of a recent species radiation: what mtDNA reveals and conceals about modes of speciation in Hawaiian crickets. Proc. Natl. Acad. Sci. USA 99: 16122–16127.

Shaw, K.L. 2005. Rapid speciation in an arthropod. Nature 433: 375–376.

Simon, C. 1987. Hawaiian evolutionary biology: an introduction. Trends Ecol. Evol. 2: 175–178.

Spaulding, A.W., and C.D. von Dohlen. 2001. Psyllid endosymbionts exhibit patterns of co-speciation with hosts and destabilizing substitutions in ribosomal RNA. Insect Mol. Biol. 10: 57–67.

Stenberg, J.A., J. Witzell, and L. Ericson. 2006. Tall herb herbivory resistance reflects historic exposure to leaf beetles in a boreal archipelago age-gradient. Oecologia 148: 414–425.

Strong, D.R., J.H. Lawton, and R. Southwood. 1984. Insects on plants. Blackwell, Oxford, UK.

Usinger, R.L. 1941. Problems of insect speciation in the Hawaiian Islands. Am. Nat. 75: 251–263.

Vandergast, A.G., R.G. Gillespie, and G.K. Roderick. 2004. Influence of volcanic activity on the population genetic structure of Hawaiian *Tetragnatha* spiders: fragmentation, rapid population growth, and the potential for accelerated evolution. Mol. Ecol. 13: 1729–1743.

Wagner, W.L., and V. Funk (eds.). 1995. Hawaiian biogeography evolution on a hot spot archipelago. Smithsonian Institution Press, Washington, DC.

Wendel, J.F., J.M. Stewart, and J.H. Rettig. 1991. Molecular evidence for homoploid reticulate evolution among Australian species of Gossypium. Evolution 45: 694–771.

Wilson, W.W., C. Mitter, R.F. Denno, and M.R. Wilson. 1994. Evolutionary patterns of host plant use by delphacid planthoppers and their relatives, pp. 7–113. In R.F. Denno and T.J. Perfect (eds.), Planthoppers: their ecology and management. Chapman and Hall, New York.

Zimmerman, E.C. 1960. Possible evidence of rapid evolution in Hawaiian moths. Evolution 14: 137–138.

TWELVE

Selection by Pollinators and Herbivores on Attraction and Defense

LYNN S. ADLER

Interactions between plants, their herbivores, and their pollinators are thought to have led to the diversification of both plants and insects. Historically, studies of plant-herbivore and plant-pollinator interactions have occurred independently. Research at both micro- and macroevolutionary levels has focused on the evolution of plant resistance in the context of herbivory, and on floral traits in the context of pollination. For example, researchers have long recognized the role of plant secondary chemicals in herbivore feeding preferences (e.g., Dethier 1941). Fraenkel (1959) proposed that the "reason for existence" of plant secondary chemicals was to attract and deter herbivores rather than as products of plant metabolism, building on earlier work by Stahl (1888) and others. Ehrlich and Raven's classic article (1964) on coevolution concluded that "the evolution of secondary plant substances and the stepwise evolutionary responses to these by phytophagous organisms have clearly been the dominant factors in the evolution of butterflies and . . . in the evolution of angiosperm subgroups" (p. 382). This publication and others in the 1970s (e.g., Feeny 1976; Rhoades and Cates 1976) led to a surge of interest in chemical defenses mediating plant-herbivore interactions. More recently, selection on plant resistance traits has been studied by manipulating herbivores as selective agents (e.g., Mauricio and Rausher 1997; Stinchcombe and Rausher 2001), and several phylogenetic studies interpret the diversification of plants and herbivores in the context of coevolution mediated by plant resistance traits (e.g., Farrell et al. 1991; Becerra 1997, 2003; Farrell and Mitter 1998; Cornell and Hawkins 2003).

By contrast, the diversity of plant floral traits has been interpreted as the result of evolution due to their obvious role in attracting pollinators and promoting efficient pollination. Sprengel's 1793 pioneering treatise (first translated to English in 1996) interpreted floral function in terms of relationships with pollinators and inspired much of the subsequent field of pollination ecology. Darwin also interpreted the floral morphological variation of heterostylous plants in terms of its role in promoting outcrossing (Darwin 1877). More recently, pollination biologists have suggested that the evolution of floral traits may be shaped by a diversity of pollinators, rather than a single pollinator or guild type (e.g., Herrera 1996; Waser et al. 1996). Recent theoretical models and manipulative studies continue to focus on the role of pollinators in shaping the evolution of floral traits (e.g., Dafni and Kevan 1997; Aigner 2001; Fenster et al. 2004) and on floral trait phylogenetic diversity in the context of pollinator attraction and efficiency (e.g., Jurgens 2004; Sargent and Otto 2004; Manning and Goldblatt 2005; Ree 2005).

Clearly, herbivores have been a major selective force in the evolution of plant defense, and pollinators have been a major selective force in the evolution of attractive floral traits. However, a growing number of studies suggest that traits that deter herbivores may affect pollinator attraction, and traits that attract pollinators may affect herbivores. Herbivores and pollinators could exert selection on plant traits either through direct interactions (a pairwise relationship between plants and insects mediated by the trait), or via indirect interactions (insect selection on plant traits that is mediated by a third species) (Wootton 1994). Over half a century ago Grant (1950) recognized that floral morphology may have evolved to protect ovules from damage by some pollinators, such as birds and beetles, as well as to promote pollen transfer. More recent phylogenetic studies demonstrate that floral traits have evolved in response to selection from both pollinators and herbivores (e.g., Armbruster 1997; Armbruster et al. 1997; Pellmyr 2003). For example, resin-secreting floral glands that defended flowers of *Dalechampia* vines were subsequently co-opted as a reward for resin-collecting pollinating bees (Armbruster 1997). Thus, the evolution of plant traits may be shaped by

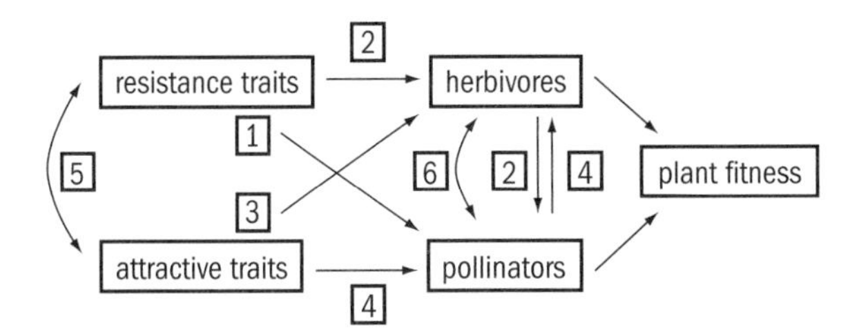

FIGURE 12.1. Paths by which plant resistance and attractive traits could affect plant fitness via interactions with herbivores and pollinators. This conceptual framework could be extended to other antagonist and mutualist interactions. Numbers refer to corresponding text sections discussing each pathway: 1, "Direct Effects of Resistance on Pollinators"; 2, "Indirect Effects of Resistance on Pollinators"; 3, "Direct Effects of Attractive Traits on Herbivores"; 4, "Indirect Effects of Attractive Traits on Herbivores"; 5, "Resistance and Attraction Traits May Not Be Independent; 6, "Herbivores and Pollinators May Not Be Independent."

simultaneous or sequential interactions with both pollinators and herbivores.

Attractive and defensive traits can be genetically correlated via linkage or pleiotropy. Thus herbivore-imposed selection on resistance may drive the evolution of floral traits and vice versa. Furthermore, herbivores and pollinators themselves are not independent of each other. Several insect taxa include species that are pollinators as adults and herbivores as larvae. Traits that attract adult pollinators therefore have the potential to increase subsequent herbivory in some systems. The consequence of these interactions for plant fitness will depend on the level of specialization and on community context. For example, a pollinating herbivore may benefit plants when other pollinators are unavailable, but reduce plant fitness when nonherbivorous pollinators are also present (Thompson and Cunningham 2002).

Abiotic factors as well as community context can alter the expression of traits and the fitness consequences of interactions. Pathogens and soil microorganisms may also play a large role in the evolution of plant traits (e.g., Agrawal et al. 1999) but are outside the scope of this book. Furthermore, many secondary compounds serve functions other than defense, such as UV protection and the oxidation of free radicals (e.g., McCloud and Berenbaum 1994; Izaguirre et al. 2003; Gould 2004). For simplicity, I consider here just the role of insects on the evolution of attractive and defensive traits, while acknowledging that other factors undoubtedly play significant roles in the evolution of these traits.

In this chapter, I review the literature on selection by pollinators and herbivores on resistance and attractive traits, with the goal of highlighting the pathways by which pollinators may affect the evolution of plant resistance, and herbivores may affect the evolution of floral attractive traits. Figure 12.1 provides a schematic diagram, with numbered paths referring to corresponding sections in the text.

Selection by Pollinators on Plant Resistance

Direct Effects of Resistance on Pollinators

Resistance to herbivores may incur a variety of costs for plants, including ecological costs of deterring other mutualists (Strauss et al. 2002). A small but growing number of studies have shown that resistance traits may have direct, negative impacts on pollinator preference. For example, *Brassica rapa* lines that were artificially selected for high myrosinase (i.e., high herbivore resistance) produced flowers with smaller petals and were less attractive to pollinators compared to low-resistance lines (Strauss et al. 1999). Such aversion could be due to the expression of defensive traits in flowers, or to defense costs resulting in reduced allocation to floral traits. Other systems have shown that resistance to floral antagonists can deter pollinators. Floral spines deterred nectar thieves but also reduced pollinator time per visit in *Centaurea solstitialis* (Agrawal et al. 2000), and nectar alkaloids in *Gelsemium sempervirens* deterred nectar robbers at a cost of reduced pollinator attraction (Adler and Irwin 2005). Additionally, if floral antagonists and pollinators prefer the same phenotypes, they may exert opposing selection on floral traits (Gomez 1993; Eriksson 1995; Ehrlen 1997). For example, floral seed predators and pollinators exerted opposing selection pressures on calyx length in *Castilleja linariaefolia* (Cariveau et al. 2004) and scape length in *Primula farinosa* (Ehrlen et al. 2002), and predispersal seed predators and pollinators may exert opposing selection on flowering phenology and inflorescence size in *Ipomopsis aggregata* (Brody 1997; Brody and Mitchell 1997). Thus, if pollinator attraction affects plant fitness, pollinators may select against resistance traits expressed in flowers or traits that cause reduced allocation to floral display or rewards.

In some cases plants may circumvent the negative effects of resistance on pollinators. For example, *Acacia* trees produced a volatile that deterred guarding ants from "protecting" young flowers against pollinators (Willmer and Stone 1997), and corollas of *Nicotiana attenuata* increased pools of the attractant benzyl acetone and decreased pools of nicotine at dusk, when *Manduca* spp. pollinators are most active (Euler and Baldwin 1996). However, when selection for increased defense in one tissue has pleiotropic consequences for expression in other tissues, plants may not be able to simultaneously evolve optimal solutions for attracting pollinators and deterring herbivores.

Indirect Effects of Resistance on Pollinators

Although herbivore resistance may directly deter pollinators, pollinators could select for higher levels of plant resistance by preferring undamaged plants. This indirect effect seems most probable when damage occurs on floral tissues, since floral cues are most likely to be used by pollinators to assess rewards. Damage to vegetative or even root tissue could also affect pollinator attraction. The effects of damage to each of these tissues (floral, vegetative, and roots) are reviewed below.

Damage to floral tissue or consumption of floral resources (collectively referred to as floral antagonism) can reduce pollinator attraction. For example, florivory reduced pollinator preference and plant male or female reproduction in several systems (Lohman and Berenbaum 1996; Krupnick and Weis 1999; Krupnick et al. 1999; Mothershead and Marquis 2000; Adler et al. 2001). Florivory can also alter sex expression (Hendrix 1984). For example, a lepidopteran herbivore preferentially consumed more exerted floral parts in the distylous vine *Gelsemium sempervirens*, so that long-styled plants became functionally male and long-filamented plants became functionally female (Leege and Wolfe 2002). Florivory can also change sex allocation in future flowers via compensatory reproduction as a tolerance mechanism (Hendrix and Trapp 1981). Nectar robbing may reduce plant fitness indirectly by deterring pollinators (e.g., Irwin and Brody 1998, 1999, 2000), although robbing is not costly in all systems (reviewed in Maloof and Inouye 2000; Irwin et al. 2001).

Although floral antagonism reduced pollinator attraction and plant reproduction in several systems, few studies have elucidated traits conferring resistance to floral antagonists. Natural and artificial flower damage in *Nemophila menziesii* induced resistance to florivory in younger flowers (McCall 2006), but the responsible traits are unknown. Defensive compounds have been detected in flowers (Detzel and Wink 1993; Euler and Baldwin 1996; Zangerl and Rutledge 1996; Adler and Wink 2001; Gronquist et al. 2001; Strauss et al. 2004; Irwin and Adler 2006), pollen, and nectar (reviewed in Adler 2000; also Gaffal and Heimler 2000; Thornburg et al. 2003). Such compounds may provide the basis for resistance to floral antagonists, but this has generally not been demonstrated (but see Gronquist et al. 2001). Other floral traits, such as corolla shape (Galen and Cuba 2001), exposure of sexual organs (Leege and Wolfe 2002), nectar concentration (Irwin et al. 2004), pollen nutritional content or defenses (Adler 2000; Roulston and Cane 2000), and escape in time or space (Irwin et al. 2001, 2004; Theis et al. 2006a), could potentially be under selection by pollinators through conferring resistance to floral antagonists. Flower number may also influence resistance to floral predators. Two hundred years after *Silene latifolia* escaped floral antagonists by invading North America from Europe, North American lineages produce more flowers than European lineages (Blair and Wolfe 2004); these lineages may also be more attractive to pollinators.

In most cases, leaf herbivory reduced plant fitness by deterring pollinators (Strauss et al. 1996; Lehtila and Strauss 1997; Strauss and Armbruster 1997; Mothershead and Marquis 2000; Hamback 2001; Poveda et al. 2003), although there can be differential effects on male compared to female fitness (Strauss et al. 2001). Leaf herbivory generally reduced floral traits associated with both male and female fitness. Simulated diabroticite beetle damage to *Cucurbita texana* branches reduced the number of male flowers, the amount of pollen per flower, and pollen siring success (Quesada et al. 1995). Vegetative herbivory in a variety of systems reduced flower number, size, height, and flowering period (Karban and Strauss 1993; Lehtila and Strauss 1997, 1999; Mothershead and Marquis 2000; Hamback 2001; Poveda et al. 2003, 2005b; Ivey and Carr 2005), and pollen and nectar production, quality, or exertion (Lehtila and Strauss 1999; Ivey and Carr 2005; Poveda et al. 2005b). Vegetative herbivory can also alter plant sex ratio (Hendrix and Trapp 1981; Hendrix 1984; Krupnick and Weis 1998; Krupnick et al. 2000; Thomson et al. 2004) and mating system (Elle and Hare 2002; Steets and Ashman 2004; Ivey and Carr 2005).

Links between belowground herbivory and pollination are only beginning to be explored. Surprisingly, root herbivory may increase pollinator attraction (Poveda et al. 2003, 2005a), although the mechanism is not clear (Poveda et al. 2005b). Belowground herbivory may also attract other aboveground mutualists. Root herbivory reduced inflorescence size in thistles, but increased attraction of both teprhitid seed predators and their parasitoids (Masters et al. 2001). Artificial and natural root herbivory in greenhouse cotton increased extrafloral nectar production, which was interpreted as an induced indirect defense to attract natural enemies (Wackers and Bezemer 2003). If these studies represent general patterns, pollinators or other aboveground mutualists have the potential to select for reduced resistance to belowground herbivory. However, very little is known about the mechanisms or genetic basis of root resistance to herbivory (but see Davis and Rich 1987; Zangerl and Rutledge 1996; Rasmann et al. 2005 for examples); this is clearly an open area for future research.

This review demonstrates that resistance traits may directly deter pollinators, but such traits can also indirectly attract pollinators by reducing herbivory. Ultimately, the net result of pollinator selection on resistance will depend on (1) the importance of pollinator service for plant fitness, and (2) the relative importance of direct deterrence versus indirect attraction of pollinators to resistant plants.

Selection by Herbivores on Floral Traits

Direct Effects of Attractive Traits on Herbivores

Attractive floral traits have the potential to directly attract herbivores as well as pollinators. This conflict is analogous to the trade-off between natural and sexual selection that is commonly studied in the animal kingdom, where individuals signal to attract mates, but such signals can also attract predators (e.g., Tuttle and Ryan 1981). In plants, floral signals and rewards may attract a variety of antagonists that consume floral resources, as well as pollinators. *Ipomopsis aggregata* plants with larger inflorescences, for example, attract higher rates of predispersal seed predation (Brody and Mitchell 1997), and *Polemonium viscosum* plants with wider corollas were more attractive to nectar-robbing ants in addition to bumblebee pollinators (Galen and Cuba 2001). Several floral traits including flower production affected resistance to the bud-clipping weevil *Anthonomus signatus* in strawberry, *Fragaria virginiana* (Ashman et al. 2004). Male plants were less resistant and more tolerant of herbivory compared to hermaphrodite plants. Within males, plants with higher pollen production had more herbivory, suggesting that pollen production could be under selection by both pollinators and herbivores. Scent also attracts herbivores as well as pollinators. *Cucurbita* species that have more fragrant flowers attract higher numbers of *Diabrotica* pollen-feeding beetles (Andersen and Metcalf 1987), and two floral volatiles attracted both pollinators and floral herbivores in *Cirsium arvense* (Theis 2006). The timing of scent emission may reflect selection to attract pollinators while avoiding florivores (Euler and Baldwin 1996; Theis and Raguso 2005; Theis et al. 2006b). In cases where floral antagonists and pollinators prefer the same floral traits, plants may experience conflicting selection pressures to maintain pollinator attraction while resisting herbivores.

In some cases, leaf herbivores may be attracted to plants by floral or extrafloral resources. This can happen when herbivores have a wide diet, including nectar or pollen in addition to leaf material. Alternatively, insects may consume nectar or pollen as adults but leaf or vegetative tissue as larvae. For example, domestic cotton varieties with extrafloral nectaries experience greater levels of herbivory from a variety of hemipteran and lepidopteran larvae whose adults feed on the nectar (Lukefahr and Rhyne 1960; Schuster et al. 1976; Flint et al. 1988; Scott et al. 1988). Such a trade-off may also exist for extrafloral nectar production in wild cotton (Rudgers 2004; Rudgers and Gardener 2004). In these cases, adults are not pollinators, and the outcome of the interaction is negative from the plant's perspective. Cases in which adult pollinators have herbivorous larvae are reviewed below (see"Herbivores and Pollinators May Not Be Independent"). In either circumstance, one of the consequences of advertising or producing floral or extrafloral rewards may be attraction of herbivores. When herbivory reduces plant fitness and attractive traits are heritable, such traits may evolve in response to herbivory as well as pollination.

Indirect Effects of Attractive Traits on Herbivores

The attraction of pollinators could benefit some herbivores such as seed predators, because seed predators require fruit set for larval survival. Seed predators that oviposit before pollination should prefer attractive flowers to increase the chances of locating a future fruit for their offspring (e.g., Brody and Morita 2000). Even seed predators that oviposit after pollination may choose plants with the highest fruit production, which may correlate with previously expressed attractive traits. For example, plants with larger inflorescence heads had higher incidences of predispersal seed predation both within and across species of Asteraceae. Thus, inflorescence size might represent a trade-off between attracting pollinators versus seed predators (Fenner et al. 2002).

Floral traits may also attract natural enemies that reduce herbivory. Many adult parasitoids feed on nectar (Kidd and Jervis 1989) and are particularly attracted to flowers with open corollas and easily accessible nectar (Patt et al. 1997; Tooker and Hanks 2000). Adding nectar sources in crop plantings alters parasitoid behavior and may help control pest herbivores (reviewed in Patt et al. 1997; Baggen and Gurr 1998). Furthermore, nectar sugar composition, scent, and accessibility can all affect parasitoid learning and preference (Patt et al. 1997, 1999; Wackers 1999, 2001), and such traits may be under selection if they reduce herbivory. However, the benefit of nectar via attracting parasitoids may be balanced by the cost of attracting herbivores. For example, access to buckwheat *(Fagopyron esculentum)* and dill *(Anethum graveolens)* flowers increased longevity and fecundity of the encyrtid wasp parasitoid *Copidosoma koehleri* and its host, the gelechiid moth *Phthorimaea operculella,* whose larvae are pests on potato. By contrast, flowers of *Phacelia* and *Nasturtium* benefited the parasitoid but not the herbivore (Baggen et al. 1999). If findings from agricultural settings hold in natural environments where communities may be more complex, then accessibility or quality of floral nectar may provide an additional benefit to plants by attracting parasitoids that reduce herbivory. Such benefits will be greatest for plants when parasitoids kill eggs or early larval stages (idiobiont parasitoids) rather than late larval or pupal stages (koinobiont parasitoids).

Finally, pollinator attraction in one generation may be linked with plant-herbivore dynamics in the next. Progeny of selfed plants of *Mimulus guttatus* had lower resistance (Carr and Eubanks 2002) and tolerance (Ivey et al. 2004) to herbivory compared to progeny of outcrossed plants. These results suggest that traits attracting pollinators could influence plant-herbivore interactions in the offspring, if pollinator attraction affects outcrossing rates.

Resistance and Attraction Traits May Not Be Independent

A growing number of studies indicate that the expression of attractive and resistance traits are not independent. Traits such as floral resins, which evolved as herbivore defenses, can be co-opted over evolutionary time as pollinator rewards (Armbruster et al. 1997). In ecological time, the same trait can serve as both pollinator attractant and defense against florivory, such as production of ultraviolet pigments in *Hypericum calycinum* or showy bracts in *Dalechampia* species (Armbruster and Mziray 1987; Armbruster 1997; Gronquist et al. 2001). Even traits such as flower color and leaf resistance may be correlated due to pleiotropy or linkage. For example, flower color polymorphism in *Ipomoea purpurea* correlated with differences in leaf herbivore resistance (Simms and Bucher 1996), although such differences may not affect damage in the field

(Fineblum and Rausher 1997). In *Raphanis sativus*, flower color morph is correlated with lower levels of indole glucosinolates in leaves, and preference and performance of a variety of leaf herbivores (Irwin et al. 2003). Such differential effects of flower color on leaf herbivory may explain why pollinator preference alone does not predict microevolution of floral color morphs (Irwin and Strauss 2005). Furthermore, alkaloid concentrations are correlated in leaves and corollas of naturally growing *Gelsemium sempervirens* (Irwin and Adler 2006), although these correlations may be due to genetic or environmental variation. The expression of nicotine and related alkaloids in nectar and leaves is phenotypically correlated across individual *Nicotiana tabacum* plants (Adler et al. 2006) and is also correlated across *Nicotiana* species (L. S. A., M. Gittinger, G. Morse, and M. Wink, unpublished data). Although not the subject of this chapter, related literature addresses the causes and consequences of toxic ripe fruit for fruit dispersers (e.g., Cipollini and Levey 1997; Cipollini 2000; Tewksbury and Nabhan 2001). Toxicity in ripe fruit may be correlated across species with toxicity of leaf defenses (Ehrlen and Eriksson 1993), providing another example where the evolution of attractive rewards (ripe fruit) may be constrained by expression of defenses in other tissues. Thus, a growing number of studies demonstrate that selection by pollinators on flower color or floral secondary compounds could drive correlated evolution of leaf traits, and selection by leaf herbivores on resistance could alter the evolution of flower color or defense (Lande and Arnold 1983). However, much work remains to elucidate the genetic basis of correlated traits across plant tissues to determine the generality of these results.

Leaf damage may alter floral traits. Optimal defense theory predicts that flowers will be constitutively defended due to their high reproductive value (McKey 1974; Rhoades and Cates 1976), and this prediction is supported by high levels of constitutive resistance in *Pastinaca sativa* flowers compared to leaves and roots (Zangerl and Rutledge 1996). However, recent studies have found that flower defense is also inducible. For example, leaf damage induced higher petal glucosinolate concentrations in anthocyanin-containing color morphs of *Raphanus sativus* (Strauss et al. 2004). In *N. attenuata*, leaf damage increased nicotine concentration in flowers (Euler and Baldwin 1996) and fruits (Baldwin and Karb 1995) and increased resistance to floral and fruit herbivory in the field (McCall and Karban 2006). Leaf damage by *Manduca sexta* induced higher levels of nectar nicotine in *N. tabacum* (Adler et al. 2006). Thus, leaf damage can affect floral traits, which may alter interactions with both pollinators and floral antagonists.

Herbivores and Pollinators May Not Be Independent

Plant interactions with herbivores and pollinators are often studied as separate and independent. However, in many systems herbivores and pollinators are the same species interacting with plants at different points in their life cycle. In some cases, pollinators oviposit into flowers or fruits that are subsequently consumed by larval seed predators; these systems can be highly obligate, such as the yucca plant–yucca moth (Pellmyr 2003) and fig plant–fig wasp interactions (Bronstein 1988; Kjellberg et al. 2001), or somewhat facultative, such as interactions between *Silene* and *Hadena* moths (Pettersson 1992; Wolfe 2002). In other cases, pollinators may oviposit leaf-feeding larvae whose success is less linked with pollinator behavior. In any case, if adults prefer plants with attractive rewards such as high nectar volumes (e.g., Real and Rathcke 1991; Hodges 1995), then attractive traits may be under conflicting selection to attract pollinators but minimize the linked cost of herbivory. I review some examples here; more complete coverage is provided by Adler and Bronstein (2004).

Members of the family Sphingidae (the hawkmoths) provide perhaps the best examples of herbivorous larvae that specialize on the same plants pollinated by adults (reviewed in Adler and Bronstein 2004). Such herbivory may represent a significant cost to plants. For example, an individual *Manduca sexta* larva can defoliate its host by the time it pupates (McFadden 1968). Among other Lepidoptera, *Pieris rapae* (Pieridae) is an efficient pollinator of *Raphanus raphanistrum* (Conner et al. 1995), and larval *P. rapae* are specialists on crucifers including this species (e.g., Agrawal 1999). Finally, several moths and butterflies that are generalist nectar-feeders as adults and generalist herbivores as larvae may incorporate certain plant species in their diets at both life-history stages; examples include *Heliothis virescens* and *Helicoverpa armigera* (Cunningham et al. 1998; De Moraes et al. 2001). This is not intended as an exhaustive list, but rather as examples that demonstrate the potential for trade-offs between attracting pollinators and experiencing increased levels of herbivory from offspring.

Only two experiments have tested the hypothesis that floral attractants could increase levels of oviposition by leaf herbivores. Adding supplemental nectar to *Datura stramonium* flowers increased the number of *M. sexta* eggs oviposited on leaves (Adler and Bronstein 2004). *Manduca sexta* is a voracious herbivore of *D. stramonium*, whose adults are also common nectar-feeding pollinators on the same plants (L. S. A., personal observations). Similarly, increasing the quality of nectar with supplemental arginine, a naturally occurring amino acid essential for egg maturation, increased *M. sexta* leaf oviposition on *N. tabacum* (A. J. Lentz and L. S. A., unpublished). Both of these studies were conducted in cages stocked with artificially high levels of *M. sexta* moths, and both removed eggs before hatching and so could not quantify the costs of herbivory. Furthermore, in these systems there was no benefit of supplemental nectar amount or quality on plant female reproduction, presumably because both *D. stramonium* and *N. tabacum* are highly selfing (Goodspeed 1954; Motten and Antonovics 1992). However, these studies represent the first steps in demonstrating that floral rewards, by attracting pollinators, may also increase leaf herbivory. Further work is needed to demonstrate whether such trade-offs occur in the field under natural insect densities, and to quantify the benefits and costs of floral traits in the context of attracting both pollinators and their herbivorous offspring.

Abiotic Factors and Geographic Variation

Selection pressures do not remain constant over time or space. Variation in the abiotic and biotic environment can alter both the expression of phenotypes and the relative importance of different selective agents. There is ample evidence that abiotic conditions mediate attractive and defensive phenotypes (e.g., Gershenzon 1984; Mattson and Haack 1987; Wyatt et al. 1992; Galen 1999b; Carroll et al. 2001; Gardener and Gillman 2001), and that selection is spatially, temporally, and environmentally heterogeneous (Boag and Grant 1981; Kalisz 1986; Stratton 1992; Dudley 1996; Stratton and Bennington 1998). However, few empirical studies have examined how selection by multiple biotic agents changes under different abiotic conditions (but see Galen 1999a; Galen and Cuba 2001; Ehrlen et al. 2002).

Although the influence of abiotic factors on selection by biotic agents has long been recognized in studies of the evolution of plant defenses (e.g., Bryant et al. 1983; Coley et al. 1985), the role of abiotic factors has only recently been studied for the evolution of floral diversity (Galen 1999b; Elle 2004), and there are few empirical tests. In two such natural experiments, pollinator selection on flower morphology changed between wet and dry years (Maad 2000; Maad and Alexandersson 2004) and across an altitude-temperature gradient (Totland 2001). Furthermore, expression of traits involved in pollinator attraction, such as flower color, may be linked with traits involved in drought tolerance (Schemske and Bierzychudek 2001), suggesting that both abiotic and biotic factors may simultaneously influence the selective advantage of plant traits. Abiotic conditions may also affect the relative significance of selection by herbivores and pollinators. In particular, resources such as water or nutrients may determine whether a plant is pollen limited and therefore the importance of pollinator attraction (Haig and Westoby 1988; Zimmerman and Pyke 1988). Reductions in pollinator visitation may not affect fitness in harsh conditions, where limited resources constrain fecundity, but may significantly reduce fitness in favorable conditions (e.g., Campbell and Halama 1993; Corbet 1998). In harsher conditions, pollinator-mediated selection may be reduced while herbivore-mediated selection remains constant or increases. Therefore, the relative importance of herbivore defense and pollinator attraction may change in different environmental contexts.

Community context may also change the pattern of selection on attractive or defensive traits. For example, the presence of an alternate pollinator changed the outcome of the interaction between a plant and pollinating seed predator from positive to negative for the plant (Thompson and Cunningham 2002). The effect of low-efficiency pollinators on plant fitness in *Campanula americana* varied from neutral to negative, depending on the abundance of high-efficiency pollinators (Lau and Galloway 2004). The presence of another plant species that competes for pollination services can also change the shape of selection on floral traits in a focal plant species (Caruso 2000, 2001). These examples demonstrate the importance of community context in shaping selection on attractive and defensive phenotypes. It is likely that community composition plays a large role in determining the magnitude and direction of interactions between plants, herbivores, and pollinators in many systems.

Future Directions

The goal of this review was to gather and synthesize a wide range of studies demonstrating the potential for herbivores to select on floral attraction, and for pollinators to select on plant resistance. Many of these studies are quite recent, demonstrating both a historical lack of attention to the potential for multispecies selection on plant traits, and a recent excitement to pursue such questions in greater depth. While these studies represent a large and growing body of work, there are clear gaps in our understanding of these interactions that should be the focus of future research.

While the traits responsible for resistance to leaf herbivores have been extensively studied, we know very little about the traits responsible for resistance to other herbivores, such as floral antagonists and root herbivores. For example, although nectar robbers can reduce plant fitness as much as leaf herbivores (Juenger and Bergelson 1997; Irwin and Brody 2000), only a handful of studies have attempted to determine what traits confer resistance to nectar robbing. Observational and manipulative studies are needed to elucidate whether the same or correlated traits are involved in attracting pollinators and deterring floral antagonists (e.g., Irwin et al. 2004).

Traits must be experimentally manipulated to isolate their effect on species interactions. Historically, mechanical (e.g., constraining floral tube shape [Galen and Cuba 2001]) or chemical (e.g., addition of hormones to induce changes [Thaler 1999]) manipulations have been used. However, the increasing feasibility of isolating and transforming or knocking out specific loci allows a greater range of manipulations and understanding of the mechanistic basis of trait expression (e.g., Kessler et al. 2004). The use of genetic modification to manipulate traits provides another powerful tool to address how such traits evolve in natural contexts.

Understanding the genetic correlations between traits expressed across tissues, such as secondary compounds in leaves and nectar, or pigmentation in flowers and vegetative tissue (e.g., Armbruster 2002), is of fundamental importance for predicting how leaf herbivores could exert correlated selection on floral traits, or how pollinators could drive the evolution of resistance traits in leaves. Furthermore, the heritability of some basic traits is poorly understood. Although nectar production and composition are critical for pollinator attraction in many systems (Dafni 1992; Pellmyr 2002) and may also attract herbivores and natural enemies, as of 2004 only seven published studies had examined the heritability of nectar traits in wild plant species (Mitchell 2004). Even less is known about the heritability of extrafloral nectar traits (but see Rudgers 2004).

One of the fundamental goals of evolutionary ecology is to understand how traits evolve in the context of their environments, but there are currently no studies that quantify the extent of selection by herbivores and pollinators on resistance or attractive traits. Determining the role of selection by herbivores or pollinators on the evolution of plant traits requires manipulating or removing the putative agent of selection and measuring changes in the pattern of selection on the traits of concern. Such studies have shown that pollinators can select on floral traits (e.g., Campbell et al. 1991; Galen 1996; Jones and Reithel 2001) and herbivores can select on plant defensive traits (e.g., Mauricio and Rausher 1997). However, as this review has shown, numerous traits may influence both attraction and defense. Little is known about the *relative* importance of selection by mutualists and antagonists on attractive and defensive traits (but see Gomez and Zamora 2000; Herrera 2000; Herrera et al. 2002). This question could be addressed by manipulating herbivores and pollinators and measuring resultant changes in the magnitude or direction of selection on both floral and defensive traits, or by examining selection in multiple populations that vary in herbivory or pollination frequency. Such studies would be intensive due to the sample sizes required to detect changes in selection (Kingsolver et al. 2001) but would be feasible in some systems. Good candidate systems would be *Brassica* and *Raphanus* species, in which the genetic basis of resistance traits is well understood (Strauss et al. 1999; Irwin et al. 2003), resistance traits reduce herbivore preference and performance (Giamoustaris and Mithen 1995; Irwin et al. 2003), leaf herbivory reduces pollinator attraction (Lehtila and Strauss 1997), pollinators are necessary for reproduction (Strauss et al. 1996), and flower traits are correlated with leaf defenses (Irwin et al. 2003; Strauss et al. 2004). Alternatively, on a macroevolutionary scale one could look for evolutionary changes in herbivore resistance that correlate with changes in pollinator mode, or vice versa, and test predictions about the conditions under which such evolutionary correlations might occur. These approaches would help to assess the relative role of pollinators and herbivores in altering the evolution of attraction and defense across microevolutionary and macroevolutionary time scales.

Acknowledgments

I thank K. Tilmon for organizing this volume and inviting me to participate, S. Halpern for her substantial contributions to the section on abiotic factors, and N. Theis, P.S. Warren, and two anonymous reviewers for their helpful comments on the manuscript. I was funded during the inspiration and writing for this chapter by the University of Massachusetts at Amherst Department of Plant, Soil, and Insect Science and by NSF DEB-0514398.

References Cited

Adler, L.S. 2000. The ecological significance of toxic nectar. Oikos 91: 409–420.

Adler, L.S., and J.L. Bronstein. 2004. Attracting antagonists: does floral nectar increase leaf herbivory? Ecology 85: 1519–1526.

Adler, L.S., and R.E. Irwin. 2005. Ecological costs and benefits of defenses in nectar. Ecology 86: 2968–2978.

Adler, L.S., and M. Wink. 2001. Transfer of alkaloids from hosts to hemiparasites in two *Castilleja-Lupinus* associations: analysis of floral and vegetative tissues. Biochem. Syst. Ecol. 29: 551–561.

Adler, L.S., R. Karban, and S.Y. Strauss. 2001. Direct and indirect effects of alkaloids on plant fitness via herbivory and pollination. Ecology 82: 2032–2044.

Adler, L.S., M. Wink, M. Distal, and A.J. Lentz, 2006. Leaf herbivory and nutrients increase nectar alkaloids. Ecology Letters 9: 960–967.

Agrawal, A.A. 1999. Induced responses to herbivory in wild radish: effects on several herbivores and plant fitness. Ecology 80: 1713–1723.

Agrawal, A.A., S. Tuzun, and E. Bent (eds.). 1999. Induced plant defenses against pathogens and herbivores: biochemistry, ecology, and agriculture. APS Press, St. Paul, MN.

Agrawal, A.A., J.A. Rudgers, L.W. Botsford, D. Cutler, J.B. Gorin, C.J. Lundquist, B.W. Spitzer, and A.L. Swann. 2000. Benefits and constraints on plant defense against herbivores: spines influence the legitimate and illegitimate flower visitors of yellow star thistle, *Centaurea solstitialis* L. (Asteraceae). Southwest. Nat. 45: 1–5.

Aigner, P.A. 2001. Optimality modeling and fitness trade-offs: when should plants become pollinator specialists? Oikos 95: 177–184.

Andersen, J.F., and R.L. Metcalf. 1987. Factors influencing distribution of *Diabrotica* spp. (Coleoptera, Chyrsomelidae) in blossoms of cultivated *Cucurbita* spp. J. Chem. Ecol. 13: 681–699.

Armbruster, W.S. 1997. Exaptations link evolution of plant-herbivore and plant-pollinator interactions: a phylogenetic inquiry. Ecology 78: 1661–1672.

Armbruster, W.S. 2002. Can indirect selection and genetic context contribute to trait diversification? A transition-probability study of blossom-colour evolution in two genera. J. Evol. Biol. 15: 468–486.

Armbruster, W.S., and W.R. Mziray. 1987. Pollination and herbivore ecology of an African *Dalechampia* (Euphorbiaceae): comparisons with New World species. Biotropica 19: 64–73.

Armbruster, W.S., J.J. Howard, T.P. Clausen, E.M. Debevec, J.C. Loquvam, M. Matsuki, B. Cerendolo, and F. Andel. 1997. Do biochemical exaptations link evolution of plant defense and pollination systems? Historical hypotheses and experimental tests with *Dalechampia* vines. Am. Nat. 149: 461–484.

Ashman, T.L., D.H. Cole, and M. Bradburn. 2004. Sex-differential resistance and tolerance to herbivory in a gynodioecious wild strawberry. Ecology 85: 2550–2559.

Baggen, L.R., and G.M. Gurr. 1998. The influence of food on *Copidosoma koehleri* (Hymenoptera: Encyrtidae), and the use of flowering plants as a habitat management tool to enhance biological control of potato moth, *Phthorimaea operculella* (Lepidoptera: Gelechiidae). Biol. Control 11: 9–17.

Baggen, L.R., G.M. Gurr, and A. Meats. 1999. Flowers in tri-trophic systems: mechanisms allowing selective exploitation by insect

natural enemies for conservation biological control. Entomol. Exp. Appl. 91: 155–161.
Baldwin, I.T., and M.J. Karb. 1995. Plasticity in allocation of nicotine to reproductive parts in *Nicotiana attenuata*. J. Chem. Ecol. 21: 897–909.
Becerra, J.X. 1997. Insects on plants: macroevolutionary chemical trends in host use. Science 276: 253–256.
Becerra, J.X. 2003. Synchronous coadaptation in an ancient case of herbivory. Proc. Natl. Acad. Sci. USA 100: 12804–12807.
Blair, A.C., and L.M. Wolfe. 2004. The evolution of an invasive plant: an experimental study with *Silene latifolia*. Ecology 85: 3035–3042.
Boag, P.T., and P.R. Grant. 1981. Intense natural selection in a population of Darwin's finches (Geospizinae) in the Galapagos. Science 214: 82–85.
Brody, A.K. 1997. Effects of pollinators, herbivores, and seed predators on flowering phenology. Ecology 78: 1624–1631.
Brody, A.K., and R.J. Mitchell. 1997. Effects of experimental manipulation of inflorescence size on pollination and pre-dispersal seed predation in the hummingbird-pollinated plant *Ipomopsis aggregata*. Oecologia 110: 86–93.
Brody, A.K., and S.I. Morita. 2000. A positive association between oviposition and fruit set: female choice or manipulation? Oecologia 124: 418–425.
Bronstein, J.L. 1988. Mutualism, antagonism, and the fig-pollinator interaction. Ecology 69: 1298–1302.
Bryant, J.P., F.S.I. Chapin, and D.R. Klein. 1983. Carbon/nutrient balance of boreal plants in relation to vertebrate herbivory. Oikos 40: 357–368.
Campbell, D.R., and K.J. Halama. 1993. Resource and pollen limitations to lifetime seed production in a natural plant population. Ecology 74: 1043–1051.
Campbell, D.R., N.M. Waser, M.V. Price, E.A. Lynch, and R.J. Mitchell. 1991. Components of phenotypic selection: pollen export and lower corolla width in *Ipomopsis aggregata*. Evolution 45: 1458–1467.
Cariveau, D., R.E. Irwin, A.K. Brody, L.S. Garcia-Mayeya, and A. von der Ohe. 2004. Direct and indirect effects of pollinators and seed predators to selection on plant and floral traits. Oikos 104: 15–26.
Carr, D.E., and M.D. Eubanks. 2002. Inbreeding alters resistance to insect herbivory and host plant quality in *Mimulus guttatus* (Scrophulariaceae). Evolution 56: 22–30.
Carroll, A.B., S.G. Pallardy, and C. Galen. 2001. Drought stress, plant water status, and floral trait expression in fireweed, *Epilobium angustifolium* (Onagraceae). Am. J. Bot. 88: 438–446.
Caruso, C.M. 2000. Competition for pollination influences selection on floral traits of *Ipomopsis aggregata*. Evolution 54: 1546–1557.
Caruso, C.M. 2001. Differential selection on floral traits of *Ipomopsis aggregata* growing in contrasting environments. Oikos 94: 295–302.
Cipollini, M.L. 2000. Secondary metabolites of vertebrate-dispersed fruits: evidence for adaptive functions. Rev. Chil. Hist. Nat. 73: 421–440.
Cipollini, M.L., and D.J. Levey. 1997. Why are some fruits toxic? Glycoalkaloids in *Solanum* and fruit choice by vertebrates. Ecology 78: 782–798.
Coley, P.D., J.P. Bryant, and F.S. Chapin III. 1985. Resource availability and plant antiherbivore defense. Science 230: 895–899.
Conner, J.K., R. Davis, and S. Rush. 1995. The effect of wild radish floral morphology on pollination efficiency by four taxa of pollinators. Oecologia 104: 234–245.
Corbet, S.A. 1998. Fruit and seed production in relation to pollination and resources in bluebell, *Hyacinthoides nonscripta*. Oecologia 114: 349–360.
Cornell, H.V., and B.A. Hawkins. 2003. Herbivore responses to plant secondary compounds: a test of phytochemical coevolution theory. Am. Nat. 161: 507–522.
Cunningham, J.P., S.A. West, and D.J. Wright. 1998. Learning in the nectar foraging behaviour of *Helicoverpa armigera*. Ecol. Entomol. 23: 363–369.
Dafni, A. 1992. Pollination ecology: a practical approach. Oxford University Press, Oxford.
Dafni, A., and P.G. Kevan. 1997. Flower size and shape: implications in pollination. Isr. J. Plant Sci. 45: 201–212.
Darwin, C. 1877. The different forms of flowers on plants of the same species. John Murray, London.
Davis, E.L., and J.R. Rich. 1987. Nicotine content of tobacco roots and toxicity to *Meloidogyne incognita*. J. Nematol. 19: 23–29.
De Moraes, C.M., M.C. Mescher, and J.H. Tumlinson. 2001. Caterpillar-induced nocturnal plant volatiles repel nonspecific females. Nature 410: 577–580.
Dethier, V.G. 1941. Chemical factors determining the choice of food plants by *Papilio* larvae. Am. Nat. 75: 61–72.
Detzel, A., and M. Wink. 1993. Attraction, deterrence or intoxication of bees *(Apis mellifera)* by plant allelochemicals. Chemoecology 4: 8–18.
Dudley, S.A. 1996. Differing selection on plant physiological traits in response to environmental water availability: a test of adaptive hypothesis. Evolution 50: 92–102.
Ehrlen, J. 1997. Risk of grazing and flower number in a perennial plant. Oikos 80: 428–434.
Ehrlen, J., and O. Eriksson. 1993. Toxicity in fleshy fruits: a non-adaptive trait? Oikos 66: 107–113.
Ehrlen, J., S. Kack, and J. Agren. 2002. Pollen limitation, seed predation and scape length in *Primula farinosa*. Oikos 97: 45–51.
Ehrlich, P.R., and P.H. Raven. 1964. Butterflies and plants: a study in coevolution. Evolution 18: 586–608.
Elle, E. 2004. Floral adaptations and biotic and abiotic selection pressures, pp. 111–118. In Q. Cronk, R. Ree, I. Taylor and J. Whitton (eds.), Plant adaptation: molecular genetics and ecology. NRC Research Press, Ottawa, ON.
Elle, E., and J.D. Hare. 2002. Environmentally induced variation in floral traits affects the mating system in *Datura wrightii*. Funct. Ecol. 16: 79–88.
Eriksson, O. 1995. Asynchronous flowering reduces seed predation in the perennial forest herb *Actaea spicata*. Acta Oecol. 16: 195–203.
Euler, M., and I.T. Baldwin. 1996. The chemistry of defense and apparency in the corollas of *Nicotiana attenuata*. Oecologia 107: 102–112.
Farrell, B.D., and C. Mitter. 1998. The timing of insect/plant diversification: might *Tetraopes* (Coleoptera: Cerambycidae) and *Asclepias* (Asclepiadaceae) have co-evolved? Biol. J. Linn. Soc. 63: 553–577.
Farrell, B.D., D.E. Dussourd, and C. Mitter. 1991. Escalation of plant defense: do latex and resin canals spur plant diversification? Am. Nat. 138: 881–900.
Feeny, P. 1976. Plant apparency and chemical defense. Recent Adv. Phytochem. 10: 1–40.
Fenner, M., J.E. Cresswell, R.A. Hurley, and T. Baldwin. 2002. Relationship between capitulum size and pre-dispersal seed predation by insect larvae in common Asteraceae. Oecologia 130: 72–77.

Fenster, C.B., W.S. Armbruster, P. Wilson, M.R. Dudash, and J.D. Thomson. 2004. Pollination syndromes and floral specialization. Annu. Rev. Ecol. Evol. Syst. 35: 375–403.

Fineblum, W.L., and M.D. Rausher. 1997. Do floral pigmentation genes also influence resistance to enemies? The W locus in *Ipomoea purpurea*. Ecology 78: 1646–1654.

Flint, H.M., N.J. Curtice, and F.D. Wilson. 1988. Development of pink bollworm populations (Lepidoptera: Gelechiidae) on nectaried and nectariless deltapine cotton in field cages. Environ. Entomol. 17: 306–308.

Fraenkel, G.S. 1959. The raison d'etre of secondary plant substances. Science 129: 1466–1470.

Gaffal, K.P., and W. Heimler. 2000. Die nektarien von herzglycosidhaltigen rachenbluetlern: eine quelle der speise fuer goetter mit herzinsuffizienz? Mikrokosmos 89: 129–138.

Galen, C. 1996. Rates of floral evolution: adaptation to bumblebee pollination in an alpine wildflower, *Polemonium viscosum*. Evolution 50: 120–125.

Galen, C. 1999a. Flowers and enemies: predation by nectar-thieving ants in relation to variation in floral form of an alpine wildflower, *Polemonium viscosum*. Oikos 85: 426–434.

Galen, C. 1999b. Why do flowers vary? The functional ecology of variation in flower size and form within natural plant populations. BioScience 49: 631–640.

Galen, C., and J. Cuba. 2001. Down the tube: pollinators, predators, and the evolution of flower shape in the alpine skypilot, *Polemonium viscosum*. Evolution 55: 1963–1971.

Gardener, M.C., and M.P. Gillman. 2001. The effects of soil fertilizer on amino acids in the floral nectar of corncockle, *Agrostemma githago* (Caryophyllaceae). Oikos 92: 101–106.

Gershenzon, J. 1984. Changes in the levels of plant secondary metabolites under water and nutrient stress, pp. 273–321. In B. N. Timmermann, C. Steelink and F.A. Loewus (eds.), Phytochemical adaptations to stress. Plenum Press, New York.

Giamoustaris, A., and R. Mithen. 1995. The effect of modifying the glucosinolate content of leaves of oilseed rape *(Brassica napus* ssp. *oleifera)* on its interaction with specialist and generalist pests. Ann. Appl. Biol. 126: 347–363.

Gomez, J.M. 1993. Phenotypic selection on flowering synchrony in a high mountain plant, *Hormathophylla spinosa* (Cruciferae). J. Ecol. 81: 605–613.

Gomez, J.M., and R. Zamora. 2000. Spatial variation in the selective scenarios of *Hormathophylla spinosa* (Cruciferae). Am. Nat. 155: 657–668.

Goodspeed, T.H. 1954. The genus *Nicotiana*: origins, relationships and evolution of its species in the light of their distribution, morphology and cytogenetics. Chronica Botanica Company, Waltham, MA.

Gould, K.S. 2004. Nature's Swiss army knife: the diverse protective roles of anthocyanins in leaves. J. Biomed. Biotechnol. 314–320.

Grant, V. 1950. The protection of the ovules in flowering plants. Evolution 4: 179–201.

Gronquist, M., A. Bezzerides, A. Attygalle, J. Meinwald, M. Eisner, and T. Eisner. 2001. Attractive and defensive functions of the ultraviolet pigments of a flower *(Hypericum calycinum)*. Proc. Natl. Acad. Sci. USA 98: 13745–13750.

Haig, D., and M. Westoby. 1988. On limits to seed production. Am. Nat. 131: 757–759.

Hamback, P. A. 2001. Direct and indirect effects of herbivory: feeding by spittlebugs affects pollinator visitation rates and seedset of *Rudbeckia hirta*. Ecoscience 8: 45–50.

Hendrix, S.D. 1984. Reactions of *Heracleum lanatum* to floral herbivory by *Depressaria pastinacella*. Ecology 65: 191–197.

Hendrix, S.D., and E.J. Trapp. 1981. Plant-herbivore interactions: insect induced changes in host plant sex expression and fecundity. Oecologia 49: 119–122.

Herrera, C.M. 1996. Floral traits and plant adaptation to insect pollinators: a devil's advocate approach, pp. 65–87. In D.G. Lloyd and S.C. H. Barrett (eds.), Floral biology: studies on floral evolution in animal-pollinated plants. Chapman and Hall, New York.

Herrera, C.M. 2000. Measuring the effects of pollinators and herbivores: evidence for non-additivity in a perennial herb. Ecology 81: 2170–2176.

Herrera, C.M., M. Medrano, P.J. Rey, A.M. Sanchez-Lafuente, M.B. Garcia, J. Guitian, and A.J. Manzaneda. 2002. Interaction of pollinators and herbivores on plant fitness suggests a pathway for correlated evolution of mutualism- and antagonism-related traits. Proc. Natl. Acad. Sci. USA 99: 16823–16828.

Hodges, S.A. 1995. The influence of nectar production on hawkmoth behavior, self pollination, and seed production in *Mirabilis multiflora* (Nyctaginaceae). Am. J. Bot. 82: 197–204.

Irwin, R.E., and L.S. Adler. 2006. Correlations among traits associated with herbivore resistance and pollination: implications for pollination and nectar robbing in a distylous plant. Am. J. Bot. 93: 64–72.

Irwin, R., and A.K. Brody. 1998. Nectar robbing in *Ipomopsis aggregata*: effects on pollinator behavior and plant fitness. Oecologia 116: 519–527.

Irwin, R.E., and A.K. Brody. 1999. Nectar-robbing bumble bees reduce the fitness of *Ipomopsis aggregata* (Polemoniaceae). Ecology 80: 1703–1712.

Irwin, R.E., and A.K. Brody. 2000. Consequences of nectar robbing for realized male function in a hummingbird-pollinated plant. Ecology 81: 2637–2643.

Irwin, R.E., and S.Y. Strauss. 2005. Flower color microevolution in wild radish: evolutionary response to pollinator-mediated selection. Am. Nat. 165: 225–237.

Irwin, R.E., A.K. Brody, and N. M. Waser. 2001. The impact of floral larceny on individuals, populations, and communities. Oecologia 129: 161–168.

Irwin, R.E., S.Y. Strauss, S. Storz, A. Emerson, and G. Guibert. 2003. The role of herbivores in the maintenance of a flower color polymorphism in wild radish. Ecology 84: 1733–1743.

Irwin, R.E., L.S. Adler, and A.K. Brody. 2004. The dual role of floral traits: pollinator attraction and plant defense. Ecology 85: 1503–1511.

Ivey, C.T., and D.E. Carr. 2005. Effects of herbivory and inbreeding on the pollinators and mating system of *Mimulus guttatus* (Phrymaceae). Am. J. Bot. 92: 1641–1649.

Ivey, C.T., D.E. Carr, and M.D. Eubanks. 2004. Effects of inbreeding in *Mimulus guttatus* on tolerance to herbivory in natural environments. Ecology 85: 567–574.

Izaguirre, M.M., A.L. Scopel, I.T. Baldwin, and C.L. Ballare. 2003. Convergent responses to stress. Solar ultraviolet-B radiation and *Manduca sexta* herbivory elicit overlapping transcriptional responses in field-grown plants of *Nicotiana longiflora*. Plant Physiol. 132: 1755–1767.

Jones, K.N., and J.S. Reithel. 2001. Pollinator-mediated selection on a flower color polymorphism in experimental populations of *Antirrhinum* (Scrophulariaceae). Am. J. Bot. 88: 447–454.

Juenger, T., and J. Bergelson. 1997. Pollen and resource limitation of compensation to herbivory in scarlet gilia, *Ipomopsis aggregata*. Ecology 78: 1684–1695.

Jurgens, A. 2004. Flower scent composition in diurnal *Silene* species (Caryophyllaceae): phylogenetic constraints or adaption to flower visitors? Biochem. Syst. Ecol. 32: 841–859.

Kalisz, S. 1986. Variable selection on the timing of germination in *Collinsia verna* (Scrophulariaceae). Evolution 40: 479–491.

Karban, R., and S.Y. Strauss. 1993. Effects of herbivores on growth and reproduction of their perennial host *Erigeron glaucus*. Ecology 74: 39–46.

Kessler, A., R. Halitschke, and I.T. Baldwin. 2004. Silencing the jasmonate cascade: induced plant defenses and insect populations. Science 305: 665–668.

Kidd, N.A.C., and M.A. Jervis. 1989. The effects of host feeding behavior on the dynamics of parasitoid-host interactions, and the implications for biological control. Res. Pop. Ecol. 31: 235–274.

Kingsolver, J.G., H.E. Hoekstra, J.M. Hoekstra, D. Berrigan, S.N. Vignieri, C.E. Hill, A. Hoang, P. Gibert, and P. Beerli. 2001. The strength of phenotypic selection in natural populations. Am. Nat. 157: 246–261.

Kjellberg, F., E. Jousselin, J.L. Bronstein, A. Patel, J. Yokoyama, and J.Y. Rasplus. 2001. Pollination mode in fig wasps: the predictive power of correlated traits. Proc. R. Soc. Lond. B 268: 1113–1121.

Krupnick, G.A., and A.E. Weis. 1998. Floral herbivore effect on the sex expression of an andromonoecious plant, *Isomeris arborea* (Capparaceae). Plant Ecol. 134: 151–162.

Krupnick, G.A., and A.E. Weis. 1999. The effect of floral herbivory on male and female reproductive success in *Isomeris arborea*. Ecology 80: 135–149.

Krupnick, G.A., A.E. Weis, and D.R. Campbell. 1999. The consequences of floral herbivory for pollinator service to *Isomeris arborea*. Ecology 80: 125–134.

Krupnick, G.A., G. Avila, K.M. Brown, and A.G. Stephenson. 2000. Effects of herbivory on internal ethylene production and sex expression in *Cucurbita texana*. Funct. Ecol. 14: 215–225.

Lande, R., and S.J. Arnold. 1983. The measurement of selection on correlated characters. Evolution 37: 1210–1226.

Lau, J.A., and L.F. Galloway. 2004. Effects of low-efficiency pollinators on plant fitness and floral trait evolution in *Campanula americana* (Campanulaceae). Oecologia 141: 577–583.

Leege, L.M., and L.M. Wolfe. 2002. Do floral herbivores respond to variation in flower characteristics in *Gelsemium sempervirens* (Loganiaceae), a distylous vine? Am. J. Bot. 89: 1270–1274.

Lehtila, K., and S. Y. Strauss. 1997. Leaf damage by herbivores affects attractiveness to pollinators in wild radish, *Raphanus raphanistrum*. Oecologia 111: 396–403.

Lehtila, K., and S.Y. Strauss. 1999. Effects of foliar herbivory on male and female reproductive traits of wild radish, *Raphanus raphanistrum*. Ecology 80: 116–124.

Lohman, D.J., and M.R. Berenbaum. 1996. Impact of floral herbivory by parsnip webworm (Oecophoridae: *Depressaria pastinacella* Duponchel) on pollination and fitness of wild parsnip (Apiaceae: *Pastinaca sativa* L.). Am. Midl. Nat. 136: 407–412.

Lukefahr, M.J., and C. Rhyne. 1960. Effects of nectariless cottons on populations of three lepidopterous insects. J. Econ. Entomol. 53: 242–244.

Maad, J. 2000. Phenotypic selection in hawkmoth-pollinated *Platanthera bifolia*: targets and fitness surfaces. Evolution 54: 112–123.

Maad, J., and R. Alexandersson. 2004. Variable selection in *Platanthera bifolia* (Orchidaceae): phenotypic selection differed between sex functions in a drought year. J. Evol. Biol. 17: 642–650.

Maloof, J.E., and D.W. Inouye. 2000. Are nectar robbers cheaters or mutualists? Ecology 81: 2651–2661.

Manning, J.C., and P. Goldblatt. 2005. Radiation of pollination systems in the Cape genus *Tritoniopsis* (Iridaceae: Crocoideae) and the development of bimodal pollination strategies. Int. J. Plant Sci. 166: 459–474.

Masters, G.J., T.H. Jones, and M. Rogers. 2001. Host-plant mediated effects of root herbivory on insect seed predators and their parasitoids. Oecologia 127: 246–250.

Mattson, W.J., and R.A. Haack. 1987. The role of drought in outbreaks of plant-eating insects: drought's physiological effects on plants can predict its influence on insect populations. BioScience 37: 110–118.

Mauricio, R., and M.D. Rausher. 1997. Experimental manipulation of putative selective agents provides evidence for the role of natural enemies in the evolution of plant defense. Evolution 51: 1435–1444.

McCall, A.C. 2006. Natural and artificial floral damage induces resistance in *Nemophila menziesii* (Hydrophyllaceae) flowers. Oikos 112: 660–666.

McCall, A.C., and R. Karban. 2006. Induced defense in *Nicotiana attenuata* (Solanaceae) fruit and flowers. Oecologia 146: 566–571.

McCloud, E.S., and M.R. Berenbaum. 1994. Stratospheric ozone depletion and plant-insect interactions: effects of UVB radiation on foliage quality of *Citrus jambhiri* for *Trichoplusia ni*. J. Chem. Ecol. 20: 525–539.

McFadden, M.W. 1968. Observations on feeding and movement of tobacco hornworm larvae. J. Econ. Entomol. 61: 352–356.

McKey, D. 1974. Adaptive patterns in alkaloid physiology. Am. Nat.108: 305–320.

Mitchell, R.J. 2004. Heritability of nectar traits: why do we know so little? Ecology 85: 1527–1533.

Mothershead, K., and R.J. Marquis. 2000. Fitness impacts of herbivory through indirect effects on plant-pollinator interactions in *Oenothera macrocarpa*. Ecology 81: 30–40.

Motten, A.F., and J. Antonovics. 1992. Determinants of outcrossing rate in a predominantly self-fertilizing weed, *Datura stramonium* (Solanaceae). Am. J. Bot. 79: 419–427.

Patt, J.M., G.C. Hamilton, and J. H. Lashomb. 1997. Foraging success of parasitoid wasps on flowers: interplay of insect morphology, floral architecture and searching behavior. Entomol. Exp. Appl. 83: 21–30.

Patt, J.M., G.C. Hamilton, and J.H. Lashomb. 1999. Responses of two parasitoid wasps to nectar odors as a function of experience. Entomol. Exp. Appl. 90: 1–8.

Pellmyr, O. 2002. Pollination by animals, pp. 157–184. In C.M. Herrera and O. Pellmyr (eds.), Plant-animal interactions: an evolutionary approach. Blackwell Science, Oxford.

Pellmyr, O. 2003. Yuccas, yucca moths, and coevolution: a review. Ann. Missouri Botanical Garden 90: 35–55.

Pettersson, M.W. 1992. Density-dependent egg dispersion in flowers of *Silene vulgaris* by the seed predator *Hadena confusa* (Noctuidae). Ecol. Entomol. 17: 244–248.

Poveda, K., I. Steffan-Dewenter, S. Scheu, and T. Tscharntke. 2003. Effects of below- and above-ground herbivores on plant growth, flower visitation and seed set. Oecologia 135: 601–605.

Poveda, K., I. Steffan-Dewenter, S. Scheu, and T. Tscharntke. 2005a. Effects of decomposers and herbivores on plant performance and aboveground plant-insect interactions. Oikos 108: 503–510.

Poveda, K., I. Steffan-Dewenter, S. Scheu, and T. Tscharntke. 2005b. Floral trait expression and plant fitness in response to below- and aboveground plant-animal interactions. Perspect. Plant Ecol. Evol. Syst. 7: 77–83.

Quesada, M., K. Bollman, and A.G. Stephenson. 1995. Leaf damage decreases pollen production and hinders pollen performance in *Cucurbita texana*. Ecology 76: 437–443.

Rasmann, S., T.G. Kollner, J. Degenhardt, I. Hiltpold, S. Toepfer, U. Kuhlmann, J. Gershenzon, and T.C.J. Turlings. 2005. Recruitment of entomopathogenic nematodes by insect-damaged maize roots. Nature 434: 732–737.

Real, L.A., and B.J. Rathcke. 1991. Individual variation in nectar production and its effect on fitness in *Kalmia latifolia*. Ecology 72: 149–155.

Ree, R.H. 2005. Phylogeny and the evolution of floral diversity in *Pedicularis* (Orobanchaceae). Int. J. Plant Sci. 166: 595–613.

Rhoades, D.F., and R.G. Cates. 1976. Toward a general theory of plant antiherbivore chemistry. Recent Adv. Phytochem. 10: 168–213.

Roulston, T.H., and J.H. Cane. 2000. Pollen nutritional content and digestibility for animals. Plant Syst. Evol. 222: 187–209.

Rudgers, J.A. 2004. Enemies of herbivores can shape plant traits: selection in a facultative ant-plant mutualism. Ecology 85: 192–205.

Rudgers, J.A., and M.C. Gardener. 2004. Extrafloral nectar as a resource mediating multispecies interactions. Ecology 85: 1495–1502.

Sargent, R.D., and S.P. Otto. 2004. A phylogenetic analysis of pollination mode and the evolution of dichogamy in angiosperms. Evol. Ecol. Res. 6: 1183–1199.

Schemske, D.W., and P. Bierzychudek. 2001. Perspective: evolution of flower color in the desert annual *Linanthus parryae*: Wright revisited. Evolution 55: 1269–1282.

Schuster, M.F., M.J. Lukefahr, and F.G. Maxwell. 1976. Impact of nectariless cotton on plant bugs and natural enemies. J. Econ. Entomol. 69: 400–402.

Scott, W.P., G.L. Snodgrass, and J.W. Smith. 1988. Tarnished plant bug (Hemiptera: Miridae) and predaceous arthropod populations in commercially produced selected nectaried and nectariless cultivars of cotton. J. Entomol. Sci. 23: 280–286.

Simms, E.L., and M.A. Bucher. 1996. Pleiotropic effect of flower color intensity on resistance to herbivory in *Ipomoea purpurea*. Evolution 50: 957–963.

Sprengel, C.K. 1996. Discovery of the secret of nature in the structure and fertilization of flowers, pp. 3–43. In D.G. Lloyd and S.C. H. Barrett (eds.), Floral biology: studies on floral evolution in animal-pollinated plants. Chapman and Hall, New York.

Stahl, E. 1888. Pflanzen und schnecken: biologische studie über die schutzmittel der pflanzen gegen schneckenfrass. Jenaische Zeitschr. Naturwiss. 22: 557–684.

Steets, J.A., and T.L. Ashman. 2004. Herbivory alters the expression of a mixed-mating system. Am. J. Bot. 91: 1046–1051.

Stinchcombe, J.R., and M.D. Rausher. 2001. Diffuse selection on resistance to deer herbivory in the ivyleaf morning glory, *Ipomoea hederacea*. Am. Nat. 158: 376–388.

Stratton, D.A. 1992. Life-cycle components of selection in *Erigeron annuus*: I. Phenotypic selection. Evolution 46: 92–106.

Stratton, D.A., and C.C. Bennington. 1998. Fine-grained spatial and temporal variation in selection does not maintain genetic variation in *Erigeron annuus*. Evolution 52: 678–691.

Strauss, S.Y., and W.S. Armbruster. 1997. Linking herbivory and pollination: new perspectives on plant and animal ecology and evolution. Ecology 78: 1617–1618.

Strauss, S.Y., J.K. Conner, and S.L. Rush. 1996. Foliar herbivory affects floral characters and plant attractiveness to pollinators: implications for male and female plant fitness. Am. Nat. 147: 1098–1107.

Strauss, S.Y., D.H. Siemens, M.B. Decher, and T. Mitchell-Olds. 1999. Ecological costs of plant resistance to herbivores in the currency of pollination. Evolution 53: 1105–1113.

Strauss, S.Y., J.K. Conner, and K.P. Lehtila. 2001. Effects of foliar herbivory by insects on the fitness of *Raphanus raphanistrum*: damage can increase male fitness. Am. Nat. 158: 496–504.

Strauss, S.Y., J.A. Rudgers, J.A. Lau, and R.E. Irwin. 2002. Direct and ecological costs of resistance to herbivory. Trends Ecol. Evol. 17: 278–285.

Strauss, S.Y., R.E. Irwin, and V.M. Lambrix. 2004. Optimal defence theory and flower petal colour predict variation in the secondary chemistry of wild radish. J. Ecol. 92: 132–141.

Tewksbury, J.J., and G.P. Nabhan. 2001. Seed dispersal: directed deterrence by capsaicin in chilies. Nature 412: 403–404.

Thaler, J.S. 1999. Jasmonate-inducible plant defenses cause increased parasitism of herbivores. Nature 399: 686–688.

Theis, N. 2006. Fragrance of Canada thistle (*Cirsium arvense*) attracts both floral herbivores and pollinators. Journal of Chemical Ecology 32(5): 917–927.

Theis, N., and R.A. Raguso. 2005. The effect of pollination on floral fragrance in thistles (*Cirsium*, Asteraceae). J. Chem. Ecol. 31: 2581–2600.

Theis, N., R.A. Raguso, and M. Lerdau. 2007. The challenge of attracting pollinators while evading floral herbivores: patterns of fragrance emission in *Cirsium arvense* and *Cirsium repandum* (Asteraceae). International Journal of Plant Sciences 168(5): 587–601.

Thompson, J.N., and B.M. Cunningham. 2002. Geographic structure and dynamics of coevolutionary selection. Nature 417: 735–738.

Thomson, V.P., A.B. Nicotra, and S.A. Cunningham. 2004. Herbivory differentially affects male and female reproductive traits of *Cucumis sativus*. Plant Biol. 6: 621–628.

Thornburg, R.W., C. Carter, A. Powell, R. Mittler, L. Rizhsky, and H. T. Horner. 2003. A major function of the tobacco floral nectary is defense against microbial attack. Plant Syst. Evol. 238: 211–218.

Tooker, J.F., and L.M. Hanks. 2000. Flowering plant hosts of adult Hymenopteran parasitoids of central Illinois. Ann. Entomol. Soc. Am. 93: 580–588.

Totland, O. 2001. Environment-dependent pollen limitation and selection on floral traits in an alpine species. Ecology 82: 2233–2244.

Tuttle, M.D., and M.J. Ryan. 1981. Bat predation and the evolution of frog vocalizations in the Neotropics. Science 214: 677–678.

Wackers, F.L. 1999. Gustatory response by the hymenopteran parasitoid *Cotesia glomerata* to a range of nectar and honeydew sugars. J. Chem. Ecol. 25: 2863–2877.

Wackers, F.L. 2001. A comparison of nectar and honeydew sugars with respect to their utilization by the hymenopteran parasitoid *Cotesia glomerata*. J. Insect Physiol. 47: 1077–1084.

Wackers, F.L., and T.M. Bezemer. 2003. Root herbivory induces an above-ground indirect defence. Ecol. Letters 6: 9–12.

Waser, N.M., L. Chittka, M.V. Price, N.M. Williams, and J. Ollerton. 1996. Generalization in pollination systems, and why it matters. Ecology 77: 1043–1060.

Willmer, P.G., and G.N. Stone. 1997. How aggressive ant-guards assist seed-set in *Acacia* flowers. Nature 388: 165–167.

Wolfe, L.M. 2002. Why alien invaders succeed: support for the escape-from-enemy hypothesis. Am. Nat. 160: 705–711.

Wootton, J.T. 1994. The nature and consequences of indirect effects in ecological communities. Ann. Rev. Ecol. Syst. 25: 443–466.

Wyatt, R., S.B. Broyles, and G.S. Derda. 1992. Environmental influences on nectar production in milkweeds *(Asclepias syriaca* and *A. exaltata).* Am. J. Bot. 79: 636–642.

Zangerl, A.R., and C.E. Rutledge. 1996. The probability of attack and patterns of constitutive and induced defense: a test of optimal defense theory. Am. Nat. 147: 599–608.

Zimmerman, M., and G.H. Pyke. 1988. Reproduction in *Polemonium*: assessing the factors limiting seed set. Am. Nat. 131: 723–738.

THIRTEEN

Adaptive Radiation: Phylogenetic Constraints and Ecological Consequences

PETER W. PRICE

A fundamental question in evolutionary biology is how adaptive radiation proceeds on continents, where most of it occurs. The question is most pressing when insects are considered, especially in phytophagous taxa, which represent over 25% of terrestrial biodiversity. Each taxon has, no doubt, followed a unique trajectory through time, but unifying themes should reveal some general patterns and processes, even if the answers recognize that with different starting points, different end points will result (cf. MacArthur 1972). Given the great diversity of insect herbivores, we should expect both divergent and convergent modes of adaptive radiation, which therefore require a pluralistic approach to their explanation.

Adaptive radiation is the relatively rapid evolutionary divergence of species in a lineage into a series of rather different adaptive zones, with each zone occupied by species with similar ecological niches. Thus, for insect herbivores, adaptive zones may involve adaptation within a single phyletic line to living in or on, and feeding upon, different plant modules such as leaves, stems, roots, flowers, or fruits. The ecological niches within any adaptive zone may be defined by the host-plant species that is exploited, so that related insect herbivore species may speciate across related host-plant species.

Adaptive radiation is a centerpiece of evolutionary biology because its study necessarily unites and integrates major aspects of the biological sciences, including ecology, evolution, behavior, systematics, and physiology. It forms a central theme in evolutionary biology. And yet, for major taxa of insect herbivores, such as the Hymenoptera, Hemiptera (Homoptera), or Lepidoptera, we have very little in the way of a conceptual framework with which to develop hypotheses and to detect patterns.

One surprising condition, perhaps, is that much of ecology has never embraced the evolutionary synthesis. Central themes in ecology such as distribution, abundance, and population dynamics have proceeded largely without the evolutionary point of view. Indeed, some ecologists would assert that population dynamics is purely an ecological subject, as if evolutionary approaches, especially employing the evolutionary background of a species or larger taxon, and the ground plan, or *Bauplan,* of the lineage, exert no influence in ecology. Although these ecologists would admit that species are adapted to their environment, they would not acknowledge that such adaptation has its own constraints (Ligon 1993), and that these constraints impact the ecology of the species. An adaptive move in one direction will limit alternative evolutionary options. For example, increasing the length of the ovipositor of a parasitoid wasp in response to more deeply concealed hosts in wood lessens the adaptive advantage for attacking unconcealed hosts. The benefits and costs of an adaptation and its attendant constraint must be kept in mind, as well as the context of adaptation represented by the ground plan of the species or group, as espoused from Darwin (1859) to Gould (2002): "recognizing organisms as products of history, rather than objects created in their present state" (Gould 2002, p. 99; see also Darwin 1859, pp. 485–486). A similar theme is adopted by Thompson (2005), who emphasizes that "species are phylogenetically conservative in their interactions" (p. 12), and "how the combination of phylogenetic conservatism and ecological opportunity shapes the structure of interaction webs remains one of the least understood aspects of community ecology" (p. 33).

The term "adaptive radiation" obviously highlights the main evolutionary advantages in a lineage through time, but inevitably constraints are involved, either explicitly or more cryptically. A balanced view of adaptations and constraints, and the explicit recognition of both, provide a blending opportunity in the field of adaptive radiation. We can embrace heretofore refractive parts of ecology into the evolutionary synthesis—in particular, distribution, abundance, and population dynamics. For, if we are to compare the adaptive radiation of lineages, and to search for pattern and mechanism, then ecological characteristics of species and higher taxa are of central concern.

The Phylogenetic Constraints Hypothesis

We have advocated an approach to adaptive radiation and insect herbivore population dynamics that we call the phylogenetic constraints hypothesis (Price et al. 1990; Price 1994, 2003). The hypothesis developed here argues that the *Bauplan* of a taxonomic group ultimately influences the ecology of that group in terms of distribution, abundance, and population dynamics. We emphasized phylogenetic constraints because we were faced with the puzzle of why a particular insect herbivore species we studied showed such relatively stable, or latent, population dynamics, when the majority of studies on variation in numbers were on insect pests exhibiting eruptive dynamics. Our ecological investigation revealed that natural enemies were unimportant for the focal species, a tenthredinid sawfly, but resources were limiting: only rapidly growing shoots of the willow host plant could be utilized. The ecological constraint of a low carrying capacity in the environment raised the question of why this should exist when most outbreaking species can consume a large proportion of the foliage in a forest. We concluded that a morphological constraint, part of the *Bauplan* of the group, dictated the use by sawflies of soft, rapidly growing willow shoots. The stilettolike ovipositor that gives the sawflies their name is used for placing eggs, a single egg at a time, into plant tissue close to the growing terminal meristem. The ovipositor is flexible but rather delicate, which limits successful oviposition to the softest shoots, which are those that are growing most rapidly. This morphology of the ovipositor therefore imparts a constraint on the sawfly, while simultaneously acting as a major adaptation for exploiting plant resources. The adaptive radiation of this sawfly group has been extensive and is best understood using a combination of explicitly recognized adaptations and constraints: The phylogenetic constraints hypothesis embraces both, as explained next.

The sawfly is the arroyo willow stem-galling sawfly, *Euura lasiolepis* Smith (Hymenoptera: Tenthredinidae), whose only host is the arroyo willow, *Salix lasiolepis* Benth. (Salicaceae). As stated above, the sawlike ovipositor acts as a phylogenetic constraint. The ovipositor is a plesiomorphic character for the sawflies, and indeed with its lepismatoid-type structure (derived from the primitive silverfish [Thysanura: Lepismatidae]), for the whole lineage of the Hymenoptera (Hunt 1999; Grimaldi and Engel 2005). Behaviorally, females of *Euura* are limited to ovipositing in soft plant tissue, right behind the growing meristem. Phenologically and physiologically this limits emergence times to early spring in the north temperate climate, when willow growth is rapid. This limits the evolution of a life cycle, making it univoltine; mature larvae in cocoons are the overwintering phase. In such confined conditions for placing an egg, typically only one is laid at a time. The time required to search for high-quality shoots and to oviposit is prolonged, so that relatively few eggs are laid, and synovigenic egg production (maturation of eggs during the female's life) is the norm. This contrasts with pro-ovigenic egg production seen in many eruptive insect species (cf. Price 2003). Therefore, we see that the morphology of the ovipositor has far-reaching consequences for many biological features in the insect's life cycle.

The sawfly has evolved with extensive adaptive responses to the constraints imposed by the ovipositor on the evolution of its life-history strategy. We have called these the adaptive syndrome, a term coined by Root and Chaplin (1976), meaning a coordinated group of adaptations that mitigate the constraint and indeed contribute to the adaptive opportunities of the lineage. These include the following. (1) Insertion of the ovipositor into plant tissue provides information on phytochemical constituents of host plants, enabling the identification of the correct host plant and long shoots on those plants (Roininen et al. 1999). (2) Such insertion also increases the chances of stimulating plant cell division and expansion to form a gall, which may provide multiple adaptive features (cf. Price et al. 1987; Stone and Schönrogge 2003). (3) Ovipositing eggs singly provides an opportunity to determine the sex of the progeny, for being haplodiploid, a basal character of the Hymenoptera, a female can deposit an unfertilized male egg or a fertilized female egg. Male eggs are placed into less favorable shoots than female eggs (Craig et al. 1992). Also, males are smaller than females, and galls on poorer shoots are smaller, so allocation of males and females maximizes the potential size of progeny in each sex. (4) Female choice of long shoots on vigorous plants maximizes larval survival; there is a strong ovipositional preference for long shoots where larval performance is high (Craig et al. 1989). (5) Females have also evolved to be philopatric in relation to the maternal willow clone, tending to oviposit on the same clone from which they emerged (Stein et al. 1994). If the last generation found favorable resources, then remaining on this willow clone is probably a safe strategy.

All these evolved traits in the phylogenetic constraint and the adaptive syndrome, which involve morphology, physiology, behavior, and evolutionary history, impose far-reaching effects on the ecology of this sawfly and are called the emergent properties: (1) Distribution over the landscape is patchy, limited to small areas where willows are growing well, often after disturbance such that regrowth is rapid or plants are young and vigorous. (2) Abundance of sawflies is generally low because of patchy resources and difficulty in finding them. (3) Population dynamics are relatively stable, or latent—much more so than for eruptive species such as outbreak species of forest Lepidoptera. Dynamics are driven from the bottom up because high soil moisture is needed for vigorous willow growth, while top-down effects from natural enemies are weak and insignificant (Hunter and Price 1998; Price and Hunter 2005).

This strong flow of influences from evolved traits to ecological relationships brings the study of the areas of distribution, abundance, and population dynamics well within the evolutionary synthesis. Phylogenetic constraints lead to mitigating evolutionary change, the adaptive syndrome, and these determine emergent properties in ecology. The hypothesis opens up avenues of inquiry into comparative studies both within related groups (those related by phylogeny), and between groups (those associated by convergent phylogenetic constraints). Comparisons can reach beyond these groups to those with different phylogenetic constraints. We can then develop a macroevolutionary, pluralistic theory on the adaptive radiation of insect herbivores, including temperate and tropical comparisons. These points of view will be explored in a later section of this chapter.

Adaptive Capture and Escape

Advocating the concept of evolutionary constraints in the phylogenetic hypothesis we have proposed places a responsibility for explaining why such evolutionary limitation endures through time. What limits evolutionary alternatives? Gould (e.g., Gould and Lewontin 1979; Gould 1980, 1989, 2002) has long been a champion for the concept of evolutionary constraints: he spends two long chapters on "the integration of constraint and adaptation" (Gould 2002). "Organisms must be analyzed as integrated wholes, with *Baupläne* so constrained by phyletic heritage, pathways of development and general architecture that the constraints themselves become more interesting and more important in delimiting pathways to change than the selective force that may mediate change when it occurs" (Gould and Lewontin 1979, p. 581). Gould (2002, p. 1025) also opened his discussion with "constraint as a positive concept," which causes "evolutionary direction and change" (p. 1028).

The validity of the existence of long-term constraints on evolution can be argued from the general case, and the specifics of the sawfly ovipositor. These are discussed in turn.

The general case for constraints has at least four bases. First, effective design precludes most alternatives and endures: the four appendages of vertebrates, the six legs of insects. Alternatives with three or five legs, or five or seven legs, respectively, have never survived, if such bizarre inventions ever emerged. Secondly, Gould (2002) uses an example of a zebra evolving the ability to fly to escape predators, but the extra appendages needed and the weight of the zebra constrain the evolution of wings and flight. "Natural selection makes nothing" by itself (p. 1029), so the raw materials of evolution are constraining in themselves. Thirdly, many traits persist over long periods of time because of their effective design, becoming hallmarks of a clade, or what we call plesiomorphic characters. Our systematic organization of taxa depends upon this conservation of characters, which must acknowledge the existence of long-term constraints on evolutionary options. Fourth, designs that are effective are frequently generally useful and adaptable. The four limbs of vertebrates have served multiple purposes, as have the six of insects.

Concerning the phylogenetic constraint of the sawfly ovipositor, the general case applies. The hymenopteran ovipositor is obviously an excellent design, having survived longer than any other form, dating back to the origin of the silverfish and firebrats (Thysanura: Lepismatidae), the basal wingless insects from the Devonian, 400 million years ago (Grimaldi and Engel 2005). The four elements in the functional, piercing part of the ovipositor are derived from two pairs of abdominal appendages; they join neatly to transfer the egg from the gonopore into the substrate, and they form a well-reinforced piercing organ. Three or five appendages would not work as well and would be counter to the ground plan of insect design. The ovipositor serves many purposes in various insect species: piercing host insects, stinging predators (as in the bees and wasps), inserting eggs into substrates (e.g., within wood, or egg sacs under plant epidermis). However, the ovipositor serves one function in the tenthredinid sawflies and relatives almost exclusively: to place eggs into plant tissue. Why such constraint?

Various kinds of constraints accompany, or group around, the design and use of the ovipositor in tenthredinid sawflies, which together canalize design. The egg physiology constraint appears to be universal in the group and a basic requirement: the egg absorbs water from the host plant and swells during embryogenesis. The eggs may swell 1.5 times their original width (Caltagirone 1964), or in another species, "an average 2.8 times because of absorption of water" (Ivanova-Kasas 1959, p. 224). The phenomenon has been noted from Swammerdam (1737) to many authors in the 1900s (e.g., Enslin 1912; Carleton 1939, Zirngiebl 1939; Benson 1950; Ivanova-Kasas 1959; Pschorn-Walcher 1982; Schedl 1991). It is a conspicuous feature of egg development observed, if not recorded, by most naturalists studying sawflies. Clearly, placement of eggs into plant tissue, and close to xylem cells, promotes osmosis of water into eggs. Were eggs to be laid onto plants, simultaneous adaptations in chorionic permeability and resistance to desiccation, and basic embryological processes would be required, having low probability of escaping the constraint. It is interesting to note that most reptile eggs also depend on water uptake during embryogenesis (Brown and Shine 2005).

The ovipositional stimulus constraint (or the adult female physiological constraint), derives from the requirement of a specific phenolic glucoside that stimulates oviposition behavior (Kolehmainen et al. 1994; Roininen et al. 1999; Price 2003). After insertion of the ovipositor and gall initiation by *E. lasiolepis*, the female may decide to lay an egg or to reject this option. Presumably, key information is derived on plant quality by receptors on the ovipositor only after insertion into the host plant (Price 2003). These could be chemoreceptors, osmoreceptors, or both. Again, all the neurophysiological and behavioral adaptations with oviposition probably result in a strong canalizing effect on any further adaptation that might result in escape from the general constraint.

The host quality constraint is critical because eggs survive only in one host-plant species, and only on vigorous growth in which osmolarity of plant tissue allows water to move into eggs. Female *E. lasiolepis* show an ovipositional preference for rapidly growing shoots with high tremulacin content, probably detected after insertion of the ovipositor (Price 2003). Utilization of the more abundant shoots of lower vigor is precluded by inadequate osmotic and ovipositional stimulant conditions, making insertion of a piercing ovipositor critical to the survival of the species.

The gall formation constraint derives from the unusual condition observed in sawflies where adult females stimulate gall growth. Injection of gall stimulants is therefore required. Escape from such a constraint, with transfer of the ability to larvae, as is the case in the majority of gall-inducing insect species, would require a sudden and large shift in adaptations, unlikely to be achieved: hence, canalization of the trait in the female.

These multiple aspects of constraints on adaptive moves in evolution reinforce the concept that canalization captures highly advantageous traits in an "evolutionary straitjacket" from which escape is highly improbable. Seldom have constraints been so well understood and researched as in the sawflies, suggesting a fruitful avenue of research in many other organisms. It may well emerge that constraints in the adaptive radiation of a group are easier to contemplate and predict than the adaptive opportunities and escape from constraints. The evidence from plesiomorphic characters used in systematics supports this view. "Too often, the adaptationist programme gave us an evolutionary biology of parts and genes, but not organisms. It assumed that all transitions could occur step by step and underrated the importance of integrated developmental blocks and pervasive constraints of history and architecture. A pluralistic view could put organisms, with all their recalcitrant, yet intelligible, complexity, back into evolutionary theory" (Gould and Lewontin 1979, p. 597).

I wish to emphasize that the ancient form of the lepismatoidlike ovipositor antedates all other traits considered, the latter clustering around the phylogenetic constraint as the Hymenoptera emerged as a group. The ovipositor, of all the characters considered, is unique in its antiquity and its complexity. It is retained in all the tenthredinid sawflies, and invariably at the oviposition site the plant epidermis is cut, allowing water to pass from the plant into the egg, so the egg swells (Alexey Zinovyev and Heikki Roininen, 2006, personal communications). This plesiomorphic character in the sawflies should be considered as *the* phylogenetic constraint, with more derived traits clustered around it, as subsequent, not coincident, stages in the evolution of the emerging sawflies.

The Adaptive Radiation of Common Sawflies

Once we understood one species well, we undertook a comparative study of as many related sawflies as we could find. While concentrating on gall-inducing sawflies we also studied some free feeders. Many show the same pattern as in *E. lasiolepis,* where vigorous shoot modules are preferred, and where it has been studied, there is a strong preference and performance linkage on rapidly growing shoots (Price 2003). We have also explored reasons why a few species have escaped the constraints imposed by the need to attack rapidly growing shoots (Price et al. 2004). Thus, the phylogenetic constraints hypothesis explains the resource use and life-history adaptation in the majority of sawflies we have studied (38 species listed by Price [2003]). A common phylogenetic background, with shared phylogenetic constraints, result in similar life histories and ecological interactions.

Further insight into the adaptive radiation of these sawflies is seen in the gradual exploitation of new resources and ways of exploiting these resources. The phylogenetic hypothesis developed by Nyman et al. (2000) shows a shift from free-feeding sawflies to leaf-folding gallers, to enclosed leaf galls, and on to stem gallers, bud gallers, and petiole gallers (Fig. 13.1). Thus the different resources, or modules, utilized—leaves, petioles, midribs, stems, and buds—constitute adaptive zones, and speciation within these zones occurs by host shifts from one willow species to another. No doubt the new adaptive zones were invaded through ovipositional mistakes because female sawflies oviposit at the terminals of young shoots where small leaves and other modules are clustered around the central meristem. Therefore, during oviposition, these modules are millimeters or less apart, providing potential adaptive zones for radiation. Even now, we can find single sawfly species with variation in oviposition, and galls formed on leaf midribs, petioles, and stems, which may ultimately result in separate species on these three module components.

Once new adaptive zones have been colonized, host shifting among willows appears to be opportunistic, probably related to the ecological juxtaposition of willow species. Hence, the phylogenies of willows and sawflies do not illustrate parallel speciation, or parallel cladogenesis (Nyman 2000).

Convergence of Constraints

We can test the phylogenetic constraints hypothesis by studying a very different group of insect herbivores, but with convergent constraints. Such a group is exemplified by the treehoppers (Hemiptera: Homoptera: Membracidae). Species are endowed with a plant-piercing ovipositor, and parents and progeny live together, feeding on the shoot that acted as an oviposition substrate. Not surprisingly, family groups of membracids are found commonly on long, vigorous shoots, with a strong ovipositional preference for the longer shoot-length classes (Price and Carr 2000).

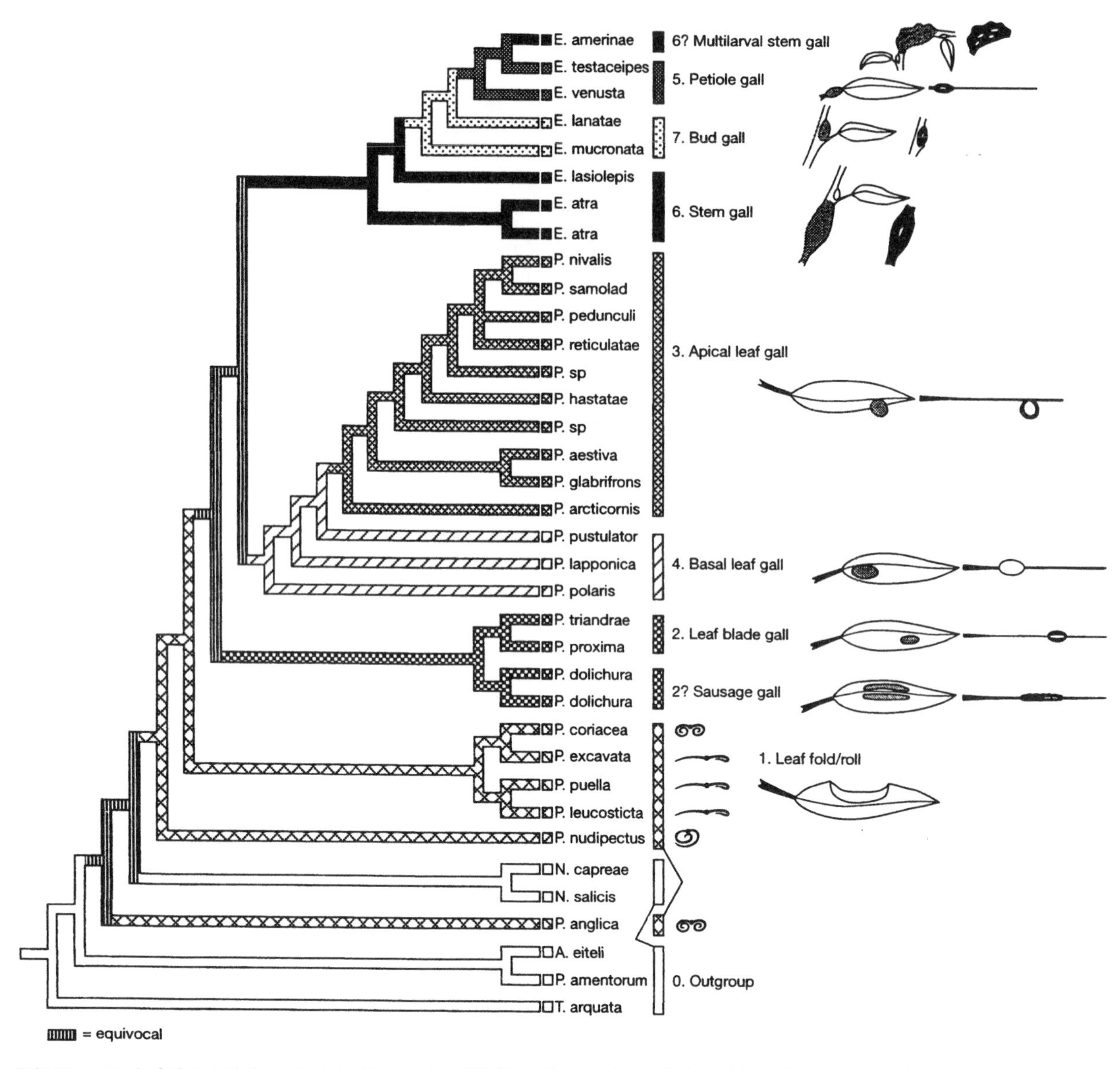

FIGURE 13.1 A phylogenetic hypothesis by Nyman et al. (2000) on the adaptive radiation of the gall-inducing tenthredinid sawflies from free-feeding progenitors. Note the general pattern of leaf galling and folding, and then enclosed leaf galls, stem, bud, and petiole galls, and finally multilarval stem galls. Latin names are provided for each species: *P, Phyllocolpa* for leaf rolls and folds; *P, Pontania* for enclosed leaf galls (sausage, leaf blade, basal leaf, and apical leaf types); and *E, Euura* for stem, bud, petiole, and multilarval stem types.

Of the four membracid species we have studied, three occur in the tropics and one in the warm temperate region. The climates in these zones are substantially different from the cool temperate climate in which sawflies thrive, and so the host plants are strongly divergent. And yet the phylogenetic constraints hypothesis explains why ecologies are similar in such divergent surroundings.

The ecology of treehoppers, in terms of distribution, abundance, and population dynamics, has not been extensively studied, but naturalists familiar with the tropics know where to find them. They frequently occur in disturbed vegetation, especially where humans have trimmed or cut back woody plants and where regrowth is rapid. They are found along trails and roads where growth is cleared, and in landscaped areas. Under natural conditions these spots with rapid or vigorous growth are patchy over the landscape, and similar in distribution to willows favored by sawflies. Membracids are widely dispersed in small populations, with a high likelihood that populations remain low. Their ecology is very similar to that of the sawflies, as are their phylogenetic constraints.

The adaptive radiation of membracids has, no doubt, followed lines similar to that of sawflies: opportunistic colonization of host plants with host shifting promoting reproductive isolation and speciation events. Reading through Wood's (1984) field notes on life-history observations on membracids in Central America we can see general patterns emerging. (1) Females often oviposit close to terminal meristems or in inflorescences. (2) Females are strongly philopatric, returning to the same shoot after disturbance. (3) Membracid species reoccur on the same host individual from one year to the next. Wood (1980) emphasized how a membracid species, *Enchenopa binotata* Say, became specialized on individual host-plant species, adapting to specific properties of each host plant to such an extent that "minimally, the species must now be considered a complex of host-plant races or incipient species" (Wood 1980, p. 159; see also Wood 1987). Adaptation to specific host-plant species involved shifts in timing of mating, ovipositional site, optimal oviposition time in the day, and preference for a specific host and mates reared from that host. Host shifting resulted in speciation events, providing a major example of sympatric speciation (cf. Berlocher and Feder 2002). Many articles provide details on the mechanisms involved with host shifting and speciation, providing insights for many other radiating groups of specialized insect herbivores (e.g., Wood and Guttman 1982; Guttman and Weigt 1989; Wood and Keese 1990; Tilmon et al. 1998; Wood et al. 1999).

The comparison of sawflies and membracids is merely one example of how the phylogenetic constraints hypothesis provides a mechanistic understanding of why the ecologies of two taxonomically divergent groups are so similar. The hypothesis also explains why similarities are observed even over strong latitudinal gradients from the north temperate to the tropics, and when host plants are so taxonomically divergent.

Divergent Constraints

We can now be quite confident that if phylogenetic constraints differ, then the ecology of the species in relation to distribution, abundance, and population dynamics will differ. Indeed, the large differences between the sawflies we studied and the outbreaking, eruptive, species of Lepidoptera in north temperate forests provided the stimulus for a search for evolutionary divergence between the groups. In north temperate forests sawfly species can show latent, or relatively noneruptive, dynamics, while forest lepidopterans can be eruptive and defoliate forests. In the same environment the distribution, abundance, and population dynamics of insect herbivores can be very different, so that evolved differences between groups are more likely to provide explanatory power than ecological factors.

Among the forest lepidopterans of north temperate America, there are 41 macrolepidopteran species that have caused visually conspicuous defoliation for at least 2 years in 20 years of survey work from 1962 to 1981 (Nothnagle and Schultz 1987). A search in the literature for relevant life-history traits for these species yielded adequate information on 33 species (Price 1994). In all cases the evidence indicated a lack of an ovipositional preference and larval performance linkage. This was because females did not choose oviposition sites in relation to larval food requirements. Eggs were frequently laid in the late summer, while larvae hatched and began feeding in the following spring. Eggs in many species were deposited in a single cluster, with hundreds of eggs per mass in some cases, and frequently egg masses were placed on branches, tree trunks, rocks, and other nonfood substrates. Some species will even oviposit on vehicles or in paper bags, illustrating a willingness to place eggs almost anywhere. Other females are wingless or so heavy with eggs that mobility is limited, making choices difficult.

In these cases of forest lepidopterans, the phylogenetic constraint is that females lack an ovipositor. Oviposition occurs from the oviduct, which vents through a pad, the cushionlike oviporus (Snodgrass 1935). There is no plant-piercing ovipositor, which results in a reduced level of information derived from the plant being available to the female. If the female cannot discriminate among the range of food quality available because of the oviporus, and because of asynchrony with the availability of larval food, then depositing eggs en masse, and on a permanent and stable substrate, such as a tree trunk, seems to be a viable strategy. Females also evolve to be pro-ovigenic, with all eggs mature and ready to deposit when the female emerges as an adult. Massed eggs also mean that larvae emerge as a group, that they are likely to remain together as a group, and that they will feed gregariously. Also, larvae must find food for themselves, even as tiny first-instar larvae, so selection pressure will be strong to accept a wide range of food quality, and even a wide range of host-plant species. Such lepidopterans are likely to be generalist feeders. The consequences for the ecology of these kinds of species are populations widely distributed in temperate forest, densities can become high, and eruptive population dynamics may be observed. With a generalized feeding ability, larvae can eat almost every green leaf in a forest, the carrying capacity of the environment is very high, with the potential to support extremely high densities of larvae. Substrates for oviposition are almost unlimited, with high egg densities possible, and the defoliation of the forest results.

All this is in stark contrast to the conditions for sawflies such as *Euura lasiolepis*, for which the carrying capacity of the environment is limited by the shortage of suitable oviposition sites, which are on vigorous shoots on willows. The chemical cues used in identification of host plants, and as oviposition stimulants, promote high host-plant fidelity and specificity, and in most cases the specialization on a single host-plant species. Such limitations prevent eruptive population dynamics, although in favorable patches populations may persist for decades and remain higher in small areas than populations of outbreak species during troughs of abundance after population crashes.

The contrast between *Euura*-like sawflies and forest lepidopterans illustrates the potential for the phylogenetic constraints hypothesis. With different starting points in evolutionary history, which we identify as phylogenetic constraints, we can predict different end points, the emergent properties of ecology: distribution, abundance, and population dynamics. The hypothesis is thus pluralistic, being able to cope with many different kinds of phylogenetic constraints, adaptive syndromes, and emergent properties (e.g., Price 1997, 2003). And we have seen that the hypothesis can encompass species from the north temperate into the tropics, and the evolved traits trump environmental variation, maintaining patterns in life histories and ecologies as lineages spread around the world. The adaptive radiation of so many groups is ruled by their shared phylogenetic constraints, which set limits on the ways in which a lineage can expand and radiate.

There is one important proviso to be emphasized with the phylogenetic constraints hypothesis. Being an evolutionary hypothesis, it describes and predicts the evolutionarily based potential for a certain set of ecological characteristics we call emergent properties. The hypothesis predicts how species in a certain lineage will behave ecologically given a certain kind of phylogenetic constraint. However, we should not expect that all members of a lineage will exhibit the same ecology, because both abiotic and biotic aspects of ecology may be superimposed on the basic evolutionary potential. The hypothesis says nothing, for example, on how natural enemies are likely to impact an insect herbivore population, and their impact may suppress the evolutionary potential for outbreak dynamics for most of the time, or even perpetually. Movement of species into extreme environments may well result in abiotic conditions too severe to permit outbreak dynamics. Therefore, the hypothesis can predict the likelihood that species will show a certain kind of ecology, which will be observed more frequently than other kinds. But prediction will take the form of a group with a certain phylogenetic constraint showing a higher probability of species with a particular ecology than another group with a different phylogenetic constraint. Unfortunately we do not have enough studies of comparative ecology within species groups to understand well why some species have eruptive population dynamics while others have stable or latent dynamics. This remains a challenge for the future.

The Similarities of Temperate and Tropical Insect Herbivores

As we saw with the comparison of temperate sawflies and tropical treehoppers, taxonomic groups with similar phylogenetic constraints exhibit similar ecologies and modes of adaptive radiation. The same is true for groups in which many species show eruptive dynamics. We can compare forest Lepidoptera in the temperate and the tropical environments and ask if phylogenetic constraints dominate patterns in ecology, or whether the large differences in climate and host-plant species override the evolved background effects.

A long tradition in ecology and evolution argues that there are dramatic differences between the ecology and evolution of species and communities in temperate and tropical environments. This started with the Victorian naturalist explorers Humboldt, Darwin, Wallace, Bates, Müller, and others who expostulated over the exuberance of tropical life and the complexity of biotic interactions. All this was encapsulated by Dobzhansky (1950) in his well-known article "Evolution in the Tropics," in which he emphasized that the ecology and evolution of a species was predominantly determined by the impact of other species in the tropics, in contrast to the overriding influence of severe abiotic factors, such as weather, in temperate regions. This view was formalized by MacArthur and Wilson (1967) in their *r*-and-*K*-selection hypothesis. They argued that in temperate environments with harsh weather, selection worked to increase the number of progeny at the expense of quality care; so the population growth, *r*, was maximized. In the tropics, competition and natural enemies were so severe that increased parental investment was selected for, reducing numbers of progeny but increasing their quality and/or the amount of parental care. They termed this *K* selection, which is selection for a reproductive strategy maximizing survival in an environment filled to capacity, *K*, with species. Such views are commonplace today, meaning that many would predict substantially different ecologies for insect herbivores in the temperate and tropical latitudes.

In fact, there is very little empirical evidence in support of such diverging ecologies among insect herbivores (Price 1991a). This is based on several lines of evidence. (1) Competition can be readily found and is commonly reported from temperate latitudes, and the majority of such studies come from these latitudes (cf. Connell 1983; Schoener 1983, 1985). (2) Insect herbivores in one taxonomic group, such as butterflies and bruchid weevils, tend to be equally specialized, independent of latitude; most of them are parasites on plants with all the constraints that select for specialist life histories associated with the parasitic life style (Price 1980). (3) There is no evidence of greater niche compression among insect herbivores in the lower latitudes, and biological interactions appear to be as effective across the latitudinal gradient, based on experiments on biological control of insect herbivore pests (cf. DeBach 1964, 1974). (4) In addition, mortality inflicted on insect herbivores by parasitoids is about the same within each herbivore feeding type in temperate and tropical latitudes, based on a meta-analysis by Hawkins (1994). However, levels of parasitism are significantly different between the ecological niches utilized by different lineages, meaning that the phylogenetic differences in niche exploitation patterns have more impact on natural enemy attack than latitude. The number of parasitoid species per insect herbivore species even increases from tropical to temperate latitudes for externally feeding herbivores, while for endophytic species, parasitoid richness remains roughly the same along the latitudinal gradient (Hawkins 1994). Cornell and Hawkins (1995) also found that based on hundreds of life-table studies, the impact of neither competition nor natural enemies on insect herbivores was significantly different between temperate and

TABLE 13.1

Comparison of Temperate and Tropical Forest Lepidoptera Families with Species Having Eruptive Population Dynamics.

Family	*Temperate North America*	*Tropical Ghana*
Nymphalidae	1	1
Pieridae	1	0
Saturniidae	3	2
Lasiocampidae	5	3
Lymantriidae	7	4
Notodontidae	4	3
Arctiidae	3	6
Geometridae	16	3
Dioptidae	1	0
Noctuidae[a]	0	9
Hesperiidae[a]	0	1
Sphingidae[a]	0	1
Limacodidae[a]	0	1
Papilionidae[a]	0	1

NOTE: Based on Nothnagle and Schultz 1987 and Wagner et al. 1991.

[a]Families also in North American forests but with species of lesser importance (cf. Baker 1972; Furniss and Carolin 1977).

tropical latitudes, although the trend was for slightly greater mortality in the tropics. In a smaller study, Dyer and Coley (2002) found significant differences in the effects of predators on prey in temperate and tropical communities.

We are forced to conclude that differences between temperate and tropical environments are not as important as originally conceived in determining the evolution of ecological interactions for insect herbivores, and that such species are actually more similar in their ecologies than they are different!

The similarities in the ecology of insect herbivores in temperate and tropical latitudes are explained, simply, by the phylogenetic constraints hypothesis. As a taxonomic group has radiated around the world, the same phylogenetic constraints have prevailed in the evolution of similar adaptive syndromes and emergent ecologies. As an example, consider the forest insects of importance in North America and Ghana. The eruptive forest Macrolepidoptera of North America are listed by Nothnagle and Schultz (1987), and those in Ghana by Wagner et al. (1991). Much of the northern temperate forest of the United States is distributed in the range of about 40 to 50° N, while Ghana's forests are in equatorial Africa between 5 and 11° N. The climates are radically different, and the host-plant families utilized by the lepidopterans are completely different, except for a couple of introduced species. Despite these differences, the families of Lepidoptera in these extremely different environments are remarkably similar (Table 13.1). Seven out of 12 families are the same in temperate and tropical habitats. The similarities are much greater when forest insects of lesser importance are included: then the overlap becomes close to 100% (cf. Baker 1972; Furniss and Carolin 1977). This demonstrates that although host-plant families have not radiated into many parts of the world, insect families have radiated. Those families with phylogenetic constraints and adaptive syndromes that promote emergent properties of eruptive population dynamics show these chains of influence both in temperate and tropical latitudes. The strength of the phlogenetic constraints overrides the influence of host-plant and climatic differences, making insect herbivores more similar over this gradient than they are different.

One pairwise comparison utilizing the magnificent emperor moths (Lepidoptera: Saturniidae) will provide an example of similar dynamics in temperate and tropical latitudes. In the northern pine forests, the pandora moth, *Coloradia pandora* Blake, is an eruptive species (Furniss and Carolin 1977), and the mopane worm, *Imbrasia belina* (Westwood), outbreaks in the mopane woodland belt across southern Africa from Namibia and Botswana to Zimbabwe, Mozambique, and South Africa (15 to 25° N). The host tree for the pandora moth is ponderosa pine, *Pinus ponderosa* Lawson, and relatives, and the leguminous mopane tree, *Colophospermum mopane* (J. Kirk ex Benth.) J. Léonard, and others for the mopane worm. So their host trees are vastly different, except for being in widespread monocultures, but the ecology and life history of the herbivores remain similar. Both lepidopteran species are eruptive, inflicting heavy to complete defoliation periodically on their host plants. Both lay eggs in clusters on leaves or bark, showing no apparent ovipositional preference for high-quality foliage beneficial to larval establishment, growth, and survival. Epidemics of the pandora moth occur in the western United States at intervals of about 20 to 30 years, lasting about 6 to 8 years per eruption (Furniss and Carolin 1977), and Native Americans are known to eat the larvae (Blake and Wagner 1987). The mopane worm appears to erupt more frequently, which is fortunate for native Africans who use the large caterpillars as an important source of protein (Skaife 1979; Scholtz and Holm 1985; Menzel and D'Aluisio 1998).

Other examples of eruptive species showing similar dynamics in a wide range of latitudes include acridid grasshoppers, whose species range from Australia, through tropical Africa, India, and into North America (reviewed in Price 1997). Most of the severely eruptive species are called locusts. Acridid grasshoppers oviposit into the soil away from nymphal feeding sites both in time and space, showing no ovipositional preference for sites in relation to food quality for first-instar nymphs. Indeed, food quality is often dictated by rainfall after oviposition, so that females could not evolve with a preference for young, rapidly growing grass and forb foliage (Nailand and Hanrahan 1993; Showler 1995).

Ovipositing into soil, as in the acridid grasshoppers, is frequently associated with eruptive dynamics. This is the case for leaf beetles (Coleoptera: Chrysomelidae) in the subfamilies of flea beetles (Alticinae), and the root worms

TABLE 13.2

Examples of Gall-Inducing Insect Herbivores from Temperate and Tropical Latitudes, All Showing a Positive Response to Module size or Plant Vigor

Insect Species (Order: Family)	*Host-Plant Species*	*Locality*	r^2	*Reference*
Temperate Latitudes				
Diplolepis fusiformans (Hymenoptera: Cynipidae)	*Rosa arizonica*	Flagstaff, Arizona	0.84	Caouette and Price 1989
Pemphigus betae (Homoptera: Aphididae)	*Populus angustifolia*	Weber R., Utah	0.88	Whitham 1978
Amblypalpis olivierella (Lepidoptera: Gelechiidae)	*Tamarix nilotica*	Caesarea, Israel	0.96	Price and Gerling 2004
Euura lappo (Hymenoptera: Tenthredinidae)	*Salix lappo*	Kevo, Finland	0.62	Price et al. 1994
Pontania amurensis (Hymenoptera: Tenthredinidae)	*Salix miyabeana*	Sapporo, Japan	0.67	Price et al. 1999
Tropical Latitudes				
Contarinia sp. (Diptera: Cecidomyiidae)	*Palicourea rigida*	Brasilia, Brazil	0.89	Vieira et al. 1996
Schizomyia bauhiniae (Diptera: Cecidomyiidae)	*Bauhinia brevipes*	Tres Marias, Brazil	up to 0.78	Fernandes 1998
Anadiplosis nr. *venusta* (Diptera: Cecidomyiidae)	*Machaerium angustifolium*	Belo Horizonte, Brazil	0.92	Price 2003
Cecidomyiid sp. (Diptera: Cecidomyiidae)	*Platypodium elegans*	Belo Horizonte, Brazil	0.81	Price 2003
Psyllid sp. (Homoptera: Psyllidae)	*Myrcia itambensis*	Belo Horizonte, Brazil	0.68	Price et al. 1995

NOTE: Adapted from Price 2003. r^2 is the square of the correlation coefficient, providing an estimate of the variation accounted for by the regression of host-plant shoot length as the independent variable and the insect numerical response to shoot length as the dependent variable. All r^2 values are significant at the $P < 0.05$ level, and most at $P < 0.01$.

and others in the Galerucinae (Price 2003). Some froghoppers (Homoptera: Cercopidae) in the tropics also oviposit into soil, making an interesting contrast to their relatives, the treehoppers. Some of these cercopids are serious pests of grasses, demonstrating eruptive dynamics in Brazil and other South American countries. Sixteen species in six genera of cercopids are listed in Pires et al. (2000), all of which oviposit into the soil, plant debris, or dead leaf sheaths. Therefore, they cannot evaluate plant quality for nymphs. Added to this is the aestivation of eggs through the dry period of the year, with dormancies lasting almost six months from late May into October (Fontes et al. 1995). Direct tests of ovipositional preference and larval performance have been conducted on one neotropical spittlebug, *Deois flavopicta* Stal, which confirm the lack of a preference-performance linkage in relation to food quality for nymphs in this eruptive species (Pires et al. 2000).

Similarities of temperate and tropical insect herbivores are also observed among the rich faunas of gall-inducing insects. These show patterns similar to *Euura lasiolepis*, with a positive response of oviposition to rapidly growing plant tissues on longer classes of shoot modules (Table 13.2). Where studied, there is an ovipositional preference and larval performance linkage, no matter where the species occur, from the cerrados, or savannas, of Brazil in tropical latitudes to Alaska, Finland, and Sweden in the north. The gall-inducing insects provide strong support for the plant vigor hypothesis (Price 1991b), which proposes that many insect herbivore species benefit more from plant vigor than from plant stress.

We also observe positive responses by various kinds of insects to vigorous plant growth in the tropics, just as in temperate regions. In northern latitudes the spring flush of shoots is favorable to many species, stimulating the proposal of the plant vigor hypothesis (Price 1991b). And after fire, a dominant factor in the cerrados of Brazil, insect herbivores quickly colonize fire-adapted plant species exhibiting rapid growth responses, while adjacent unburned areas support little or no fresh herbivore activity (e.g., Prada et al. 1995; Seyffarth et al. 1996; Vieira et al. 1996). Thus, we see a flush of insect herbivore activity after fire, and a patchy distribution because fire is local, resulting in a mosaic over the landscape of vegetation in different stages of response to fire, which is not synchronous, and an equivalent checkerboard pattern of insect herbivore distribution and abundance. Also, similar kinds of insects are observed to colonize rapid plant regrowth after fire, or after clear-cutting on the whole latitudinal gradient: aphids, leafhoppers, leaf beetles, and gall-inducers, to mention a few groups.

Repeatedly, we observe close similarities among insect herbivores in temperate and tropical latitudes. In general, species and higher taxa appear to be much more similar than they are different in relation to the insects' responses to plant traits and their ecology (distribution, abundance, and population dynamics). I took a sabbatical leave in Brazil with the expectation that I would find patterns that contrasted with those I had observed in temperate regions. I thought that such contrasts would provide a new perspective that might enable an expansion of the phylogenetic constraints hypothesis. However, in every case I learned of, or studied myself, conformity to the hypothesis was the norm.

In the past, the effects of insect herbivores on plants have been compared across latitudes, perhaps more than studying the life histories and ecological interactions of the actual herbivores. When plant damage is compared, significant differences have been found (Coley and Aide 1991; Coley and Barone 1996; Dyer and Coley 2002). Thus, there is an interesting contrast between the insect characteristics themselves, and the impact that they have on the vegetation.

Adaptive Radiation in Temperate and Tropical Environments

We can now see that the adaptive radiation of insect herbivores has proceeded to populate most of the world with related species inextricably linked to associated evolutionary and ecological traits. As examples, eruptive moths (Lepidoptera), gall-inducing midges (Diptera: Cecidomyiidae), leaf-mining flies (Diptera: Agromyzidae), moths (e.g., Gelechiidae, Gracillariidae), and grasshoppers (Orthoptera: Acrididae) have all carried with them the phylogenetic constraints of their lineage and therefore have the evolutionary potential to perform in similar ways. While the plant lineages have diverged substantially in evolutionary time, resulting from their more limited dispersal, the flying insects have colonized plants very broadly across plant groups.

However, the ways in which adaptive radiation may have proceeded in the insect herbivores could be very different. In the species with high preference-performance linkage and latent dynamics (e.g., adaptive radiation of common sawflies), shifts into new adaptive zones may involve simple ovipositional errors, and colonization of new ecological niches could be opportunistic and within the same habitat. Both processes in adaptive radiation could be relatively rapid and may well involve sympatric speciation. This is suggested by the presence of many sibling, or cryptic, species in the gall-inducing sawflies (Roininen et al. 1993; Kopelke 2000, 2001; Nyman 2002), and the large size of genera in this group. The galling sawflies are represented by few genera (*Phyllocolpa*, *Pontania*, and *Euura*), but many species per genus, suggesting that speciation has been rapid without enough morphological divergence to justify generic distinctions. In contrast, during geographic speciation, we would expect large differences between species and the evolution of more genera with fewer species. When species simply shift from one related host plant to another, as the sawflies have shifted among willow species, apparently very little selection is imposed on morphological change, making the identification of good species and their taxonomy problematic. These kinds of patterns and problems are seen in many specialist plant-feeding insect groups such as the fruit flies (Diptera: Tephritidae) (Smith and Bush 2000) and membracids (Wood 1980), in which host shifts have been shown to result in speciation (e.g., Schwarz et al. 2005). Many other cases are discussed by Berlocher and Feder (2002), where host shifts have resulted in morphologically similar but evolutionarily distinct insect populations. Roininen et al. (2005) estimated that the number of species of gall-inducing sawflies per genus ranges from tens to over a hundred (*Phyllocolpa* 35–80 species, *Pontania* 80–230 species, *Euura* 40–100 species).

The scenario for adaptive radiation among the forest Lepidoptera is likely to be very different from that for those species with latent population dynamics. Generalist feeders, adapted to extensive northern temperate forests and with wide geographic ranges, are more likely to speciate allopatrically, with host shifts not playing such an important role in the speciation process. Many genera occupy the same adaptive zone of free-feeding larvae on forest tree leaves, and many tree species may provide suitable food. For example, the forest tent caterpillar, *Malacosoma disstria* Hubner, "feeds on a wide variety of hardwoods" (Baker 1972, pp. 330–331), as does the fall cankerworm, *Alsophila pometaria* (Harr). Several genera exist, but they are small and often have only one to six species per genus, suggesting slow speciation but significant divergence into new genera as radiation proceeds (e.g., in North America: *Alsophila,* one species; *Lymantria,* one species; *Operophtera,* four species; *Orgyia,* four species; and *Malacosoma,* six species [Arnett 1993]).

Similar patterns in species per genus hold in the tropics. In the gall midges, for example, genera again are often large, spreading from the tropics into temperate regions (e.g., *Asphondylia,* 260 described species and 82 Neotropical species; *Contarinia,* 300 species; *Rhopalomyia,* 135 species including 9 members so far described in the tropics [Gagné

1994]). While some of these taxa are "catchall" genera, the necessity for such agglomerations of species indicates that much speciation occurs without the establishment of distinctive generic boundaries, and this supports the concept of the rapid generation of new species (cf. Gagné 1989). Most species "are narrowly adapted and synchronized to their hosts" (Gagné 1989, p. 37), implying that host shifts could result in sympatric speciation due to allochronic isolation. At the other extreme are the moth families and genera in which eruptive species are found. In the group of greater emperor moths to which the mopane moth belongs, the mean genus size is 3 species, with 17 genera included (cf. Pinhey 1972). In Australia, with 70 species in 16 genera in the tussock moths (Lymantriidae) the mean number per genus is about 4 species (cf. Common 1990). This pattern may be widespread, as the same families of macrolepidopterans appear in various parts of the tropics as they do in Ghana (Table 13.1). A short list provided by Speight and Wylie (2001) includes an arctiid in South America, geometrids in China and Brazil, a lasiocampid in China, Vietnam, and Thailand, a lymantriid in Papua New Guinea, a noctuid in India, Southeast Asia, Australia, and Oceania, a notodontid in Brazil, and a pierid in Africa, Southeast Asia, Australia, and Oceania. This list includes many families noted in Table 13.1. Forest tree pests in Indonesia include the families Pieridae, Lymantriidae, Geometridae, and Noctuidae, which are also present in Ghana (Table 13.1) (Nair et al. 1996; Nair 2000). Around the world, the same families are represented by eruptive species, and these are in small genera.

When emphasizing the two contrasting kinds of species, one group with high preference-performance linkage and latent population dynamics, and the other with no preference-performance linkage and the potential for eruptive population dynamics, only the extremes of a continuum between the two have been discussed. Many species, and perhaps most, occur somewhere along the continuum between these archetypes of evolutionary history and ecology. Filling in the middle ground is challenging because, as gradations in phylogenetic constraints and adaptive syndromes occur, subtle differences probably exert important effects on ecological interactions. Furthermore, it is these subtleties that are rarely recorded in the literature, meaning that specific and directed research is needed to address questions on the traits resulting in intermediate kinds of ecology on the continuum. Particularly critical is the female's morphology and behavior when placing eggs, and its relationship to available high-quality food and a place to live for larvae. This is because the establishment of immatures in a safe and nutritious feeding site is a critical episode in the life history for the majority of insect species. Unfortunately, the details of oviposition behavior, oviposition cues and their variation in host plants, the range in substrates utilized, and the time and distance to food for first-instar nymphs or larvae are all pertinent but are known for very few species. In addition, these details need to be known on a broad species basis so that the roles of each facet of female behavior, and their consequences can be compared.

In the meantime, we have the phylogenetic constraints hypothesis, which provides a basis for comparison, pattern detection, and theoretical development and has the following benefits, which were formerly absent and/or poorly developed in the literature: (1) The hypothesis provides a strongly comparative approach to the central themes of ecology on distribution, abundance and population dynamics, with high potential for detecting patterns in nature. (2) There is a strong evolutionary emphasis, centered on the basic evolved attributes of individuals and species, especially female morphology and behavior. (3) The macroevolutionary scenarios developed explain the macroecological patterns observed. (4) The ecological characteristics of species and groups, especially population dynamics, are brought into the bosom of the evolutionary synthesis for the first time. (5) The approaches adopted are applicable at all latitudes, from the temperate to the tropical, because phylogenetic patterns are so strong and predictable and evidently have more influence than variations in climate and host-plant species and families. (6) The hypothesis is pluralistic, being able to encompass various starting points in the phylogenetic constraints and a range of end points in latent and eruptive dynamics. (7) The hypothesis is predictive, because once the constraints are known, likely adaptive syndromes can be envisaged, as well as their resulting emergent properties: we can predict which kinds of species and groups are likely to be latent and which eruptive. (8) There is good potential for pattern detection and the mechanistic explanation of pattern, permitting the development of theory that accounts for a wide variety of ecological patterns in herbivorous insects, especially distribution, abundance, and population dynamics.

Acknowledgments

I am most grateful to Kelley Tilmon for her invitation to contribute to the symposium and book, and for her editorial comments. Alexey Zinovyev and Heikki Roininen provided invaluable expertise on sawfly ovipositors and oviposition, and Hubert Pschorn-Walcher contributed references on the swelling of sawfly eggs. Reviews provided by Rex Cocroft, Timothy Craig, Heikki Roininen, Tommi Nyman, and an anonymous reviewer contributed importantly to this chapter. I am grateful to Tommi Nyman for permission to use his phylogeny of the gall-inducing sawflies.

References Cited

Arnett, R. H. 1993. American insects: a handbook of insects of America north of Mexico. Sandhill Crane Press, Gainesville, FL.

Baker, W. L. 1972. Western forest insects. U.S. Dept. Agr. For. Serv. Misc. Pub. 1175. U.S. Government Printing Office, Washington, DC.

Benson, R. B. 1950. An introduction to the natural history of British sawflies (Hymenoptera: Symphyta). Trans. Soc. Brit. Entomol. 10: 45–142.

clades of herbivores undergo speciation more rapidly than their nonherbivore counterparts (Mitter et al. 1988). The bottom line is that herbivore biology strongly promotes evolutionary diversification (Funk et al. 2002).

Sequential Radiation

What has been little explored is whether diversification of herbivores in response to their host plants causes differentiation of their natural enemies. Does the genetic differentiation of an herbivorous insect create a new resource that when exploited by a natural enemy causes that natural enemy to undergo genetic and/or behavioral differentiation itself? In this chapter, we explore five examples in which the genetic diversification of herbivores has created new resource opportunities that have been exploited by natural enemies, which subsequently have undergone differentiation via host-race formation themselves.

We have previously termed the process of herbivore differentiation causing natural-enemy divergence "sequential radiation" (Abrahamson et al. 2003). In this sense, sequential radiation is characterized by a diversification of taxa (e.g., natural enemies) farther up a trophic web in response to diversification of their hosts (in this case, of herbivorous insects), which differentiated as a consequence of shifts to novel host plants.

However, sequential radiation could also be used to identify instances in which the diversification of an organism at one trophic level results in diversification of organisms that are lower in a trophic web (e.g., endosymbionts of herbivores or mycorrhizae of host plants). For example, sequential radiation could conceivably describe the reasonably well-understood process of cospeciation (e.g., Clayton et al. 2003; Hafner et al. 2003; Page 2003), such as with the vertically transmitted endosymbiont *Buchnera aphidicola*, which is differentiating in response to its pea aphid *(Acyrthosiphon pisum)* host's specialization (Baumann et al. 1995; Via 1999, 2001; Shigonebu et al. 2000; Simon et al. 2003). The pea aphid has formed genotypic races, which attack different host plants (e.g., pea, alfalfa, clover) (Via 1999; Simon et al. 2003). In response to the aphid's host-race formation, *B. aphidicola* is differentiating in parallel. The evolution of these organisms is linked by their mutualistic association, as neither can reproduce in the absence of the other (Baumann et al. 1995). *Buchnera aphidicola* is so specialized that it cannot survive outside the eukaryotic cell; it resembles an organelle. As a consequence of the tight mutualism, which includes both vertical transmission and obligate endosymbiosis, cospeciation results (Shigonebu et al. 2000).

The situation is quite different for free-living organisms, where the influence of the host is less overwhelming. The parthenogenetic pea aphids are not known to cross-infect each other with *Buchnera,* so there is no option for host shifting. However, the radiation of differences up the trophic levels in free-living organisms is more complex and less inevitable.

In this chapter, we focus on sequential radiation as it describes an escalation of biodiversity up a trophic system based on differentiation of an herbivore via host shifts among host plants and subsequent differentiation by a natural enemy. Hence, the examples that we present are not simply cases of cospeciation in which the speciation process occurs in two interacting species in parallel. Each example involves a three-trophic-level interaction, and in all but one example, one of the trophic levels (herbivore) has constructed a new niche and/or "engineered its ecosystem" with subsequent exploitation of created resources and host-race formation.

In addition to examining five examples of natural-enemy differentiation in response to the diversification of their host herbivore, we also explore one example that appears similar, yet no differentiation farther up the trophic web has occurred. Comparison of cases where sequential radiation is present versus absent should shed light on the conditions necessary for escalation of biodiversity. Unfortunately, we have few examples at this point to draw upon, and as a result our ability to determine the conditions that promote reproductive isolation, host-race formation, and speciation are limited. As a consequence, the conclusions we reach must be viewed as tentative as they require testing in additional studies. Nonetheless, our exploration will illustrate a potentially common source of biodiversity and will emphasize the contribution that evolutionary ecologists and population biologists can make to explain "what determines species diversity." Our goals in reviewing these examples are (1) to illustrate a potentially common source of biodiversity among insects and (2) to identify the conditions that promote sequential radiation.

Conditions for Host-Race Formation

The circumstances that are probably relevant to host-race formation in natural enemies are likely to be the conditions that apply to host-race formation in general. Studies of sympatric speciation among insects have examined the setting of host-race formation. Jaenike (1981), Diehl and Bush (1984), Tauber and Tauber (1989), Bush (1993), Itami et al. (1997), and Abrahamson et al. (2001) have summarized a number of conditions that are often present in host races:

1. Host-race formation and speciation involve a shift to a new habitat or resource. In moving to a new host, species can modify their own niche and occupy formerly unoccupied niche space (niche construction) or, alternatively, modify the environment ("ecosystem engineering"), unintentionally facilitating the production of niches for other species. When host races form via shifts to new resources, gene flow among host-associated populations can be reduced, and genetic differentiation between sympatric, host-associated populations becomes more likely.

2. Habitat selection and fidelity to a host are under genetic control, although in some examples, host fidelity is subject to modification through conditioning to the host.
3. Females determine the larval host and discriminate among hosts for oviposition. Preferences in oviposition on different hosts by host races are essential to a successful host shift and subsequent host-race formation.
4. Insect phenology is related to host-plant phenology. Differences in the phenologies of hosts can promote reproductive isolation among organisms associated with different hosts. Host-associated emergence differences of only a few days can dramatically affect the availability of mates from the same or alternative hosts because many adult insects are short-lived.
5. Fitness is related to differences in host-associated traits. For example, individuals in a population on a derived host may have a fitness advantage over individuals on the ancestral host in spite of poor adaptation to the new host if individuals on the derived host have shifted into enemy-reduced space or to a better-quality resource.
6. Mate choice is dependent on habitat or resource selection. Specifically, insect males and females use their hosts as a rendezvous for courtship and mating. The coupling of mating and habitat or host choice facilitates assortative mating via host selection and can rapidly establish ecological isolation between the ancestral and the derived populations. Herbivorous insects in which mate choice is directly dependent on habitat or resource choice represent some of the best-documented cases of evolutionary diversification and sympatric speciation.

In the following section, we explore five cases of natural-enemy differentiation and one case of no differentiation to identify which, if any, of the above conditions, determined from studies of herbivores, are also present during the differentiation of natural enemies.

Cases of Sequential Radiation via Host-Race Formation

Case 1: Host-Habitat Fidelity

Herbivore Radiation *Rhagoletis pomonella* (Diptera: Tephritidae) and its sibling species, *R. mendax, R. zephyria,* and *R. cornivora*, are morphologically similar yet they exhibit strong association with different host plants (Frey et al. 1992; Feder et al. 1994; Berlocher 2000; Linn et al. 2003, 2004; Schwarz et al. 2005). For example, the blueberry maggot *R. mendax* attacks the fruits of ericaceous shrubs such as highbush blueberry *(Vaccinium corymbosum),* while the apple maggot *R. pomonella* attacks hawthorn (*Crataegus* spp.) and apples *(Malus pumila)* (Table 14.1). The most-often cited example of sympatric speciation via host shifts involves host-race formation within the apple maggot. During the past 150 years, this fly has formed genetically divergent host races as a consequence of a host-plant shift from its ancestral hawthorn host to introduced, domestic apples (Bush 1969, 1994; Feder et al. 1993). The entire *Rhagoletis* species complex is believed to have diversified in sympatry through shifts to new hosts. This diversification has been facilitated by their striking host fidelity, which produces premating reproductive isolation among diverging and existing species. Host fidelity in *Rhagoletis* is based on the fly's strong host-associated mating and oviposition behaviors, both of which occur on unabscised host fruits (Feder et al. 1994).

The Natural Enemy The wasp *Diachasma alloeum* (Hymenoptera: Braconidae) is a specialist endoparasitoid that is known to attack only the third-instar larvae of the apple maggot (Glas and Vet 1983) and second-instar larvae of the blueberry maggot (Stelinski et al. 2004). *Diachasma alloeum* seems to be tracking both the herbivore and its host fruit. Experimental studies using hawthorn fruits suggest that *D. alloeum* finds its host by locating unabscised fruit and that visual cues are important to the process (Glas and Vet 1983). However, cues from the host insect are also important. Studies using apple maggot–infested as well as uninfested hawthorn fruits show that the time spent on a fruit and the level of fruit-probing activity by mated *D. alloeum* females are strongly affected by the presence of an apple maggot larva within the fruit. *Diachasma alloeum* females exhibited nonrandom probing when they explored hawthorn fruits containing actively gnawing apple maggot larvae; in contrast, *D. alloeum* females showed random probing on fruits that contained parasitized and hence paralyzed apple maggot larvae (Glas and Vet 1983). Similarly, field observations of host foraging by *D. alloeum* on highbush blueberry fruits showed that this endoparasitoid alighted more frequently on clusters of fruits infested with blueberry maggot larvae than on uninfested clusters or mechanically damaged fruit (Stelinski et al. 2004).

Stelinski et al. (2004) posited that ovipositing *D. alloeum* females are attracted by either plant volatile compounds released by blueberry maggot–infested blueberries but not mechanically damaged fruits, or acoustic signals given off by larvae as they gnaw and tunnel through infested fruits. The timing of the host was also important. The majority of *D. alloeum* females were attracted to host-infested blueberries 15 to 21 days after *R. mendax* females had oviposited into fruit. Such natural-enemy attentiveness to plant cues, alteration of host-plant traits, or host resources is likely a frequent correlate of differentiation of natural enemies in response to their host's differentiation.

Sequential Radiation *Diachasma alloeum* wasps appear to have formed host races by tracking the speciation of their hosts, apple maggots on one hand or blueberry maggots on the other. Conditions that might lead to assortative mating live beetle emergence hardly occurred in the gall beetles on sympatric *S. altissima* host plants.

Beetles from the two hosts have shown behavioral evidence of reproductive isolation (Eubanks et al. 2003). In no-choice experiments, pairs of beetles reared from different host plants were significantly less likely to mate than pairs reared from the same host. In choice experiments, beetles preferred mates from their own host. These experiments were conducted in the absence of the host plants and thus lead to no conclusions about one of the traits expected to lead to host-race formation: the linking of mate choice and host choice through mating on the host. Where the beetles mate in nature is unknown. Assortative mating in the absence of the host is a surprising finding in host races whose slight genetic differentiation suggests that their impediments to gene flow are recent.

Mordellistena convicta females choose the oviposition site, a trait common to insects that form host races. There are suggestions that females prefer their natal host for egg laying. When beetles of both sexes were caged on galls of either their host or nonhost and allowed to mate and lay eggs, offspring were recovered the following spring only from host galls. When beetles were allowed to mate and oviposit while caged with both host and nonhost plants, significantly more beetles were produced from host-plant galls: 84% of the *S. altissima* beetle cages and 82% of the *S. gigantea* cages produced offspring only from the host galls (Eubanks et al. 2003). These experiments likely indicate ovipositional host preference, although host-related offspring performance may also be involved.

Ecosystem engineering by the gall flies provided new niches for two rounds of beetle differentiation: between the stem beetles and a subset of that population that shifted to galls and also between the populations in the two host galls. Differentiation between the gall fly host races has radiated up the trophic web to the beetles that have also formed host races in the *Eurosta* galls. Gall-niche-related selection seems to be the driving force behind gall-beetle differentiation on *S. altissima* and *S. gigantea*: beetle phenology is related to host-gall phenology, and fitness is related to differences in host-gall-influenced traits such as final beetle mass. Similar to the situation with the kleptoparasitic *Koptothrips* who also inhabit another's gall, the same host-related selection that drives gall-inducer host-race formation drives the process in the beetle. Once the gall-inducer alters the plant environment, others are attracted to the new niches and differentiate as the result of the niches being on separate plants.

Case 4: A More Distant Relationship with the Host Gall

The Herbivore *Rhopalomyia solidaginis* (Diptera: Cecidomyiidae) and its closely related sister species *R. capitata* form galls on *S. altissima* and *S. gigantea*, respectively, as do the host races of *Eurosta solidaginis* (McEvoy 1988; Stireman et al. 2005) (Table 14.1). Analysis of mtDNA sequences from several sympatric populations of *Rhopalomyia* on the two goldenrods in the Upper Midwest of the United States revealed host-related differentiation at the COI locus consistent with cryptic sibling species (10.2% net sequence divergence) (Stireman et al. 2005). As in *Eurosta*, the *S. gigantea* host form seems to be the derived one because it exhibits far less genetic variation. Although female *Rhopalomyia* from the two host species are indistinguishable, there are interspecific differences in male genitalia and gall morphology. On both plant species, there are two generations of *Rhopalomyia*, one emerging in spring and one in late summer (McEvoy 1988; Netta Dorchin, 2005, personal communication). Adults of the spring-emerging generation lay their eggs in the apical meristem of the host plants to induce the summer galls, whereas adults of the late-summer generation lay their eggs in the ground near the plant (McEvoy 1988).

The Natural Enemy On both host plants, eggs laid by the spring generation in the apical meristems are attacked by *Platygaster variabilis* (Hymenoptera: Platygastridae), a polyembryonic egg parasitoid. Whether *P. variabilis* attacks the eggs of the following generation is yet to be determined (Netta Dorchin, 2005, personal communication). Not much is known about *P. variabilis* behavior, including whether this species mates on the herbivore's host plant. Since *P. variabilis* is an egg parasitoid, the ovipositing females need to orient to midge eggs in the apical meristem prior to the formation of the gall.

Sequential Radiation *Platygaster variabilis* appears to be differentiating along the same lines as *Rhopalomyia*. Maximum likelihood analysis of mtDNA from wasp broods collected from *Rhopalomyia* apical galls in Minnesota, Iowa, Nebraska, and South Dakota indicate that *P. variabilis* is divided into two clades, one attacking *Rhopalomyia* on *S. altissima* and one on *S. gigantea* (Stireman et al. 2006). As is the case with its midge host, within-clade genetic diversity is much higher for the *S. altissima* clade of *P. variabilis* than for the *S. gigantea* clade. Likewise, the degree of differentiation suggests morphologically cryptic sibling species. Stireman et al. (2006) describe the differentiation of the *P. variabilis* parasitoid in response to the differentiation of its host *Rhopalomyia* as "cascading host-associated differentiation." Although these authors restrict cascading host-associated differentiation to the host-related diversification of parasitoids, the pattern depicted is the same pattern we describe as sequential radiation.

Little is known about factors in *P. variabilis* that might lead to host-race formation. The species does have at least one known host-race related trait: the ovipositing female chooses the host egg so that a genetic change in the female's host preference would be sufficient to initiate a host shift. Similar to *Diachasma alloeum*, the female *P. variabilis* must orient to the correct host plant to find her prey. Thus, chemical differences between the host-plant species could influence host-race formation in this parasitoid. *Platygaster variabilis* phenology is likely influenced by the timing of egg laying in *Rhopalomyia*, but how narrowly constrained *Rhopalomyia* egg laying is by plant phenology is unknown.

Unlike *M. convicta* and *Koptothrips* species, which also attack gall-inducers, *P. variabilis* larvae do not consume gall tissue. As an endoparasitoid, *P. variabilis* contacts the host gall only indirectly through the herbivorous *Rhopalomyia* larva that is eating the plant as it is itself being eaten. Thus it need not adapt to eating tissue from two different hosts. Although it does not eat the gall, *P. variabilis* benefits from niche construction and ecosystem engineering by the gall midge. As an endoparasitoid of the midge larva, it needs the gall for development of its prey and therefore itself. But unlike the inquiline *M. convicta* and the kleptoparasite *Koptothrips* species, this parasitoid is not attracted by the gall itself, and gall features are unlikely to affect its differentiation.

Case 5: Recurring Local Sequential Radiation

The Herbivore The moth *Gnorimoschema gallaesolidaginis* (Lepidoptera: Gelechidae) also forms galls on both *S. altissima* and *S. gigantea* (Table 14.1). Females lay eggs in the autumn on the underside of goldenrod leaves, and larvae overwinter in the eggs from which they hatch in the spring. They crawl to a new goldenrod shoot and burrow laterally into the terminal bud and down into the stem where their feeding stimulates gall formation (Leiby 1922a). Analysis of 12 allozyme loci from six sites in the Midwest revealed host-associated differentiation consistent with reduced gene flow between the moth populations on *S. altissima* and *S. gigantea* (Nason et al. 2002). An F_{ST} estimate of 0.16 between the two host-associated populations lies within the range common to insect herbivore host races. Nason et al. (2002) speculate that the two populations are either well-established host races or young cryptic species.

The Natural Enemy *Gnorimoschema gallaesolidaginis* is parasitized on both goldenrod plants by *Copidosoma gelechiae* (Hymenoptera: Encyrtidae), a polyembryonic parasitoid wasp that attacks the moth eggs in the autumn (Leiby 1922a). Multiple *C. gelechiae* larvae (about 165 per moth larva) develop within the overwintering moth larva and remain within it when it forms its gall. Larval development and pupation in the parasitoid is closely aligned to development in the moth (Leiby 1922b). When the moth dies in its final instar, *C. gelechiae* larvae pupate in the moth mummy until late summer, when they emerge as adults and gnaw through the gall to parasitize the new generation of moth eggs (Patterson 1915; Leiby 1922b).

Sequential Radiation Analysis of nine polymorphic enzyme loci revealed host-associated genetic differentiation in large samples of *C. gelechiae* from *Gnorimoschema* galls in intermixed stands of *S. altissima* and *S. gigantea* from three geographically distant areas: New Brunswick, Ontario, and Minnesota (Stireman et al. 2006). Within each of the sympatric sites, the wasp exhibited host-related genetic differentiation. However, the divergence was caused by a different group of allozyme loci at each site, suggesting that differentiation was local and developed independently in each area. Also, the differentiation at each site was minor ($F_{ST\ site}$ = 0.002–0.05), which indicates recent divergence. So although the herbivore has only two host forms spread throughout the sampled area, sequential radiation, or as termed by Stireman et al. (2006), cascading host-associated differentiation, in *C. gelechiae* seems to have produced multiple local host-related forms in the recent past. Repeated local host differentiation that affects allozymes suggests an almost inevitable process; rather than occurring only once and spreading, differentiation seems to occur repeatedly.

Like *P. variabilis*, *C. gelechiae*, except when chewing its way out as a newly emerged adult, is not required to digest plant tissue from two divergent herbivore hosts. And the emergence chewing is brief because *C. gelechiae* adults use the exit tunnel created by the gall moth larva and are merely required to nibble through the final tissue layer to the outside (Leiby 1922b). Thus, again like *P. variabilis*, *C. gelechiae* benefits from environmental engineering only insofar as the gall offers it protection during development. But *C. gelechiae* does have a relationship to the herbivore's host plant. Like *P. variabilis*, it must find the host plants on which the gall moth has laid its eggs, most likely through chemical tracking, a process requiring plant recognition. The *C. gelechiae* female, like the *D. alloeum* parasitoid that attacks the apple and blueberry maggots, has considerable contact with the herbivore's host plant, crawling about the goldenrod leaves and stems, searching assiduously until her antennae contact an egg (Leiby 1922b). The need to find and search the correct plant may lead to the host fidelity characteristic of differentiating host races. But determination of the mechanisms of divergence require behavioral and ecological data not yet available (Stireman et al. 2006).

Example Showing No Differentiation

Case 6: Goldenrod Gall Flies and a Parasitoid Wasp—Probably No Host-Race Formation

The parasitoid *Eurytoma gigantea* (Hymenoptera: Eurytomidae) oviposits into the central chambers of fully grown galls induced by the goldenrod gall fly *Eurosta solidaginis* (Weis and Abrahamson 1985) (Table 14.1). As described above, the gall fly has formed genetically and behaviorally distinct races on its host plants *S. altissima* and *S. gigantea*. Cronin and Abrahamson (2001) tested the hypothesis that *E. gigantea* has formed host races in direct response to the host shift and subsequent host-race formation of *E. solidaginis* by determining emergence times, mating preference, and female oviposition preference for parasitoids derived from galls of each gall fly host race.

Glas, P. C. G., and L. E. M. Vet. 1983. Host-habitat location and host location by *Diachasma alloeum* Muesebeck (Hym.; Braconidae), a parasitoid of *Rhagoletis pomonella* Walsh (Dipt.; Tephritidae). Neth. J. Zool. 33: 41–54.

Hafner, M. S., J. W. Demastes, T. A. Spradling, and D. L. Reed. 2003. Cophylogeny between pocket gophers and chewing lice, pp. 195–220. In R. D. M. Page (ed.), Tangled trees: phylogeny, cospeciation, and coevolution. University of Chicago Press, Chicago.

Hess, M. D., W. G. Abrahamson, and J. M. Brown. 1996. Intraspecific competition in the goldenrod ball-gallmaker *(Eurosta solidaginis):* larval mortality, adult fitness, ovipositional and host-plant response. Am. Midl. Nat. 136: 121–133.

Itami, J. K., T. P. Craig, and J. D. Horner. 1997. Factors affecting gene flow between the host races of *Eurosta solidaginis*, pp. 375–407. In S. Mopper and S. Y. Strauss (eds.), Genetic structure and local adaptation in natural insect populations: effects of ecology, life history, and behavior. Chapman and Hall, New York.

Jaenike, J. 1981. Criteria for ascertaining the existence of host races. Am. Nat. 117: 830–834.

Jones, C. G., J. H. Lawton, and M. Shachak. 1997. Positive and negative effects of organisms as physical ecosystem engineers. Ecology 78: 1946–1957.

Kranz, B. D., M. P. Schwarz, L. C. Giles, and B. J. Crespi. 2000. Split sex ratios and virginity in a gall-inducing thrips. J. Evol. Biol. 13: 700–708.

Leiby, R. W. 1922a. Biology of the goldenrod gall-maker *Gnorimoschema gallaesolidaginis* Riley. J. New York Entomol. Soc. 30: 81–94.

Leiby, R. W. 1922b. The polyembryonic development of *Copidosoma gelechiae* with notes on its biology. J. Morph. 37: 194–249.

Linn, C. E. Jr., J. L. Feder, S. Nojima, H. R. Dambroski, S. H. Berlocher, and W. L. Roelofs. 2003. Fruit odor discrimination and sympatric host race formation in *Rhagoletis*. Proc. Natl. Acad. Sci. USA 100: 11490–11493.

Linn, C. E. Jr., H. R. Dambroski, J. L. Feder, S. H. Berlocher, S. Nojima, and W. L. Roelofs. 2004. Postzygotic isolating factor in sympatric speciation in *Rhagoletis* flies: reduced response of hybrids to parental host-fruit odors. Proc. Natl. Acad. Sci. USA 101: 17753–17758.

McEvoy, M. V. 1988. The gall insects of goldenrod (Compositae: *Solidago*) with a revision of the species of *Rhopalomyia* (Diptera: Cecidomyiidae). M.S. thesis, Cornell University, Ithaca, NY.

McPheron, B. A., D. C. Smith, and S. H. Berlocher. 1988. Genetic differences between host races of *Rhagoletis pomonella*. Nature 336: 64–66.

Mitter, C., B. Farrell, and B. Wiegmann. 1988. The phylogenetic study of adaptive zones: has herbivory promoted insect diversification? Am. Nat. 132: 107–128.

Morris, D. C., and L. A. Mound. 2002. Thrips as architects: modes of domicile construction on *Acacia* trees in arid Australia, pp. 279–282. In R. Marullo and L. A. Mound (eds.), Thrips and tospoviruses: proceedings of the 7th international symposium on Thysanoptera. Australian National Insect Collection, Canberra, Australia.

Morris, D. C., L. A. Mound, M. P. Schwarz, and B. J. Crespi. 1999. Morphological phylogenetics of Australian gall-inducing thrips and their allies: the evolution of host-plant affiliations, domicile use and social behaviour. Syst. Entomol. 24: 289–299.

Nason, J. D., S. B. Heard, and F. R. Williams. 2002. Host-associated genetic differentiation in the goldenrod elliptical-gall moth, *Gnorimoschema gallaesolidaginis* (Lepidoptera: Gelechiidae). Evolution 56: 1475–1488.

Odling-Smee, F. J., K. N. Laland, and M. W. Feldman. 2003. Niche construction: the neglected process in evolution. Princeton University Press, Princeton, NJ.

Page, R. D. M. 2003. Tangled trees: phylogeny, cospeciation, and coevolution. University of Chicago Press, Chicago.

Patterson, J. T. 1915. Observations on the development of *Copidosoma gelechiae*. Biol. Bull. 29: 291–305.

Pennisi, E. 2005. What determines species diversity? Science 309: 90.

Perry, S. P., T. W. Chapman, M. P. Schwarz, B. J. Crespi. 2004. Proclivity and effectiveness in gall defense by soldiers in five species of gall-inducing thrips: benefits of morphological caste dimorphism in two species *(Kladothrips intermedius* and *K. habrus)*. Behav. Ecol. Sociobiol. 56: 602–610.

Poff, A. C., M. Szymanski, D. Back, M. A. Williams, and J. T. Cronin. 2002. Bird predation and the host-plant shift by the goldenrod stem galler, *Eurosta solidaginis* (Diptera: Tephritidae). Can. Entomol. 134: 215–227

Schwarz, D., B. M. Matta, N. L. Shakri-Botteri, and B. A. McPheron. 2005. Host shift to an invasive plant triggers rapid animal hybrid speciation. Nature 436: 546–549.

Shigonebu, S., H. Watanabe, M. Hattori, Y. Sakaki, and H. Ishikawa. 2000. Genome sequence of the endocellular bacterial symbiont of aphids *Buchnera* sp. APS. Nature 407: 81–86.

Simon, J.-C., S. Carré, M. Boutin, N. Prunier-Leterme, B. Sabater-Muñoz, A. Latorre, and R. Bournoville. 2003. Host-based divergence in populations of the pea aphid: insights from nuclear markers and the prevalence of facultative symbionts. Proc. R. Soc. Lond. B 270: 1703–1712.

Smith, P. T., K. Krager, J. T. Cronin, and S. Kambhampati. 2002. Mitochondrial DNA variation among host races of *Eurosta solidaginis* Fitch (Diptera: Tephritidae). Mol. Phylogenet. Evol. 25: 372–376.

Smith, P. T., K. Krager, J. T. Cronin, and S. Kambhampati. 2003. Erratum to "Mitochondrial DNA variation among host races of *Eurosta solidaginis* Fitch (Diptera: Tephritidae)." Mol. Phylogenet. Evol. 29: 648.

Stelinski, L. L., and O. E. Liburd. 2005. Behavioral evidence for host fidelity among populations of the parasitic wasp, *Diachasma alloeum* (Muesebeck). Naturwissenschaften 92: 65–68.

Stelinski, L. L., K. S. Pelz, and O. E. Liburd. 2004. Field observations quantifying attraction of the parasitic wasp, *Diachasma alloeum* (Hymenoptera: Braconidae) to blueberry fruit infested by the blueberry maggot fly, *Rhagoletis mendax* (Diptera: Tephritidae). Fla. Entomol. 87: 124–129.

Stinner, B. R., and W. G. Abrahamson. 1979. Energetics of the *Solidago canadensis*-stem gall insect-parasitoid guild interaction. Ecology 60: 918–926.

Stireman, J. O., J. D. Nason, and S. B. Heard. 2005. Host-associated genetic differentiation in phytophagous insects: general phenomenon or isolated exceptions? Evidence from a goldenrod-insect community. Evolution 59: 2573–2587.

Stireman, J. O., J. D. Nason, S. B. Heard, and J. M. Seehawer. 2006. Cascading host-associated genetic differentiation in parasitoids of phytophagous insects. Proc. R. Soc. Lond. B 273: 523–530.

Strong, D. R., J. H. Lawton, and T. R. E. Southwood. 1984. Insects on plants. Harvard University Press, Cambridge, MA.

Tauber, C. A., and M. J. Tauber. 1989. Sympatric speciation in insects: perception and perspective, pp. 307–344. In D. Otte and J. A.

Endler (eds.), Speciation and its consequences. Sinauer, Sunderland, MA.

Uhler, L. D. 1961. Mortality of goldenrod gall fly *Eurosta solidaginis* in vicinity of Ithaca, New York. Ecology 42: 215–216.

Via, S. 1999. Reproductive isolation between sympatric races of pea aphids. I. Gene flow restriction and habitat choice. Evolution 53: 1446–1457.

Via, S. 2001. Reproductive isolation in animals: the ugly duckling grows up. Trends Ecol. Evol. 16: 381–390.

Waring, G. L., W. G. Abrahamson, and D. J. Howard. 1990. Genetic differentiation among host-associated populations of the gallmaker *Eurosta solidaginis* (Diptera: Tephritidae). Evolution 44: 1648–1655.

Weis, A. E., and W. G. Abrahamson. 1985. Potential selective pressures by parasitoids on a plant-herbivore interaction. Ecology 66: 1261–1269.

Weis, A. E., and W. G. Abrahamson. 1986. Evolution of host-plant manipulation by gall makers: ecological and genetic factors in the *Solidago-Eurosta* system. Am. Nat. 127: 681–695.

Wood, T. K. 1980. Divergence in the *Enchenopa binotata* Say complex (Homoptera: Membracidae) effected by host plant adaptation. Evolution 34: 147–160.

Wood, T. K. 1993. Speciation of the *Enchenopa binotata* complex (Insecta: Homoptera: Membracidae), pp. 299–317. In D. R. Lees and D. Edwards (eds.), Evolutionary patterns and processes. Academic Press, San Diego.

FIFTEEN

The Oscillation Hypothesis of Host-Plant Range and Speciation

NIKLAS JANZ AND SÖREN NYLIN

From a humble beginning in the early Cretaceous, angiosperm plants have quickly conquered the earth so that they now make up one of the most ubiquitous and species-rich groups (Crane et al. 1995; Wikström et al. 2001; Stuessy 2004; Friis et al. 2005). Likewise, the pioneer insects that once colonized this novel resource have multiplied to such an extent that they have become an ecologically dominating group in all terrestrial ecosystems (Mitter et al. 1988; Farrell 1998). Together, the seed plants and the insects that feed on them make up a good half of all described species, and their diversification is among the most remarkable in the history of life, both in terms of magnitude and relative speed. Consequently, if we want to understand the processes that generate biodiversity on earth, this is a good place to start.

Even though it is clear that both seed plants and plant-feeding insects have undergone rapid diversification to a much larger extent than their respective sister groups, we have a surprisingly poor understanding of why this has happened. The analogous diversification of these groups has led to the idea that they must have diversified together, reciprocally influencing each other's cladogenesis (Ehrlich and Raven 1964). However, apparent disparity in the relative timing of the diversification of the two groups has led many to suggest that they may have undergone rapid speciation for reasons that are not necessarily connected (Jermy 1984; Janz and Nylin 1998). Furthermore, while explanations for how insects could have influenced the diversification of flowering plants have typically focused on pollination (Pellmyr 1992; Waser 1998; Dodd et al. 1999; Grimaldi 1999), explanations for the potential impact of plants on the diversification of insects have often revolved around herbivory (Mitter et al. 1988; Farrell 1998; Kelley et al. 2000).

Here, we will primarily deal with the latter question: how the utilization of flowering plants as food resources could have promoted speciation rates in insects. Seed plants are different from many other food resources. As already mentioned, the resource itself has quickly diversified, and a typical consequence of this seems to the development of high resource specialization among the insects that feed on them. Relative host specialists dominate most groups of plant-feeding insects, and there are indications that many polyphagous species often show considerable geographic specialization (Thompson 2005). Flowering plants may be an abundant resource, but different plant species offer very different chemical challenges, and it appears difficult to be able to cope with more than a few plant species at a time. These difficulties can be both metabolic (Via 1991; Joshi and Thompson 1995; Mackenzie 1996; Agrawal 2000) and neurological (Bernays and Wcislo 1994; Janz and Nylin 1997; Bernays 2001).

This widespread specialization is a natural candidate in the search for mechanisms behind the high speciation rates among these insects and has often figured in such discussions (Jaenike 1990; Futuyma 1991; Thompson 1994; Kelley et al. 2000; Hawthorne and Via 2001). However, understanding the reasons for host specialization is just the beginning. The proposed role of specialization is that it can aid population fragmentation, either by the formation of host races (Bush 1975; Feder et al. 1988; Carroll and Boyd 1992; Hawthorne and Via 2001) or by an increased likelihood of geographic fragmentation (Peterson and Denno 1998; Kelley et al. 2000). Under both these scenarios, specialization acts as a pruning process; it increases the likelihood of population subdivision and speciation by removing plants from the repertoire. Thus, by itself, this process would very soon run out of fuel.

The next challenge then, is to understand what processes generate novel variation in host use, the processes that have driven the spread of plant-feeding insects across the phylogeny of flowering plants. Even though there is a considerable conservatism in host-plant use among groups of related herbivorous insects, plant-feeding insects have, to various extents, conquered all major groups of flowering plants. How and why does the *interaction* diversify? What are the relative roles of host-range expansions and host shifts?

This is an important part of our hypothesis for how plant diversity generates diversity in the insects that feed on them, but a final, and fundamental, part remains: does this diversification of the interaction indeed lead to elevated speciation rates? It is the purpose of this chapter to suggest a comprehensive explanation for the generation of variation in host use, the subsequent pruning of this variation, and how it can influence the diversification rates of plant-feeding insects.

The Oscillation Hypothesis

We will first briefly outline our general hypothesis for how diversity of phytophagous insects may be promoted by oscillations in host-plant range, before going into the process in more detail in later sections. The hypothesis is related to both the biogeographical concept of "taxon pulses" first described by Erwin (1981) and the smaller-scale oscillations described by M. C. Singer (this volume) as "peristaltic evolution."

Most phytophagous insects are specialists on a group of plants, at the species level or at a higher taxonomical level (Fig. 15.1 box a). As noted above, such specialization has a limited potential to create diversity, unless the process of specialization is at least occasionally reversed.

For an increase in host-plant range (Fig. 15.1 box b), not only the plant preferences of females need to change, but also the ability of juveniles to feed on more than one plant taxon, unless the expansion involves only very similar plants or plants to which the insect is for some other reason preadapted (see "Colonizations and Host-Range Expansions"). Once a wider host-plant range has evolved, it may lead to speciation through the formation of more specialized host-plant races, in sympatry or parapatry (Fig. 15.1, the direct route from box b to box e). A more general process may perhaps occur via a geographical range expansion (Fig. 15.1, box c), facilitated by the ability to feed on more than one plant group. The evidence for a link between host-plant range and geographical range is reviewed in "Diet Breadth and Geographical Range," below. The initial colonizers during such a range expansion may well be members of populations or species that are prone to dispersal and show high levels of gene flow, preventing local adaptation. However, a strong dispersive tendency should often be a short-lived evolutionary stage. As was noted by Futuyma (1998, pp. 575–576) "natural selection against alleles that promote dispersal is almost inevitable." This is because alleles for dispersal tend to remove themselves automatically from local populations, because dispersion is risky and mortality during this phase high, and because alleles for staying put keep organisms in a habitat where they have a chance to become locally adapted.

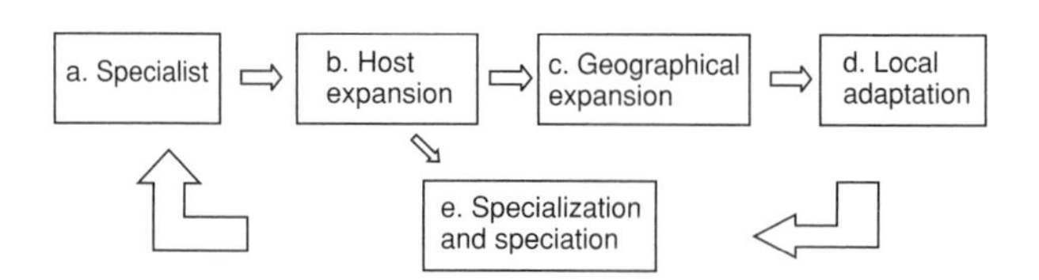

FIGURE 15.1. A scheme for oscillations in host-plant range promoting higher rates of speciation.

Local adaptation is bound to follow as soon as gene flow is no longer ubiquitous (Fig. 15.1, box d). Life histories and other seasonal adaptations will evolve to fit local conditions, and so will host-plant utilization. Local host-plant specialization should often reoccur (Fig. 15.1, box e; see "Specialization and Fragmentation") because females are expected to evolve to prefer the plants that maximize their fitness, given the local patterns of seasonality, the local abundances of potential host plants, the life-history consequences of feeding on one plant or the other, *and* the fitness consequences of these particular life histories under the local patterns of seasonality and climate. Speciation may be facilitated by the resulting geographic variation in host-plant preferences and life histories (to some extent this will be a coadapted package), through processes of ecological speciation (see "Speciation").

For an understanding of the processes governing host-plant range (Fig. 15.1, boxes a, b, and e) it is necessary to study the preferences of females (including the genetics and mechanistic aspects, e.g., oviposition stimulants), as well as the performance of offspring when reared on different host plants, in other words, the adaptive and nonadaptive plasticity of offspring—especially in fitness-related life-history traits. To address the process generating biodiversity (Fig. 15.1, boxes c through e) we need to make comparisons between populations and species, including phylogenetic studies.

It should be noted that in this chapter we use the concept of plasticity in a way similar to West-Eberhard (2003), that is, to refer to a potential to follow different developmental pathways. This concept of developmental plasticity is related to both phenotypic and evolutionary plasticity, because it may be expressed in either way. In the context of host plants, a potential in an insect lineage to develop on more than one host can be expressed either as oligo- or polyphagy, or as frequent evolutionary shifts between hosts in the potential range, or both.

Colonizations and Host-Range Expansions

As plant-feeding appears to have repeatedly led to increased species richness among insects (Mitter et al. 1988; Farrell 1998), a logical conclusion would be that there is something about plants as a resource that increases the likelihood of speciation (or decreases the likelihood of extinction). So, what is it that is so special about plants?

As mentioned in the introduction, the most striking aspect of the seed plants is that they are themselves such a highly diverse group. They are diverse in terms of species numbers but also in terms of internal chemistry and external physical structures. The seed plants have conquered all major habitats of the world and are, indeed, defining features of many of them, such as the rain forests, the grassland prairies or steppes, and so on. It would seem then that an important key to understanding the diversification of plant-feeding insects lies in understanding the diversification of host-plant use among these insects. What has caused them to spread across the seed plants to such an extent? Taking the butterflies as an example, Janz and Nylin (1998) showed that the ancestor of the butterflies probably colonized a relatively derived plant within the clade that now contains Fabaceae, Rosaceae, and the "urticalean rosids" (Urticaceae and relatives). If this is true, the present patterns of host-plant use must have been caused by colonizations from these ancestral plants onto an already diverse group; there is no indication that the process of plant diversification has in itself promoted the spread of butterflies onto them.

Many studies have pointed out the conservative aspect of host-plant use among plant-feeding insects, where related insects tend to feed on related plants (e.g., Ehrlich and Raven 1964; Futuyma and Mitter 1996; Janz and Nylin 1998; Thompson 2005). Yet, in spite of this, these insects have managed to colonize a substantial portion of the seed plants in a relatively short time. A closer look at what was going on among the butterflies within the tribe Nymphalini revealed an almost chaotic picture, with a large number of shifts between distantly related plants (Janz et al. 2001). Furthermore, there are examples of truly rapid changes in host use, with host shifts occurring in observable time as responses to, for example, changes in human land use (Strong 1974; Strong et al. 1977; Tabashnik 1983; Singer et al. 1993). Apparently, the available data point in different directions; host-plant use cannot possibly be conservative and opportunistic at the same time. Or can it?

Futuyma and colleagues found that host shifts within the beetle genus *Ophraella* tended to involve the plants that were already used by other congeneric species (Futuyma et al. 1993, 1994, 1995). Hence, even if host shifts could potentially be numerous, they were not randomly distributed across plant taxa but appeared to be constrained to plants that were used by related species. A similar pattern was later seen in the butterfly tribe Nymphalini (Janz et al. 2001), as well as in a number of other groups (Nosil 2002). Within the Nymphalini, a dozen plant families were reconstructed to have been independently colonized 29 to 37 times within a phylogeny of less than 30 terminal species, some families as many as 5 to 9 times. This is a truly remarkable number, considering the size of the butterfly clade under investigation, and certainly not something that could be characterized as a conservative host use. This led to the conclusion that many of these colonizations were in fact not independent; the defining trait that allowed the colonization, the "preadaptation," was probably more ancestral than the actual colonization events themselves. This ancestral event could have been an actual colonization of the plant that was followed by a secondary loss from the repertoire, but where much of the essential "machinery" needed to utilize the plant was retained. But it could also be a more literal preadaptation, where an evolved trait (like a new enzyme) makes the colonization of a given plant possible. A series of larval establishment tests on a wide range of plants (known hosts of the group as well as a sample of non-host-plant groups) also showed that the larvae were often capable of feeding and surviving on a wider range of plants than were actually used as hosts by a species. Moreover, the additional plants were invariably plants that had, according to the phylogenetic reconstruction, been used as a host by an ancestral butterfly. Hence, the study contrasted the actual host range, the range of plants that are actually oviposited on and used as hosts at a given time, with the potential host range, meaning the total number of plants that an insect is capable of using (or where there exists genetic variation for feeding within the population). This distinction between actual and potential host ranges should be kept in mind when reading this chapter, as it will be fundamental to understanding our further reasoning.

Apparently, host use can be very dynamic and even opportunistic, but only within a restricted set of plants that appear to be shared among a group of insects. How, then, do genuinely novel associations evolve? Janz et al. (2001) found that most of the changes in host use seemed to be concentrated in periods of expanded host ranges. There was a strong tendency for colonizations to lead to periods of multiple host use (host-range expansions), rather than to immediate host shifts. "Immediate" in this context means before the next speciation event, so even the few examples of so-called immediate host shifts might have involved extended periods of multiple host use. During these periods, further colonizations were more common than during periods of specialized host use. Furthermore, the few examples of radical host shifts in the tribe, to plants outside the typical set of plants used by the group, seemed to be happening during these polyphagous phases. This pattern has recently been confirmed and further elaborated by Weingartner et al. (2006). In that study, the colonization of novel hosts did not appear to be caused by chemical breakthroughs in the sense of Ehrlich and Raven (1964), but rather by a slower accumulation of hosts during times of increased plasticity in resource use (cf. West-Eberhard 2003). In both studies on the Nymphalini butterflies, there were also clear indications that the plastic phases of expanded host ranges were evolutionarily transient; there were clear tendencies to respecialize on either the ancestral host or one of the newly incorporated hosts (Janz et al. 2001; Weingartner et al. 2006).

Reconstructing ancestral host range is difficult for several reasons (cf. Stireman 2005) and will require very detailed data on host use and well-resolved complete phylogenies to

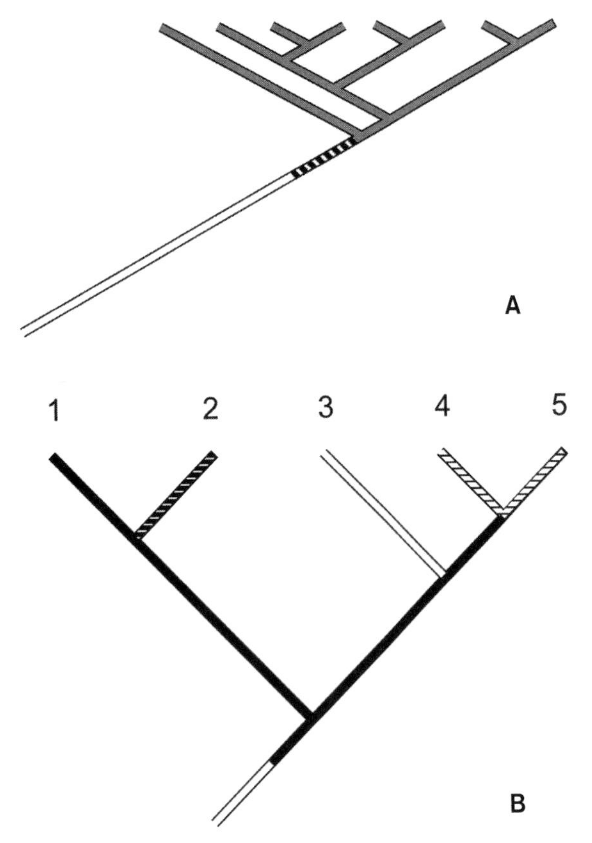

FIGURE 15.2. Schematic representations of the "explosive radiation scenario" (A) and "plasticity scenario" (B). Under the former we expect a short stage of shifting from the ancestral host (white) to the novel (gray) over a stage where perhaps only one of these two hosts were used (striped). Under the latter we expect a long-lasting stage (black) of at least potential ability to feed on several plant taxa, with respecializations on the ancestral host (white) as well as on different novel hosts (all striped) in some species.

have any chance of success. Moreover, a lesson from the Nymphalini studies is that even so, ancestral host ranges will tend to be underestimated because of the nature of host-range evolution (Janz et al. 2001; Weingartner et al. 2006). In many cases, the phylogenetic traces of a host-range expansion will be lost with time; a period of expanded host range followed by specialization on one of the novel hosts will in retrospect appear as a clean host shift, once all traces of the additional hosts are lost (cf. species 2, 4, and 5 in (Fig. 15.2B). Indeed, a respecialization on the ancestral host will often be impossible to detect at all once enough time has passed, and it will thus look as if nothing has happened at all (cf. species 3 in Fig. 15.2B). In these cases, establishment tests might provide additional clues to the reconstruction of ancestral host ranges (Janz et al. 2001); or, if this is not possible, the combined host ranges of a clade appears to be a relatively accurate estimation of past host-range expansions, even if it is not possible to tell whether there has been one or several expansions (Janz et al. 2001, 2006).

To sum up, it appears that host-plant colonizations typically lead to host-range expansions rather than to immediate host shifts, and that such periods of expanded host ranges facilitate further colonization events, including the inclusion of truly novel hosts into the repertoire. As a consequence, host expansions are probably an important part of the process that has allowed plant-feeding insects to spread across the seed plants.

Diet Breadth and Geographical Range

If host expansions are to contribute to increased diversification rates among plant-feeding insects, they must somehow increase the likelihood of a species to subdivide into distinct subpopulations. At a first glance, this might not seem very likely. After all, multiple-host use implies increased behavioral and physiological plasticity, which should buffer against environmental heterogeneity and thus, rather, act to counterbalance local adaptations and population fragmentation. What host expansions can accomplish, however, is to help set the stage for other factors, which do lead to population fragmentation, to start acting.

Many factors may influence the geographic distribution of a species, but the distribution of its available resources sets the ultimate boundaries for where the insect can live and reproduce. For a plant-feeding insect, the occurrence of acceptable larval host plants will determine its possible limits of distribution. As two plant species will not be likely to have the exact same distribution, a wider host range should allow a plant-feeding insect to spread over a wider geographic area. There is some evidence for this, especially from the British Isles, where patterns of butterfly distribution are particularly well documented (Quinn et al. 1998; Dennis et al. 2000). Regional studies from Germany (Brändle et al. 2002), Japan (Kitahara and Fujii 1994), Finland (Päivinen et al. 2005), and the United States (Nitao et al. 1991) also confirm this picture. All these studies were limited in geographical and/or taxonomical scope, but until a more comprehensive study of a more global scope is conducted, the weight of evidence is definitely in favor of a general positive correlation between the diet breadths of insect species and the size of their geographical ranges.

There are at least two mechanisms by which such a correlation could arise. First, an alternative host with a geographical distribution that is partly overlapping with the ancestral host may be colonized. This would allow the insect to expand its geographical range into the areas where the novel host grows, but where the ancestral host does not. In this scenario, the novel host is colonized within the original area of distribution, and the geographical range expansion is a secondary process. The colonization of the novel host has simply opened up a larger area of suitable habitat.

Alternatively, a novel host could be colonized as a consequence of active dispersal into areas where the ancestral host does not exist. This, however, would require a preexisting capacity to both recognize the novel plant as a host (so that the female will accept it for oviposition), as well as a corresponding preexisting capacity of the larvae to grow and survive on the plant. It may not seem like a very plausible scenario, but there are some features of the interaction that may make it more likely. As described in the previous section, most colonizations and host shifts seem to involve plants that either are used by close relatives or have been used as hosts in the past. Such plants may share important characteristics with the presently used host, which makes them more likely to be recognized as hosts and to support at least some larval survival. Alternatively, the insect may be literally preadapted to the plant, because it has already been used historically, which means that many of the adaptations needed to cope with the plant are already present. Similarly, the many examples of parallel colonizations of the same plant group documented in the butterfly tribe Nymphalini (Janz et al. 2001; Weingartner et al. 2006) can be explained by a shared plesiomorphic trait that facilitates the colonization of a particular plant.

This process, where a geographical range expansion leads to a colonization or host shift with negligible initial evolutionary change, has been called "ecological fitting" (Janzen 1985). When a species arrives at a new location, its survival there will be determined by its fit into the local ecosystem. If it cannot find suitable resources or habitat to exploit, it will simply not survive to be a part of the local community.

Ecological fitting has recently received renewed attention in biogeography, where it has become clear that species interactions within a community are often not evolved in place but have instead evolved elsewhere and exist in the present context due to secondary dispersal and ecological fitting (Wilkinson 2004; Brooks and Ferrao 2005). The vegetation of Ascension Island serves as a good example of the process of ecological fitting. When Darwin passed by the island on the Beagle on his way back to England, he described the island as a barren rock, "entirely destitute of trees" (Darwin 1839). Today its slopes are covered by a cloud forest, a complex, species-rich ecosystem that has formed in less than 200 years (Wilkinson 2004). In this case, plant dispersal to this remote island has to a large extent been carried out by humans, which has certainly increased the speed of the process dramatically and thus given the processes of local adaptation and coevolution less of a chance. Nevertheless, it shows that a complex community *can* form almost entirely through a series of ecological fittings (Wilkinson 2004). It is clear that species with wider environmental tolerance will be more likely to succeed in fitting into a novel community. Hence, the capacity to (potentially) utilize a wider range of hosts should facilitate successful geographical range expansions.

We do not mean to say that coevolution and local adaptation are not important processes in the shaping of species interactions, but these processes can start acting only *after* the interaction has come into existence in the first place, through biotic expansion and ecological fitting. A good example of this can be seen in the recent host shift of some populations of the prodoxid moth *Greya politellla* in the mountains of Idaho. Throughout most of its geographical range, it feeds on plants of the genus *Lithophragma* (Saxifragaceae). In central Idaho, some populations have shifted to the related plant *Heuchera grossularifolia*, a biogeographically complex set of diploid and autopolyploid populations (Segraves et al. 1999). Although the moths attack plants of both ploidies, they preferentially attack tetraploids over diploids (Thompson et al. 1997). When presenting diploid and tetraploid *Heuchera* plants along with the native *Lithophragma* plants to a population that has had no known prior contact with the novel host species, these moths laid almost all eggs in their native host (Janz and Thompson 2002). Clearly, the moths were not well adapted to the novel host. Even so, they did lay some eggs in the novel host, which showed that the potential for a host colonization was there, should the local plant community change, or should they disperse to a nearby locality where only the novel host was present. In this example, the initial colonization of the novel plant is made possible through ecological fitting, while the subsequent increase in preference and survival, as well as the higher preference for tetraploid variants, must have evolved locally, after the colonization.

Specialization and Fragmentation

The general tendency among plant-feeding insects to evolve toward increased specialization in the host-plant utilization is well documented (Thompson 1994). As mentioned in the introduction, this pervasive trend toward increased specialization has been suggested to play an important role in the ongoing diversification of plant-feeding insects (Bush 1975; Feder et al. 1988; Carroll and Boyd 1992; Peterson and Denno 1998; Kelley et al. 2000; Hawthorne and Via 2001; Morse and Farrell 2005).

The main question we face for the general hypothesis outlined in this chapter is whether such specialization can decrease gene flow among populations of geographically widespread species, by selecting for different patterns of host use in different parts of an insect's geographical range.

Local populations often differ in the degree of specialization of their species interactions, and indeed with what species they interact, leading to geographic selection mosaics across the distributions of interacting species (Thompson 1994, 2005). The importance of such selection mosaics in this context is that they provide the opportunity for evolution to follow diverging paths in different parts of a species' geographical range. There can be several reasons for such divergent selection. Even on small geographical scales, habitat fragmentation may be enough to cause differences in local availability of potential hosts (Singer et al. 1989; Kuussaari et al. 2000); and, depending on connectivity between habitat patches, such differences may lead relatively quickly to genetic differences in host preference and performance (Singer et al. 1993; Singer and Thomas 1996). In other cases, the mere presence or absence of additional species, such as competitors, predators, and parasites, can cause differences in local selection pressures. Such is the case in the interaction between the *Greya* moths mentioned in the previous section and their saxifragaceous hosts, where the local presence or absence of additional pollinators determines whether the interaction is mutualistic or parasitic, and hence the direction of selection on both the insect and the plant (Thompson et al. 2004).

Obviously, the larger the geographical range, the larger the chance that a species will experience different conditions in different parts of its range. Members of a widespread species may find themselves in diverse climatic conditions, each requiring specific local adaptations. In the temperate zones, conditions and requirements change dramatically with latitude, and it is not surprising that patterns of host use and specialization do too (Scriber 1973). Each potential plant will have its specific life-history consequences (and these too may change with latitude), and what is favorable in a strictly univoltine northern population may not be favored in a multivoltine southern population.

A good example of this is the host use within the butterfly species complex *Papilio glaucus/canadensis* in the eastern United States. Host preference and detoxification abilities differ between populations, with a sharp cline in a hybrid zone in the North American Great Lakes region (Scriber 1988, 1994; Scriber et al. 1991, this volume). This highly polyphagous butterfly feeds on several trees, and while some hosts are shared across most of its geographical range, the southern *P. glaucus* variant mainly uses *Magnolia* and *Liriodendron*, while the northern *P. canadensis* variant mainly feeds on *Populus*. The two variants are now considered true species (Hagen et al. 1991), and there are good indications that divergent host utilization played a part in the speciation process (Scriber 1994).

Even where relative host-plant ranking is the same across a latitudinal cline, degree of specialization on the preferred host can change due to differences in voltinism (Nylin 1988; Scriber and Lederhouse 1992). While northern univoltine populations of the Eurasian butterfly *Polygonia c-album* regularly use hosts from seven families in five orders, bi- or multivoltine southern populations feed almost exclusively on the preferred host *Urtica dioica*, which can support the higher growth rates needed to fit several generations into the favorable season (Nylin 1988; Janz et al. 1994; Janz and Nylin 1997). These differences in specificity are genetically determined, and like in the *Papilio* example above, the differences in host preference are largely sex-linked (Janz 1998, 2003; Nylin et al. 2005). Sex-linked genes are disproportionately represented among the reported genetic differences between closely related species in the Lepidoptera (Sperling 1994; Prowell 1998), suggesting a role in speciation. *Polygonia c-album* is a widespread, polyphagous species in which current gene flow probably prevents further fragmentation. However, there is a potential for further divergence and speciation should these populations become more isolated in the future.

All in all, there is good evidence that local populations of widespread species will often evolve toward specialization on a subset of the original host range. Moreover, different requirements throughout the distributional range will often lead to variable outcomes, so that populations are likely to specialize on different plants.

Speciation

At this stage we need to ask ourselves how host plants might be directly or indirectly connected to speciation events in phytophagous insects. The different modes of speciation are traditionally classified according to geography, from sympatric speciation (where the new species form without physical or geographic isolation, within a single ancestral population) to allopatric speciation (where the incipient species are first completely isolated from each other). Many other modes have been suggested as part of this general scheme; the most important of these (for our purposes) is the intermediate mode: parapatric speciation (where the incipient species occur in different parts of the ancestral range and are thus partially genetically isolated by distance).

Recently, it has been suggested that it may be more useful to classify speciation models according to the mechanisms responsible for the evolution of reproductive isolation (Schluter 2000, 2001; Via 2001). In such schemes, the major distinction is according to whether reproductive isolation evolves primarily because of chance events (such as genetic drift) or because of selection. A further distinction can be made according to whether barriers to gene flow evolve as a result of ecologically based selection differing between environments ("ecological speciation" [Schluter 2000, 2001; Rundle and Nosil 2005]) or other forms of selection (such as sexual selection) where reproductive isolation is not primarily due to ecological differences between environments. It is in the category of ecological speciation that host plants are likely to have their most direct effects, and this category cuts across all of the geographical categories. The role of host plants in speciation has been reviewed by Berlocher and Feder (2002), Drès and Mallet (2002), and Funk et al. (2002) (and see also several other chapters in this volume). However, these treatments typically focus on the role of host-plant specialization in speciation, rather than on host-plant range, so we will briefly outline how we believe a wide range of potential hosts may promote speciation under different modes of speciation.

SYMPATRIC SPECIATION

This mode of speciation has always been controversial, but there has recently been a strong reappraisal of the importance of sympatric speciation in nature (Via 2001). It is interesting to note, however, that in the latest major synthesis of the field, Coyne and Orr (2004) conclude that sympatric speciation is theoretically possible, but that necessary conditions are more stringent than for allopatric speciation, which, in their opinion, should thus remain the null model for speciation.

They also critically review the evidence from nature and find only a few cases where sympatric speciation seems "plausible." All of them involve differences in resource use, suggesting that ecologically based divergent selection may well be important for this mode of speciation. None of the suggested cases of sympatric speciation involving "host races" survive their scrutiny, usually because an allopatric phase during the formation of the host races cannot be ruled out. In line with the recent trend toward classification according to isolating mechanism rather than geography, we can still conclude that the cases reviewed by Berlocher and Feder (2002[AuQ2]) and Drès and Mallet (2002) (such as the famous *Rhagoletis* system) represent evidence that differences in host-plant utilization between incipient species can contribute strongly toward their isolation in sympatry and *perhaps* can result in speciation, even without an allopatric phase.

What is usually not recognized enough, however, is the simple fact that host-plant races cannot form at all if the ancestral species is strictly specialized on a single host plant. In the *Rhagoletis* system (see Berlocher and Feder 2002 for references), the ancestral race of *R. pomonella* evidently fed on hawthorn, and a new race formed on introduced apple. But where did the ability to feed on apples come from? Clearly, it must have been part of a potential range of hosts even before apples were introduced. There must have been at least some ability to recognize and feed on this host, an ability that could later be fine tuned in a process of "genetic accommodation" (sensu West-Eberhard 2003). The probability of such an evolutionary event should increase with a wider host-plant range. Furthermore, assortative mating is necessary for reproductive isolation to evolve in sympatry. In the case of host-plant-driven speciation this could be due to mate choice being influenced by plant-derived substances, a tendency to mate on or near the host plant, differences in plant phenology isolating the populations in time, or differences in plant habitat isolating them in space. All of these possibilities would seem more likely the more different the plants are, and with a wider ancestral host-plant range the differences are likely to be larger.

In our own study organisms, the butterflies, host-plant-driven sympatric speciation (or speciation involving a sympatric phase where the reproductive isolation is strengthened by selection) seems less likely than in some other phytophagous insects. Butterflies are typically very mobile as adults, making strong isolation due to differences in plant habitat requirements unlikely (but not impossible, since even a mobile insect can choose to stay in one habitat if this is favored by selection). They do not typically mate on the host plant, and indeed the adult phase is often only very weakly linked to the larval host plant. Adults do not feed on the host plant used by larvae, in contrast to many phytophagous beetles. We do not rule out the possibility of sympatric speciation, however, especially in sedentary butterflies such as some blues (Lycaenidae), or in cases where plant-derived pheromones are of strong importance for mating. Plant-derived toxic substances could also play a role via aposematic coloration, mimicry, and sexual selection (Willmott and Mallet 2004).

PARAPATRIC SPECIATION

Populations that are isolated by distance may diverge because of environmental differences and divergent selection, even if they exchange genes, and reproductive isolation can evolve as a by-product of the differentiation (Coyne and Orr 2004). It is clearly conceivable that differences in host-plant utilization could contribute to such speciation events, because different host plants may be optimal to use in different parts of the geographical range (see "Specialization and Fragmentation"), and genes for using a particular host (including both preference traits and performance traits) could be part of a whole coadapted set of genes that also includes life-history and diapause characteristics. In the *Rhagoletis* case, for instance, the host-plant races differ in their acceptance of fruits for oviposition (but not in survival, so both hosts are "potential" hosts in our terminology) and also in having pupal diapause regimes that adapt them to the fruiting times of their respective hosts.

Host-plant-driven parapatric speciation should leave a signal in the form of sister species with abutting distributions, differing in host-plant use. As resolved phylogenies become available for butterflies and other phytophagous insects, it will be of interest to search for such evidence. The challenge, as usual for speciation studies, will be to determine whether such pairs of sister species really formed in parapatry or allopatry, but this distinction is of less importance for our purposes. It should also be added that a wider host-plant range could affect rates of parapatric speciation in a more general way: a wider host range may increase the geographical range of the species (see "Diet Breadth and Geographical Range") and hence set the scene for any environmental factor that varies over this range to contribute to divergence between subpopulations and eventually speciation.

TWO-STAGE MODELS

It may well be that the geographical extremes of sympatric and allopatric speciation are rare in their "ideal" form; perhaps real speciation events typically occur between these extremes (Rundle and Nosil 2005). One such intermediate form would be when differentiation starts in allopatry, but reproductive isolation is completed during a stage of secondary contact between the populations. In the allopatric stage there can be divergent selection between the different environments. In the later sympatric or parapatric stage the resulting ecological differences between the incipient species may be of great importance in preventing too-high levels of gene flow, and selection may also act to increase the differences further. This could occur, for instance, because of selection to reduce competition between the two forms, or through selection against the production of unfit hybrids, perhaps even leading to incipient species preferring to mate with their own kind (reinforcement).

It can easily be seen how differences in host-plant utilization that have evolved in allopatry may be involved in such a process. For instance, Forister (2005) has shown how correlations between preference and performance on particular host plants that have evolved in allopatry are broken by hybridization in the lycaenid butterfly *Mitoura* species complex, which may reduce gene flow between host races. Under this two-stage scenario, a wider host-plant range in the ancestral population should increase the probability that differences in host-plant use will evolve during the allopatric stage, forming the raw material for later completion of the ecological speciation.

ALLOPATRIC SPECIATION

The general importance of this mode of speciation is not denied even by the most ardent champions of sympatric speciation. Allopatric speciation is a likely mode for two reasons: first, the populations involved are physically/geographically isolated from each other to begin with and are thus free to evolve in different directions without gene flow; second, this means that *both* divergent selection and chance processes can contribute to the differentiation and eventual reproductive isolation. Thus, allopatric speciation is possible even when the habitats are identical.

We would like to stress that host-plant-driven speciation is not ruled out even if speciation is always strictly allopatric. The simplest scenario would be when a wider host-plant range increases the geographical range of a species, creating opportunities for later geographical isolation, and for a speciation process that does not further involve the host plants (this would not qualify as ecological speciation in the strict sense). Colonization of a single novel host, which is found in many habitats, such as the colonization of grasses by satyrine butterflies, could promote allopatric speciation in a similar way. It is worth pointing out that such a colonization event always involves *at least* a short stage of using both the old and the new plant, and probably typically involves a more polyphagous stage (see "Colonizations and Host-Range Expansions" and " Synthesis: Speciation Rates"). It is also clearly possible that differential use of host plants in the incipient species could contribute further toward their eventual reproductive isolation. The host plant is a key part of the life of any phytophagous insect, and isolation could evolve as a by-product of divergent selection that involves the host plants in one way or the other, causing hybrids to be unfit for extrinsic reasons (sensu Coyne and Orr 2004; hybrids viable but lack a suitable niche) or intrinsic reasons (hybrids inviable or sterile, e.g., due to epistatic interactions among parental genes).

Synthesis: Speciation Rates

How can host-plant range affect speciation rates? First of all, a distinction has to be made between rates of speciation per se and net speciation rates—the sum of both extinctions and speciation events. The latter measure is the one that is (more or less explicitly) applied when species numbers are simply compared between two sister clades, the standard method to assess the effect of some extrinsic or intrinsic factor on speciation rates. Effects of such factors on extinction risk may thus confound the comparisons, a problem that can only be avoided with methods that explicitly model the probabilities of both speciation and extinction through evolutionary time (Barraclough and Nee 2001). Large, reasonably complete phylogenies and data on branch lengths are necessary for such studies.

Although this has not yet been studied phylogenetically, we find it likely that host-plant range affects not only speciation probability but also extinction risk. Ecological data showing that populations of specialist butterflies have declined over the last decades, at the same time as populations of more polyphagous species have been able to increase and expand their ranges (see "Diet Breadth and Geographical Range") are suggestive in this respect, although not proof of actual species extinctions. If present, this effect of a wider host-plant range on extinction risk would affect the number of species in a clade positively, namely, in the same direction as the effect that we predict on speciation probabilities. Here, we focus the discussion on the latter aspect, but the available data is on species numbers (net speciation rates), and it should be kept in mind that differences in extinction risk may also have contributed to the observed patterns.

As noted above, there is evidence that phytophagous insect taxa are more species rich than their nonphytophagous sister taxa (Mitter et al. 1988), so that the plant-feeding habit evidently somehow promotes higher net speciation rates. This could be due to the fact that feeding at a lower trophic level increases the amount of available resources, or it could be due to the greater diversity of the plant resources as compared to, for example, animal tissues as food (Mitter et al. 1988). The first explanation cannot easily be ruled out as a contributing factor, but it is possible to investigate whether plant diversity has had an effect on insect diversity—over and above any effects of resource abundance—by studying the effects of host-plant range and host-plant shifts on speciation rates within a phytophagous insect taxon. Importantly, Farrell (1998) showed that angiosperm-feeding beetle clades have more species than their sister clades, evidence that angiosperm diversity is part of the explanation for beetle diversity.

At least implicitly, this is the reasoning behind many insect-plant studies, since Ehrlich and Raven's inspiring paper on butterfly-plant interactions (1964). These authors reasoned that plant taxa differ strongly in their (particularly chemical) properties, as a result of selection to defend themselves against herbivory. A plant species that invents a new defence may be free of herbivory and as a result be able to speciate at a higher rate, creating a whole clade of chemically similar species. Because of the chemical diversity among such clades, it is a challenge for an insect to colonize a new host-plant taxon, but if it manages to do so it may be able to speciate at a higher rate in this new adaptive zone. The pattern that we observe today, where related butterflies feed on related plants, could have created by this rather diffuse form of coevolution (Ehrlich and Raven 1964) or simply through a combination of host-plant conservatism and some insect speciation occurring between relatively rare host-plant shifts, without evoking coevolution (Janz and Nylin 1998). In any case, it would seem that such phylogenetic patterns, where host-plant records have a tendency to follow insect taxonomical groupings, suggest that there is some sort of link between plant variation and insect diversity.

However, it is not clear what the link is (Weingartner et al. 2006). Is it that phytophagous insects tend to quickly radiate into many species when they shift from an old to a new plant taxon? Let's call this scenario (which is consistent with Ehrlich and Raven's [1964], with or without coevolution) the "explosive adaptive radiation scenario." Versions of this scenario are generally invoked to explain the diversity of angiosperm-feeding insects, and often in case studies of host-plant shifts (e.g., Farrell 1998; Cook et al. 2002; Marvaldi et al. 2002). Or is it, rather, that what matters for high net speciation rates in an insect taxon is that it has a wide range of potential hosts (i.e., with several taxonomically distant plant taxa), whether or not they are actually used as hosts at a given evolutionary moment? Let's call this the "plasticity scenario," because a wide potential range means that some "machinery" for plant recognition and metabolization must be present (albeit perhaps often in an imperfect form) even for plants not currently utilized. Such species must be both more evolutionarily plastic and more phenotypically plastic than extreme specialists. The connection is obvious to West-Eberhard's idea that developmental plasticity may promote diversification, by providing more types of "building blocks" for selection to act upon (West-Eberhard 1989, 2003).

In a series of butterfly studies at different taxonomical levels, we are investigating the relative importance of these—not mutually exclusive—scenarios. They may be difficult to separate in practice using phylogenetic techniques, but we have proposed a few possible criteria (Weingartner et al. 2006; Nylin and Wahlberg, in press). First, the explosive radiation process would seem more likely if the plant and insect clades involved are of similar age. Explicit reasons are seldom given for why insects should speciate when colonizing a novel host taxon. It is often said that the new host is a new adaptive zone, but this only means that different ecological niches on this host will be more similar to each other than to niches on other hosts (cf. Futuyma 1998). This could even lower the rate of speciation, if the ancestral host or range of hosts was ecologically more diverse than the new zone. We see three possibilities for the observed patterns of radiation on novel host taxa. The colonization may have occurred early in the evolution of a radiating plant taxon and the insects then rode on this wave of plant divergence and diverged along with the formation of new niches (as suggested for beetles colonizing angiosperms at an early stage of angiosperm radiation [Farrell 1998]). Such a radiation would also be explosive to the extent that the plant radiation was rapid. There may instead (or also) be "empty niches" in the new zone because it is largely free from competition from other phytophagous animals (this was the process envisaged by Ehrlich and Raven [1964]). This possibility is also more likely if the clades are of similar age, because if an insect colonizes a plant clade that has been in existence for a long time, there is no strong reason to expect it to be free of competitors. A plant clade that diverged recently could, in contrast, have been successful in its radiation because of a new defence that excludes herbivores (Ehrlich and Raven 1964; Farrell et al. 1991).

For the plasticity scenario, age of the plant clades is not as strong an issue, and a third possibility to explain radiation after colonizations is actually that a host-plant shift only looks like a shift from one zone (plant A) to another (plant B) on a phylogeny, because we are studying it long after it happened. The insect may have colonized an ancient and already well-diversified plant taxon, and done so during a stage when it was using both plant A and B and several others as well (a polyphagous species would seem more likely to encounter, accept, and survive on a novel host). The general drive toward specialization after stages of more generalist feeding (Janz et al. 2001; Nosil 2002) will tend to obscure such patterns. Perhaps it has also sometimes obscured the fact that much of the insect diversification occurred in a stage when species in the clade had a wide range of potential hosts, as a result of having a wide zone rather than a novel and/or empty one?

A second criterion could thus be the timing of diversification. Both scenarios postulate an increased net diversification rate in the insect clade following colonization, but the pattern of diversification over time can be expected to differ. An explosive radiation process of ecological release from competition should lead to a fast diversification over a short time span immediately following the colonization. The plasticity scenario, rather, predicts a higher than earlier, but not necessarily much higher, rate of diversification, sustained over a relatively long time span. Thirdly, the pattern of host-plant usage over time is expected to differ between the scenarios (Fig. 15.2). Under the explosive radiation scenario we expect to see a shift from one plant clade to the other, with only a short period of using both hosts, because fitness is expected to be much higher on the new host if it is free of competition or superior in another way that supposedly explains the radiation. Such shifts are likely to look instantaneous when traced onto a phylogeny long after the event. Under the plasticity scenario we may instead see phylogenetic traces of a relatively long-lasting potentially polyphagous stage. This could be in the form of polyphagous species even today using both the ancestral and the novel plants, higher taxa where some species use the novel plant clade and others use the ancestral hosts

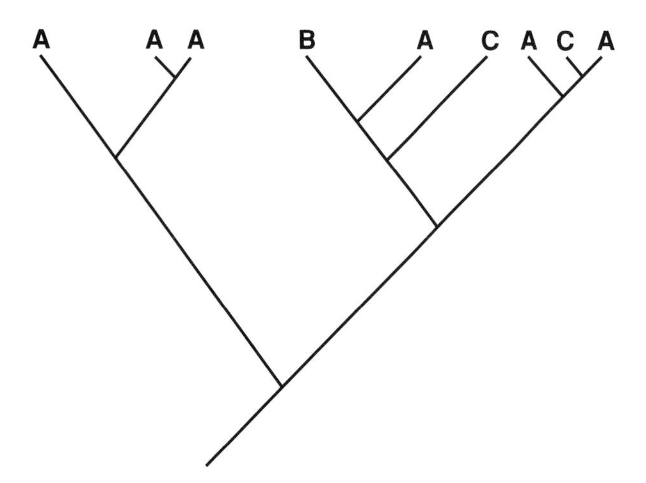

FIGURE 15.3. A hypothetical insect phylogeny with the host plants of the species in the recent environment represented by letters above terminal branches. Although all of the species are specialists, the total range of hosts used in the right-hand clade is wider. This pattern indicates that there must have been several shifts between plants A, B, and C, and that most likely the species at the internal nodes of the phylogeny, the ancestors of recent species, were at least potentially able to feed on more than one of these plants.

(suggesting recolonization), and species that have colonized yet other plant clades, such events presumably being facilitated by the wide potential range of hosts. Finally, testing juveniles for the ability to survive on nonhosts is a fourth key piece of evidence for distinguishing between scenarios (Janz et al. 2001).

To begin with, it should be determined whether diversity of plant utilization is really strongly linked to insect diversity within clades of phytophagous insects. If this is so, a clade where the total range of plants used (all of the involved species could actually be specialists today; cf. (Fig. 15.3) is wider than in the sister clade should on average contain more species. A similar logic was applied in Farrell's beetle study (1998) mentioned above, contrasting ancestral feeding habits (cycads and conifers) with derived feeding on the more diverse angiosperms. In such a study on nymphalid butterflies (Janz et al. 2006), we made use of recent advances in the knowledge of phylogeny of the family Nymphalidae to create sister-group comparisons where the clades differed in diversity of host-plant use, taking into account both number of plant families used and the number of different orders to which they belong.

We found 22 sister-group comparisons differing in diversity of host-plant use, and in 18 of these the number of species was larger in the clade with the most diverse host-plant use, a significant correlation in accordance with our prediction. Host plants can hardly be the only factor that drives speciation rates, and this is demonstrated by the fact that if sister groups were instead chosen in the reciprocal manner, according to whether they differed in the number of species, the correlation was weaker; 16 out of 24 contrasts had a more diverse host use in the more species-rich clade (not significant in a sign test). Importantly, this also shows that the causality behind our observed pattern is unlikely to have been the opposite one, that is, species-rich clades do not automatically have a more diverse host-plant use.

The reported pattern could conceivably have also resulted at least in part from colonizations of novel hosts followed by radiation, although we find it likely that diverse host-plant use in a clade generally represents a wide potential range in the clade, in line with the plasticity scenario (Fig. 15.3). Furthermore, the explosive radiation scenario should often produce the opposite pattern, where the happy colonizers have radiated dramatically on a single novel host clade, and a smaller group of species in the sister group is doing more poorly on a more diverse set of ancestral hosts.

In any case, there is a need for more detailed studies. In one such (Nylin and Wahlberg, in press), host-plant shifts in the subfamily Nymphalinae were studied, using a recent phylogeny (Wahlberg et al. 2005). Traditionally, there are three tribes in the subfamily: Nymphalini, Kallimini, and Melitaeini. Nymphalini was found to be the sister group to Kallimini plus Melitaeini, but Kallimini was found to be a paraphyletic "grade," with some genera being closer to Melitaeini than others. The "urticalean rosid" clade (Urticaceae and relatives) is the ancestral host taxon for Nymphalini (Janz et al. 2001), and it was found to also be the likely ancestral host clade for Nymphalinae as a whole. A major host-plant shift to the order Lamiales is superficially evident at the base of Kallimini plus Melitaeini, and a second major colonization of a novel order—Asterales—occurred early in the evolution of Melitaeini. This would at first glance seem consistent with the explosive radiation scenario and the general scheme of Ehrlich and Raven (1964). However, a closer investigation reveals that in the basal parts of the Kallimini grade there are species feeding on both urticalean rosids and Lamiales, genera where different species specialize on one or the other of these orders, and several other respecializations on urticalean rosids after the "shift" to Lamiales. In fact, even though this is a very ancient event (probably about 60 million years ago [Wahlberg, 2006]) we can still see traces of what looks like a long-lasting stage of a wider potential host range. Most likely this range once included other plants besides urticalean rosids and Lamiales, but no traces now remain of these more minor hosts.

A long-lasting stage with a wider host-plant range is even more evident in the case of the more recent colonization of Asterales, which is used alongside Lamiales in many genera and even some species throughout the Melitaeini. Work on dating these phylogenies (Wahlberg, 2006) gives further insights. We find both cases of colonizations of (at the time) young plant clades, and other cases where it is clear that the plant clades involved are much older than the butterfly clades (Nylin and Wahlberg, in press).

Finally, returning to the case of Nymphalini (Janz et al. 2001; sections above), but focusing more closely on species numbers (Weingartner et al. 2006), we find that although only a limited number of contrasts can be made (and some of them are not phylogenetically independent of each other), all of them support our predictions and the general pattern of host use is consistent with the plasticity scenario. The genus *Aglais*, whose larvae feed on the ancestral urticalean rosids, has fewer species than the sister clade *Nymphalis* plus *Polygonia,* whose members use a much wider range of host plants. Based on the fact that the same set of host plants have been repeatedly recolonized in the latter clade, and the fact that juveniles show some capacity to feed on many nonhosts within this set but not on other plants (Janz et al. 2001), we believe that the "backbone" of this clade has a wide potential host-plant range. Within *Polygonia*, there seems to have been several independent respecializations on one or the other family of urticalean rosids, and interestingly this seems to halt the higher speciation rates. In all such instances, there are more species in the sister group, which includes species that use the novel host families.

Conclusion

Host-plant range is dynamic over evolutionary time (Janz et al. 2001; Nosil 2002). We suggest that it is precisely because host specialization is *not* a "dead end" that the guild of plant-feeding insects has been able to become so rich in species. Many previously investigated insect-plant systems could benefit from looking at them again with fresh eyes, not only at the important and very interesting phase where specialization on different plants does its work and may promote speciation, but also at the earlier phase where new host plants were added to the potential range, creating the fuel for host-plant-driven speciation.

Acknowledgments

This work was supported by grants from the Swedish Research Council to N.J. and S.N. We are grateful to M.C. Singer and an anonymous reviewer for valuable comments on the manuscript.

References Cited

Agrawal, A.A. 2000. Host-range evolution: adaptation and trade-offs in fitness of mites on alternative hosts. Ecology 81: 500–508.

Barraclough, T.G., and S. Nee. 2001. Phylogenetics and speciation. Trends Ecol. Evol. 16: 391–399.

Berlocher, S.H. and J.L. Feder. 2002. Sympatric speciation in phytophagous insects: moving beyond controversy? Annu. Rev. Entomol. 47: 773–815.

Bernays, E.A. 2001. Neural limitations in phytophagous insects: implications for diet breadth and evolution of host affiliation. Annu. Rev. Entomol. 46: 703–727.

Bernays, E.A., and W.T. Wcislo. 1994. Sensory capabilities, information processing, and resource specialization. Q. Rev. Biol. 69: 187–204.

Brändle, M., S. Ohlschlager, and R. Brandl. 2002. Range sizes in butterflies: correlation across scales. Evol. Ecol. Res. 4: 993–1004.

Brooks, D.R., and A.L. Ferrao. 2005. The historical biogeography of co-evolution: emerging infectious diseases are evolutionary accidents waiting to happen. J. Biogeogr. 32: 1291–1299.

Bush, G.L. 1975. Sympatric speciation in phytophagous parasitic insects, pp. 187–206. In P.W. Price (ed.), Evolutionary strategies of parasitic insects and mites. Plenum, London.

Carroll, S.P., and C. Boyd. 1992. Host race formation in the soapberry bug: natural history with the history. Evolution 46: 1052–1069.

Cook, J.M., A. Rokas, M. Pagel, and G.N. Stone. 2002. Evolutionary shifts between host oak sections and host-plant organs in *Andricus* gallwasps. Evolution 56: 1821–1830.

Coyne, J.A., and H.A. Orr. 2004. Speciation. Sinauer, Sunderland, MA.

Crane, P.R., E.M. Friis, and K.R. Pedersen. 1995. The origin and early diversification of angiosperms. Nature 374: 27–33.

Darwin, C. 1839. Journal of researches into the geology and natural history of the various countries visited by H.M.S. Beagle under the command of Captain Fitzroy R.N. from 1832 to 1836. Henry Colburn, London.

Dennis, R.L.H., B. Donato, T.H. Sparks, and E. Pollard. 2000. Ecological correlates of island incidence and geographical range among British butterflies. Biodivers. Conserv. 9: 343–359.

Dodd, M.E., J. Silvertown, and M.W. Chase. 1999. Phylogenetic analysis of trait evolution and species diversity variation among angiosperm families. Evolution 53: 732–744.

Drès, M., and J. Mallet. 2002. Host races in plant-feeding insects and their importance in sympatric speciation. Phil. Tran. R. Soc. Lond. B 357: 471–492.

Ehrlich, P.R., and P.H. Raven. 1964. Butterflies and plants: a study in coevolution. Evolution 18: 586–608.

Erwin, T.L. 1981. Taxon pulses, vicariance, and dispersal: an evolutionary synthesis illustrated by carabid beetles,. pp. 159–196. In G. Nelson and D. E. Rosen (eds.), Vicariance biogeography: a critique. Columbia University Press, New York.

Farrell, B.D. 1998. "Inordinate fondness" explained: why are there so many beetles? Science 281: 555–559.

Farrell, B.D., D.E. Dussourd, and C. Mitter. 1991. Escalation of plant defense: do latex and resin canals spur plant diversification? Am. Nat. 138: 881–900.

Feder, J.L., C.A. Chilcote, and G.L. Bush. 1988. Genetic differentiation between sympatric host races of the apple maggot fly *Rhagoletis pomonella.* Nature 336: 61–64.

Forister, M.L. 2005. Independent inheritance of preference and performance in hybrids between host races of *Mitoura* butterflies (Lepidoptera: Lycaenidae). Evolution 59: 1149–1155.

Friis, E.M., K.R. Pedersen, and P.R. Crane. 2005. When Earth started blooming: insights from the fossil record. Curr. Opin. Plant Biol. 8: 5–12.

Funk, D.J., K.E. Filchak, and J.L. Feder. 2002. Herbivorous insects: model systems for the comparative study of speciation ecology. Genetica 116: 251–267.

Futuyma, D.J. 1991. Evolution of host specificity in herbivorous insects: genetic, ecological, and phylogenetic aspects, pp. 431–454. In P.W. Price, T.M. Lewinsohn, G.W. Fernandes, and W.W. Benson (eds.), Plant-animal interactions: evolutionary ecology in tropical and temperate regions. John Wiley and Sons, New York.

Futuyma, D.J. 1998. Evolutionary biology, 3d ed. Sinauer, Sunderland, MA.

Futuyma, D.J., and C. Mitter. 1996. Insect-plant interactions: the evolution of component communities. Phil. Tran. R. Soc. Lond. B 351: 1361–1366.

Futuyma, D.J., M.C. Keese, and S.J. Scheffer. 1993. Genetic constraints and the phylogeny of insect-plant associations: responses of *Ophraella communa* (Coleoptera, Chrysomelidae) to host plants of its congeners. Evolution 47: 888–905.

Futuyma, D.J., J.S. Walsh, T. Morton, D.J. Funk, and M.C. Keese. 1994. Genetic variation in a phylogenetic context: responses of two specialized leaf beetles (Coleoptera, Chrysomelidae) to host plants of their congeners. J. Evol. Biol. 7: 127–146.

Futuyma, D.J., M.C. Keese, and D.J. Funk. 1995. Genetic constraints on macroevolution: the evolution of host affiliation in the leaf beetle genus *Ophraella*. Evolution 49: 797–809.

Grimaldi, D. 1999. The co-radiations of pollinating insects and angiosperms in the Cretaceous. Ann. Mo. Bot. Gardens 86: 373–406.

Hagen, R.H., R.C. Lederhouse, J.L. Bossart, and J.M. Scriber. 1991. *Papilio canadensis* and *P. glaucus* (Papilionidae) are distinct species. J. Lepidopt. Soc. 45: 245–258.

Hawthorne, D.J., and S. Via. 2001. Genetic linkage of ecological specialization and reproductive isolation in pea aphids. Nature 412: 904–907.

Jaenike, J. 1990. Host specialization in phytophagous insects. Annu. Rev. Ecol. Syst. 21: 243–273.

Janz, N. 1998. Sex-linked inheritance of host-plant specialization in a polyphagous butterfly. Proc. R. Soc. Lond. B 265: 1675–1678.

Janz, N. 2003. Sex-linkage of host plant use in butterflies, pp. 229–239. In C.L. Boggs, P.R. Ehrlich, and W.B. Watt (eds.), Butterflies: ecology and evolution taking flight. University of Chicago Press, Chicago.

Janz, N., and S. Nylin. 1997. The role of female search behaviour in determining host plant range in plant feeding insects: a test of the information processing hypothesis. Proc. R. Soc. Lond. B 264: 701–707.

Janz, N., and S. Nylin. 1998. Butterflies and plants: a phylogenetic study. Evolution 52: 486–502.

Janz, N., and J.N. Thompson. 2002. Plant polyploidy and host expansion in an insect herbivore. Oecologia 130: 570–575.

Janz, N., S. Nylin, and N. Wedell. 1994. Host plant utilization in the comma butterfly: sources of variation and evolutionary implications. Oecologia 99: 132–140.

Janz, N., S. Nylin, and K. Nyblom. 2001. Evolutionary dynamics of host plant specialization: a case study of the tribe Nymphalini. Evolution 55: 783–796.

Janz, N., S. Nylin, and N. Wahlberg. 2006. Diversity begets diversity: host expansions and the diversification of plant-feeding insects. BMC Evol. Biol. 6: 4.

Janzen, D.H. 1985. On ecological fitting. Oikos 45: 308–310.

Jermy, T. 1984. Evolution of insect/host plant relationships. Am. Nat. 124: 609–630.

Joshi, A., and J.N. Thompson. 1995. Trade-offs and the evolution of host specialization. Evol. Ecol. 9: 82–92.

Kelley, S.T., B.D. Farrell, and J.B. Mitton. 2000. Effects of specialization on genetic differentiation in sister species of bark beetles. Heredity 84: 218–227.

Kitahara, M., and K. Fujii. 1994. Biodiversity and community structure of temperate butterfly species within a gradient of human disturbance: an analysis based on the concept of generalist vs. specialist strategies. Res. Popul. Ecol. 36: 187–199.

Kuussaari, M., M. Singer, and I. Hanski. 2000. Local specialization and landscape-level influence on host use in an herbivorous insect. Ecology 81: 2177–2187.

Mackenzie, A. 1996. A trade-off for host plant utilization in the black bean aphid, *Aphis fabae*. Evolution 50: 155–162.

Marvaldi, A.E., A.S. Sequeira, C.W. O'Brien, and B.D. Farrell. 2002. Molecular and morphological phylogenetics of weevils (Coleoptera, Curculionoidea): do niche shifts accompany diversification? Syst. Biol. 51: 761–785.

Mitter, C., B. Farrell, and B. Wiegmann. 1988. The phylogenetic study of adaptive zones: has phytophagy promoted insect diversification? Am. Nat. 132: 107–128.

Morse, G.E., and B.D. Farrell. 2005. Interspecific phylogeography of the *Stator limbatus* species complex: the geographic context of speciation and specialization. Mol. Phyl. Evol. 36: 201–213.

Nitao, J.K., M.P. Ayres, R.C. Lederhouse, and J.M. Scriber. 1991. Larval adaptation to lauraceous hosts: geographic divergence in the spicebush swallowtail butterfly. Ecology 72: 1428–1435.

Nosil, P. 2002. Transition rates between specialization and generalization in phytophagous insects. Evolution 56: 1701–1706.

Nylin, S. 1988. Host plant specialization and seasonality in a polyphagous butterfly, *Polygonia c-album* (Nymphalidae). Oikos 53: 381–386.

Nylin, S., G.H. Nygren, J.J. Windig, N. Janz, and A. Bergström. 2005. Genetics of host-plant preference in the comma butterfly *Polygonia c-album* (Nymphalidae), and evolutionary implications. Biol. J. Linn. Soc. 84: 455–765.

Nylin, S., and N. Wahlberg. Does plasticity drive speciation? Host plant shifts and diversification of nymphaline butterflies during the tertiary. Biol. J. Linn. Soc., in press.

Päivinen, J., A. Grapputo, V. Kaitala, A. Komonen, J.S. Kotiaho, K. Saarinen, and N. Wahlberg. 2005. Negative density-distribution relationship in butterflies. BMC Biol. 3: 5.

Pellmyr, O. 1992. Evolution of insect pollination and angiosperm diversification. Trends Ecol. Evol. 7: 46–49.

Peterson, M.A., and R.F. Denno. 1998. The influence of dispersal and diet breadth on patterns of genetic isolation by distance in phytophagous insects. Am. Nat. 152: 428–446.

Prowell, D.P. 1998. Sex linkage and speciation in Lepidoptera, pp. 309–319. In S. Berlocher and D. Howard (eds.), Endless forms: species and speciation. Oxford University Press, Oxford.

Quinn, R.M., K.J. Gaston, and D.B. Roy. 1998. Coincidence in the distributions of butterflies and their foodplants. Ecography 21: 279–288.

Rundle, H.D., and P. Nosil. 2005. Ecological speciation. Ecol. Lett. 8: 336–352.

Schluter, D. 2000. The ecology of adaptive radiation. Oxford University Press, Oxford.

Schluter, D. 2001. Ecology and the origin of species. Trends Ecol. Evol. 16: 372–380.

Scriber, J.M. 1973. Latitudinal gradients in larval feeding specialization of the world Papilionidae (Lepidoptera). Psyche 80: 355–373.

Scriber, J.M. 1988. Tale of the tiger: beringial biogeography, binomial classification, and breakfast choices in the *Papilio glaucus* complex of butterflies, pp. 241–301. In K. C. Spencer (ed.), Chemical mediation of coevolution. Academic Press, New York.

Scriber, J.M. 1994. Climatic legacies and sex chromosomes: latitudinal patterns of voltinism, diapause, body size, and host-plant selection on two species of swallowtail butterflies at their hybrid zone. In H.V. Danks (ed.), Theory, evolution and ecological consequences for seasonality and diapause control. Kluwer Academic Publishers, Dordrecht, Netherlands.

Scriber, J.M., and R.C. Lederhouse. 1992. The thermal environment as a resource dictating geographic patterns of feeding specialization of insect herbivores, pp. 429–466. In M.R. Hunter, T. Ohgushi, and P.W. Price (eds.), Effects of resource distribution on animal-plant interactions. Academic Press, New York.

Scriber, J.M., B.L. Giebink, and D. Snider. 1991. Reciprocal latitudinal clines in oviposition behaviour of *Papilio glaucus* and *P. canadensis* across the Great Lakes hybrid zone: possible sex-linkage of oviposition preferences. Oecologia 87: 360–368.

Segraves, K.A., J.N. Thompson, P.S. Soltis, and D.E. Soltis. 1999. Multiple origins of polyploidy and the geographic structure of *Heuchera grossulariifolia*. Mol. Ecol. 8: 253–262.

Singer, M.C., and C.D. Thomas. 1996. Evolutionary responses of a butterfly metapopulation to human- and climate-caused environmental variation. Am. Nat. 148: S9–S39.

Singer, M.C., C.D. Thomas, H.L. Billington, and C. Parmesan. 1989. Variation among conspecific insect populations in the mechanistic basis of diet breadth. Anim. Behav. 37: 751–759.

Singer, M.C., C.D. Thomas, and C. Parmesan. 1993. Rapid human-induced evolution of insect-host associations. Nature 366: 681–683.

Sperling, F.A.H. 1994. Sex-linked genes and species differences in Lepidoptera. Can. Entomol. 126: 807–818.

Stireman, J.O. 2005. The evolution of generalization? Parasitoid flies and the perils of inferring host range evolution from phylogenies. J. Evol. Biol. 18: 325–336.

Strong, D.R. 1974. Rapid asymptotic species accumulation in phytophagous insect communities: the pests of cacao. Science 185: 1064–1066.

Strong, D.R., E.D. McCoy, and J.R. Rey. 1977. Time and the number of herbivore species: the pests of sugarcane. Ecology 58: 167–175.

Stuessy, T.F. 2004. A transitional-combinational theory for the origin of angiosperms. Taxon 53: 3–16.

Tabashnik, B.E. 1983. Host range evolution: the shift from native legume hosts to alfalfa by the butterfly *Colias philodice eriphyle*. Evolution 37: 150–162.

Thompson, J.N. 1994. The coevolutionary process. University of Chicago Press, Chicago.

Thompson, J.N. 2005. The geographic mosaic of coevolution. University Of Chicago Press, Chicago.

Thompson, J.N., B.M. Cunningham, K.A. Segraves, D.M. Althoff, and D. Wagner. 1997. Plant polyploidy and insect/plant interactions. Am. Nat. 150: 730–743.

Thompson, J.N., S.L. Nuismer, and K. Merg. 2004. Plant polyploidy and the evolutionary ecology of plant/animal interactions. Biol. J. Linn. Soc. 82: 511–519.

Via, S. 1991. The population structure of fitness in a spatial network: demography of pea aphid clones from two crops in a reciprocal transplant. Evolution 45: 827–852.

Via, S. 2001. Sympatric speciation in animals: the ugly duckling grows up. Trends Ecol. Evol. 16: 381–390.

Wahlberg, N., A.V.Z. Brower, and S. Nylin. 2005. Phylogenetic relationships and historical biogeography of tribes and genera in the subfamily Nymphalinae (Lepidoptera: Nymphalidae). Biol. J. Linn. Soc. 86: 227–251.

Wahlberg, N. 2006. That awkward age for butterflies: insights from the age of the butterfly subfamily Nymphalinae (Lepidoptera: Nymphalidae). Syst. Biol., 55: 703–714.

Waser, N.M. 1998. Pollination, angiosperm speciation, and the nature of species boundaries. Oikos 82: 198–201.

Weingartner, E., N. Wahlberg, and S. Nylin. 2006. Dynamics of host plant use and species diversity in *Polygonia* butterflies (Nymphalidae). J. Evol. Biol. 19: 483–491.

West-Eberhard, M.J. 1989. Phenotypic plasticity and the origins of diversity. Annu. Rev. Ecol. Syst. 20: 249–278.

West-Eberhard, M.J. 2003. Developmental plasticity and evolution. Oxford University Press, New York.

Wikström, N., V. Savolainen, and M.W. Chase. 2001. Evolution of the angiosperms: calibrating the family tree. Proc. R. Soc. Lond. B 268: 2211–2220.

Wilkinson, D.M. 2004. The parable of Green Mountain: ascension island, ecosystem construction and ecological fitting. J. Biogeogr. 31: 1–4.

Willmott, K.R., and J. Mallet. 2004. Correlations between adult mimicry and larval host plants in ithomiine butterflies.Proc. R. Soc. Lond. B 271: S266–S269.

SIXTEEN

Coevolution, Cryptic Speciation, and the Persistence of Interactions

JOHN N. THOMPSON

We are faced with three seemingly conflicting observations regarding the diversification of plant-feeding insects. Insects can evolve at astoundingly rapid rates when confronted with new selection pressures, as shown in hundreds of studies in recent decades. Nevertheless, most insect lineages remain highly conservative in the range of species with which they interact. Occasionally, though, insects make great phylogenetic jumps, even jumping between eudicotyledonous and monocotyledonous plant taxa. It is the juxtaposition of these three observations that has historically created the conceptual tension between fields of study that focus on current selection, such as evolutionary ecology and population genetics, and fields that focus on higher-level patterns in the diversification of life, such as systematics and paleobiology.

Studies of the geographic mosaic of coevolution, specialization, and population divergence have begun to resolve these apparently conflicting observations. In this chapter I discuss how our developing knowledge of the geographic mosaic of coevolution and cryptic speciation may help us better understand how plant-insect interactions persist for millions of years despite ongoing rapid evolution in ever-changing environments. Throughout the chapter I focus on interactions between prodoxid moths in the genus *Greya*, their host plants, and parasitoids as exemplars of how our understanding of the dynamics of interactions involving phytophagous insects is advancing as researchers integrate results from multiple subdisciplines.

A Blending of Perspectives: Populations, Species, and Species Interactions

One of the most important results we have obtained from detailed study of insect species over the past several decades is that insect populations may often become adapted to their local host species (Craig et al. 2000; Boggs et al. 2003; Singer 2003; Ehrlich and Hanski 2004; Fischer and Foitzik 2004; Gotthard et al. 2004; Nielsen and de Jong 2005). Moreover, it is becoming evident that local adaptation may often occur at a finer geographic scale than is evident in phylogeographic analyses on the scale of population differentiation identified by using neutral molecular markers. For example, multiple *Greya* moth species, which are close relatives of yucca moths, occur in the northern U.S. Rocky Mountains (Thompson 1985; Davis et al. 1992; Pellmyr et al. 1996). Populations of some of these species differ geographically in their use of plant species, even within regions in which there is little evidence of neutral molecular differentiation among populations (Brown et al. 1997; Thompson et al. 1997; Janz and Thompson 2002; Thompson and Cunningham 2002). In turn, some wasps in the braconid genus *Agathis*, which attack larvae of *Greya* moths by searching within and among host plants, differ geographically in their search behaviors in ways that match the local availability of *Greya* larvae (Althoff and Thompson 2001). As with the moths, the geographic scale of differences in searching behavior is smaller than the scale of differentiation found in concomitant analyses of neutral molecular markers (Althoff and Thompson 1999, 2001). Hence, there is a complex mosaic of ecological differentiation in these plant-insect-parasitoid interactions that is underestimated by analyses of neutral molecular differentiation among populations.

Differences in the geographic scale of population differentiation obtained by these different kinds of analysis should not be surprising. These moths and parasitoids are living in post-Pleistocene environments that have been available for only about the past 10,000 years, allowing little time for molecular differentiation in genes that are diverging through mutation and random genetic drift alone. Genes under strong natural selection should be expected to evolve much faster than molecular regions that are diverging through mutation and random genetic drift alone, especially in fairly large populations. Hence, the scale of population differentiation is much more apparent in genes subject to selection than in molecular regions assessed using neutral markers. The important point is that insect populations probably are often more differentiated from one another than is evident in the burgeoning number of studies of molecular differentiation among insect populations.

Moreover, we know that evolutionary changes in local adaptation can sometimes occur very quickly, reshaping insect populations within decades, as shown in the adaptation of some insect populations to the introduction of novel plant species. Examples include rapid biochemical adaptation of lepidopteran larvae to the novel defensive plant chemistry of invasive plants (Berenbaum and Zangerl 1998; Zangerl and Berenbaum 2003), modification of the length of the piercing mouthparts and other traits of hemipterans to gain access to seeds of introduced plants (Carroll et al. 1998, 2001), and changes in oviposition preference in novel environments containing introduced plants (Singer and Thomas 1996; Forbes et al. 2005). Collectively, studies of local adaptation and of rapid evolutionary response to novel hosts have shown that insect species are collections of genetically distinct populations that are continually reshaped by natural selection.

These differences among populations in local selection and adaptation may often result from a geographic selection mosaic in which a particular insect genotype may interact with plants, and also with its parasitoids, in different ways in different environments. That is, in the parlance of population genetics, interspecific interactions may exhibit a genotype by genotype by environment interaction. Each interacting insect population and plant population is therefore a potentially novel coevolutionary experiment.

Within this geographic selection mosaic, coevolution is likely to occur only in some environments, because reciprocal selection will not be equally strong in all environments. These coevolutionary hotspots, in which selection is truly reciprocal, will often be embedded in a broader geographic matrix of coevolutionary coldspots, where selection may act on one of the species but not on the other. For example, in some environments a local insect population may depend solely on a particular plant species without affecting the fitness of that host species, whereas in other environments herbivory or pollination by a particular insect may have large effects on plant fitness. In addition, through gene flow, random genetic drift, and metapopulation dynamics, novel traits will continue to appear, spread, and disappear, creating an ever-changing mix of adaptation and counter-adaptation in coevolving species. The result is an ever-changing mosaic of adaptation and coadaptation across landscapes, sometimes imposed upon larger-scale clines.

The combined effects of geographic selection mosaics, coevolutionary hotspots, and trait remixing compose the geographic mosaic theory of coevolution (Thompson 1994, 2005). Mathematical models of this view of coevolution have shown that such spatially structured coevolution can lead to rapid and ongoing change in interacting species, while allowing long-term persistence of the interaction (Gomulkiewicz et al. 2000; Nuismer et al. 2000, 2003; Gandon and Michalakis 2002). During that process, some local populations persist, whereas others go extinct, as the interaction continues to evolve in different ways in different environments. The "interaction" between an insect species and a plant species is therefore often a collection of interactions that may differ across populations and environments in the traits subject to selection and in ecological outcome (e.g., antagonism, commensalism, mutualism).

Most of these differences among populations driven by coevolution are likely local adaptations that do not lead to long-term directional change. They are, in effect, coevolutionary meanderings, moving populations around in a restricted arena of possible phenotypes and patterns of specialization. These ongoing evolutionary and coevolutionary changes, however, are of great short- and long-term ecological importance, because they keep the players in the evolutionary game. In a very rough simile, they are like two people dancing along a mountain road. Their dancing is bounded by the road. Unless they can dance back and forth along that road, they will eventually get hit by a truck. Similarly, unless populations and interactions continually adapt to locally changing conditions, they are likely to go extinct.

Evidence of geographic selection mosaics and coevolutionary hotspots in coevolving interactions has been accumulating in recent years for an increasingly wide range of interactions (Benkman et al. 2001; Brodie et al. 2002; Benkman 2003; Zangerl and Berenbaum 2003; Fischer and Foitzik 2004; Laine 2005; Nielsen and de Jong 2005). For example, the prodoxid moth *Greya politella* is a major pollinator of woodland stars, *Lithophragma parviflorum* (Saxifragaceae), in some, but not all, populations within the northern U.S. Rocky Mountains (Thompson and Pellmyr 1992; Pellmyr and Thompson 1996; Thompson and Fernandez 2006). A female *G. politella* lays eggs in the flowers of *Lithophragma* by inserting her abdomen through the corolla and piercing the ovary with her ovipositor. During oviposition, pollen adhering to the base of her abdomen from previous flowers she visited rubs off onto the stigma. In populations where there are few copollinators, plants depend upon these ovipositing *Greya* females for pollination. In other populations, however, copollinators are common.

Long-term ecological studies across the complex mountainous landscapes of the northern Rockies have shown that some local interactions between *Greya* and *Lithophragma* are consistently mutualistic, whereas others are commensalistic or antagonistic (Thompson and Cunningham 2002; Thompson and Fernandez 2006). In these studies, mutualism means that the moths completely depend on that host plant and that flowers receiving moth eggs are more likely to develop mature seeds than flowers without eggs. Antagonism means that flowers with *Greya* eggs were more likely to be aborted than flowers without eggs.

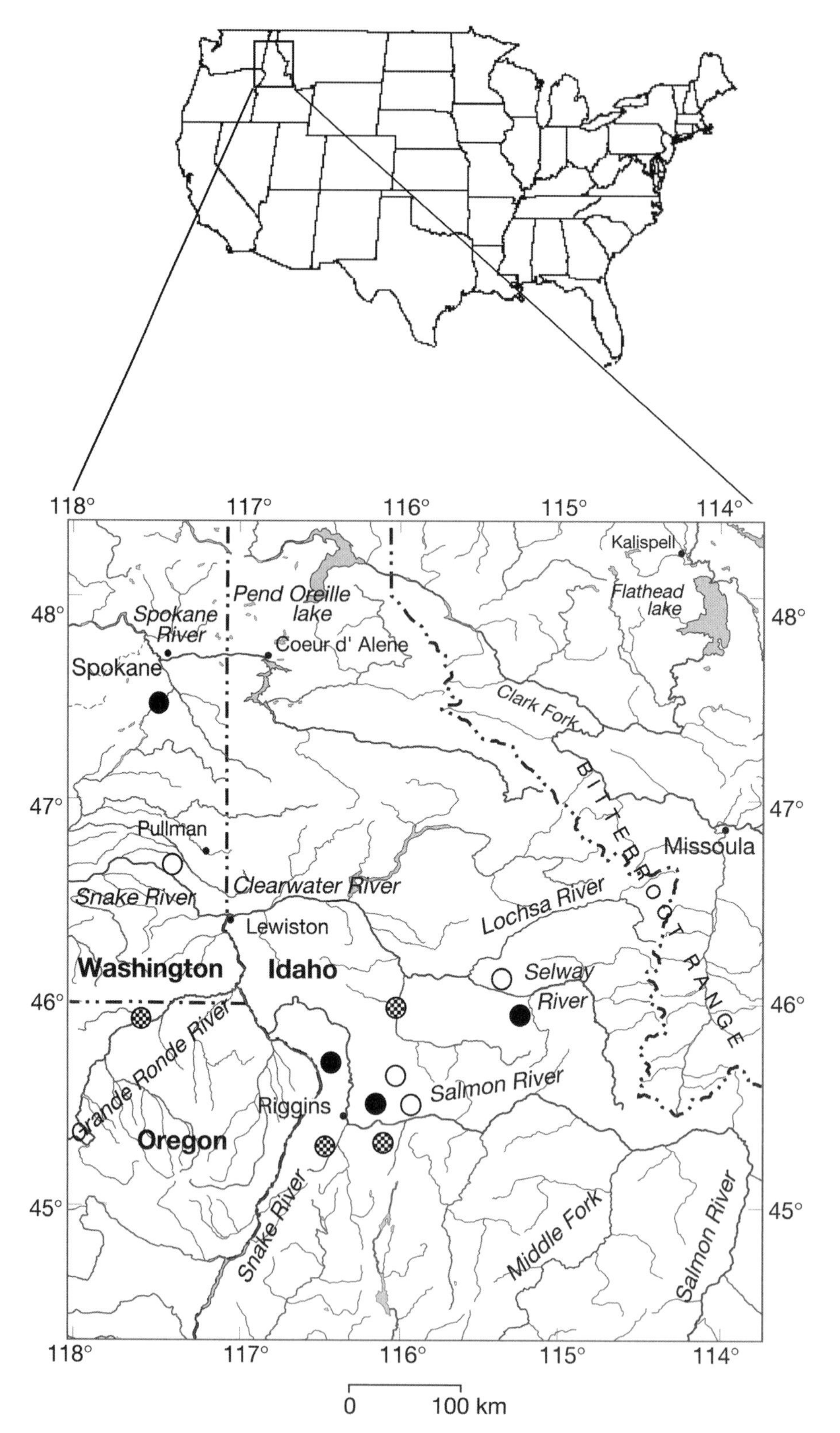

FIGURE 16.1. Distribution of sampled populations of the plant *Lithophragma parviflorum* that either rely upon the moth *Greya politella* for pollination in at least some years (black circles), selectively abort floral capsules containing *Greya* larvae (checked circles), or have no effect on the probability that a flower will develop mature seeds in any year studied (open circles). All populations shown have been studied for at least three years, and some have been studied for nine years. From Thompson (2005), modified from Thompson and Cunningham (2002).

Moreover, these different ecological outcomes are distributed in a mosaic pattern across these mountainous habitats (Fig. 16.1). During seven years of study of the interaction at Turnbull National Wildlife Refuge in Washington State, the interaction was mutualistic in five years and commensalistic in two years (Thompson and Fernandez 2006). In contrast, during the same seven years, the interaction at Rapid River, Idaho, was antagonistic in two years and commensalistic in five years. Direct pairwise comparison of flowers containing *Greya* eggs to those lacking *Greya* eggs on the same plants in two populations during four years showed that these populations differ strongly and consistently in the range of

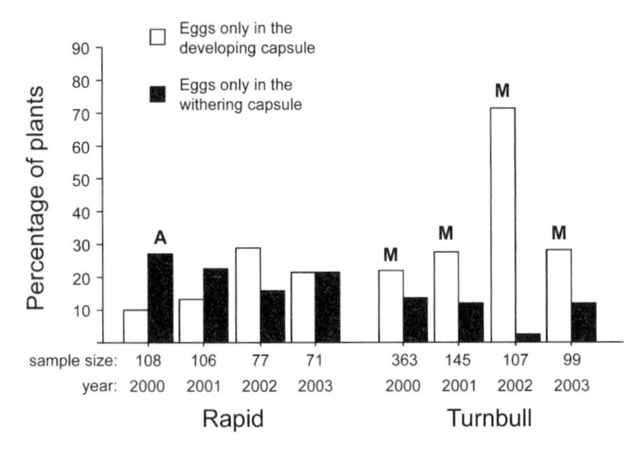

FIGURE 16.2. Ecological outcome of the interaction between *Greya politella* and *Lithophragma parviflorum* at Turnbull National Wildlife Refuge in Washington State and Rapid River, Idaho. Each pair of bars indicates whether the developing flower or the aborted flower is more likely to have *Greya* eggs within pairs of developing and withered flowers collected from the same plant. Each comparison is restricted to plants in which only one of the two flowers has eggs, because only those comparisons provide information for the analysis. Sample sizes indicate the number of plants from which pairs were drawn. Significant differences ($P < 0.05$) were determined using a paired sign test for each population and year and are shown as either A, significantly antagonistic, or M, significantly mutualistic. Comparisons without A or M indicate that the interactions are ecologically commensalistic. From Thompson and Fernandez (2006).

ecological outcomes between the plants and the moths (Fig. 16.2). Overall, the results of these ecological studies in multiple habitats have suggested that this interaction exhibits mutualistic coevolutionary hotspots, antagonistic coevolutionary hotspots, and commensalistic coevolutionary coldspots across the landscapes of the northern U.S. Rocky Mountains.

Evidence of geographic selection mosaics is now known for some other interactions between insects and plants. For example, in Denmark, populations of the crucifer *Barbarea vulgaris* differ geographically in the presence of a compound toxic to many insects (Nielsen 1997). Some populations of the flea beetle *Phyllotreta nemorum*, however, have a set of alleles that allow them to overcome these defenses (de Jong and Nielsen 1999, 2002; Nielsen 1999), creating a geographic mosaic of plant and beetle populations with or without these defenses and counterdefenses. The ability of beetles to use the resistant plants is controlled by a major dominant gene. The frequency of beetles harboring this gene varies among populations, depending upon the local presence of alternative host plants in different sites and years (Nielsen and de Jong 2005). Hence, the geographic mosaic of this interaction depends upon the geographic distribution of host plants available to the beetles.

There is now even evidence that coevolutionary mosaics can form very quickly. In North America, introduced wild parsnips *(Pastinaca sativa)* produce a mix of furanocoumarins that act as effective chemical defenses against herbivores. However, the specialist oecophorid moth *Depressaria pastinacella* has chemical counterdefenses against these compounds. Studies of multiple populations throughout the midwestern United States have shown that the ratios of different furanocoumarins differ among parsnip populations, and the ratios of the counterdefenses in the moths also vary in ways that match the plant populations (Berenbaum and Zangerl 1998; Zangerl and Berenbaum 2003). Related studies have shown that these two species have significant effects on each other's Darwinian fitness (Berenbaum and Zangerl 1994; Zangerl and Berenbaum 1997). These studies therefore suggest that wild parsnips and parsnip webworms have been rapidly coevolving across the landscapes of the New World during the past several hundred years since these plants were first introduced and escaped from gardens.

Thus, there is a growing realization that interactions between insects and plants do not simply coevolve over millions of years in a slow and stately way through long-term directional selection. They are continually evolving in multiple ways across ever-changing landscapes, sometimes over time scales that we can observe within human lifetimes. Each coevolving plant and insect species is a collection of small coevolutionary experiments, most of which will fail but some of which result in novel adaptations that allow persistence of the interaction over time in a subset of landscapes. The various degrees of population divergence across the continuum from local populations to species therefore fuel ongoing coevolutionary change between plants and insects.

Cryptic Speciation in Insects

An almost inevitable consequence of the geographic mosaic of coevolution, at least in widespread species, is the formation of one or more highly divergent populations that have the potential to form new species. The potential for the development of host races in insects, either as allopatric or sympatric populations, has a long (see review in Berlocher and Feder 2001) and continuing history of study in insect evolutionary biology (Abrahamson et al. 2001; Emelianov et al. 2001; Drès and Mallet 2002; Blair et al. 2005; Schwarz et al. 2005). As more insect species have been studied across multiple populations, it is becoming evident that insect populations often differ geographically in their relative oviposition preference for particular plant species, or their ability to feed on different plant species. Populations or diverging lineages of sibling species within familiar taxa such as *Euphydryas*, *Polygonia*, and *Papilio* butterflies all show geographic differences among populations in their relative preference or performance among plant species (Thompson 1998; Janz 2003; Singer 2003; Ehrlich and Hanski 2004; Nylin et al. 2005; Scriber and Ording 2005), as do species in multiple other insect taxa (Sezer and Butlin 1998; Craig et al. 2000).

Relative preference for hosts often varies within narrow bounds among populations within any one species, but it can vary almost as a continuum among populations within

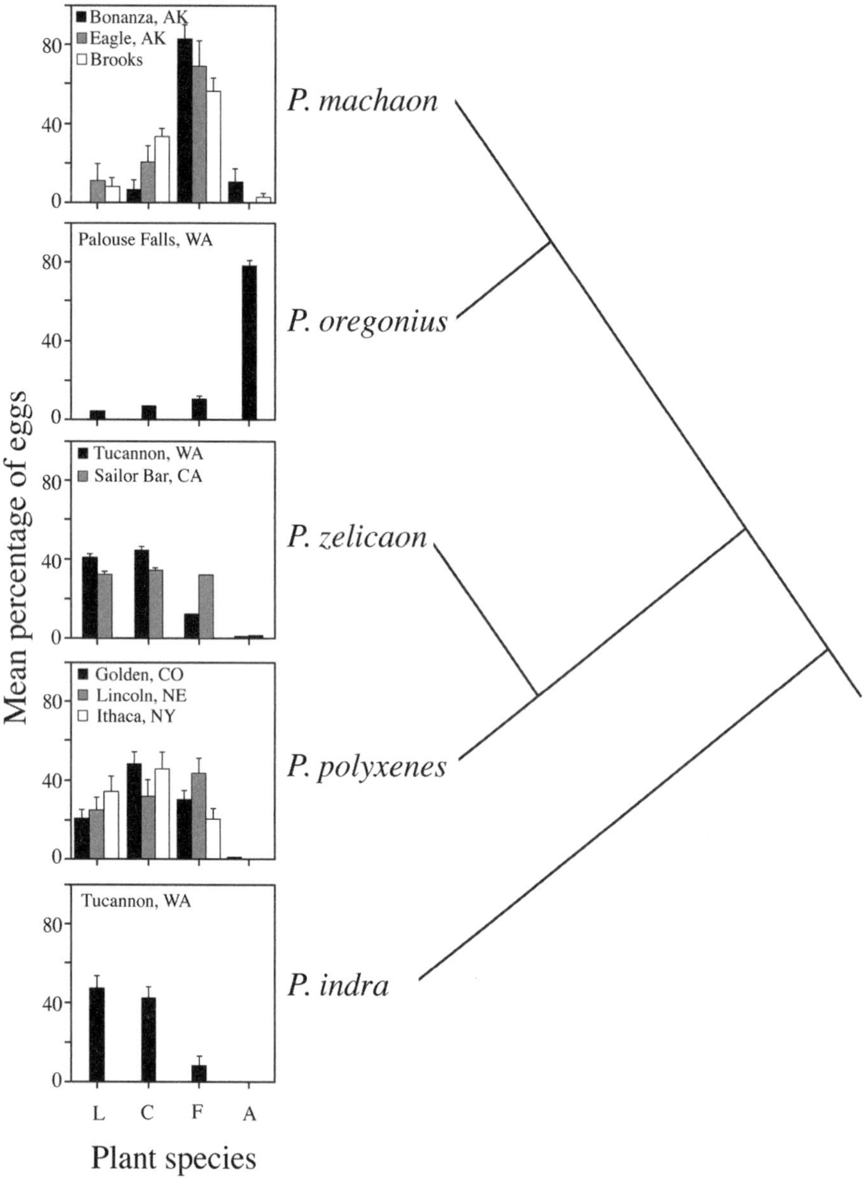

FIGURE 16.3. Variation in oviposition preference hierarchies among populations and species within the *Papilio machaon* species complex of swallowtail butterflies in North America. L, *Lomatium grayi* (Apiaceae); C, *Cymopterus terebinthinus* (Apiaceae); F, *Foeniculum vulgare* (Apiaceae); A, *Artemisia dracunculus* (Asteraceae). Preference hierarchies differ significantly among these species and also among populations within species, based upon multivariate analysis of variance. Extreme populations are shown for species in which multiple populations have been tested. Modified from Thompson (1998).

a group of closely related species. A continuum—or something approaching a continuum—of differentiation can sometimes be observed if multiple populations within a monophyletic group of insect species are tested experimentally using a standard set of host plants (Fig. 16.3). For example, in the *Papilio machaon* species complex in North America, each species is distinct in its relative preferences for a standard reference set of four plant species (Thompson 1998). However, within each species there are some geographic differences among populations. For example, the Sailor Bar, California, population of the western anise swallowtail *(P. zelicaon),* differs from all other populations that have been studied of this species in its relatively high preference for fennel *(Foeniculum vulgare),* which is an introduced plant now common in this part of California (Thompson 1993). This population differs significantly from 13 populations studied in the Pacific Northwest (Wehling and Thompson 1997), of which one is shown in Fig. 16.3. Other populations within the *P. machaon* group show a wide range of relative preferences among these four plant species.

These differences in relative preference become the basis for new tips on phylogenetic branches in the diversification of insects, and the boundary between differentiated populations and species becomes obscure as populations diverge from one another at different rates. Studies of the moth *Greya politella* have provided evidence for how cryptic species may develop as insect and plant populations diversify, adding another level to the geographic mosaic of interactions among plant and insect lineages. In addition to the differences in ecological outcome found among interacting populations of *Greya* and *Lithophragma* in the northern U.S. Rocky Mountains, a few *Greya* populations have colonized a closely related plant species, *Heuchera grossulariifolia*. The two plant species occur parapatrically, and sometimes sympatrically, along some of the major river drainages of northern Idaho. *Lithophragma* and *Heuchera* moths differ in their preference for laying eggs on the other host. Moreover, when sympatric, the two hosts are often phenologically separated. Hence, the *Lithophragma*-feeding moth populations and the *Heuchera*-feeding moth populations function effectively as separate species or almost separate species. The degree of gene flow between these moth populations has not yet been assessed quantitatively, but the differences in phenology of the plants suggest that the moth populations are unlikely to be panmictic.

Similar, or even stronger, cases of demonstrated or potential cryptic speciation in phytophagous insects are accumulating as molecular and ecological data are identifying divergence among populations once thought to be single species (Jordal et al. 2004; Scriber and Ording 2005). Among the most astonishing are the results of molecular analyses of the neotropical skipper butterfly *Astraptes fulgerator*, which have shown that this species is a group of at least 10 species (Hebert et al. 2004). Most of these 10 species differ within Costa Rica in food plants used by the caterpillars and in ecosystem preferences. Similarly, isozyme analyses of the tumbling flower beetle *Mordellistena convicta* have indicated that this species is, in fact, a complex of at least six cryptic species, each with fewer hosts than have been reported for the nominal species (Blair et al. 2005). Evidence of cryptic speciation is also accumulating for insect taxa often considered to be less "specialized" in their interactions with other species, such as predatory insects including lacewings (Henry et al. 1999) and damselflies (Stoks et al. 2005).

Cryptic species complexes add a previously hidden level to the process of adaptation, speciation, and adaptive radiation. Complexes of reproductively isolated cryptic species driven by specialization to different hosts are an extension of the geographic mosaic of evolving interactions between insects and plants. Some of these novel interactions, such as those between *Greya* moths and *Heuchera* plants in northern Idaho, may become failures over the long term. Alternatively, these interactions at the northern periphery of the species ranges could persist while the interactions between *Greya* and *Lithophragma* in other geographic areas become extinct. In general, the evolution of divergent populations or cryptic species on different, but related, hosts adds another level to the ongoing dynamics of the geographic mosaic of evolving insect-plant interactions.

Cryptic Speciation in Plants and Its Effect on Interactions with Insects

Cryptic speciation within plant lineages contributes yet another potentially important dimension to the dynamics of adaptation, coevolution, and speciation in insect-plant interactions. Most plant lineages include at least some polyploid taxa, and many plant lineages include species differing in level of ploidy. Hence, genome duplication has been a major mechanism by which plants have evolved. Beginning in the 1990s, it started to become evident that repeated evolution of polyploidy was much more common within plant lineages than previously suspected (Soltis and Soltis 1995, 1999; Soltis et al. 2004). Some plant species with a single Latin binomial name turned out to be complexes of diploid and polyploid populations, with polyploids arising multiple times from the diploid ancestor (Segraves et al. 1999). Since reproduction between diploid and polyploid populations may often be reduced, at least some of these diploid-polyploid complexes function as cryptic species pairs or swarms.

This realization created the opportunity to begin asking how the repeated evolution of polyploid populations may shape the ecology and evolution of plant-insect interactions. The initial studies again involved the moth *Greya politella* and its host plants. Studies of the populations of *H. grossulariifolia* used by *G. politella* in northern Idaho showed that these plant populations differ considerably in levels of *Greya* attack. The plant family Saxifragaceae, within which *Heuchera* occurs, is well known for harboring many polyploid complexes of species, and *H. grossulariifolia* was known to include diploids and autotetraploids. Further study showed that some, but not all, naturally occurring sympatric diploids and tetraploids could be separated by eye, and that that polyploid populations had arisen multiple times within these plant species (Segraves et al. 1999). Most importantly, *G. politella* distinguished between these plants, preferring tetraploids to diploids when ovipositing (Thompson et al. 1997; Nuismer and Thompson 2001; Janz and Thompson 2002).

Further studies have shown that other insect herbivores associated with *H. grossulariifolia* also differentiate between diploids and tetraploids. Two other major herbivores attack these *Heuchera* populations and both are lepidopterans: *G. piperella* and the geometrid moth *Eupithecia misturata*. Both of these species preferentially attack diploid plants rather than neighboring tetraploid plants (Nuismer and Thompson 2001).

In addition, the suites of floral visitors differ between diploids and tetraploids, as shown in studies in which the pattern of visits has been recorded to diploid and tetraploid plants growing within a meter of each other. Along the Salmon River in Idaho, the rate of insect visitation to diploid and tetraploid plants does not differ, but some species such as the bee *Lasioglossum* spp. visit diploid plants more often, whereas other species such as the bombyliid fly *Bombylius major* visit tetraploid plants more often. In fact, one bee species, *Bombus centralis*, differs among castes in its pattern of visitation. Queens visit tetraploids more often than diploids, whereas workers visit diploids (Segraves and Thompson 1999; K. Merg and J.N.T., unpublished data). Some of these differences in visitation reflect phenological differences in the flowering time of diploids and tetraploids, and recent studies have suggested that sympatric diploid and tetraploid populations are under variable selection for divergence from each other in flowering time (Nuismer and Cunningham 2005). Nevertheless, even when comparisons are restricted to neighboring diploid and tetraploid plants flowering at the same time, plants of different ploidy have different patterns of visitation.

Similar differences in insect visitation to diploids and polyploids have also been shown for insects visiting fireweed, *Chamerion angustifolium*. In one study that analyzed the patterns of pollinator movement in a mixed population of diploids and tetraploid plants, bees visited tetraploids more than diploids (Husband and Schemske 2000; Husband and Sabara 2003). Some of the differential use of tetraploids, however, may have resulted from differences in clumping among diploid and tetraploid plants, which may affect foraging behavior in some insect species. Regardless of the cause, this study provides another indication that the evolution of polyploidy in plants may have important impacts on the ecology and evolution of plant-insect interactions.

Nevertheless, the studies of *Heuchera* and *Chamerion* are the only long-term, ongoing analyses of how plant diversification through polyploidy has shaped the adaptation and the diversification of plant-insect interactions. It is probably one of the largest gaps in our understanding of the ecology and evolution of insect-plant interactions. More studies of other species are now underway in multiple laboratories, but there are still very few. This is remarkable, given the pervasiveness of polyploidy among plant taxa and the crucial role of polyploidy in agriculture. Most of the most important crop plants on earth are polyploid. Hence, we need to understand much better the role of plant polyploidy in shaping the ecology and diversification of insect-plant species through adaptation and cryptic speciation.

Implications

Our increased appreciation of the geographic mosaic of coevolution, cryptic speciation in insects, and cryptic speciation in plants calls for a more hierarchical view of the evolution and diversification of insect-plant interactions. Evolution and coevolution are not glacially slow processes that gradually shift the genetic structure of species over millions of years. Insect-plant interactions—and all other interspecific interactions—are constantly evolving in multiple ways across landscapes, creating novel evolutionary experiments in different environments. These experiments involve genetically differentiated populations and, sometimes, complexes of cryptic species with little or no gene flow between them. Over time, natural selection and other evolutionary processes sort out these experiments. Most of these evolutionary experiments fail, whereas others blend together through gene flow among populations and through hybridization among cryptic species. Collectively, most of these experiments remain bounded within a small region of morphological and biochemical space, creating the illusion of little current and ongoing evolution until studied in detail across multiple populations using multiple ecological and molecular approaches.

Occasionally, however, one of the experiments breaks new ground as an insect population or cryptic species colonizes a host plant that is phylogenetically remote from its ancestors. For example, knowing what we know about the extensive, but sometimes cryptic, geographic differentiation in the use of plant species by *Greya* species, it becomes easier to understand how at various points throughout the evolutionary history of the Prodoxidae new clades have evolved to feed on different plant families and even on monocots rather than on eudicots. Each tip (i.e., species) on a phylogenetic branch is not a single bud; it is a cluster of rapidly evolving populations with various degrees of reproductive isolation among them.

The tools of molecular biology and ecology have improved considerably in the past decade. We have multiple ways of assessing the patterns of genetic connection among populations. We also have multiple interactions that have now been studied in sufficient detail that they can form the basis of new studies that evaluate how interactions vary in traits, ecological outcomes, and forms of selection across landscapes. That makes this the most promising time ever for us to get robust answers to questions of how interactions between insects and plants diversify through coevolution and speciation and sometimes persist for millions of years.

Getting robust answers to these questions is becoming more important as we fragment and fundamentally change ecosystems worldwide. If we are to conserve species and interactions for the future, we need to understand the processes that have allowed them to persist in the past. That requires an increase in the number of studies that analyze in detail the genetic and ecological structure of interactions throughout their geographic ranges. We need a much better understanding of the geographic scale of adaptation, differentiation, and cryptic speciation in multiple interactions as a guide for designing conservation strategies that capture the true ecological and evolutionary structure of species networks.

Moreover, we need studies now that use the earth's last remaining wilderness areas to get precious data on the geographic structure of interactions, including adaptation and cryptic speciation. All of the earth's environments are changing quickly, and even highly protected areas are increasingly subject to the spread of invasive species. We can maintain the presence of native species and even restore some species once they become locally extinct, but we can never extract the knowledge of long-term coevolution once these communities are highly altered. Yet, we need that information now to test our increasingly sophisticated hypotheses and models against real data. Short-term experiments in highly altered environments cannot tell us how interactions evolve over 5,000 years or 50,000 years. The highly coevolved interactions distributed within and among the earth's shrinking wilderness areas are the among the most precious and free research-and-development resources that we have for developing a science of conservation that preserves the evolutionary processes that continually reshape interspecific interactions.

Acknowledgments

I thank Carol Boggs, Paul Ehrlich, Catherine Fernandez, and an anonymous reviewer for very helpful comments on the manuscript. This work was supported by NSF grant DEB-344147.

References Cited

Abrahamson, W.G., C.P. Blair, M.D. Eubanks, and A.V. Whipple. 2001. Gall flies, inquilines, and goldenrods: a model for host-race formation and sympatric speciation. Am. Zool. 41: 928–938.

Althoff, D.M., and J.N. Thompson. 1999. Comparative geographic structures of two parasitoid-host interactions. Evolution 53: 818–825.

Althoff, D.M., and J.N. Thompson. 2001. Geographic structure in the searching behaviour of a specialist parasitoid: combining molecular and behavioural approaches. J. Evol. Biol. 14: 406–417.

Benkman, C.W. 2003. Divergent selection drives the adaptive radiation of crossbills. Evolution 57: 1176–1181.

Benkman, C.W., W.C. Holimon, and J.W. Smith. 2001. The influence of a competitor on the geographic mosaic of coevolution between crossbills and lodgepole pine. Evolution 55: 282–294.

Berenbaum, M.R., and A.R. Zangerl. 1994. Costs of inducible defense: protein limitation, growth, and detoxification in parsnip webworms. Ecology 75: 2311–2317.

Berenbaum, M., and A. Zangerl. 1998. Chemical phenotype matching between a plant and its insect herbivore. Proc. Natl. Acad. Sci. USA 95: 13743–13748.

Berlocher, S.H., and J.L. Feder. 2001. Sympatric speciation in phytophagous insects: moving beyond controversy? Annu. Rev. Entomol. 47: 773–815.

Blair, C.P., W.G. Abrahamson, J.A. Jackman, and L. Tyrrell. 2005. Cryptic speciation and host-race formation in a purportedly generalist tumbling flower beetle. Evolution 59: 304–316.

Boggs, C.L., A.D. Watt, and P.R. Ehrlich (eds.). 2003. Butterflies: ecology and evolution taking flight. University of Chicago Press, Chicago.

Brodie, E.D., Jr., B.J. Ridenhour, and E.D. Brodie III. 2002. The evolutionary response of predators to dangerous prey: hotspots and coldspots in the geographic mosaic of coevolution between newts and snakes. Evolution 56: 2067–2082.

Brown, J.M., J.H. Leebens-Mack, J.N. Thompson, O. Pellmyr, and R. G. Harrison. 1997. Phylogeography and host association in a pollinating seed parasite *Greya politella* (Lepidoptera: Prodoxidae). Mol. Ecol. 6: 215–224.

Carroll, S.P., S.P. Klassen, and H. Dingle. 1998. Rapidly evolving adaptations to host ecology and nutrition in the soapberry bug. Evol. Ecol. 12: 955–968.

Carroll, S.P., H. Dingle, T.R. Famula, and C.W. Fox. 2001. Genetic architecture of adaptive differentiation in evolving host races of the soapberry bug, *Jadera haemotoloma*. Genetica 112/113: 257–272.

Craig, T.P., J.K. Itami, C. Shantz, W.G. Abrahamson, J.D. Horner, and J.V. Craig. 2000. The influence of host plant variation and intraspecific competition on oviposition preference and offspring performance in the host races of *Eurosta solidaginis*. Ecol. Entomol. 25: 7–18.

Davis, D.R., O. Pellmyr, and J.N. Thompson. 1992. Biology and systematics of *Greya* Busck and *Tetragma*, new genus (Lepidoptera: Prodoxidae). Smithsonian Contrib. Zool. 524: 1–88.

de Jong, P.W., and J.K. Nielsen. 1999. Polymorphism in a flea beetle for the ability to use an atypical host plant. Proc. R. Soc. Lond. B. 266: 103–111.

de Jong, P.W., and J.K. Nielsen. 2002. Host plant use of *Phyllotreta nemorum:* do coadapted gene complexes play a role? Entomol. Exp. Appl. 104: 207–215.

Drès, M., and J. Mallet. 2002. Host races in plant-feeding insects and their importance in sympatric speciation. Philos. Trans. R. Soc. B 357: 471–492.

Ehrlich, P.R., and I. Hanski (eds.). 2004. On the wings of checkerspots: a model system for population biology. Oxford University Press, Oxford.

Emelianov, I., M. Drès, W. Baltensweiler, and J. Mallet. 2001. Host-induced assortative mating in host races of the larch budmoth. Evolution 55: 2002–2010.

Fischer, B., and S. Foitzik. 2004. Local co-adaptation leading to a geographic mosaic of coevolution in a social parasite system. J. Evol. Biol. 17: 1026–1034.

Forbes, A.A., J. Fisher, and J.L. Feder. 2005. Habitat avoidance: overlooking an important aspect of host-specific mating and sympatric speciation? Evolution 59: 1552–1559.

Gandon, S., and Y. Michalakis. 2002. Local adaptation, evolutionary potential and host-parasite coevolution: interactions between migration, mutation, population size and generation time. J. Evol. Biol. 15: 451–462.

Gomulkiewicz, R., J.N. Thompson, R.D. Holt, S.L. Nuismer, and M.E. Hochberg. 2000. Hot spots, cold spots, and the geographic mosaic theory of coevolution. Am. Nat. 156: 156–174.

Gotthard, K., N. Margraf, and M. Rahier. 2004. Geographic variation in oviposition choice of a leaf beetle: the relationship between host plant ranking, specificity, and motivation. Entomol. Exp. Appl. 110: 217–224.

Hebert, P.D.N., E.H. Penton, J.M. Burns, D.H. Janzen, and W. Hallwachs. 2004. Ten species in one: DNA barcoding reveals cryptic species in the neotropical skipper butterfly *Astraptes fulgerator*. Proc. Natl. Acad. Sci. USA 101: 14812–14817.

Henry, C.S., M.L.M. Wells, and C.M. Simon. 1999. Convergent evolution of courtship songs among cryptic species of the *Carnea* group of green lacewings (Neuroptera: Chrysoperla). Evolution 53: 1165–1179.

Husband, B., and H.A. Sabara. 2003. Reproductive isolation between autotetraploids and their diploid progenitors in fireweed, *Chamerion angustifolium* (Onagraceae). New Phytol. 161: 703–713.

Husband, B.C., and D.W. Schemske. 2000. Ecological mechanisms of reproductive isolation between diploid and tetraploid *Chamerion angustifolium*. J. Ecol. 88: 689–701.

Janz, N. 2003. Sex linkage of host plant use in butterflies, pp. 229–239. In C.L. Boggs, W.B. Watt, and P.R. Ehrlich (eds.), Butterflies: ecology and evolution taking flight. University of Chicago Press, Chicago.

Janz, N., and J.N. Thompson. 2002. Plant polyploidy and host expansion in an insect herbivore. Oecologia 130: 570–575.

Jordal, B.H., L.R. Kirkendall, and K. Harkestad. 2004. Phylogeny of a Macronesian radiation: host-plant use and possible cryptic speciation in *Liparthrum* bark beetles. Mol. Phylogenet. Evol. 31: 554–571.

Laine, A.-L. 2005. Spatial scale of local adaptation in a plant-pathogen metapopulation. J. Evol. Biol. 18: 930–938.

Nielsen, J.K. 1997. Variation in defences of the plant *Barbarea vulgaris* and in counteradaptations by the flea beetle *Phyllotreta nemorum*. Entomol. Exp. Appl. 82: 25–35.

Nielsen, J.K. 1999. Specificity of a Y-linked gene in the flea beetle *Phylotreta nemorum* for defences in *Barbarea vulgaris*. Entomol. Exp. Appl. 91: 359–368.

Nielsen, J.K., and P.W. de Jong. 2005. Temporal and host-related variation in frequencies of genes that enable *Phyllotreta nemorum* to utilize a novel host plant, *Barbarea vulgaris*. Entomol. Exp. Appl. 115: 265–270.

Nuismer, S.L., and B.M. Cunningham. 2005. Selection for phenotypic divergence between diploid and autotetraploid *Heuchera grossulariifolia*. Evolution 59: 1928–1935.

Nuismer, S.L., and J.N. Thompson. 2001. Plant polyploidy and non-uniform effects on insect herbivores. Proc. R. Soc. Lond. B 268: 1937–1940.

Nuismer, S.L., J.N. Thompson, and R. Gomulkiewicz. 2000. Coevolutionary clines across selection mosaics. Evolution 54: 1102–1115.

Nuismer, S.L., J.N. Thompson, and R. Gomulkiewicz. 2003. Coevolution between species with partially overlapping geographic ranges. J. Evol. Biol. 16: 1337–1345.

Nylin, S., G.H. Nygren, J.J. Windig, N. Janz, and A. Bergstrom. 2005. Genetics of host-plant preference in the comma butterfly *Polygonia c-album* (Nymphalidae), and evolutionary implications. Biol. J. Linn. Soc. 84: 755–765.

Pellmyr, O., and J.N. Thompson. 1996. Sources of variation in pollinator contribution within a guild: the effects of plant and pollinator factors. Oecologia 107: 595–604.

Pellmyr, O., J.N. Thompson, J.M. Brown, and R.G. Harrison. 1996. Evolution of pollination and mutualism in the yucca moth lineage. Am. Nat. 148: 827–847.

Schwarz, D., B.M. Matta, N.L. Shakir-Botteri, and B.A. McPheron. 2005. Host shift to an invasive plant triggers rapid animal hybrid speciation. Nature 436: 546–549.

Scriber, J.M., and G.J. Ording. 2005. Ecological speciation without host plant specialization: possible origins of a recently described cryptic *Papilio* species. Entomol. Exp. Appl. 115: 247–263.

Segraves, K.A., and J.N. Thompson. 1999. Plant polyploidy and pollination: floral traits and insect visits to diploid and tetraploid *Heuchera grossulariifolia*. Evolution 53: 1114–1127.

Segraves, K.A., J.N. Thompson, P.S. Soltis, and D.E. Soltis. 1999. Multiple origins of polyploidy and the geographic structure of *Heuchera grossulariifolia*. Mol. Ecol. 8: 253–262.

Sezer, M., and R.K. Butlin. 1998. The genetic basis of oviposition preference differences between sympatric host races of the brown planthopper *(Nilaparvata lugens)*. Proc. R. Soc. Lond. B 265: 2399–2405.

Singer, M.C. 2003. Spatial and temporal patterns of checkerspot butterfly-host plant association: the diverse roles of oviposition preference, pp. 207–228. In C.L. Boggs, W.B. Watt, and P.R. Ehrlich (eds.), Butterflies: ecology and evolution taking flight. University of Chicago Press, Chicago.

Singer, M.C., and C.D. Thomas. 1996. Evolutionary responses of a butterfly metapopulation to human- and climate-caused environmental variation. Am. Nat. 148: S9-S39.

Soltis, D.E., and P.S. Soltis. 1995. The dynamic nature of polyploid genomes. Proc. Natl. Acad. Sci. USA 92: 8089–8091.

Soltis, D.E., and P.S. Soltis. 1999. Polyploidy: recurrent formation and genome evolution. Trends Ecol. Evol. 14: 348–352.

Soltis, D.E., P.S. Soltis, J.C. Pires, A. Kovarik, J.A. Tate, and E. Mavrodiev. 2004. Recent and recurrent polyploidy in *Tragopogon* (Asteraceae): cytogenetic, genomic and genetic comparisons. Biol. J. Linn. Soc. 82: 485–501.

Stoks, R., J.L. Nystrom, M.L. May, and M.A. McPeek. 2005. Parallel evolution in ecological and reproductive traits to produce cryptic damselfly species across the Holarctic. Evolution 59: 1976–1988.

Thompson, J.N. 1985. Postdispersal seed predation in *Lomatium* spp. (Umbelliferae): variation among individuals and species. Ecology 66: 1608–1616.

Thompson, J.N. 1993. Preference hierarchies and the origin of geographic specialization in host use in swallowtail butterflies. Evolution 47: 1585–1594.

Thompson, J.N. 1994. The coevolutionary process. University of Chicago Press, Chicago.

Thompson, J.N. 1998. The evolution of diet breadth: monophagy and polyphagy in swallowtail butterflies. J. Evol. Biol. 11: 563–578.

Thompson, J.N. 2005. The geographic mosaic of coevolution. University of Chicago Press, Chicago.

Thompson, J.N., and B.M. Cunningham. 2002. Geographic structure and dynamics of coevolutionary selection. Nature 417: 735–738.

Thompson, J.N., and C.C. Fernandez. 2006. Temporal dynamics of antagonism and mutualism in a geographically variable plant-insect interaction. Ecology 87: 103–112.

Thompson, J.N., and O. Pellmyr. 1992. Mutualism with pollinating seed parasites amid co-pollinators: constraints on specialization. Ecology 73: 1780–1791.

Thompson, J.N., B.M. Cunningham, K.A. Segraves, D.M. Althoff, and D. Wagner. 1997. Plant polyploidy and insect/plant interactions. Am. Nat. 150: 730–743.

Wehling, W.F., and J.N. Thompson. 1997. Evolutionary conservatism of oviposition preference in a widespread polyphagous insect herbivore, *Papilio zelicaon*. Oecologia 111: 209–215.

Zangerl, A.R., and M.R. Berenbaum. 1997. Cost of chemically defending seeds: furanocoumarins and *Pastinaca sativa*. Am. Nat. 150: 491–504.

Zangerl, A. R., and M. R. Berenbaum. 2003. Phenotypic matching in wild parsnip and parsnip webworms: causes and consequences. Evolution 57: 806–815.

SEVENTEEN

Cophylogeny of Figs, Pollinators, Gallers, and Parasitoids

SUMMER I. SILVIEUS, WENDY L. CLEMENT,
AND GEORGE D. WEIBLEN

Cophylogeny provides a framework for the study of historical ecology and community evolution. Plant-insect cophylogeny has been investigated across a range of ecological conditions including herbivory (Farrell and Mitter 1990; Percy et al. 2004), mutualism (Chenuil and McKey 1996; Kawakita et al. 2004), and seed parasitism (Weiblen and Bush 2002; Jackson 2004). Few examples of cophylogeny across three trophic levels are known (Currie et al. 2003), and none have been studies of plants, herbivores, and their parasitoids. This chapter compares patterns of diversification in figs *(Ficus* subgenus *Sycomorus)* and three fig-associated insect lineages: pollinating fig wasps (Hymenoptera: Agaonidae: Agaoninae: *Ceratosolen*), nonpollinating seed gallers (Agaonidae: Sycophaginae: *Platyneura*), and their parasitoids (Agaonidae: Sycoryctinae: *Apocrypta*). Molecular phylogenies of each participant in this tritrophic interaction can illuminate histories of ancient association ranging from codivergence to host switching. We distinguish cospeciation, the simultaneous speciation of a host and an associate (Page 2003), from coevolution, strictly defined as reciprocal evolutionary change between lineages acting as agents of selection upon each other (Stearns and Hoekstra 2000). Tests of cospeciation and alternative hypotheses, including host switching, can evaluate whether evolutionary change in host use was associated with speciation (Coyne and Orr 2004).

Nonpollinating fig wasps have not received as much attention as the pollinating wasps, and so little is known about their evolutionary history (Ulenberg 1985; Bronstein 1991; Lopez-Vaamonde et al. 2001; Cook and Rasplus 2003). Weiblen and Bush (2002) argued from phylogeny that nonpollinating seed gallers are less closely cospeciated with figs than pollinators sharing the same hosts. It is not known if the same is true for fig wasp parasitoids.

Robust phylogeny estimates inform our inference of past evolutionary processes from patterns of diversity seen in the present. Topological congruence between the phylogenies of host organisms and their associated lineages is the first line of evidence for cospeciation. On the other hand, phylogenetic incongruence may indicate other historical patterns of association, including host switching. When host and associate topologies and divergence times are more closely congruent than expected by chance (Page 1996), ancient cospeciation may have occurred. Incongruence between phylogenies requires more detailed explanation, including the possibility that error is associated with either phylogeny estimate. Ecological explanations for phylogenetic incongruence include extinction, "missing the boat," host switching, and host-independent speciation (Page 2003). "Missing the boat" refers to the case where an associate tracks only one of the lineages following a host-speciation event. For most plant-insect interactions, these other events are more common than cospeciation (Farrell and Mitter 1990). A few highly specialized interactions, including obligate pollination mutualisms (Weiblen and Bush 2002; Kawakita et al. 2004), show evidence of cospeciation.

We examined patterns of historical association across three trophic levels in the fig microcosm, comparing mitochondrial DNA phylogenies for pollinators, gallers, and parasitoids to multigene phylogenies of the host figs. Given that nonpollinators play no direct role in fig reproductive isolation, we tested the prediction that gallers and parasitoids show less evidence of cospeciation with their hosts than do pollinators influencing patterns of host gene flow. Some background on fig pollination is needed to understand this prediction in more detail.

Background

Fig Pollination

More than 800 fig species (*Ficus*, Moraceae) occur in the tropics and subtropics worldwide (Berg 1989). Fig species exhibit a wide array of growth forms including shrubs, trees, climbers, and hemi-epiphytic stranglers. The genus is characterized by an obligate mutualism with pollinating fig wasps (Chalcidoidea: Agaonidae: Agaoninae) and a specialized inflorescence called a syconium, which comprises an enclosed cavity filled with numerous, highly reduced unisexual flowers (Bronstein 1992). At the apex of the syconium is a bract-lined opening, or ostiole, that is the point of entry for pollinating fig wasps. Pollinators depend on figs for the rearing of their offspring, and figs depend on wasps for pollination. Pollinators are very small, approximately a millimeter in body length, and often bear specialized thoracic pockets for the transport of pollen grains (Weiblen 2002). Most fig species are pollinated by unique fig wasp species (Wiebes 1979), but not all cases involve one-to-one species specificity. Molbo et al. (2003) found cryptic pollinator species on the same host in sympatry, whereas Rasplus (1994) reported instances of peripatric pollinator species on widespread hosts with intervening zones of contact (Michaloud et al. 1996).

Chemical volatiles are the primary cues that attract unique pollinators species to receptive figs (Hossaert-McKey et al. 1994; Grison-Pige et al. 2002). Female wasps crawl through the bract-lined ostiole to reach the fig cavity where egg laying takes place. Oviposition involves piercing the style of a pistillate flower and depositing a single egg inside the fig ovule to form a gall (Galil and Eisikowitch 1971). Most Agaoninae actively pollinate fig flowers during oviposition by removing pollen from their thoracic pockets and spreading it across the synstigma, a surface formed by numerous, closely packed pistillate flowers (Jousselin and Kjellberg 2001). Wasp larvae consume fig seeds, a single larva feeding on the contents (typically the endosperm) of a single galled flower (Weiblen 2002). Sexual dimorphism reflects the specialized roles of male and female fig wasps. Females are mated in their galls by males who are typically wingless, blind, and less pigmented than the winged females. The emergence of females from their galls into the fig cavity coincides with anthesis, when males chew an opening in the syconium and females collect pollen from staminate flowers as they exit. The search for receptive syconia, followed by pollination and oviposition, marks the completion of the pollinator life cycle (Jousselin et al. 2001b).

Stability of this mutualism depends on the relative allocation of floral resources to pollen, seeds, and pollinators. Given that pollinators also eat seeds, there is potential evolutionary conflict between seed production and seed consumption (Cook and Rasplus 2003). Fig lineages have resolved this conflict independently by different means (Kameyama et al. 1999). This chapter focuses on the subgenus *Sycomorus*, which is in large part functionally dioecious, segregating the production of seeds and pollinators in two types of figs called "gall figs" and "seed figs" on separate plants (Weiblen et al. 2001). Gall figs contain staminate and short-styled pistillate flowers, whereas seed figs contain only long-styled pistillate flowers (Weiblen and Bush 2002; Weiblen 2004). Because differences in flower morphology affect oviposition success, the former produce pollinators and pollen while the latter produce only seed. There are obvious negative fitness consequences for pollinators of seed figs given that these "tomb blossoms" have ostiole bracts that prevent pollinators from exiting. Nonetheless, seed figs regularly deceive pollinators into visiting despite the absence of any reproductive reward (Grafen and Godfray 1991).

Ficus Subgenus *Sycomorus*

Ficus subgenus *Sycomorus* has attracted the attention of evolutionary biologists interested in explaining the stability of obligate pollination mutualism (Kameyama et al. 1999; Harrison and Yamamura 2003). The group is monophyletic and includes approximately 140 species occurring across Africa, Asia, and the Indo-Pacific (Weiblen 2000). All species of *Ficus* subgenus *Sycomorus* are pollinated by the monophyletic genus *Ceratosolen* (Weiblen 2001). A diverse assemblage of nonpollinating wasps that utilize fig resources is also associated with *Sycomorus* (Kerdelhue et al. 2000). The absence of pollen pockets and the presence of extremely long ovipositors with which to pierce the fig wall distinguish nonpollinators from *Ceratosolen*. However, a few nonpollinator species in Africa (Kerdelhue and Rasplus 1996) and Asia (Jousselin et al. 2001a) with short ovipositors are known to crawl into the fig and oviposit internally.

Nonpollinating Fig Wasps

Nonpollinators occupy two trophic levels in the fig microcosm, as gallers and parasitoids. *Sycomorus* figs are attacked by gallers of the genus *Platyneura* (Sycophaginae). The name *Platyneura* takes priority over *Apocryptophagus* (J. Y. Rasplus, 1997, personal communication) and includes all *Sycomorus* gallers known from Indo-Pacific, the center of *Sycomorus* fig diversity. *Platyneura* are essentially seed parasites. Oviposition in a fig ovule induces an abnormal growth such that *Platyneura* galls are much larger than those of *Ceratosolen*. Each galler occupies a single flower (Galil et al. 1970) and reduces fig fecundity by exactly one seed, while negatively affecting pollinators through competition for floral resources (Weiblen et al. 2001).

Nonpollinators occupying the third trophic level in the fig microcosm are parasitoids of other fig wasps. Three genera of parasitoids (Agaonidae: Sycoryctinae) associated with *Sycomorus* figs are *Sycoscapter*, *Philotypesis*, and *Apocrypta*. They are distinguished by ovipositors associated with three different means of piercing the fig wall. *Sycoscapter* and *Philotypesis* attack *Ceratosolen*, whereas *Apocrypta* are specifically associated with *Platyneura* gallers (Ulenberg 1985). The focus of this chapter is the three-way interaction between *Sycomorus*, *Platyneura*, and *Apocrypta*. *Apocrypta* females pierce the fig wall and deposit solitary eggs in *Platyneura* galls, where these parasitoids kill, consume, or outcompete galler larvae (Weiblen et al. 2001). *Apocrypta* is not polyembryonic, and, as with the rest of the fig wasp community, a single adult is the product of each gall. Little is known about the systematics of *Apocrypta* apart from that provided by Ulenberg (1985), and no molecular data have been published until now.

Phylogenetic Knowledge

Previous studies on the phylogenetic relationships of *Sycomorus* figs (Weiblen 2000) and pollinators (Weiblen 2001) suggest a history of cospeciation and extreme host specificity. These findings are in broad agreement with predictions from pollinator life history. Cospeciation is expected when associates are vertically transmitted (Herre et al. 1999), and the transportation of fig pollen from cradle to grave by pollinating wasps *(Ceratosolen)* closely approximates this mode of transmission because pollinators tend to select the same host species as their parents. Pollinators rarely make mistakes when choosing host figs, and, in the event that a pollinator enters the wrong host, there is a chance of pollen incompatibility or the failure to lay eggs due to a morphological mismatch between ovipositor and fig flowers (Weiblen and Bush 2002). Only some of these conditions apply to externally ovipositing nonpollinators; the prediction that nonpollinator speciation should be less closely tied to *Sycomorus* speciation is supported by evidence from molecular phylogeny (Weiblen and Bush 2002). Nonpollinator host use may have more to do with the size of the fig and the thickness of the fig wall in relation to their ovipositor length (Weiblen 2004). Phylogenetic conservatism or convergent evolution (Jousselin et al. 2003) in fig size could affect patterns of speciation and host use in nonpollinators and lead to departures from simple expectations of cospeciation.

Overview

This chapter describes a sympatric assemblage of *Sycomorus* figs and the associated fig wasps from a lowland rain forest in Papua New Guinea. New Guinea is the global center of *Sycomorus* fig diversity, with over 60 species on the island. We focus on a set of sympatric and closely related host species because the co-occurrence of close relatives provides opportunities for host switching and departures from one-to-one species specificity. Phylogenies of sympatric *Sycomorus* figs and associated wasps based on multiple gene regions were used to explain the host specificity of nonpollinators and to test hypotheses of cospeciation between figs, pollinators, gallers, and parasitoids. In addition to topological comparisons of host and associate phylogenies, tests of temporal congruence were performed. Different predictions about the timing of divergence are drawn from the contrasting life histories of fig pollinators and nonpollinators. Complete reproductive interdependence of the mutualists, and the observation that extreme pollinator specificity enforces premating reproductive isolation between sympatric fig species, leads to the prediction that divergence of fig and pollinator clades should be temporally congruent. On the other hand, these conditions do not apply to nonpollinators or their hosts, and there is less reason to expect temporal congruence among interacting lineages when the mode of transmission is horizontal (Herre et al. 1999).

Sampling and DNA Sequencing

The study area was located in moderately disturbed lowland rain forest around the Ohu Conservation Area (145°41(E, 5°14″ S, ca. 50–200 m) in the Madang district of Papua New Guinea. The climate at the study area was perhumid (a wet climate with humidity index values above 100) with average annual rainfall of 3558 mm and annual average temperature of 26.5°C (McAlpine et al. 1983). Wasps were reared from 19 sympatric fig species by parataxonomists at the New Guinea Binatang Research Center between 1995 and 2005 (Table 17.1). These fig species are functionally dioecious (Weiblen et al. 2001) as described above. Wasps were reared from gall figs by enclosing each individual ripe fig in a container covered by nylon mesh held with a rubber band. Following the emergence of wasps from their galls and the fig, the entire community from an individual fig was preserved in 70% ethanol. Communities were sorted to genus and morphospecies using a dissecting microscope and digitally photographed.

Table 17.1 lists the 19 *Sycomorus* species and the associated pollinators, gallers, and parasitoids that were examined. Whereas most of the *Ceratosolen* species are known to science, the nonpollinator species have yet to be described. We recognized nonpollinator morphospecies on the basis of body size, ovipositor length, host use, and mitochondrial DNA sequence divergence (S.I.S., unpublished data). *Ceratosolen* pollinators in the study area are extremely specialized and are involved in one-to-one relationships with particular host species (Weiblen et al. 2001). It has been assumed that nonpollinators are similarly specialized (Ulenberg 1985; Weiblen and Bush 2002), but the dissertation research of Silvieus (unpublished data) suggests otherwise.

A molecular phylogeny for 19 sympatric *Sycomorus* host species was inferred from three gene regions: the internal transcribed spacer region of nuclear ribosomal DNA (ITS), glyceraldehyde 3-phosphate dehydrogenase (G3PDH), and granule-bound starch synthase (GBSS or waxy). Amplification of ITS (Weiblen 2000), G3PDH (Strand et al. 1997), and waxy followed published protocols (Mason-Gamer et al. 1998; Evans et al. 2000). Two *Ficus* outgroups were included for rooting purposes: *F. virens*, representing subgenus *Urostigma*, and *F. wassa*, representing subgenus *Sycidium* (Berg and Corner 2005). DNA was extracted from dried leaves using the DNeasy plant mini kit (Qiagen, Valencia, CA). ITS and G3PDH were sequenced directly from polymerase chain reaction (PCR) products, whereas waxy was cloned using the TOPO-TA PCR cloning kit (Invitrogen,

TABLE 17.1

Sympatric *Ficus* Subgenus *Sycomorus* Species and Associated *Ceratosolen* Pollinators, *Platyneura* Gallers, and *Apocrypta* Parasitoids from a New Guinea Lowland Rainforest

Ficus Species	*Ficus Section*	*Trees Sampled*	*Figs Sampled*	*Ceratosolen*	*Platyneura*	*Apocrypta*
adelpha	*Sycocarpus*	9	21	ex *F. adelpha*	*ex* F. adelpha[a]	ex *F. adelpha, bernaysii* & cf. *ternatana*
adenosperma	*Adenosperma*	5	5	*adenospermae*[b]	ex *F. adenosperma* sp. A ex *F. adenosperma* sp. B	
arfakensis	*Sycocarpus*	5	14	*solitarius*	ex *F. arfakensis* & cf. *ternatana*	ex *F. arfakensis*
bernaysii	*Sycocarpus*	21	51	*hooglandi*	ex *F. bernaysii* sp. A ex *F. bernaysii* sp. B	ex *F. adelpha, bernaysii* & cf. *ternatana*
botryocarpa	*Sycocarpus*	21	39	*corneri*	ex *F. botryocarpa*	
congesta	*Sycocarpus*	18	36	*notus*	ex *F. congesta* ex *F. congesta* & *hispidioides*	ex *F. congesta* & *hispidioides*
dammaropsis	*Adenosperma*[c]	18	16	*abnormis*		
hispidioides	*Sycocarpus*	19	27	*dentifer*	ex *F. hispidioides*[a] ex *F. congesta* & *hispidioides*	ex *F. congesta* & *hispidioides*
mollior	*Adenosperma*	5	16	*medlarianus*	ex *F. mollior* sp. A ex *F. mollior* sp. B	
morobensis	*Sycocarpus*	3	5	ex *F. morobensis*		

nodosa	*Sycomorus*	19	23	*nexilis*	ex *F. nodosa* sp. A ex *F. nodosa* sp. B	ex *F. nodosa*
pachyrrhachis	*Sycocarpus*	7	10	ex *F. pachyrrhachis*	ex *F. pachyrrhachis*	
pungens	*Sycocarpus*[d]	23	35	*nanus*		
robusta	*Sycomorus*	3	6	*grandii*[e,g]	ex *F. robusta*[a,e]	
semivestita	*Sycomorus*[f]	1	1	*grandii*[e,g]	ex *F. semivestita*[e]	
septica	*Sycocarpus*	18	32	*bisulcatus*		
subcuneata	*Adenosperma*	3	3	ex *F. subcuneata*[f]	ex *F. subcuneata*	
cf. *ternatana*[h]	*Sycocarpus*	2	2	ex *F.* cf. *ternatana*	ex *F. arfakensis* & cf. *ternatana*[a,e]	ex *F. adelpha, bernaysii* & cf. *ternatana*
variegata	*Sycomorus*	17	27	*appendiculatus*	ex *F. variegata* sp. A ex *F. variegata* sp. B	ex *F. variegata*

NOTE: Total numbers of trees and figs examined for nonpollinators are listed. Nonpollinator morphospecies are named according to the host fig species with which they are associated. All hosts are listed for nonpollinator morphospecies attacking multiple host species.

[a]Cytochrome B sequence unavailable.

[b]400 bp at 3′ end of cytochrome oxidase I unavailable.

[c]Molecular and morphological phylogenetic inferences support the placement of *F. dammaropsis* in section *Adenosperma*. Section *Dammaropsis* sensu Berg and Corner (2005) is embedded within section *Adenosperma* and should not be recognized.

[d]Weiblen (2000) placed *F. pungens* in section *Sycocarpus* based on morphological and molecular phylogenetic evidence. Section *Bosscheria* sensu Berg and Corner (2005) is embedded within section *Sycocarpus* and should not be recognized.

[e]400 bp at 5′ end of cytochrome oxidase I unavailable.

[f]Contrary to Weiblen (2000, 2001), *F. semivestita* is a member of section *Sycomorus* sensu Berg and Corner (2005). Weiblen (2000) included misidentified specimens of *F. subcuneata* in molecular and morphological analysis, erroneously suggesting the transfer of *F. semivestita* to section *Adenosperma*. The pollinator of *F. subcuneata* was misidentified on the basis of the host as *Ceratosolen grandii* in Weiblen (2001) and Weiblen and Bush (2002).

[g]According to morphology and COI sequences *Ceratosolen grandii* was reared from both *F. semivestita* and *F. robusta*.

[h]*Ficus* cf. *ternatana* is an undescribed New Guinea species evidently related to *F. ternatana* of the Moluccas.

Carlsbad, CA). Ten clones were screened for inserts, and plasmids were isolated from four of these using the Qiagen plasmid prep kit. Sequencing followed standard protocols for Big Dye v.3 (Applied Biosystems, Foster City, CA) with an ABI 377 automated DNA sequencer (PE Biosystems, Foster City, CA). Nucleotide sequences for each region were aligned using Clustal X (Thompson et al. 1997), followed by manual editing. Multiple copies of waxy are known in the Rosales (Evans et al. 2000), and therefore it was necessary to ensure that phylogeny reconstruction was performed with orthologous copies. A preliminary survey detected two copies in Moraceae, GBSS1 and GBSS2, which were easily distinguished on the basis of size and intron alignment (W.L.C., unpublished data). Analyses were based solely on GBSS1 because GBSS2 was encountered less commonly in *Sycomorus* figs.

Mitochondrial genes cytochrome oxidase I (COI) and cytochrome B (cytB) were sequenced from pollinating and nonpollinating wasps associated with sympatric *Sycomorus* species. Mitochondrial genes in fig wasps have very high rates of molecular evolution, so they are variable enough to resolve and support relationships among closely related species but are still conservative enough to align without ambiguity (Lunt et al. 1996; Lopez-Vaamonde et al. 2001; Machado et al. 2001; Weiblen 2001; Molbo et al. 2003). Although most morphospecies (hereafter species) were reared from multiple trees and multiple figs per tree (Table 17.1), a single exemplar from each species was included in phylogenetic analysis to reduce computation time. Genomic DNA representing each morphospecies was isolated by pooling up to 10 individuals reared from the same fig in extractions with the DNeasy tissue kit (Qiagen). Amplification and direct sequencing of COI and cytB followed standard protocols for fig wasps (Kerdelhue et al. 1999; Weiblen 2001). Taxon sampling of fig wasps was restricted to the local assemblage of *Sycomorus* associates plus a non-fig-associated chalcid, *Anaphes nitens* (Mymaridae), as an outgroup (C. Lopez-Vaamonde, unpublished data). Sequencing included ~800 bp of COI and ~600 bp of cytB, except in cases noted in Table 17.1 where PCR products were not obtained. Sequences from *Apocrypta*, *Ceratosolen*, and *Platyneura* were analyzed simultaneously, as the genera are reciprocally monophyletic and their relative phylogenetic positions were not the object of this community study.

Phylogenetic Analysis

Phylogenetic analysis was performed with PAUP*4.0b (Swofford 2001) under parsimony and maximum likelihood (ML) criteria. Our experience teaches that the application of multiple optimality criteria in phylogenetic analysis can identify strong inferences that are robust to different assumptions about underlying processes of molecular evolution (Datwyler and Weiblen 2004). Under parsimony, heuristic searches with 1000 random addition sequence replicates explored the possibility of multiple islands of most parsimonious (MP) trees (Maddison 1991). The relative strength of clade support was assessed using nonparametric bootstrap resampling (Felsenstein 1985a) with 1000 replicates and 100 random addition sequence replicates per bootstrap replicate. Although ML is computationally more demanding than parsimony, it assumes explicit models of nucleotide substitution that can be compared statistically. To expedite ML searches, MP topologies were used as starting trees to optimize substitution models, parameter estimates, and search for optimal ML topologies. Substitution models were selected using the Aikake information criterion as implemented in Modeltest (Posada and Crandall 1998; Posada and Buckley 2004), and the best-fitting models with the fewest additional parameters were used in heuristic searches to identify topologies that maximized the likelihood of the data.

Gene regions were analyzed separately and statistical tests of phylogenetic congruence were conducted before combining data sets in simultaneous analyses. MP topologies resulting from searches of separate data sets were compared under ML using the best-fitting substitution model and model parameters for all genes combined (Shimodaira and Hasegawa 1999). If topologies resulting from separate analyses of regions were not significantly different, the results of combined analyses were accepted as the best available estimate of phylogeny.

Regions from *Sycomorus* figs and fig wasps were tested for rate constancy using likelihood ratio tests of models with and without the assumption of a molecular clock (Felsenstein 1985b). When rate constancy was rejected, we applied a nonparametric rate smoothing method to ML branch lengths (Sanderson 1997) as implemented in the program TreeEdit (A. Rambant and M. Charleston, University of Oxford, Oxford, UK) to obtain ultrametric trees for the purpose of molecular dating. As there are no fossils attributed to *Sycomorus* or *Sycomorus*-associated wasps, we used a geological time constraint to estimate divergence times. *Ficus* sections *Papuasyce*, *Dammaropsis*, and *Adenosperma* (Weiblen 2000; Berg and Corner 2005) compose a clade that is endemic to New Guinea and the Solomon Islands. Since neither the island of New Guinea nor its antecedents existed more than ~40 million years from the present (Axelrod and Raven 1982), this age was applied as a maximum constraint for this *Sycomorus* clade and its associated *Ceratosolen* pollinators. The dated fig wasp phylogeny was split into the three component genera for separate reconciliation analyses with the dated host-fig phylogeny.

Reconciliation Analysis

Hypotheses of topological and temporal congruence among the lineages as implied by cospeciation were tested by comparison of dated molecular phylogenies for *Sycomorus* figs, *Ceratosolen* pollinators, *Playneura* gallers, and *Apocrypta* parasitoids. Host and associate phylogenies were reconciled under event-based parsimony and tested for cocladogenesis using TreeMap software (Page 1995). Phylogenetic reconciliation infers the minimal number of evolutionary events needed to fit an associate phylogeny to a host phylogeny. Cocladogenesis is inferred if phylogenies of hosts and associates are topologically and temporally congruent. Deviations from perfect congruence suggest other evolutionary events including host switching (Percy et al. 2004). It is assumed that associates are restricted to a single host at any point in time, and, if there is a host switch, speciation is assumed with one daughter shifting to the new host (Ronquist 1998). Randomization tested whether reconciled host and associate tree topologies show significantly higher levels of cospeciation than expected by chance. Both host and associate phylogenies, randomized 1000 times using the proportional-to-distinguishable model in TreeMap, generated a null distribution of maximum cospeciation. Failure to reject the null hypothesis suggests that other evolutionary events besides cospeciation are needed to account for host associations in the present day (Hafner and Page 1995). Dates of divergence for *Sycomorus* clades and their inferred ancestral associates were then compared to assess the temporal plausibility of cospeciation. Significant correlation of the age of congruent, ancestral host and associate lineages provides evidence that speciation might have been synchronous. On the other hand, if divergence time estimates are wildly asynchronous, cospeciation is unlikely, and alternative scenarios such as host switching, speciation in the associate but not the host, speciation in the host but not the associate, may be invoked (Percy et al. 2004).

Phylogenies of Figs and Wasps

Sequencing of 19 *Sycomorus* species and two outgroups for three gene regions produced a total of 3040 aligned nucleotide positions. GBSS1 (waxy) provided the most phylogenetic information, with 179 parsimony-informative characters out of 1677 aligned positions (11%). G3PDH and ITS provided 61 and 114 parsimony informative characters (10% and 15%) out of 611 and 752 aligned positions, respectively. Separate heuristic searches of each gene region yielded 216, 10, and 495 MP trees for waxy, ITS, and G3PDH, respectively. Shimodaira-Hasegawa tests indicated that trees resulting from searches of individual gene regions were not significantly incongruent when compared in a likelihood framework incorporating information from all three genes. Likelihood and parsimony analyses were then conducted on the combined data sets. The best-fitting model of nucleotide substitution according to the Aikake information criterion was general time reversible (GTR) with additional parameters for heterogeneity in substitution rates across sites (Γ) and the proportion of invariant sites (I). Parsimony yielded 48 MP trees, and the ML topology was identical to one of these. Model parameters and branch lengths were estimated for the ML tree under GTR + I + Γ. Nonparametric bootstrapping under parsimony supported only half of the clades in the ML tree. Even three congruent gene regions providing 354 parsimony-informative characters failed to provide robust estimates of relationships among recently diverged, New Guinea endemic species in *Ficus* sections *Adenosperma* and *Sycocarpus* (Fig. 17.1). Deeper divergences among sections and subsections, however, were generally well supported according to bootstrap values >70%.

Sequencing of 19 *Ceratosolen*, 20 *Platyneura*, 5 *Apocrypta*, and an outgroup species for two mitochondrial genes provided a total of 1168 aligned nucleotide positions. Cytochrome oxidase I (COI) comprised the larger data set with 436 parsimony informative characters (55%) out of 787 aligned positions, compared to cytochrome B (cytB) with 212 parsimony informative characters (56%) out of 381 positions in total. Heuristic searches of COI alone yielded a single tree, while cytB alone yielded 60 MP trees. Shimodaira-Hasegawa tests indicated that all 60 cytB topologies were significantly less likely ($P < 0.001$) than the COI tree when compared in a likelihood framework incorporating information from both genes. We attribute significant incongruence between nonrecombining mitochondrial genes to homoplasy and the lower number of informative characters from cytB. Heuristic searches of concatenated cytB and COI assuming the best-fitting model of nucleotide substitution (GTR + Γ + I) yielded a topology that was significantly more likely ($P < 0.001$) than either data set alone. Heuristic searches of the two genes in combination under parsimony yielded six trees including a topology identical to the ML tree. We present the ML tree based on the combined data set as the best phylogeny estimate for sympatric *Sycomorus* fig wasps from our study area. The tree was broken along the backbone to yield the individual topologies for pollinators, gallers, and parasitoids shown in Figs. 17.1, 17.2, and 17.3. Nonparametric bootstrap values supported relationships among species groups associated with *Sycomorus* sections and subsections, but clade support within species groups was generally lacking. For example, well-supported *Ceratosolen* clades included pollinators associated with *Ficus* section *Adenosperma*, section *Sycomorus*, and section *Sycocarpus*.

Host Specificity of Nonpollinating Fig Wasps

Gallers and parasitoids exhibited different patterns of association with *Sycomorus* than pollinators (Figs. 17.1, 17.2, and 17.3). *Platyneura* species were as numerous as *Ceratosolen* but did not conform to the pattern of one-to-one specificity that characterizes the mutualists. Departures from one-to-one species specificity included four parasite-free *Sycomorus* species, seven *Sycomorus* species attacked by multiple *Platyneura* gallers, and two *Platyneura* attacking multiple hosts. *Sycomorus* species were commonly attacked by two *Platyneura* that differed in ovipositor length and oviposition timing (Kerdelhue and Rasplus 1996). Species with short ovipositors lay eggs prior to pollination when figs are small in diameter, whereas species with long ovipositors lay eggs after pollination when figs are larger (Weiblen and Bush 2002). Two cases in which sister species of *Platyneura* parasitized the same host are discussed later as examples of adaptive divergence. One *Apocrypta* parasitoid

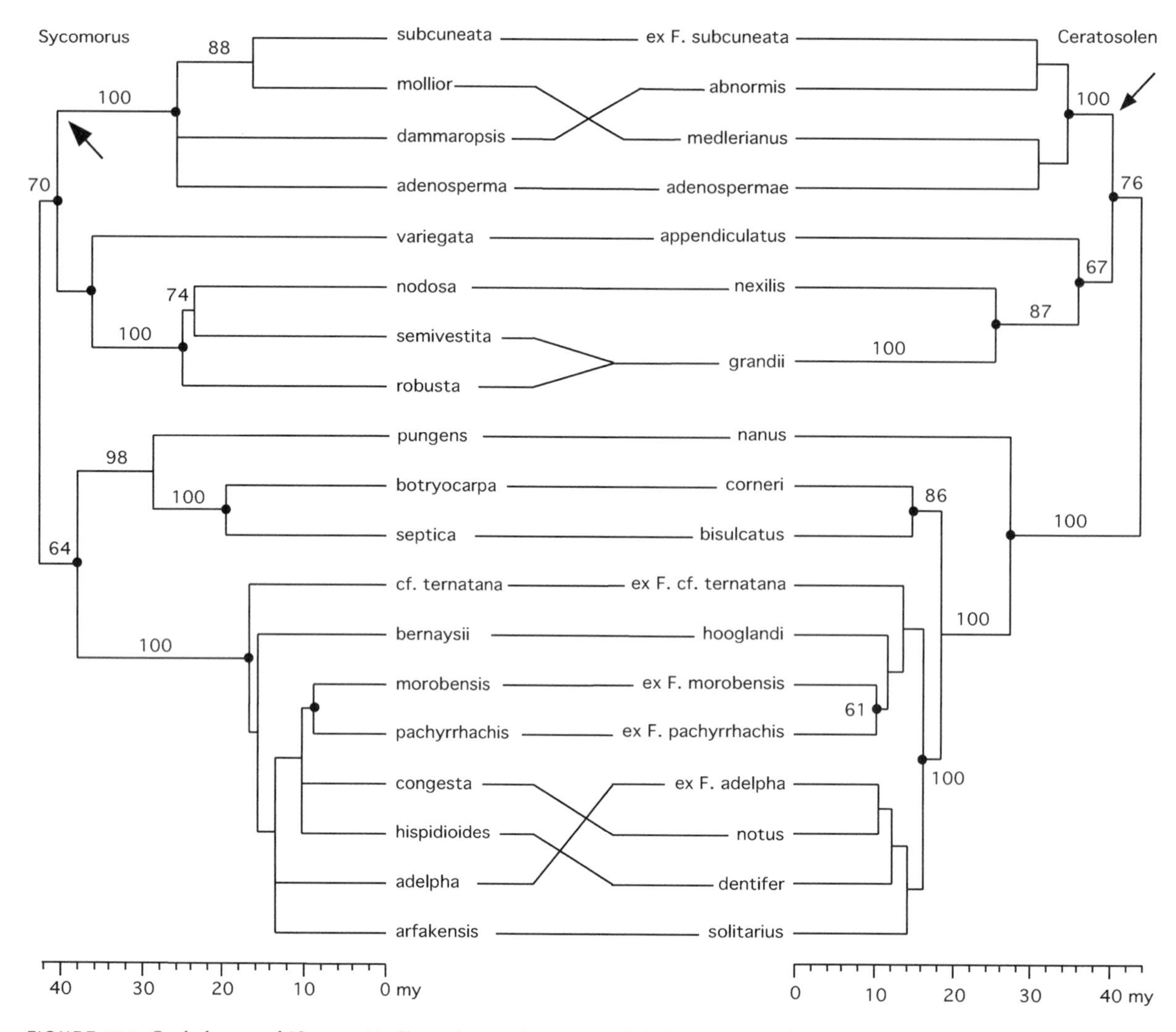

FIGURE 17.1. Cophylogeny of 19 sympatric *Ficus* subgenus *Sycomorus* and their species-specific *Ceratosolen* pollinators. Nonparametric bootstrap values >50% are shown above the nodes. Nonparametric rate smoothing of maximum likelihood topologies with general time reversible plus invariant sites plus heterogeneity in substitution rates across sites (GTR + I + Γ) branch lengths yielded ultrametric trees. Maximum ages for the New Guinea endemic clade (40 million years), represented by *F. adenosperma*, *F. dammaropsis*, *F. mollior*, and *F. subcuneata*, and their respective pollinators (indicated with arrows) were used to calibrate molecular phylogenies with respect to divergence times. The fig wasp community phylogeny was split into three component genera with *Platyneura* and *Apocrypta* shown in Figs. 17.2 and 17.3. Divergence times of congruent fig and pollinator clades marked by dots are compared in Fig. 17.4.

species exhibited broader host ranges than any pollinator or galler. Some parasitoids attacked closely related *Platyneura* on multiple fig species, and other species attacked multiple *Platyneura* on a single *Sycomorus* host. In the most extreme case, a single *Apocrypta* species attacked *Platyneura* on three closely related fig species from section *Sycocarpus*.

Double Dating of Figs and Fig Wasps

Figs and Pollinating Fig Wasps

Previous reconciliation analysis indicated that the extent of codivergence inferred for *Sycomorus* figs and their *Ceratosolen* pollinators is greater than expected by chance (Weiblen and Bush 2002). The same was true for the sympatric, interacting species pairs from New Guinea shown in Fig. 17.1. Heuristic searches maximizing codivergence and minimizing other evolutionary events yielded a reconciled tree with 13 out of 18 possible cospeciation events, four duplications, one host switch, and 16 lineage sorting events. Duplications refer to cases in which a pollinator lineage underwent speciation when the host-fig lineage did not, and lineage sorting refers to the case in which pollinators "missed the boat" and failed to colonize one of the fig lineages resulting from a host-speciation event. The maximum number of inferred cospeciation events was significantly greater than the null expectation based on pairs of randomized trees ($P < 0.001$). Rejecting this rather weak null hypothesis is not terribly informative, however, as nonrandom patterns of historical association could result from processes other than cospeciation. The calibration of molecular phylogenies with respect to divergence times based on

nize that phenological shifts in oviposition habit and the evolution of reproductive isolation could also have occurred in allopatry. In any event, the temporal incongruence of *Platyneura* and *Sycomorus* is consistent with molecular dating evidence from other plant-insect interactions (Percy et al. 2004), indicating that herbivores and their hosts have not cospeciated.

Parasitoids

Sycomorus parasitoids are younger, fewer in species, and less host specific than pollinators and parasites (Fig. 17.3). Abrahamson and Blair (Chapter 14) point out that gallers engineer niches for their use and parasitoids exploit these niches, fueling sequential radiations wherein herbivore speciation drives the speciation of natural enemies. *Apocrypta* reared from different *Ficus* species have very similar mitochondrial DNA sequences (<2% divergence), indicating that fig parasitoids have broader host range than previously thought (Ulenberg 1985; Lopez-Vaamonde et al. 2001). Temporal evidence is consistent with a relatively recent invasion of the *Sycomorus* fig wasp community followed by stepwise colonization of existing host lineages. But why is *Apocrypta* less host specific than other fig wasp lineages? A possible explanation is population size. *Platyneura* gallers are orders of magnitude less abundant than pollinators, and *Apocrypta* parasitoids are similarly rare when compared to *Platyneura*. Many of the figs we sampled did not contain *Platyneura* and few contained *Apocrypta*. Levels of parasitism are possibly limited because parasitoid oviposition search time is wasted on figs lacking *Platyneura* hosts (Weiblen et al. 2001). Species-specific parasitoids may face a higher risk of extinction than less-specialized parasitoids that maintain larger population sizes by utilizing a broader resource base. We have yet to encounter multiple *Apocrypta* species on a single host, and it is thought that a single parasitoid species can attack multiple *Platyneura* species in the same fig. It will be possible in the future to examine the relationship between population size and host range by developing a species concept and molecular phylogeny for the *Sycoscapter* parasitoids of *Ceratosolen*. Since pollinators are orders of magnitude more abundant than gallers, they are the easier host for parasitoids to locate, and we predict that parasitoids of pollinators are more abundant than those of gallers. Host-range differences between parasitoids of pollinators and gallers associated with population size could support the hypothesis that extreme specialization is a dead end (Kelley and Farrell 1998).

Detailed historical inferences in this system will require more robust phylogeny estimates for the interacting lineages, as support for relationships in many species groups was weak even with the inclusion of multiple genes. Nuclear DNA sequence data for fig wasps would be a significant addition. Other possibilities for future study include detailed investigations of particular evolutionary scenarios including host switching, using approaches more sophisticated than maximum cospeciation analysis (Charleston 1998).

Conclusions

The fig microcosm provides a uniquely controlled and bounded environment in which to compare the impact of life-history variation on modes of diversification. Phylogenies for sympatric *Sycomorus* figs and associated fig wasps from New Guinea supported predictions about the extent of codivergence drawn from the contrasting life histories of fig pollinators, gallers, and parasitoids. Complete reproductive interdependence of the mutualists and extreme pollinator specificity enforces premating reproductive isolation between sympatric fig species such that many fig and pollinator clades are of equal age. These conditions do not apply to nonpollinators that appear to have diversified more commonly by processes other than cospeciation, including host switching. Externally ovipositing nonpollinators may colonize new hosts more readily than pollinators because they do not have to enter the fig cavity to lay eggs. Not only must host-shifting pollinators be morphologically compatible with the fig opening and flowers, but there are the additional challenges of pollen compatibility and competition with locally adapted resident pollinators. These considerations combined with the fundamental role that pollinators play in fig reproductive isolation account for the striking differences in modes of speciation for insect lineages that use the same plant resources in very different ways.

Acknowledgments

We thank Brus Isua and the staff of the New Guinea Binatang Research Center for field collecting, Warren Abrahamson, Catherine Blair, and an anonymous reviewer for helpful comments, and the National Research Institute, Forest Research Institute, and Department of Environment of Conservation of Papua New Guinea for permission to conduct the work. This material is based upon work supported by the National Science Foundation under grant DEB 0128833.

References Cited

Axelrod, D.I., and P.H. Raven. 1982. Paleobiogeography and origin of the New Guinea flora, pp. 919–941. In J. L. Gressitt (ed.), Biogeography and ecology of new guinea. Junk, The Hague, Netherlands.

Berg, C.C. 1989. Classification and distribution of *Ficus*. Experientia 45: 605–611.

Berg, C.C., and E.J.H. Corner. 2005. Moraceae *(Ficus)*. Flora Malesiana 17: 1–702.

Bronstein, J. 1992. Seed predators as mutualists: ecology and evolution of the fig/pollinator interaction, pp. 1–47. In E.A. Bernays (ed.), Insect-plant interactions. CRC Press, Boca Raton, FL.

Bronstein, J.L. 1991. The nonpollinating wasp fauna of *Ficus pertusa*: exploitation of a mutualism? Oikos 61: 175–186.

Bush, G.L. 1975. Modes of speciation in animals. Annu. Rev. Ecol. Syst. 6: 339–364.

Charleston, M.A. 1998. Jungles: a new solution to the host/parasite phylogeny reconciliation problem. Math. Biosci. 149: 191–223.

Chenuil A., and D.B. McKey. 1996. Molecular phylogenetic study of a myrmecophyte symbiosis: Did *Leonardoxa* ant associations diversify via cospeciation? Mol. Phylog. Evol. 6: 270–286.

Clayton, D.H., S. Al-Tamimi, and K.P. Johnson. 2003. The ecological basis of coevolutionary history, pp. 310–350. In R.D.M. Page (ed.), Tangled trees: phylogeny, cospeciation, and coevolution. University of Chicago Press, Chicago.

Cook, J.M., and J.-Y. Rasplus. 2003. Mutualists with attitude: coevolving fig wasps and figs. Trends Ecol. Evol. 18: 241–248.

Coyne, J.A., and H.A. Orr. 2004. Speciation. Sinauer, Sunderland, MA.

Currie, C.R., A.N. M. Bot, and J.J. Boomsma. 2003. Experimental evidence of a tripartite mutualism: bacteria protect ant fungus gardens from specialized parasites. Oikos 101: 91–102.

Datwyler, S.L., and G.D. Weiblen. 2004. On the origin of the fig: phylogenetic relationships of Moraceae from *ndhF* sequences. Am. J. Bot. 91: 767–777.

Dieckmann, U., and M. Doebeli. 1999. On the origin of species by sympatric speciation. Nature 400: 354–357.

Evans, R.C., L.A. Alice, C.S. Campbell, E.A. Kellogg, and T.A. Dickinson. 2000. The granule-bound starch synthase (GBSSI) gene in the Rosaceae: multiple loci and phylogenetic utility. Mol. Phylogenet. Evol. 17: 388–400.

Farrell, B., and C. Mitter. 1990. Phylogenesis of insect/plant interactions: have *Phyllobrotica* leaf beetles (Chrysomelidae) and the Lamiales diversified in parallel? Evolution 44: 1389–1403.

Felsenstein, J. 1985a. Confidence limits on phylogenies: an approach using the bootstrap. Evolution 39: 783–791.

Felsenstein, J. 1985b. Confidence limits on phylogenies with a molecular clock. Syst. Zool. 34: 152–161.

Galil, J., and D. Eisikowitch. 1971. Studies on mutualistic symbiosis between syconia and sycophilous wasps in monoecious figs. New Phytol. 70: 773–787.

Galil, J., R. Dulberger, and D. Rosen. 1970. The effects of *Sycophaga sycomori* L. on the structure and development of the syconium in *Ficus sycomorus* L. New Phytol. 69: 102–111.

Gibbons, J.R.H. 1979. A model for sympatric speciation in *Megarhyssa* (Hymenoptera: Ichneumonidae): competitive speciation. Am. Nat. 114: 719–741.

Grafen, A., and H.C.J. Godfray. 1991. Vicarious selection explains some paradoxes in dioecious fig-pollinator systems. Proc. R. Soc. Lond. B 245: 73–76.

Grison-Pige, L., J.M. Bessiere, and M. Hossaert-McKey. 2002. Specific attraction of fig-pollinating wasps: role of volatile compounds released by tropical figs. J. Chem. Ecol. 28: 283–295.

Hafner, M.S., and R.D.M. Page. 1995. Molecular phylogenies and host-parasite cospeciation: gophers and lice as a model system. Phil. Trans. R. Soc. Lond. 350: 77–83.

Harrison, R.D., and N. Yamamura. 2003. A few more hypotheses for the evolution of dioecy in figs (*Ficus*, Moraceae). Oikos 100: 628–635.

Herre, E.A. 1989. Coevolution of reproductive characteristics in 12 species of New World figs and their pollinator wasps. Experientia 45: 637–647.

Herre, E.A., N. Knowlton, U.G. Mueller, and S.A. Rehner. 1999. The evolution of mutualisms: exploring the paths between conflict and cooperation. Trends Ecol. Evol. 14: 49–53.

Hossaert-McKey, M., M. Gibernau, and J.E. Frey. 1994. Chemosensory attraction of fig wasps to substances produced by receptive figs. Entomol. Exp. Appl. 70: 185–191.

Jackson, A.P. 2004. Cophylogeny of the *Ficus* microcosm. Biol. Rev. 79: 751–768.

Johnson, P., F. Hoppensteadt, J. Smith, and G.L. Bush. 1996. Conditions for sympatric speciation: a diploid model incorporating habitat fidelity and non-habitat assortative mating. Evol. Ecol. 10: 187–205.

Jousselin, E. and F. Kjellberg. 2001. The functional implications of active and passive pollination in dioecious figs. Ecol. Lett. 4: 151–158.

Jousselin, E., J.Y. Rasplus, and F. Kjellberg. 2001a. Shift to mutualism in parasitic lineages of the fig/fig wasp interaction. Oikos 94: 287–294.

Jousselin, R., M. Hossaert-Mckey, D. Vernet, and F. Kjellberg. 2001b. Egg deposition patterns of fig pollinating wasps: implications for studies on the stability of the mutualism. Ecol. Entomol. 26: 602–608.

Jousselin, E., J.Y. Rasplus, and F. Kjellberg. 2003. Convergence and coevolution in mutualism: evidence from a molecular phylogeny of *Ficus*. Evolution 57: 1255–1269.

Kameyama, T., R. Harrison, and N. Yamamura. 1999. Persistence of a fig wasp population and evolution of dioecy in figs: a simulation study. Res. Popul. Ecol. 41: 243–252.

Kawakita, A., A. Takimura, T. Terachi, T. Sota, and M. Kato. 2004. Cospeciation analysis of an obligate pollination mutualism: have *Glochidion* trees (Euphorbiaceae) and pollinating *Epicephala moths* (Gracillariidae) diversified in parallel? Evolution 58: 2201–2214.

Kelley, S.T., and B. D. Farrell. 1998. Is specialization a dead end? The phylogeny of host use in *Dendroctonus* bark beetles (Scolytidae). Evolution 52: 1731–1743.

Kerdelhue, C. and J.Y. Rasplus. 1996. Non-pollinating Afro-tropical fig wasps affect the fig-pollinator mutualism in *Ficus* within the subgenus *Sycomorus*. Oikos 75: 3–14.

Kerdelhue, C., I. Le Clainche, and J. Rasplus. 1999. Molecular phylogeny of the *Ceratosolen* species pollinating *Ficus* of the subgenus *Sycomorus sensu stricto*: biogeographical history and origins of the species-specificity breakdown cases. Mol. Phylogenet. Evol. 11: 401–414.

Kerdelhue, C., J.P. Rossi, and J.Y. Rasplus. 2000. Comparative community ecology studies on old world figs and fig wasps. Ecology 81: 2832–2849.

Kondrashov, A.S., and F. Kondrashov. 1999. Interactions among quantitative traits in the course of sympatric speciation. Nature 400: 351–354.

Lopez-Vaamonde, C., J.Y. Rasplus, G.D. Weiblen, and J.M. Cook. 2001. DNA-based phylogenies of fig wasps: partial co-cladogenesis of pollinators and parasites. Mol. Phylogenet. Evol. 21: 55–71.

Lunt, D.H., D.X. Zhang, J.M. Szymura, and G.M. Hewitt. 1996. The insect cytochrome oxidase I gene: evolutionary patterns and conserved primers for phylogenetic studies. Insect Mol. Biol. 5: 153–165.

Machado, C.A., E. Jousselin, F. Kjellberg, and S.G. Compton. 2001. Phylogenetic relationships, historical biogeography and character evolution of fig-pollinating wasps. Proc. R. Soc. Lond. B 268: 1–10.

Machado, C. A., N. Robbins, T.P. Gilbert, and E.A. Herre 2005. Critical review of host specificity and its coevolutionary implications in the fig/fig-wasp mutualism. Proc. Natl. Acad. Sci. USA 102: 6558–6565.

Maddison, D.R. 1991. The discovery and importance of multiple islands of most-parsimonious trees. Syst. Zool. 40: 315–328.

Mason-Gamer, R.J., C.F. Weil, and E.A. Kellogg. 1998. Granule-bound starch synthase: structure, function, and phylogenetic utility. Mol. Biol. Evol. 15: 1658–1673.

McAlpine, J.R., G. Keig, and R. Falls. 1983. Climate of Papua New Guinea CSIRO. Australian National University Press, Canberra.

Michaloud, G., S. Carriere, and M. Kobbi. 1996. Exceptions to the one: one relationship between African fig trees and their fig wasp pollinators: possible evolutionary scenarios. J. Biogeogr. 23: 521–530.

Molbo, D., C.A. Machado, J.G. Sevenster, L. Keller, and E.A. Herre. 2003. Cryptic species of fig-pollinating wasps: implications for the evolution of the fig-wasp mutualism, sex allocation, and precision of adaptation. Proc. Natl. Acad. Sci. USA 100: 5867–5872.

Page, R.D.M. 1995. Maps between trees and cladistic analysis of historical associations among genes, organisms, and areas. Syst. Biol. 443: 58–77.

Page, R.D.M. 1996. Temporal congruence revisited: comparison of mitochondrial DNA sequence divergence in cospeciating pocket gophers and their chewing lice. Syst. Biol. 45: 151–167.

Page, R.D.M. 2003. Introduction, pp. 1–21. In R.D.M. Page (ed.), Tangled trees: phylogeny, cospeciation, and coevolution. University of Chicago Press, Chicago.

Percy, D.M., R.D.M. Page, and Q.C.B. Cronk. 2004. Plant-insect interactions: double-dating associated insect and plant lineages reveals asynchronous radiations. Syst. Biol. 53: 120–127.

Posada, D., and T. Buckley. 2004. Model selection and model averaging in phylogenetics: advantages of akaike information criterion and bayesian approaches over likelihood ratio tests. Syst. Biol. 53: 793–808.

Posada, D., and K. A. Crandall. 1998. Modeltest: testing the model of DNA substitution. Bioinformatics 14: 817–818.

Ramadevan, S., and M.A.B. Deakin. 1990. The Gibbons speciation mechanism. J. Theor. Biol. 145: 447–456.

Rasplus, J.Y. 1994. The one-to-one specificity of the *Ficus*-Agaoninae mutualism: how casual? pp. 639–649. In L.J.G. van der Maesen, X.M. van der Burgt, and J.M. van Medenbach de Rooy (eds.), The biodiversity of African plants. Kluwer Academic Publisher, Dordrecht, Netherlands.

Ronquist, F. 1998. Three-dimensional cost-matrix optimization and maximum cospeciation. Cladistics 14: 167–172.

Rønsted, N., G.D. Weiblen, J.M. Cook, N. Salamin, C.A. Machado, and V. Savolainen. 2005. 60 million years of co-divergence in the fig-wasp symbiosis. Proc. R. Soc. Lond. B 272: 2593–2599.

Sanderson, M.J. 1997. A nonparametric approach to estimating divergence times in the absence of rate constancy. Mol. Biol. Evol. 14: 1218–1231.

Shimodaira, H., and M. Hasegawa. 1999. Multiple comparisons of log-likelihoods with applications to phylogenetic inference. Mol. Biol. Evol. 16: 114–116.

Stearns, S.C., and R.F. Hoekstra. 2000. Evolution: an introduction. Oxford University Press, Oxford.

Strand, A.E., J. Leebens-Mack, and B.G. Milligan. 1997. Nuclear DNA-based markers for plant evolutionary biology. Mol. Ecol. 6: 113–118.

Swofford, D.L. 2001. PAUP*. Phylogenetic analysis using parsimony (*and other methods). Sinauer, Sunderland, MA.

Thompson, J.D., T.J. Gibson, F. Plewmiak, F. Jenamougin, and D.G. Higgins. 1997. The Clustal X Windows interface: flexible strategies for multiple sequence alignment aided by quality analysis tools. Nucl. Acids Res. 24: 4876–4882.

Ulenberg, S.A. 1985. The systematics of the fig wasp parasites of the genus *Apocrypta* Coquerel. North-Holland Publishing Company, Amsterdam.

Van Noort, S., and S.G. Compton. 1996. Convergent evolution of Agaonine and Sycoecine (Agaonidae, Chalcidoidea) head shape in response to the constraints of host fig morphology. J. Biogeog. 23: 415–424.

Weiblen, G.D. 2000. Phylogenetic relationships of functionally dioecious *Ficus* (Moraceae) based on ribosomal DNA sequences and morphology. Am. J. Bot. 87: 1342–1357.

Weiblen, G.D. 2001. Phylogenetic relationships of fig wasps pollinating functionally dioecious Ficus based on mitochondrial DNA sequences and morphology. Syst. Biol. 50: 243–267.

Weiblen, G.D. 2002. How to be a fig wasp. Annu. Rev. Entomol. 47: 299–330.

Weiblen, G.D. 2004. Correlated evolution in fig pollination. Syst. Biol. 53: 128–139.

Weiblen, G.D., and G.L. Bush. 2002. Speciation in fig pollinators and parasites. Mol. Ecol. 11: 1573–1578.

Weiblen, G.D., D.W. Yu, and S.A. West. 2001. Pollination and parasitism in functionally dioecious figs. Proc. R. Soc. Lond. B 268: 651–659.

Wiebes, J.T. 1979. Co-evolution of figs and their insect pollinators. Annu. Rev. Ecol. Syst. 10: 1–12.

EIGHTEEN

The Phylogenetic Dimension of Insect-Plant Interactions: A Review of Recent Evidence

ISAAC S. WINKLER AND CHARLES MITTER

The dramatic expansion of research on insect-plant interactions prompted by Ehrlich and Raven's (1964) essay on coevolution focused at first mainly on the proximate mechanisms of those interactions, especially the role of plant secondary chemistry, and their ecological consequences. Subsequently, in parallel with the resurgence of phylogenetics beginning in the 1970s and 1980s, there arose increasing interest in the long-term evolutionary process envisioned by Ehrlich and Raven (e.g., Benson et al. 1975; Zwölfer 1978; Berenbaum 1983; Mitter and Brooks 1983; Miller 1987). Since the early 1990s, spurred in part by the increasing accessibility of molecular systematics, there has been a happy profusion of phylogenetic studies of interacting insect and plant lineages. The results so far have reinforced skepticism about the ubiquity of the particular macroevolutionary scenario envisioned by Ehrlich and Raven, now commonly termed "escape and radiation" coevolution (Thompson 1988). However, this model continues to inspire and organize research on the evolution of insect-plant assemblages because it embodies several themes of neo-Darwinism, each of interest in its own right, which have been taken up anew in the modern reembrace of evolutionary history. In this chapter we attempt to catalog some of the postulates about phylogenetic history derivable from Ehrlich and Raven's essay and evaluate their utility for explaining the structure of contemporary insect-plant interactions.

The escape and radiation model (reviewed in Berenbaum 1983) tacitly assumes, first, that the traits governing species' interactions, such as insect host-plant preference, are phylogenetically conserved due to constraints such as limited availability of genetic variation. Such constraints create time lags between successive insect and plant counteradaptations, allowing the lineage bearing the most recent innovation to increase its rate of diversification. Second, a related general implication is that, because of genetic or other constraints on evolutionary response to new biotic surroundings, the structure of present-day insect-plant interactions (e.g., who eats whom) will be governed more by long-term evolutionary history than by recent local adaptation. This postulate parallels a broader recent shift in thinking about community assembly, from a focus on equilibrium processes to a greater appreciation of the role of historical contingency (Webb et al. 2002; Cattin et al. 2004; DiMichele et al. 2004). Third, the radiation component of escape and radiation perfectly encapsulates the "new synthesis" view, lately enjoying a revival (Schluter 2000), that diversification is driven primarily by ecological interactions. Insect-plant interactions have figured prominently in the modern reexamination of all three of these broad postulates.

This chapter surveys the recent evidence on the phylogeny of insect-plant interactions, focusing chiefly on among-species differences in larval host-plant use by herbivorous insect lineages (largely neglecting pollinators, which are treated by Adler in this volume), and organized around the themes sketched above. We draw mostly on literature of the past dozen years, that is, subsequent to early attempts at a similar survey (e.g., Mitter and Farrell 1991; Farrell and Mitter 1993). Given the great diversity of phytophage life histories and feeding modes, full characterization of host-use evolution will require, in addition to hypothesis tests in particular groups, the estimation of relative frequencies of alternative evolutionary patterns across a broad sampling of lineages. Our emphasis here is on the latter approach. A complete catalog is no longer feasible, but we have made a concerted and continuing effort to compile as many phylogenetic studies of phytophagous insect groups as possible. These are entered into a database that at this writing contained over 1000 entries, many of which were obtained from the Zoological Record database. Our analyses and conclusions are based chiefly on approximately 200 of these reports that contain both a phylogenetic tree and information on

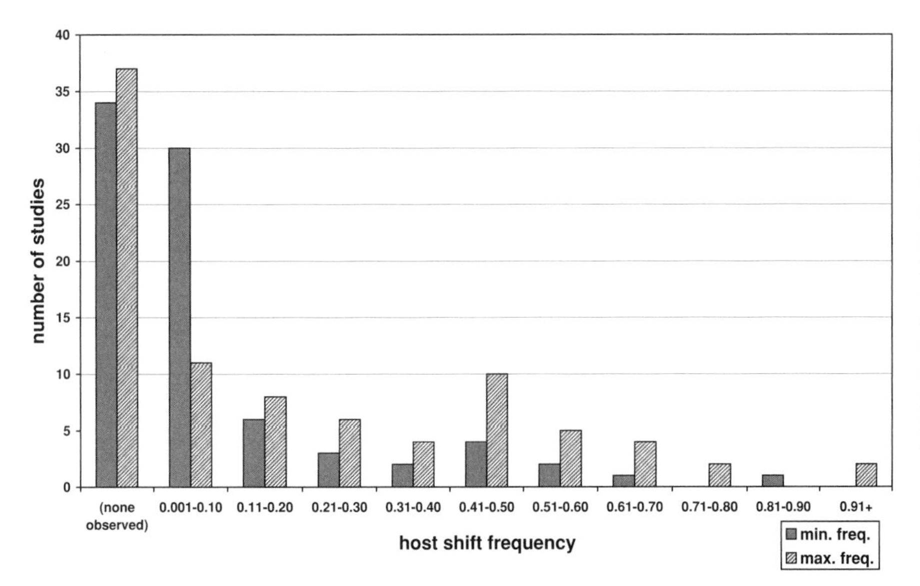

FIGURE 18.1. Frequency of host shifts per speciation event for 93 phytophagous insect phylogenies, calculated by dividing number of host-family shifts observed on phylogeny by number of included in-group species (solid bars, maximum host-shift frequency), and by total number of described species in the ingroup clade (hatched bars, minimum host-shift frequency). For references and taxa included, see Online Supplementary Table S2

host-plant use. Many of the phylogenies are based on DNA sequences, while for others the chief evidence is morphology. This database, intended as a community resource to promote further synthesis, is available at www.chemlife.umd.edu/entm/mitterlab, as are the data compilations and other supplementary materials mentioned in the text. Our nomenclature follows APGII (2003) for angiosperm families and higher groups, and Smith et al. (2006) for ferns.

Conservatism of Host-Plant Use

Full understanding of the influence of evolutionary history on insect-plant associations will require a broad accounting of the degree to which the different dimensions of the feeding niches of phytophagous insects are phylogenetically conserved. Much evidence on some aspects of this question has accumulated in the past decade.

Conservation of Host-Taxon Associations

The strongest generalization that can be made about the evolution of host-plant use is that related insect species most often use related hosts. This long-standing conclusion is now supported by numerous studies in which the history of host-taxon use has been reconstructed, most often under the parsimony criterion, on an insect phylogeny inferred from other characters. An early compilation (Mitter and Farrell 1991) of the few phylogenetic studies then available (~25) suggested that on average, less than 20% of speciation events were accompanied by a shift to a different plant family; strictly speaking, the compilation was of the fraction of branches subtended by the same node on the phylogeny that have diverged in host-family use, as inferred under the parsimony criterion. We have now repeated that calculation using essentially all applicable phylogenies we could find, totaling 93 (27 Coleoptera, 28 Hemiptera, 19 Lepidoptera, 12 Diptera, 5 Hymenoptera, and 1 each of Thysanoptera and Acari [honorary insects for the purposes of this chapter]). Some of the uncertainty in host-shift estimates comes from incomplete sampling of species. In the earlier compilation, host-shift frequency was calculated as the total number of host-family shifts inferred under the parsimony criterion, divided by one less than the number of sampled species with known hosts. This should be an unbiased estimate of the actual frequency of host shifts, if the included species are a random subset of the clade sampled. However, sampling in phylogenetic studies is often deliberately overdispersed across subclades (e.g., genera within a tribe), which should tend to inflate the average evolutionary distance among sampled species and hence the apparent frequency of host shifts. To evaluate the importance of this effect, we also calculated a corrected frequency estimate, dividing the number of shifts detected on the phylogeny by the total number of species with known hosts, including ones not included in the phylogenetic study. We will refer to these two estimates, in the order here described, as maximum versus minimum. In further contrast to the earlier tabulation, this one excluded the relatively few polyphagous species (defined here as those using more than two plant families); several phylogenies including a high proportion of polyphagous species were excluded, as well. A detailed tabulation of the phylogenies is given in the Online Supplementary Table S2, while the results are summarized in (Fig. 18.1).

The histogram of Fig. 18.1 shows a result very similar to that of the earlier tabulation, underscoring the prevalence of host conservatism. The distributions of host-family shift frequencies, strongly right-skewed, have medians of 0.08 (maximum frequency) and 0.03 (minimum frequency). Statistical tests of the hypothesis of nonrandom phylogenetic conservatism in host-genus or host-family use have now become routine within studies of the kind tabulated here. These most often use the so-called PTP test (permutation tail probability [Faith and Cranston 1991]), in which the null distribution is generated by random redistribution of the observed host-family associations across the insect phylogeny. Significant "phylogenetic signal" has been detected in nearly every instance (e.g., Table 18.2). In addition, several authors have used randomization tests on frequencies of shift among different host families or groups thereof to show that these preferentially involve related high-rank host taxa (Janz and Nylin 1998; Ronquist and Liljeblad 2001); conservatism at the level of major angiosperm clades (APGII 2003) is probably common as well.

It is widely accepted that conserved host-taxon associations primarily reflect conserved recognition of and other adaptations to plant secondary chemistry, but this assumption has been difficult to test because of the generally close correlation of chemistry with plant taxonomy. Several cases of mismatch between host-chemical and taxonomic similarity have now been examined phylogenetically and shown closer correspondence of insect phylogeny to chemistry than plant relatedness (Becerra 1997; Wahlberg 2001; Kergoat et al. 2005). Recent studies include reexamination of classic examples (Dethier 1941; Feeny 1991) of repeated shifts by lepidopterans between unrelated host families bearing similar secondary compounds (e.g., Lauraceae, Rutaceae, and/or Apiaceae [Berenbaum and Passoa 1999; Zakharov et al. 2004; Berenbaum, this volume]). This subject is by no means exhausted, as many more such syndromes surely await documentation. It should be noted, however, that herbivore groups feeding on plants without distinctive chemical defenses or on undefended plant parts can also show similarly specialized, conserved host associations (e.g., leafhoppers [Nickel 2003]).

Variation in Rates of Major Host Shift

Although conservatism is pervasive, phylogenetic studies continue to document great variation among phytophage lineages in the frequency of "major" host shifts (e.g., to different plant families). Establishing patterns to this variation will be a key step toward understanding the constraints on diet evolution. Many predictors for differential host-shift rates have been advanced (reviewed in Mitter and Farrell 1991), some invoking properties of plant taxa and/or communities, others invoking traits of the phytophages. Attempts to test these, however, remain few, and the subject seems ripe for further synthesis. In one of the few explicit analyses, Janz and Nylin (1998) present evidence that among butterflies, shifts among major angiosperm clades are less frequent in herb feeders than tree feeders. Nyman et al. (2006) found that internally feeding nematine sawfly clades have colonized significantly fewer plant families than their externally feeding sister groups. Radiations on oceanic islands have been suggested to undergo exaggerated divergence in niches, including host-plant use, compared to continental relatives (e.g., Schluter 1988). In the only test for phytophages, the eight genera of delphacid planthoppers endemic to various Pacific islands were found to have a significantly higher mean rate of host-family shift (two times higher), and frequency of polyphagy, than the 52 continental genera (Wilson et al. 1994); systematic work in progress will permit reanalysis with better control for phylogeny. Possible explanations for elevated host-shift rates on islands include limited availability of preferred hosts of colonizers, lesser chemical distinctiveness among host species due to relaxed herbivore pressure, and absence of continental competitors and/or insect natural enemies (reviewed in Wilson et al. 1994). Further comparisons to insular radiations may help to identify causes of the prevailing host specificity and conservatism of mainland phytophages.

Compilations of host-shift rates as in Online Supplementary Table S2 should permit further tests of hypotheses about differential host conservatism. Following Fagan et al. (2002), we used phylogenies from the literature to concatenate all the groups in the table into a single metaphylogeny (presented in Online Supplementary Fig. S3). One can then map onto the phylogeny the inferred host-shift frequencies plus the distribution of traits postulated to affect them (e.g., internal versus external feeding). The metatree can then be divided into a maximal number of independent regions (contrasts), each consisting of a set of contiguous branches and containing an inferred evolutionary change in the putative predictor trait. For each contrast, a single response measure is calculated (e.g., the difference in mean host-shift frequency between groups having the opposing states of the predictor variable). Paired comparisons are then used to test for a consistent effect of the predictor variable on host-shift frequency. In a first analysis, strong support was found for elevated mean frequency of host-family shifts inferred from just the oligophagous species (i.e., polyphages not scored) in lineages that include one or more polyphagous species, as opposed to lineages lacking polyphages (12/12 contrasts differing in the same direction; $P < 0.0001$, sign test). This finding supports the conjecture (e.g., Janz and Nylin, this volume) that rapid shift among host taxa and polyphagy of individual species are related phenomena.

It has often been suggested (e.g., Farrell and Mitter 1990) that dependence on host-derived toxins for larval and/or adult phytophage defense should reduce the likelihood of major host shifts. This postulate has had no formal comparative test. However, recent phylogenetic evidence suggests that use of such defenses itself is in general not so conservative, or so intimately tied to larval diet, as might be supposed (Dobler et al. 1996; Dobler 2001), probably because herbivores often have multiple defenses. Thus, in the chrysomelid beetle subtribe Chrysomelina (Termonia et al. 2001; Kuhn et al. 2004) the ancestral larval defense is entirely autogenous, but there have been two independent origins, within Salicaceae-feeding lineages, of dependence on host-derived salicin. Within one of these groups there has been subsequent addition of a second type of defense, based on a combination of autogenous and host-derived pathways, followed by multiple host shifts to another family (Betulaceae) from which salicin is not available. Availability of more than one defense-metabolism pathway may likewise have facilitated repeated host-family shifts in other groups, such as the tropical chrysomeline genus *Platyphora* (Termonia et al. 2002). Moths of the typically aposematic family Arctiidae are one of several groups that have converged on defensive use of plant-derived pyrrholizidine alkaloids (PAs), while producing endogenous other toxins as well. A recent phylogeny for arctiids implies a single origin of larval feeding on PA-containing plants and sequestration of PAs that are retained into the adult stage (Weller et al. 1999). In a species-rich subclade of the ancestrally PA-plant-feeding lineage, there have been repeated shifts to non-PA larval hosts, implying lack of constraint by chemical dependence. Adult defense, however, shows strong apparent phylogenetic inertia, as adults in this subclade have evolved to actively collect and use PAs. A similar "constraint" explanation was proposed for the propensity of adults in one African and one New World galerucine chrysomelid subtribe to feed on, and use in courtship and defense, toxic cucurbitacins from Cucurbitaceae, which are at present fed on by larvae in just a single genus in each subtribe. Recent phylogenetic evidence (Gillespie et al. 2003, 2004), however, strongly supports independent New World and Old World origins for both larval and adult use of cucurbits, and points, albeit less strongly, to adult use arising first.

Other Conserved Aspects of Host Use

Most discussion of the impact of host-plant use on insect diversification has focused on host-taxon differences, but other conserved dimensions of the feeding niche have also been recognized (e.g., Powell 1980; Powell et al. 1998), including host growth form and habitat, plant part exploited, mode of insect feeding, and phenology of oviposition and feeding. Most herbivorous insects are specialized to particular host tissues, such as leaves, flowers, fruits, seeds, stems, or roots, in addition to particular host taxa. On any one plant part, moreover, insects are typically specialized for one of a great variety of feeding modes. For example, a partial list of feeding behaviors exhibited by insects that eat leaves includes galling, mining, leaf rolling or tying, and external folivory. The relative rates of evolution of the various niche dimensions are fundamental to assessing their roles in phytophage diversification.

Several authors have begun to quantify these rates and their variation. Cook et al. (2002) used a maximum likelihood approach to show that a genus of cynipid gall wasps shifts among host-plant organs more often than among sections of their host genus, oaks. Farrell and Sequeira (2004) used similar methods to demonstrate, conversely, that in chrysomeloid beetles, shift among major host clades outpaces shift among host tissues. Other reports reinforce this latter trend at the host species level (Condon and Steck 1997; Favret and Voegtlin 2004). However, studies of gallers are mostly consistent in finding rapid shift among host tissues (e.g., Yang and Mitter 1994; Plantard et al. 1997; Nyman et al. 2000; Dorchin et al. 2004); shifts in gall location, shape, and timing, often on the same host species, may be important facilitators of galler speciation. Host growth form (i.e., trees versus herbs) often shows very strong phylogenetic conservatism relative to host clade (Ronquist and Liljeblad 2001; Bucheli et al. 2002; Lopez-Vaamonde et al. 2003), but not always (Janz and Nylin 1998; Schick et al. 2003). Timing of oviposition or development with respect to host phenology is another dimension of host use that may frequently contribute to speciation, either on the same host or on a novel host (e.g., Wood 1993; Pratt 1994; Whitcomb et al. 1994; Harry et al. 1998; Filchak et. al. 2000; Weiblen and Bush 2002; Sachet et al. 2006).

A special form of oft-conserved host use, occurring in some groups of aphids and gall wasps (Cynipidae), is obligate alternation between different host taxa in successive generations. Host alternation may have originated multiple times in aphids (Moran 1988, 1992; von Dohlen and Moran 2000; von Dohlen et al. 2006), though this inference rests mostly on differences in the mode of host alternation and other life-history features, as the phylogenetic evidence cannot adequately distinguish between gains and losses of host alternation per se. Regardless, this kind of complex host association has clearly evolved only a few times, while the loss of one or the other host has occurred repeatedly within ancestrally host-alternating lineages (Moran 1992; see also Cook et al. 2002). The degree to which host alternation (as opposed to simply shifting to a different host) reflects constraint versus adaptation has been debated (Moran 1988, 1990, 1992; Mackenzie and Dixon 1990).

Parallelism, Reversal, and Genetic Constraints on Host Shift

Although conservatism of host-use traits can suggest the influence of phylogenetic "constraint" or "inertia" (Blomberg and Garland 2002), this interpretation is not automatic, as stabilizing selection is a plausible alternative (Hansen and Orzack 2005). The constraint interpretation would receive powerful support if one could demonstrate limitations on within-population genetic variation, for traits determining host use, that corresponded to the actual history of shifts undergone by the larger clade to which the test populations belonged. In a series of studies deserving wide emulation, Futuyma and colleagues (reviewed in Futuyma et al. 1995; see also Gassman et al. 2006) reconstructed the history of host use in oligophagous *Ophraella* leaf beetles, then screened four species for genetic variation in larval and adult ability to feed and survive on the hosts (various genera of Asteraceae, in several tribes) fed on by their congeners. In only 23 of 55 tests (species by host) was there any detectable genetic variation for ability to use the alternative host. Such variation as did appear was mainly for use of hosts of closely related beetle species; these plants were themselves closely related to the normal host. Thus, lack of available variation for use of alternative hosts is probably much of the explanation for the conserved association of this genus with Asteraceae. Other lines

TABLE 18.1

Summary of Host and Distribution Overlap versus Nonoverlap for 145 Sister-Species Pairs from 45 Phytophagous Insect Phylogenies

	Total Species Pairs	*Host Species Overlap*	*Host Species Overlap*	*Host Species Disjunct*	*Host Species Disjunct*	*Total Hosts Disjunct*	*Total Distributions Disjunct*
		Distributions Overlap	Distributions Disjunct	Distributions Overlap	Distributions Disjunct		
All pairs	145	27	48	26	44	48%	63%
Continental pairs only	101	22	27	22	30	52%	56%
Island pairs only	44	5	21	4	14	41%	80%

NOTE: Host species overlap, members of pair sharing at least one host species; host species disjunct, sharing no host species; distributions overlap, with >10% areal overlap in geographic distribution; distributions disjunct, with <10% overlap in geographic distribution. Details including sources in Online Supplementary Table S4.

of evidence, less direct, point to an analogous conclusion for other clades and traits. Many authors have noted (e.g., Janz and Nylin 1998; Hsiao and Windsor 1999; Janz et al. 2001; Swigoňová and Kjer 2004; Zakharov et al. 2004) that host-family use is often highly homoplasious (i.e., showing multiple independent origins of the same habit), sometimes with repeated colonizations of a single plant family inferred to be an ancestral host. Janz et al. (2001) tested the long-standing hypothesis that such a propensity reflects retained ability to use former hosts, finding that Nymphalini butterfly larvae of most species were willing to feed on the ancestral host (*Urtica*), regardless of what host they normally fed on. Some specific kinds of phylogenetic pattern also strongly suggest genetic constraint. Thus, in several unrelated groups of galling insects, it has been found that features such as gall structure or gall position on the plant follow an ordered multistep progression on the phylogeny, for example from simple to successively more complex (Nyman et al. 2000; Ronquist and Liljeblad 2001). If the evolution of such traits were not limited by genetic variation, it is hard to see why it should nonetheless follow the presumptive path of "genetic least resistance" (Schluter 2000). The nature and extent of genetic constraints, critical to a full understanding of host-use evolution, is an underexplored subject on which modern genetic/genomic approaches hold promise for rapid progress (e.g., Berenbaum and Feeny, this volume).

Conservatism, Host Shifts, and Speciation

Given the pervasive conservatism of higher-host-taxon use, one might wonder whether diet conservatism on a finer scale has been underestimated, and shifts to different host species consequently assigned too large a role in phytophagous insect speciation. One requisite for answering this question is a broad estimate of the proportion of speciation events that are accompanied by a change in host species. To our knowledge, no such survey has been published. We provide an estimate based on 145 presumptive sister species pairs found within 45 phylogenies of phytophagous insect genera or species groups in our database for which information about hosts and geographic distribution was available. Taxa other than confirmed species (e.g., host races or unconfirmed sibling species) were excluded. Each species pair was scored as sharing a host-plant species or not; pairs were also scored as having hosts from the same genus, family, or higher angiosperm clade (defined in APGII 2003). To contrast the frequency of host differences to that of differences in distribution, each sister pair was also scored as having distributions overlapping by 10% or more (subjectively estimated) versus <10%. No characterization of the accuracy of these phylogenies was attempted. A possible source of bias is that island radiations, which show a somewhat greater frequency of allopatry between sister species than continental forms and (surprisingly) a somewhat lower mean proportion of host differences, comprise over 25% of our data set. Therefore, we also present results with and without island lineages. Our tabulation and its sources are given in Online Supplementary Table S4, and the results are summarized in (Table 18.1).

Overall, about 48% of the divergence events we tabulated are associated with an apparent change in host species. This is our best estimate of the fraction of speciation events that could have been driven by host shifts (though of course we have no way of knowing whether the host differences actually accompanied speciation, rather than arising prior to or after speciation). Our results are consistent with a major role for host shifts in phytophage speciation, but not a ubiquitous one; we estimate that about half of all speciation events are unaccompanied by a host shift. Of course, many of the latter could have involved change in tissue fed upon or other aspects of host use.

Greater circumspection is required in interpreting our compilation of differences in distribution, which potentially bear on the controversial question of sympatric speciation

(Lynch 1989). The utility of phylogenetic evidence on this issue has been doubted, even dismissed, because species' distributions can shift rapidly (Barraclough and Vogler 2000; Losos and Glor 2003; Fitzpatrick and Turelli 2006). Thus, the proportion of sister species that are sympatric might reflect dispersal ability rather than frequency of sympatric speciation (Chesser and Zink 1994; Losos and Glor 2003). Indeed, allopatric speciation has recently been suggested to play a prominent role even in the *Rhagoletis pomonella* group, the poster child for sympatric speciation (Barraclough and Vogler 2000). Nonetheless, we follow Berlocher (1998) in holding the comparative approach worthy of further exploration. Berlocher suggested that there should be a higher frequency of sympatry between sister species in host shifting than in non-host-shifting taxa, if host differences are commonly important in allowing species to originate, or at least remain distinct, in sympatry. In our compilation, however, extant sister species using different host species were sympatric only slightly (and not significantly) more often than those not differing in host, 37% (N = 70) versus 36% (N = 75). This result seems to cast doubt on the ubiquity of divergence by sympatric host shift, but that interpretation may be too conservative. For example, among-group variation in dispersal ability, which we did not correct for, might obscure the "signal" for host-associated sympatric divergence in our tabulation. Moreover, the probability of sympatric divergence may depend strongly on how different the hosts are. Thus, sister species that differ in host genus used show a markedly higher frequency of sympatry (50%) than pairs whose hosts are congeneric if they differ at all (33%), though this difference was not statistically significant ($P = 0.189$, χ^2 test). This observation is at least consistent both with a role for "major" host differences in promoting sympatric divergence, and with the postulate that shifts to distantly related hosts are more likely in sympatry, which allows for prolonged prior adaptation (Percy 2003). We should note, finally, that the study of phytophagous insect speciation and host-shift mechanisms is being revolutionized by, among other advances, the advent of fine-scale, intraspecific molecular phylogenetics including phylogeography sensu Avise (2000). These applications of phylogenies are treated elsewhere in this volume.

Phylogenesis of Host Range

Special attention has focused on the evolution of diet breadth, namely, the diversity of host plants fed on by a single herbivore species. Restriction to a small subset of the available plants is a dominant feature of phytophagous insect ecology. In addition to demanding an explanation in its own right (Bernays and Chapman 1994), it has made herbivorous insects a leading exemplar for investigating the ecological and evolutionary consequences of specialization (Schluter 2000; Funk et al. 2002). Phylogenies can potentially serve three roles in the study of host range. First, they delimit independent contrasts for identifying traits or geographic and other circumstances whose occurrence is correlated with evolutionary changes in host range, facilitating both comparative and experimental studies of the adaptive significance and consequences of those changes. Second, the rate and direction of changes in host range inferred on a phylogeny can point to genetic/phylogenetic constraints or lack thereof on host-range evolution. Third, phylogenies can in principle detect differential effects of broad versus narrow host range on diversification rates. Analyses of the second and third kinds could potentially support nonadaptive, macroevolutionary explanations for the predominance of host specificity, such as more frequent speciation in specialists than in generalists, in contrast to hypotheses invoking a prevailing individual advantage (Futuyma and Moreno 1988).

The study of host-range evolution is still something of a conceptual and methodological tangle. A fundamental question is how to define host range. Although broad, somewhat arbitrary categories of relative specialization may often suffice to reveal evolutionary patterns (e.g., Janz et al. 2001), objective, quantitative measures may yield greater statistical power and allow more meaningful comparisons across studies (Symons and Beccaloni 1999). However it is defined, host range is surely a composite feature likely to reflect different combinations of (typically unknown) adult and immature traits in different groups. It is probably subject to a heterogeneous mix of influences that vary in relative strength with the scale of comparison. Small-scale changes in host range might reflect behavioral plasticity or local adaptation in response to differences in host abundance or quality, or host-associated assemblages of competitors, predators, or parasitoids (e.g., Singer et al. 2004; Bernays and Singer 2005). Such changes could also represent short-lived intermediate steps in the evolution of new specialist species (e.g., Hsiao and Pasteels 1999; Janz et al. 2001, 2006). In contrast, changes evident mainly on longer time scales, and spanning a greater range of diet breadths, could reflect less frequent but more pervasive evolutionary shifts involving multiple component adaptations. At any scale of examination, broader host range could result from different causes in different lineages.

Given the heterogeneity of potential causes, evolutionary patterns of host range are likely to differ widely among groups. Phylogenetic evidence has begun to accumulate, but we are far from having an adequate characterization of that variation, let alone an explanation. The most useful studies will be those in which (1) unambiguous distinctions are evident in host range, reflecting intrinsic differences among species (not higher taxa as in Berenbaum and Passoa [1999], contra Nosil [2002] and Nosil and Mooers [2005]), and (2) taxon sampling is dense enough to permit detection of evolutionary trends if these exist. Only a handful of the studies in our database appear to meet these criteria. We summarize the nine that we judged to come closest in (Table 18.2). No criticism is implied of any work not included in this somewhat subjective selection, particularly

TABLE 18.2

Synopsis of Nine Recent Studies Bearing on Phylogenetic Patterns of Host Range

Insect Taxon	*Taxon Sample*	*No. Specialists vs. no. Generalists*	*Criterion for Specialist vs. Generalist*	*Forms of Directionality Reported*	*Significant Host-Taxon Conservation?*	*Significant Host-Range Conservation?*	*Source*
Coleoptera: Chrysomelidae: *Oreina*	12/24 spp., spanning all ecological variation in genus	6 vs. 6	1 vs. > 1 host tribe	None	Host family and tribe conserved, $P = 0.01$ PTP	No: $P = 0.47$, PTP	Dobler et al. 1996
Coleoptera: Curculionidae: Scolytinae: *Dendroctonus*	18/19 spp. (position of remaining species taken from literature)	6 vs. 13	Using $<½$ vs. $>½$ of available host species	Specialists limited to tips of phylogeny	Host genus conserved, $P = 0.01$, PTP (authors)	No: $P = 1.00$, PTP (authors)	Kelley and Farrell 1998
Coleoptera: Bruchidae: *Stator*	21/22 spp. with known hosts	4 vs. 16	1 vs. > 1 host tribe	Generalists derived	Use of 1 of 4 host genera conserved, $P = 0.03$, PTP (authors)	No: $P = 1.00$, multiple tests (authors)	Morse and Farrell 2005
Lepidoptera: Nymphalidae: Melitaeini	10/10 gen., 65/250 spp.; sparse sampling in one large Neotropical clade	51 vs.14	1 vs. >1 host family	Host-family gains > losses	Host-family use conserved, $P = 0.003$ (author)	No for gen. vs. spec. ($P = 0.34$ PTP; yes ($P = 0.02$) for number of host families (1–6)	Wahlberg 2001

Lepidoptera: Nymphalidae: Nymphalini	27/70 spp.; most taxa in some nearctic lineages	17 vs. 10	1 vs. >1 host order	Specialist ancestor; host gains > losses	Host-family use conserved, $P = 0.01$, PTP (authors)	Marginal: $P = 0.06$, PTP	Janz et al. 2001
Lepidoptera: Nymphalidae: *Polygonia*	14/16 spp.	7 vs. 5	1 vs. >1 host order	Specialist ancestor; host gains > losses	Host-order use conserved, $P = 0.001$, PTP	No for gen. vs. spec. and number of host orders (1–5; $P > 0.30$, PTP)	Weingartner et al. 2006
Lepidoptera: Saturniidae: *Hemileuca*	22 populations in 17 spp./28 spp. total; excluded taxa may be synonyms	15 vs. 7	Using primarily 1 vs. >1 host family	None	Host-family use conserved, $P = 0.02$, PTP	No: $P = 1.00$, PTP (authors)	Rubinoff and Sperling 2002
Diptera: Tephritidae: *Tomoplagia*	19/59 spp.; sampling limited to part of Brazil	11 vs. 8, or 14 vs. 5, or 15 vs. 4	1 vs. >1 host genus, or subtribe, or tribe	Depends on criterion	No: $P > 0.18$ (PTP) for host subtribe use	No: $P > 0.50$, PTP, all criteria	Yotoko et al. 2005
Phasmida: Timematidae: *Timema*	14 spp. (17 taxa)/21 spp. (remainder described subsequently)	11 vs. 6	1 (95% of records) vs. >1 host genus	Generalist ancestor	Host genus conserved, $P < 0.04$ (authors)	No: $P = 0.59$, PTP	Crespi and Sandoval 2000

NOTE: See "Table References" for source information. Gen., generalist; PTP, permutation tail probability test; spec., specialist; spp., species.

since the tracing of diet breadth has only rarely been an explicit goal.

The strongest generalization evident so far is that host range is quite evolutionarily labile, much more so than use of particular host taxa. As a gauge of that lability, we tabulated the results of PTP tests (Faith and Cranston 1991) on degree of host specificity treated as a binary character with changes in the two directions equally weighted (one versus more than one host family, or other criteria specified by the authors or otherwise appropriate to the study group; about half these analyses were performed by the authors). In seven of nine cases, this test cannot reject a random distribution of host range on the phylogeny, whereas in each case but one, use of individual host taxa is significantly conserved. As several authors have noted, host range is clearly not subject to strong forms of phylogenetic constraint or "inertia" (Blomberg and Garland 2002), such as absolute irreversibility (Nosil and Mooers 2005; Yotoko et al. 2005). In fact, the paucity of obvious phylogenetic signal may complicate further characterization of host-range evolution by limiting the utility of some standard strategies of phylogenetic character analysis. Thus, when a two-state likelihood model is applied to estimate the relative rates of transition to and from specialization, the rates can most often be closely predicted from just the proportions of specialists and generalists among the terminal taxa (Nosil 2002; Nosil and Mooers 2005). This outcome, intuitively expected if the states are distributed randomly on the tree, might be taken to suggest that phylogenies have little to contribute to the understanding of host-range evolution. And indeed, it is possible that much of the variation in host range analyzed so far is in fact phylogenetically "random" in the sense of reflecting idiosyncratic local fluctuation, for example in the availability of, and/or selective advantage of using, particular hosts. This may be especially true when all the species within the study group are specialists in the broad sense of feeding on plants in, for instance, the same family.

As several authors have noted, however, it is plausible that larger-scale phylogenetic regularities remain to be discovered through the elaboration of more detailed, process-oriented models of host-range evolution (Stireman 2005). Multiple approaches can be distinguished. Thus, host range might be thought of as a trait phylogenetically ephemeral in itself, but with probabilities of change predictable from the states of other, more conserved features, inviting use of the "comparative method." For example, distribution of the use of two versus more than two tribes of legumes appears by itself to be random on a phylogeny of the seed beetle genus *Stator*. Closer inspection, however, shows that independent origins of broader host range are significantly concentrated in lineages that oviposit on predispersal seeds, rather than on intact seed pods or dispersed seeds (Morse and Farrell 2005).

An alternative approach focuses on the genetic and ecological mechanisms by which host range changes. Thus, Crespi and Sandoval (2000; see also Nosil et al. 2003) conclude that host specialization in *Timema* walking sticks comes about when host-associated color polymorphism in polyphagous ancestors is converted into species differences under disruptive selection by predators. Phylogenetic evidence by itself is consistent with but does not strongly establish ancestral polyphagy. However, that interpretation is supported by abundant experimental and other evidence. Similar logic is reflected in the elaboration of a novel hypothesis about butterfly host range (e.g., Janz et al. 2001; Weingartner et al. 2006; Janz and Nylin, this volume). A phylogeny for the nymphalid tribe Nymphalini suggests ancestral restriction to Urticales followed by repeated host-range expansions as well as contractions, with multiple ostensibly independent colonizations of a set of disparate plant families. Complementary experiments show that larvae of many species are able to feed on hosts not presently used by that species, but characteristic of their inferred ancestors and/or extant relatives. Retained latent feeding abilities may help to explain rapid expansions (and hence observed lability) of host range. Polyphagy may also facilitate radical host shifts (and/or further broadening of host range), given that less specialized species seem to generally make more oviposition mistakes (Janz et al. 2001), and has been suggested to thereby promote diversification (Janz et al. 2006; Weingartner et al. 2006; Janz and Nylin, this volume). This postulate stands in direct contrast to the prediction that specialization promotes faster speciation, for which evidence is currently lacking (see below).

Several of the foregoing hypotheses may apply to a broad phylogenetic pattern of host range in the noctuid moth subfamily Heliothinae (Mitter et al. 1993; Cho 1997; Fang et al. 1997; S. Cho, A. Mitchell, C. Mitter, J. Reiger, M. Matthews, submitted). A paraphyletic basal assemblage, species-rich and almost entirely oligophagous or monophagous (80% on Asteraceae), contrasts sharply with an advanced "*Heliothis* clade" containing a much higher proportion of polyphages. Host range is correlated with phylogeny, albeit weakly, but the most dramatic difference is in its much higher rate of change in the *Heliothis* clade. That lineage appears to have a set of conserved life-history features (higher fecundity, body size, and other traits) that are relatively permissive of changes in host range, while the low fecundity, small size, low vagility, and other traits of the more basal species may strongly disfavor host-range expansion. Phylogenetically controlled analyses of the life-history correlates of diet breadth are still too few, but the number is growing (e.g., Beccaloni and Symons 2000) and further synthesis seems imminent (Jervis et al. 2005).

With so many promising recent leads at hand, we can look forward to rapid progress in understanding of the phylogenetic patterns of host-range evolution.

Signatures of Long-Term History in Extant Insect-Plant Interactions

Strong conservatism of host taxon or other aspects of host-plant use raises the possibility that the current distribution of insects across plant species reflects some form of long-term synchrony in the diversification of those associates. One extreme form of synchronous evolution would be strict parallel phylogenesis or cospeciation, in which descendant lineages of the insect ancestor maintain continuous and exclusive association with the descendants of the ancestral plant species; the expected signature is a characteristic form of correspondence between the phylogenetic relationships, and the absolute ages, of the extant associates. Extensive methodological and empirical work on this general issue over the past 15 years, in many groups of organisms, has established that strict or nearly strict parallel phylogenesis is almost entirely limited to parasites and other symbionts that are directly transmitted between host parent and offspring individuals (e.g., Page 2003). However, variants of this scenario more likely for free-living phytophages have also been envisioned, involving intermittent and/or less specific association of insect species with particular host-plant taxa, and producing corresponding forms of incomplete phylogeny matching. Under escape and radiation coevolution, for example, the closest match is expected not between phylogenies per se, but between phylogenetic sequences of escalating plant defenses and insect counteradaptations (Mitter and Brooks 1983). The marks of other forms of shared evolutionary history might lie primarily elsewhere. For example, it has been proposed that differences in the predominant host associations of major phytophagous insect clades reflect differences in which plant groups dominated the global flora in the different eras in which those phytophages arose (Zwölfer 1978). The critical evidence on such postulates will often be absolute datings. For the full range of questions considered in this section, a combined approach from phylogenetics and paleontology is proving especially powerful (reviewed in Labandeira 2002; see also Grimaldi and Engel 2005). There is currently a surge of interest in molecular dating studies, driven in part by the increasing sophistication of methods for combining evidence from fossils and molecular divergence (reviewed in Magallón 2004; Welch and Bromham 2005), though the reliability of such datings is still poorly understood.

In this section we attempt to sketch out and evaluate the evidence for several forms of historical imprint on insect-plant associations. Such inquiry matters for two reasons. First, traces of shared long-term history imply that there has been at least the opportunity for prolonged reciprocal evolutionary influence—coevolution in a broad sense—and may even provide evidence on the nature and extent of that coevolution. Second, from the ecological point of view, unique marks of history imply that the assembly of extant insect-plant communities cannot be fully explained by just the current properties of local or regional species pools or even the evolutionary propensities of these; one may need also to invoke the contingent historical sequence in which particular insect and plant lineages appeared on earth (Farrell and Mitter 1993).

Early in the current era of phylogenetic studies, there was much interest in the possibility of parallel phylogenesis between insect and host-plant clades. There is now enough evidence to state with confidence that correspondence of phytophagous insect and host phylogeny is rare on the taxonomic scale at which it has most often been examined, namely within and among related insect genera. Even groups involved in obligate pollination mutualisms show much less correspondence with host phylogeny than previously assumed (Pellmyr 2003; Kawakita et al. 2004; Machado et al. 2005). An early compilation (Mitter and Farrell 1991) examined 14 studies, in only one of which was there unambiguous support for parallel phylogenesis. Here we tabulate a subset of 18 of the many relevant studies appearing since then, limited to papers in which the authors themselves drew conclusions about parallel cladogenesis (Table 18.3). In the great majority of these, there is little evidence, from either cladogram concordance or datings, for parallel diversification. Our sample undoubtedly underestimates the true prevalence of such negative evidence, as we did not include the many papers in which parallel cladogenesis is implicitly ruled out at the start. One exception to the rule is particularly instructive: a group of psyllids showed significant phylogeny concordance with its legume hosts, but molecular clock and fossil datings indicate that host diversification was likely complete before the group was colonized by these phytophages (Percy et al. 2004). Presumably, host shifts in these herbivores have been governed by plant traits correlated with plant phylogeny; it is less clear why colonization should start at the base of the host phylogeny. In light of this finding, it seems especially important that newly discovered instances of possible cladogram match, for example, as reported for a group of gracillariid moths that obligately pollinate their hosts (Kawakita et al. 2004), be investigated for equivalence of ages.

The few plausible cases for both cladogram match and equivalence of ages include two genera of herb-feeding beetles (leaf beetles on skullcap mints [Farrell and Mitter 1990]; longhorn beetles on milkweeds [Farrell and Mitter 1998; Farrell 2001]). The vast assemblage of figs and their mutualist wasp pollinators, the subject of many recent phylogenetic studies (Silvieus et al., this volume), shows clear elements of parallel diversification, although it now appears that host specificity and parallel speciation are much less strict than was formerly thought (Machado et al. 2005).

Datings based on fossils, molecular clocks, and biogeography also continue to identify other patterns suggesting long-continued, not necessarily coevolutionary interactions (e.g., von Dohlen et al. 2002). One of the most elaborate apparent historical interaction signatures involves *Blepharida* alticine leaf beetles and related genera. Beetle phylogeny shows only tenuous concordance with that of the chief hosts, *Bursera* and relatives (Burseraceae/Anacardiaceae), but much stronger match to a phenogram of leaf extract gas chromatography profiles (compounds not specified) (Becerra 1997). Shared geographic disjunction between the New World and African tropics implies comparable overall

TABLE 18.3

Synopsis of 18 Recent Studies Testing for Parallel Insect-Plant Phylogenesis at Lower Taxonomic Levels

Insect Order and Family	*Insect Clade*	*Host Clade(s)*	*Overall Phylogeny Correspondence Plausible?*	*Equivalent Ages Plausible?*	*Sources*
Coleoptera: Cerambycidae	*Tetraopes*	*Asclepias* (Apocynaceae)	Yes, significant cladogram similarity	Yes	Farrell 2001, Farrell and Mitter 1998
Coleoptera: Chrysomelidae	*Ophraella*	Asteraceae	No, cladograms do not match	No, beetles younger than hosts	Funk et al. 1995
Coleoptera: Chrysomelidae	*Blepharida*	Burseraceae	Maybe, depends on analysis	Yes	Bercerra 1997, 2003
Coleoptera: Curculionidae	*Anthonomus grandis* grp.	*Hampea* (Malvaceae)	No, cladograms do not match	Not tested	Jones 2001
Hymenoptera: Tenthredinidae	*Euurina* (Nematinae)	*Salix* (Salicaceae)	No, cladograms do not match	Not tested	Nyman et al. 2000, Roininen et al. 2005
Hymenoptera: Cynipidae	Major lineages of cynipids	Asteraceae, Lamiaceae, Fagaceae, Rosaceae, Papaveraceae	No, cladograms not significantly similar	Maybe (based on fossils, biogeography)	Ronquist and Liljeblad 2001
Hymenoptera: Agaonidae	Agaoninae	*Ficus* (Moraceae)	Yes,[a] but correspondence not universal	Yes	Machado et al. 2005
Hymenoptera: Agaonidae	*Apocryptophagus* (nonpollinators)	*Ficus* (Moraceae)	No, cladograms not significantly similar	Not tested	Weiblen and Bush 2002
Diptera: Tephritidae	*Urophora*	Cardueae (Asteraceae)	No, cladograms not significantly similar	No, flies younger than hosts	Brändle et al. 2005
Lepidoptera: Gracillariidae	*Epicephala*	*Glochidion* (Phyllanthaceae)	Maybe, depends on type of analysis	Not tested	Kawakita et al. 2004
Lepidoptera: Gracillariidae	*Phyllonorycter*	>30 families of angiosperms	No, cladograms not significantly similar	No (individual moth/host radiations tested)	Lopez-Vaamonde et al. 2003, 2006

Lepidoptera: Geometridae	Lithinini	Ferns, multiple families	No, multiple shifts to distantly related hosts	No, moths younger than hosts	Weintraub et al. 1995
Hemiptera: Aphididae	*Uroleucon*	Asteraceae	No, multiple shifts to distantly related hosts	No, aphids much younger than hosts	Moran et al. 1999
Hemiptera: Psyllidae	Arytaininae	Fabaceae: Genisteae of Macaronesia	Yes, significant cladogram similarity	No, psyllids much younger than hosts	Percy et al. 2004
Hemiptera: Psyllidae	*Calophya, Tainarys*	*Schinus* (Anacardiaceae)	Maybe, depends on group and analysis	Not tested	Burckhardt and Basset 2000
Hemiptera: Delphacidae	Tribes of delphacids	Various monocots	Little evidence for cladogram match	Maybe (ages uncertain)	Wilson et al. 1994
Hemiptera: Delphacidae	*Nesosydne*	Hawaiian silverswords (Asteraceae: Heliantheae, 3 genera)	Maybe, depends on analysis (sampling incomplete)	Yes	Roderick 1997
Acari: Eriophyidae	*Cecidophyopsis*	*Ribes* (Grossulariaceae)	No, cladograms do not match	No, mites younger than hosts	Fenton et al. 2000

NOTE: See "Table References" for source information.

[a]Machado et al. (2005) found fig and pollinator wasp phylogeny congruence to be nonsignificant and point out the paucity of evidence for cladogram matching of figs and their pollinating wasps at lower levels, as well. However, it is evident that substantial overall codivergence has occurred (Rønsted et al. 2005), and widespread (but not universal) congruence at lower levels still seems plausible (see also Weiblen and Bush 2002; Silvieus et al., this volume, and references therein).

ages (112 million years; but see Davis et al. 2002) for the interacting clades, and molecular clocks point to similar, younger ages for two associated beetle and plant subsets marked by corresponding innovations in resin canal defense and counterdefense (Becerra 2003). This case, an exemplar of the broad syndrome of parallel origins of resin/latex canal defenses and counteradaptations thereto (Farrell et al. 1991), is perhaps the most detailed to date for long-term insect-plant "arms race" sequences as envisioned by Ehrlich and Raven (1964; but see Berenbaum 2001), though evidence for the accelerated diversification expected with each innovation is lacking.

We digress here to note that such putative escalations of plant defense are underinvestigated and possibly rare. Aside from resin/latex canals, the two most strongly stated hypotheses involve evolutionary trends toward chemical complexity in coumarins and other secondary compounds in Apiaceae (reviewed in Berenbaum 2001) and in cardenolides of milkweeds (*Asclepias*; reviewed in Farrell and Mitter 1998). Although the modern revolution in plant phylogeny has underscored the conservatism of some major secondary chemistry types (e.g., Rodman et al. 1998), phylogenetic studies directed explicitly at the evolution of plant defense are still few (but see Armbruster 1997; Wink 2003; Rudgers et al. 2004). Agrawal and Fishbein (2006) mapped an array of putative defense traits that included total cardenolides (though not the hypothesized arms race aspects thereof) onto a molecular phylogeny for 24 *Asclepias* species. Rather than reflecting plant phylogeny, these traits appear to define three distinct, convergently evolved defense syndromes, each possibly optimal in the right circumstances. This implicit optimality/equilibrium view of plant defense is very different from the historically contingent view inherent in the arms race hypothesis. Under the latter, we expect some lineages to have acquired novel defenses that confer, at least temporarily, a ubiquitous fitness advantage over relatives lacking those innovations. The relative applicability of these two views of defense evolution across the diversity of plants and their defensive traits has yet to be determined.

Reinforcing the view that ancient host associations may have left widespread, if not numerically dominant, traces on contemporary assemblages is the increasing evidence for broad-scale correspondence between the ages of currently associated insect and plant groups, over time spans encompassing major evolutionary changes in the global flora. The case for this long-standing postulate (see Zwölfer 1978) is best developed for the beetle clade Phytophaga (Chrysomeloidea + Curculionoidea, ~135,000 species), whose hosts span the chief lineages of seed plants (Farrell 1998; Marvaldi et al. 2002; Farrell and Sequeira 2004). Recent phylogeny estimates show most of the basal phytophagous lineages in both superfamilies to feed exclusively on conifers or cycads, the most basal seed plants. The five gymnosperm-associated clades, totaling about 220 species, have apparently Gondwanan-relict distributions, and several are known as Jurassic fossils from the same deposits as are members of their present-day host groups. Within both superfamilies, moreover, there are early splits between monocot and (eu)dicot feeders, possibly established during the early divergence between these two main lineages of angiosperms (Farrell 1998). A similar pattern is evident, in abbreviated form, in the Lepidoptera, first known from the early Jurassic (Grimaldi and Engel 2005). Larvae of the most basal lineage (Micropterigidae) inhabit riparian moss and liverwort beds, apparently feeding on these and/or other plant materials. Their habits match those of the inferred common ancestor of Lepidoptera and their sister group Trichoptera (Kristensen 1997). Recent morphological and molecular phylogenies (Kristensen 1984; Wiegmann et al. 2000, 2002) firmly establish that the most basal lineage of the remaining Lepidoptera, which are otherwise mostly restricted to advanced angiosperms, consists of two Australasian species that feed inside cones of the conifer *Araucaria*. This association, which parallels basal gymnosperm feeding (specifically within reproductive structures) in Phytophaga (Farrell 1998), is quite plausibly viewed as predating the availability (or at least the dominance) of angiosperm hosts. It is, however, the only obvious such relictual habit in Lepidoptera. While other primitive lineages also have apparent Gondwanan-relict distributions, suggesting mid-Mesozoic ages, they feed on advanced (mainly eurosid) dicots, and their phylogenetic relationships correspond not at all to those of their chief host-plant taxa (Powell et al. 1998). Host use appears to evolve considerably faster in Lepidoptera than in Phytophaga, thus traces of earlier feeding habits are probably more quickly obliterated.

Ancient host associations in other phytophagous lineages that date to the early Mesozoic and before, less well characterized, await clarification by modern studies. Recent progress on phylogeny of sawflies (basal hymenopterans) (e.g., Schulmeister 2003), modern families of which date to the early Jurassic or even Triassic, should permit elucidation of the degree to which the multiple conifer feeding lineages, totaling several hundred species, represent ancestral habits. We can hope for similar enlightenment about the Aphidomorpha (aphids and relatives), probably Triassic in age, in which the phylogenetic positions of the few extant gymnosperm-associated lineages are still obscure (Heie 1996; Normark 2000; von Dohlen and Moran 2000; Ortiz-Rivas et al. 2004). Moreover, documentation of such deep-level relictual host associations may prompt reexamination of some younger groups for which synchronous diversification with hosts seems at first glance implausible. Thus, analysis of the more than 1000 species of cynipid gall wasps detected no significant overall phylogeny match with their host-plant families, mostly woody rosids and herbaceous asterids (Ronquist and Liljeblad 2001). However, recently discovered taxa have raised the possibility that the ancestral gall wasps, like one basal extant lineage, fed on Papaveraceae, a member of the most basal eudicot lineage, Ranunculales (but see Nylander 2004; Nylander et al. 2004). Fossils date the gall wasps to at least the late Cretaceous, thus it is possible that this habit has been retained since before the rise to prominence of the host groups commonly used today (Ronquist and Liljeblad 2001). A similar history is possible for some genera of leaf-mining agromyzid flies (Spencer 1990).

Aphids, agromyzids, and other groups may participate in another broad historical pattern that is receiving increased attention. Insect groups whose chief diversity is associated with modern (especially poaceous or euasterid) herbaceous plants in temperate regions might well have diversified in parallel with the great Tertiary expansion of open habitats and herbaceous vegetation, driven by global cooling, drying, and latitudinal climate stratification trends (Behrensmeyer et al. 1992; Graham 1999). This postulate, in need of rigorous test, shares some elements with escape-and-radiation coevolution, including the ascription of diversification to ecological opportunity, and the distribution of insect lineages across plants to long-term historical trends. The hypothesis predicts that phylogenies of these herbivores should exhibit trends toward use of successively younger host groups (and/or perhaps from trees to herbs), and subclade ages should roughly match those of their hosts and/or biomes (Dietrich 1999; von Dohlen and Moran 2000; von Dohlen et al. 2006). One among many candidate lineages is the so-called trifine Noctuidae (Noctuidae sensu stricto; more than 11,000 species). Trifines have a markedly higher ratio of temperate-to-tropical species than any other large family of Macrolepidoptera and, unlike those families, are mostly herb feeders instead of tree feeders. Recent phylogenies confirm that the trifine groups most closely adapted to open, boreal habitats, which are often ground-dwelling "cutworms" as larvae, are among the most derived (Holloway and Nielsen 1998; Mitchell et al. 2006).

Diversification of Phytophagous Insects

The extraordinary species richness of plant-feeding insects is a salient feature of terrestrial biodiversity (Strong et al. 1984). It is therefore not surprising that insect-plant interactions have been a prominent model in the modern revival of interest in diversification (Wood 1993; Schluter 2000; Coyne and Orr 2004). Full understanding of the diversification of phytophagous insects will require both detailed analysis of speciation (and extinction) mechanisms, and comparative study of broad diversification patterns. These enterprises are of course intertwined, and phylogeny is relevant to both. Our review, however, will focus mainly on the comparative aspect.

A fundamental question to be asked is whether the apparent exceptional diversity of phytophagous insects is actually the result of consistent clade selection (Williams 1992), rather than a coincidental impression created by a few groups whose hyperdiversity could reflect some other cause. Sister-group comparisons between independently originating phytophagous insect clades and their nonphytophagous sister groups, which control for clade age and other traits possibly influencing diversification rate, show that phytophages have consistently elevated diversities (Mitter et al. 1988). This conclusion is at least consistent with the results of an analysis screening for significant variation in diversification rate across the insect orders (Mayhew 2002). It should be noted that the finding rests at present on only a small fraction of the potential evidence, as the phylogenetic positions of most originations of insect phytophagy are only now beginning to be resolved. Thus, further test of this hypothesis is desirable.

Why should phytophagous insects have elevated diversification rates? Several broad hypotheses have been advanced. One possibility is adaptive radiation (Simpson 1953), redefined loosely by Schluter (2000) as "evolution of ecological diversity in a rapidly multiplying lineage" (p. 1). Vascular plants might constitute an "adaptive zone" providing an extraordinary diversity of underutilized, distinct resources on which insect specialization is possible. A contributing factor might be that more niches supporting a sustainable population size are available at the primary consumer level than to higher levels or to decomposers, no matter how those niches are filled. Diversification could be accelerated still further if plant diversity continually increases due to coevolution sensu Ehrlich and Raven (1964). In a contrasting, though complementary, hypothesis (Price 1980), phytophage diversity reflects instead a broad propensity of the "parasitic lifestyle" for rapid diversification, due in part to the ease with which populations of small, specialized consumers can be fragmented by the patchy distribution of hosts.

Some progress has been made toward sorting out these alternatives. The finding that insect groups parasitic on animals are, if anything, less diverse than their nonparasitic sister groups (Wiegmann et al. 1993) casts strong doubt on the primacy of the parasitic lifestyle hypothesis. The leading hypothesis, adaptive radiation, makes two chief predictions. One of these, the subject of a vigorous area of research (Via 2001; Berlocher and Feder 2002; Rundle and Nosil 2005; other chapters this volume), is that shifts to new plant resources should be a major contributor to the origin of new species. Earlier, we estimated that about 50% of speciation events in phytophagous insects involve shifts to a different host-plant species. This is an underestimate of the importance of plant resource diversity to speciation, because niche shifts within the same host-plant species (e.g., to different host organs or tissues) and changes in host range (with retention of at least one previous host) are not included. Comparative data, then, are at least consistent with a major role for host-related divergence in phytophage diversification. It should be noted that ecological differences between sister species can arise by multiple mechanisms before, during, or after speciation (Futuyma 1989; Schluter 2000). Even if host-related differences were incidental to speciation, however, a broad form of the adaptive zone or radiation hypothesis could be said to hold, if those differences produced higher net diversification rate by forestalling extinction due to competition for resources or enemy-free space. As the foregoing suggests, hypotheses attributing diversification to ecological differentiation have rarely been explicit about which of the many possible mechanisms are involved (reviewed in Allmon 1992). Ongoing ecological study of the importance of competition and natural enemies to phytophage fitness and host use (e.g., Denno et al. 1995; Murphy 2004) should help to distinguish among plausible candidate mechanisms.

A second prediction of the adaptive radiation hypothesis is that the diversification rate of a phytophagous lineage should be correlated with the number of plant resource niches available to it. The strongest evidence on this question so far comes from studies of the beetle clade Phytophaga. In each of 10 contrasts identified so far (Farrell 1998; Farrell et al. 2001), beetle groups feeding on conifers or other gymnosperms were less diverse than their angiosperm-feeding sister groups. To these can be added the contrast in Lepidoptera between the basal conifer-feeding lineage Agathiphagidae (two species) and its almost entirely angiosperm-feeding sister group Heterobathmiidae + Glossata (~160,000 spp.; Wiegmann et al. 2000). Although exceptions will undoubtedly be found (e.g., probably lachnine aphids [Normark 2000]; xyelid sawflies [Blank 2002]), elevated diversity of angiosperm feeders seems likely to remain one of the strongest diversification effects known (Coyne and Orr 2004) as the numerous additional contrasts are examined. Ascription of this trend to the much greater taxonomic and chemical diversity of flowering plants, rather than some unique historical circumstance or the global biomass difference between angiosperms and gymnosperms, gains credibility from the great variation in ages and geographic distributions among the contrasted lineage pairs, and the fact that some represent secondary return to gymnosperms (Farrell et al. 2001). It will now be of great interest to determine whether association of enhanced insect diversification with more diverse host groups holds on smaller plant-taxonomic scales as well.

Ehrlich and Raven (1964) speculated that diversification of the angiosperms was promoted by their novel and diverse secondary chemistry, which improved protection from herbivores. Correspondingly greater diversity in angiosperm-feeding insects than in related relict gymnosperm feeders is at least consistent with their hypothesis. Broad-scale escape and radiation coevolution is also lent credence by recent evidence that adaptations to and interaction with insects (and other organisms) have marked influence on plant diversification rates. Plant clades bearing latex or resin canals, one of the most elaborate plant defense syndromes known, were shown to be consistently more diverse than sister groups lacking such canals (Farrell et al. 1991). More recently, several types of innovations in reproductive structures, affecting pollinator fidelity or fruit dispersal, have also been shown to be associated with more rapid plant diversification (Sargent 2004; Bolmgren and Erikkson 2005; reviewed in Coyne and Orr 2004). Thus, mounting evidence supports a central tenet of the new synthesis, implicit in escape and radiation coevolution, namely that adaptations to biotic interactions have major influence on diversification.

While substantial progress has been made in establishing phytophage diversification patterns at the broadest scale, countless questions remain, particularly at shorter evolutionary time scales. There is almost no unambiguous evidence on whether repeated counteradaptations to plant defenses have accelerated insect diversification, as predicted under escape-and-radiation coevolution (but see Farrell et al. [2001] regarding mutualism with ambrosia fungi in bark beetles; parallel examples of fungal mutualism in cecidomyiid gall midges discussed by Bisset and Borkent [1988] and Gagné [1989] await further phylogenetic study). Numerous other causes have been postulated for differential diversification of phytophages, including, among others, species richness, secondary chemical diversity, growth form, and geographic distribution of the host group (e.g., Price 1980; Strong et al. 1984; Lewinsohn et al. 2005); mode of feeding, including plant tissue attacked, internal versus external feeding, and gall-making (and advanced forms thereof) (Ronquist and Liljeblad 2001); trenching and other forms of herbivore "offense" (Karban and Agrawal 2002); degree of food plant specialization; host-shift frequency; and various traits (often host-use-related) rendering phytophages less susceptible to natural enemies (Singer and Stireman 2005). Indeed, just about any trait that might be conserved on phylogenies becomes a plausible candidate. Ideally, one would like to determine the relative importance of and interactions among these factors, and compare them to other types of influence on diversification. In the Lepidoptera, for example, the most pervasive differential influence on diversification may prove to be the repeated evolution of ultrasound detectors allowing adults to avoid bat predation (e.g., Yack and Fullard 2000), rather than any "bottom-up" factor having to do with host plants.

Progress on testing such hypotheses has been quite limited so far, probably for several reasons. First, although phylogenies are accumulating rapidly, the detailed phylogenetic resolution needed to detect correlates of diversification rates is still lacking within most families of phytophagous insects; in some cases, even species diversities are not yet well characterized. Second, we are only beginning to understand the phylogenetic distributions of most candidate traits. Many of these appear to be much more evolutionarily labile than the relatively conserved features reviewed earlier. Rapid trait evolution can frustrate estimation of ancestral states, particularly when life-history information is incomplete, making reliable sister-group comparisons hard to identify. For example, our scan of published studies uncovered essentially no unambiguous contrasts between lineages with broader versus narrower species host ranges, though sister clades often differed in average host range. Moreover, the groups characterized by labile traits, when identifiable, will often be so recent that dissecting deterministic from stochastic influences on diversification would require a large number of comparisons. Sister-group comparisons remain the most robust and straightforward method for detecting traits correlated with diversification rate (Vamosi and Vamosi 2005). But, unless traits that vary mostly at lower taxonomic levels are to be dismissed as unlikely to influence diversification rates, additional approaches will be needed (Ree 2005).

Fortunately, there is now a diverse, rapidly growing literature on diversification rate analysis, a full survey of which is beyond the scope of this chapter. Any of several approaches might prove useful for testing the association of relatively labile traits with diversification rates, depending on the nature of the data. If the chief difficulty is that inferred trait origins do not clearly define sister-group comparisons, one might identify comparisons a priori, then score sister groups simultaneously for diversity and some appropriate measure of frequency of the predictor trait. To select potentially informative comparisons, one might employ one of the various model-based methods proposed for identifying significant shifts in rates of diversification (Sanderson and Donoghue 1994; Magallón and Sanderson 2001; Moore et al. 2004); possible drawbacks include the need for well-resolved phylogenies and high variance of trait frequency estimates in extremely asymmetrical comparisons. For quantitative predictor variables (e.g., average host range) a variant of the independent contrasts method is available (Isaac et al. 2003). When lack of deeper-level phylogeny resolution limits identification of sister groups, one might make independent comparisons among groups of different ages, using estimates of absolute or relative diversification rates (Purvis 1996; Bokma 2003; application in Nyman et al. 2006). For relatively recent radiations, average time between speciation events may be a more sensitive estimator of diversification rate than species numbers per se (Ree 2005). Clock-based temporal analyses of diversification can in principle also detect changes in diversification rate over time (e.g., Nee et al. 1992, 1996; Paradis 1997), allowing test of such refinements of the adaptive zone hypothesis as the postulated slowing of diversification as niches are filled (Simpson 1953; Schluter 2000). Recently, this and other approaches have been used to identify periods of accelerated insect diversification and correlate these with potential causes such as radiation of particular plant clades, or particular biogeographic events (e.g., McKenna and Farrell 2006; Moreau et al. 2006; but see Brady et al. 2006).

While phytophage diversification rate variation at lower levels is a daunting problem, even the analysis of relatively conserved traits remains underdeveloped. To underscore this point, we end with a summary of progress on one much-discussed issue that bears on the puzzle of phytophagous insect diversity, namely, the macroevolutionary consequences of internal versus external feeding. Both habits are widespread, although their frequencies differ markedly across insect phylogeny. Most hemimetabolous insect herbivores, in orders such as Orthoptera, Phasmida, Hemiptera, and Thysanoptera, are free-living external feeders, though some (e.g., thrips) may hide in flowers or other plant structures; the chief exceptions are gall-formers, which have evolved repeatedly in the piercing/sucking lineages. In contrast, larvae of a large fraction of phytophagous Holometabola, including the basal members of nearly all the major lineages, actively bore or mine inside living plants. External phytophagy has arisen infrequently in most holometabolous orders, or not at all (e.g., higher Diptera), while return to endophagy has occurred somewhat more often. Overall, the opposing traits seem sufficiently conserved, yet also sufficiently labile, to permit replicated sister-group comparisons.

Opposing predictions have been made about diversification under these contrasting feeding modes, drawing on broader theories about ecological specialization (reviewed in Wiegmann et al. 1993; Yang and Mitter 1994). Although analyses controlled for phylogeny are needed (Nyman et al. 2006), internal feeders appear to be more host specific than external feeders (e.g., Gaston et al. 1992). Greater specialization, as argued earlier, could promote speciation by increasing the strength of population subdivision and diversifying selection (e.g., Miller and Crespi 2003). Internal feeding could also be viewed as an adaptive zone providing escape from pathogens and some parasites, and desiccation or other physical stresses (Connor and Taverner 1997). Conversely (Powell et al. 1998; Nyman et al. 2006), one could predict that external feeding, by providing release from constraints on body size, voltinism, and leaf excision, might typically increase individual and (thereby) clade fitness. Moreover, by lowering the barriers to colonization of alternative hosts and habitats, exophagy might open more opportunities for speciation.

Sister-group contrasts between internal and external feeders are potentially numerous. For example, there is strong evidence for several to many independent transitions between internal and external larval feeding within Lepidoptera (Powell et al. 1998), Coleoptera-Phytophaga (Marvaldi et al. 2002; Farrell and Sequiera 2004), and basal Hymenoptera (sawflies), and between galling and free-living habits within Aphidoidea (von Dohlen and Moran 2000), Coccoidea (Cook and Gullan 2004), Psylloidea (Burckhardt 2005), and Thysanoptera (Morris et al. 1999). Surprisingly, however, from our literature survey we are able to extract at most eight unambiguous comparisons (Table 18.4). The only phylogenetic analysis study directed specifically at this question is that of Nyman et al. (2006); others are clearly needed.

Disregarding the one tie, five of the seven sister-group comparisons we identified show the external-feeding lineage to be more diverse than its internal-feeding closest relatives. Nyman et al. (2006), in a nonoverlapping set of comparisons within the sawfly subfamily Nematinae (Tenthredinidae), found external feeders to be more diverse in 10 of 13 sister-group contrasts. Taken together, these compilations yield a result just significant by a two-tailed sign test (external feeders more diverse in 15 of 20 pairs,

TABLE 18.4
Sister Group Diversity Comparisons Between Endo- and Exophytophage Lineages

Higher Taxon	*Internally Feeding Clade*	*Diversity*	*Externally Feeding Clade*	*Diversity*	*Sources*
Coleoptera: Chrysomelidae	Bruchinae + Sagriinae	3,300	Chrysomelinae + Criocerinae + others, minus secondary internal feeders	10,000	Farrell and Sequeira 2004
Hymenoptera	Cephidae + Siricidae + Anaxyelidae + Xiphydriidae, with parasitic subclade Vespina excluded	280	Pamphilidae + Megalodontesidae	350	Brown 1989, Heitland 2002, Schulmeister 2003
Hymenoptera	Blasticotomidae	9	Remaining Tenthredinoidea	7,000+	Nyman et al. 2006, Schulmeister 2003
Hymenoptera: Xyelidae	Xyelinae	71	Macroxyelinae	11	Blank 2002, Schulmeister 2003
Lepidoptera	Cossoidea	1,873	Zygaenoidea	2,115	Powell et al. 1998
Lepidoptera	Obtectomera minus Macrolepidoptera (part or all)	<22,0000	Macrolepidoptera	87,000	Powell et al. 1998
Lepidoptera: Heliodinidae	*Lamprolophus* + 9 genera	56	*Epicroesa* + *Philocoristis*	6	Hsu and Powell 2004
Thysanoptera: Phlaeothripidae	*Kladothrips*	22	*Rhopalothripoides* (+ 5 possibly related genera)	22	Crespi et al. 2004, Morris et al. 2002

NOTE: Compilation excludes nematine tenthredinid sawflies, studied by Nyman et al. (2006). See "Table References" for source information.

$P = 0.042$), corroborating the trend in an earlier, more limited compilation by Connor and Taverner (1997).

Although progress is evident, continued study of this question is desirable. The statistical significance of the observed trend is still marginal; several of the comparisons in Table 18.4 are based on provisional phylogenies, and in several the diversity differences are small; it will also be of much interest to separately test the effects of different categories of internal feeding (e.g., gallers versus miners), and of gains versus losses of external feeding. At the least, however, the current evidence appears to firmly reject the hypothesis of consistently faster diversification by internal feeders. The result parallels previous rejection of the hypothesis of higher diversification in animal-parasitic than free-living insects due to their exceptionally specialized lifestyles (Wiegmann et al. 1993). Together, these observations suggest that, even if phytophages are more ecologically specialized in some sense that other insects, specialization per se is an unlikely explanation for their exceptional diversity. Rather, the evidence increasingly points to the importance of the sheer diversity of niches available to insects feeding on plants, particularly flowering plants.

Synopsis and Conclusions

In this chapter we have attempted to compile and synthesize the recent literature (mainly since 1993) treating aspects of the phylogenesis of associated insects and plants. We have focused on phylogenies at the among-species level and higher, mostly for insects, and on their bearing on three general questions posed implicitly by Ehrlich and Raven's hypothesis of coevolution. These are (1) the degree to which the various traits governing use of host plants are conserved during phylogenesis; (2) the degree to which contemporary associations show evidence, from phylogenies and other sources, of long-continued interactions between particular insect and plant lineages; and (3), the degree to which evolution in traits affecting their interactions affects the diversification rates of interacting insect and plant lineages. Our main conclusions are as follows:

1. Ubiquitous conservation of plant higher taxon use during insect phylogenesis is confirmed and quantified in a compilation of 93 phylogenies of mostly

oligophagous insect groups. The median frequency of shift to a different plant family is estimated to be about 0.03 to 0.08 per speciation event. Important initial insights have been gained on the reasons for this conservatism.

2. There are many hypotheses to explain among-clade variation in the frequency of among-plant-family shift, but few quantitative tests. The strongest evidence to date is for more frequent host shifting in tree feeders than in herb feeders among butterflies, and among oligophages within lineages that contain one or more polyphagous species than in lineages that do not (across 95 insect phylogenies). Recent case studies suggest that reliance on plant-derived compounds for insect defense poses less of a barrier to larval host shift than was formerly thought.
3. In contrast to the prevailing broad-scale host conservatism, shifts to a different host species have accompanied about 50% of 145 phytophage speciation events tabulated, consistent with a substantial but not universal role for host shifts in phytophage speciation. There is a suggestive but not statistically significant tendency for greater host differentiation between sympatric than allopatric species pairs.
4. The as yet limited evidence on phylogenetic patterns of host-plant range provides no support for directionality or other strong constraints but suggests an important distinction between ephemeral, phylogenetically random fluctuation, and larger-scale trends interpretable using experimental approaches combined with phylogenetic "comparative methods."
5. It is now clear that with very few exceptions, the host-use variation within and among phytophagous insect genera, in contrast to that in some vertically transmitted parasites and symbionts, reflects colonization of already-diversified hosts rather than any form of strict parallel phylogenesis. At the same time, however, evidence is increasing that associations established in the distant past, especially the Mesozoic, have left widespread if not numerically dominant marks on contemporary insect-plant assemblages; the full range of such historical "signatures" is only beginning to be explored.
6. Because phylogenetic studies directed specifically at plant defense evolution are still few, we do not yet know whether that evolution is characterized more by sequential coevolutionary "escalations," or by stably coexisting syndromes reflecting optimal adaptations for differing environments.
7. Replicated sister-group comparisons have established elevated diversification rates for phytophagous over nonphytophagous insects and for angiosperm over nonangiosperm feeders among phytophages, both at least consistent with diffuse insect-plant coevolution sensu Ehrlich and Raven (1964). Recent studies on plant diversification rates demonstrate a role for interaction with insects and other animals, likewise consonant with that theory, though most examples do not involve defense. Evidence on most phytophage diversification hypotheses (including "offense" innovations), however, has been slow to accumulate, and diversification studies at finer taxonomic scales, mostly lacking, may face methodological obstacles. A progress report on sister-group comparisons of internal versus external feeders effectively negates the hypothesis of faster radiation by endophages, thought to be more specialized, and strongly suggests the opposite trend.

Given the range of questions mapped out, the tools available, and the cornucopia of phylogenetic studies now ongoing in nearly all major herbivorous insect groups and their host plants, we can look forward to spectacular near-future advances in understanding of the evolution of insect-plant interactions, with increasing integration between phylogenetic and other perspectives.

Author's Note: For online supplementary tables and figures, go to www.chemlife.umd.edu/entm/mitterlab. These include the following:

- Item S1: Database of insect/plan phylogeny studies. FileMaker, Access formats.
- Item S2: Table compiling host-shift frequencies on phylogenies. Excel format.
- Item S3: Figure showing metaphylogeny of taxa included in table S2 for comparative analysis of host-shift frequency versus host range. PDF format.
- Item S4: Table compiling host and distribution differences for speciation events. Excel format.

Acknowledgments

We are grateful to Anurag Agrawal, Brian Farrell, Kathleen Pryer, Stephan Blank, and their coauthors for sharing unpublished material, to the editor and reviewers for helpful comments on the manuscript, and to the National Science Foundation and U.S. Department of Agriculture for financial support.

Table References

Becerra, J. X. 1997. Insects on plants: macroevolutionary chemical trends in host use. Science 276: 253–256.

Becerra, J. X. 2003. Synchronous coadaptation in an ancient case of herbivory. Proc. Natl. Acad. Sci. USA 100: 12804–12807.

Blank, S. M. 2002. Biosystematics of the extant Xyelidae with particular emphasis on the Old World taxa (Insecta: Hymenoptera). Dissertation, Freie Universität Berlin.

Brändle, M., S. Knoll, S. Eber, J. Stadler, and R. Brandl. 2005. Flies on thistles: support for synchronous speciation? Biol. J. Linn. Soc. 84: 775–783.

Brown, W.L. 1989. Hymenoptera, pp. 652–680. In S.P. Parker (ed.), Synopsis and classification of living organisms. McGraw Hill, New York.

Burckhardt, D., and Y. Basset. 2000. The jumping plant-lice (Hemiptera, Psylloidea) associated with *Schinus* (Anacardiaceae): systematics, biogeography and host plant relationships. J. Nat. Hist. 34: 57–155.

Crespi, B.J., and C.P. Sandoval. 2000. Phylogenetic evidence for the evolution of ecological specialization in *Timema* walking-sticks. J. Evol. Biol. 13: 249–262.

Crespi, B.J., D.C. Morris, and L.A. Mound. 2004. Evolution of ecological and behavioral diversity: Australian *Acacia* thrips as model organisms. CSIRO, Canberra, Australia.

Dobler, S., P. Mardulyn, J.M. Pasteels, and M. Rowell-Rahier. 1996. Host-plant switches and the evolution of chemical defense and life history in the leaf beetle genus *Oreina*. Evolution 50: 2373–2386.

Farrell, B.D. 2001. Evolutionary assembly of the milkweed fauna: cytochrome oxidase I and the age of *Tetraopes* beetles. Mol. Phylogenet. Evol. 18: 467–478.

Farrell, B.D., and C. Mitter. 1998. The timing of insect/plant diversification: might *Tetraopes* (Coleoptera: Cerambycidae) and *Asclepias* (Asclepiadaceae) have co-evolved? Biol. J. Linn. Soc. 63: 553–577.

Farrell, B.D., and A.S. Sequeira. 2004. Evolutionary rates in the adaptive radiation of beetles on plants. Evolution 58: 1984–2001.

Fenton, B., A.N.E. Birch, G. Malloch, P.G. Lanham, and R.M. Brennan. 2000. Gall mite molecular phylogeny and its relationship to the evolution of plant host specificity. Exp. Appl. Acarol. 24: 831–861.

Funk, D., D.J. Futuyma, G. Orti, and A. Meyer. 1995. A history of host associations and evolutionary diversification for *Ophraella* (Coleoptera: Chrysomelidae): new evidence from mitochondrial DNA. Evolution 49: 1008–1017.

Heitland, W. 2002. U. Ord. Symphyta. Pflanzenwespen, sawflies. www.faunistik.net/DETINVERT/HYMENOPTERA/symphyta.html.

Hsu, Y.-F., and J.A. Powell. 2004. Phylogenetic relationships within Heliodinidae and systematics of moths formerly assigned to *Heliodines* Stanton. University of California Publications in Entomology, Vol. 24.

Janz, N., K. Nyblom, and S. Nylin. 2001. Evolutionary dynamics of host-plant specialization: a case study of the tribe Nymphalini. Evolution 55: 783–796.

Jones, R.W. 2001. Evolution of the host plant associations of the *Anthonomus grandis* species group (Coleoptera: Curculionidae): phylogenetic tests of various hypotheses. Ann. Entomol. Soc. Am. 94: 51–58.

Kawakita, A., A. Takimura, T. Terachi, T. Sota, and M. Kato. 2004. Cospeciation analysis of an obligate pollination mutualism: have *Glochidion* trees (Euphorbiaceae) and pollinating *Epicephala* moths (Gracillariidae) diversified in parallel? Evolution 58: 2201–2214.

Kelley, S.T., and B.D. Farrell. 1998. Is specialization a dead end? The phylogeny of host use in *Dendroctonus* bark beetles (Scolytidae). Evolution 52: 1731–1743.

Lopez-Vaamonde, C., H.C.J. Godfray, and J.M. Cook. 2003. Evolutionary dynamics of host-plant use in a genus of leaf-mining moths. Evolution 57: 1804–1821.

Lopez-Vaamonde, C., N. Wikström, C. Labandeira, H.C.J. Godfray, S.J. Goodman, and J.M. Cook. 2006. Fossil-calibrated molecular phylogenies reveal that leaf-mining moths radiated millions of years after their host plants. J. Evol. Biol. 19: 1314–1326.

Machado, C.A., N. Robbins, M.T.P. Gilbert, and E.A. Herre. 2005. Critical review of host specificity and its coevolutionary implications in the fig/fig-wasp mutualism. Proc. Natl. Acad. Sci. USA 102: 6558–6565.

Moran, N.A., M.E. Kaplan, M.J. Gelsey, T.G. Murphy, and E.A. Scholes. 1999. Phylogenetics and evolution of the aphid genus *Uroleucon* based on mitochondrial and nuclear DNA sequences. Syst. Entomol. 24: 85–93.

Morris, D.C., M.P. Schwarz, S.J.B. Cooper, and L.A. Mound. 2002. Phylogenetics of Australian *Acacia* thrips: the evolution of behaviour and ecology. Mol. Phylogenet. Evol. 25: 278–292.

Morse, G.E., and B.D. Farrell. 2005. Ecological and evolutionary diversification of the seed beetle genus *Stator* (Coleoptera: Chrysomelidae: Bruchinae). Evolution 59: 1315–1333.

Nyman, T., A. Widmer, and H. Roininen. 2000. Evolution of gall morphology and host-plant relationships in willow-feeding sawflies (Hymenoptera: Tenthredinidae). Evolution 54: 526–533.

Nyman, T., B.D. Farrell, A.G. Zinovjev, and V. Vikberg. 2006. Larval habits, host-plant associations, and speciation in nematine sawflies (Hymenoptera: Tenthredinidae). Evolution 60: 1622–1637.

Percy, D.M., R.D.M. Page, and Q.C.B. Cronk. 2004. Plant-insect interactions: double-dating associated insect and plant lineages reveals asynchronous radiations. Syst. Biol. 53: 120–127.

Powell, J.A., C.M. Mitter, and B. Farrell. 1998. Evolution of larval feeding habits in Lepidoptera, pp. 403–422. In N.P. Kristensen (ed.), Handbook of zoology, Lepidoptera. Vol. 1: Systematics and evolution. W. de Gruyter, Berlin.

Roderick, G.K. 1997. Herbivorous insects and the Hawaiian silversword alliance: co-evolution or co-speciation? Pacific Sci. 51: 440–449.

Roininen, H., T. Nyman, and A. Zinovjev. 2005. Biology, ecology, and evolution of gall-inducing sawflies (Hymenoptera: Tenthredinidae and Xyelidae), pp. 467–494. In A. Raman, C.W. Schaefer, and T.M. Withers (eds.), Biology, ecology, and evolution of gall-inducing arthropods. Science Publishers, Enfield, NH.

Ronquist, F., and J. Liljeblad. 2001. Evolution of the gall wasp-host plant association. Evolution 55: 2503–2522.

Rønsted, N., G.D. Weiblen, J.M. Cook, N. Salamin, C.A. Machado, and V. Savolainen. 2005. 60 million years of co-divergence in the fig-wasp symbiosis. Proc. R. Soc. Lond. B 272: 2593–2599.

Rubinoff, D., and F.A.H. Sperling. 2002. Evolution of ecological traits and wing morphology in *Hemileuca* (Saturniidae) based on a two-gene phylogeny. Mol. Phylogenet. Evol. 25: 70–86.

Schulmeister, S. 2003. Simultaneous analysis of basal Hymenoptera (Insecta): introducing robust-choice sensitivity analysis. Biol. J. Linn Soc. 79: 245–275.

Wahlberg, N. 2001. The phylogenetics and biochemistry of host-plant specialization in melitaeine butterflies (Lepidoptera: Nymphalidae). Evolution 55: 522–537.

Weiblen, G.D., and G.L. Bush. 2002. Speciation in fig pollinators and parasites. Mol. Ecol. 11: 1573–1578.

Weingartner, E., N. Wahlberg, and S. Nylin. 2006. Dynamics of host plant use and species diversity: a phylogenetic investigation in *Polygonia* butterflies (Nymphalidae). J. Evol. Biol. 19: 483–491.

Weintraub, J.D., J.H. Lawton, and M.J. Scoble. 1995. Lithinine moths on ferns: a phylogenetic study of insect-plant interactions. Biol. J. Linn. Soc. 55: 239–250.

Wilson, S., C. Mitter, R.F. Denno, and M.R. Wilson. 1994. Evolutionary patterns of host plant use by delphacid planthoppers and their relatives, pp. 7–113. In R. F. Denno and J. Prefect (eds.), The ecology and management of planthoppers. Chapman and Hall, New York.

Yotoko, K.S.C., P.I. Prado, C.A.M. Russo, and V.N. Solferini. 2005. Testing the trend towards specialization in herbivore-host plant associations using a molecular phylogeny of *Tomoplagia* (Diptera: Tephritidae). Mol. Phylogenet. Evol. 35: 701–711.

References Cited

Agrawal, A.A., and M. Fishbein. 2006. Plant defense syndromes. Ecology 87: S132–S149.

APGII. 2003. An update of the angiosperm phylogeny group classification for the orders and families of flowering plants: APG II. Bot. J. Linn. Soc. 141: 399–436.

Allmon, W.D. 1992. A causal analysis of stages in allopatric speciation, pp. 219–257. In D. Futuyma and J. Antonovics (eds.), Oxford surveys in evolutionary biology, vol. 8. Oxford University Press, Oxford.

Armbruster, W.S. 1997. Exaptations link the evolution of plant-herbivore and plant-pollinator interactions: a phylogenetic inquiry. Ecology 78: 1661–1674.

Avise, J.C. 2000. Phylogeography: the history and formation of species. Harvard University Press, Cambridge, MA.

Barraclough, T.G., and A.P. Vogler. 2000. Detecting the geographical pattern of speciation from species-level phylogenies. Am. Nat. 155: 419–434.

Beccaloni, G.W., and F.B. Symons. 2000. Variation of butterfly diet breadth in relation to host-plant predictability: results from two faunas. Oikos 90: 50–66.

Becerra, J.X. 1997. Insects on plants: macroevolutionary chemical trends in host use. Science 276: 253–256.

Becerra, J.X. 2003. Synchronous coadaptation in an ancient case of herbivory. Proc. Natl. Acad. Sci. USA 100: 12804–12807.

Behrensmeyer, A.K., J.D. Damuth, W.A. DiMichele, R. Potts, H.D. Sues, and S.L. Wing (eds.). 1992. Terrestrial ecosystems through time: evolutionary paleoecology of terrestrial plants and animals. University of Chicago Press, Chicago.

Benson, W.W., K.S. Brown, and L.E. Gilbert. 1975. Coevolution of plants and herbivores: passion flower butterflies. Evolution 29: 659–680.

Berenbaum, M.R. 1983. Coumarins and caterpillars: a case for coevolution. Evolution 37: 163-179.

Berenbaum, M.R. 2001. Chemical mediation of coevolution: phylogenetic evidence for Apiaceae and associates. Ann. Missouri Bot. Gard. 88: 45–59.

Berenbaum, M.R., and S. Passoa. 1999. Generic phylogeny of North American Depressariinae (Lepidoptera: Elachistidae) and hypotheses about coevolution. Ann. Entomol. Soc. Am. 92: 971–986.

Berlocher, S.H. 1998. Can sympatric speciation via host or habitat shift be proven from phylogenetic and biogeographic evidence? pp. 99–113. In D.J. Howard and S.H. Berlocher (eds.), Endless forms: species and speciation. Oxford University Press, Oxford.

Berlocher, S.H., and J.L. Feder. 2002. Sympatric speciation in phytophagous insects: moving beyond controversy? Annu. Rev. Entomol. 47: 773–815.

Bernays, E.A., and R.F. Chapman. 1994. Host-plant selection by phytophagous insects. Chapman and Hill, New York.

Bernays, E.A., and M.S. Singer. 2005. Insect defences: taste alteration and endoparasites. Nature 436: 476.

Bisset, J., and A. Borkent. 1988. Ambrosia galls: the significance of fungal nutrition in the evolution of the Cecidomyiidae (Diptera), pp. 203–225. In K.A. Pirozynski and D.L. Hawksworth (eds.), Coevolution of fungi with plants and animals. Academic Press, London.

Blank, S.M. 2002. Biosystematics of the extant Xyelidae with particular emphasis on the Old World taxa (Insecta: Hymenoptera). Dissertation, Freie Universität Berlin.

Blomberg, S.P., and T. Garland. 2002. Tempo and mode in evolution: phylogenetic inertia, adaptation and comparative methods. J. Evol. Biol. 15: 899–910.

Bokma, F. 2003. Testing for equal rates of cladogenesis in diverse taxa. Evolution 57: 2469–2474.

Bolmgren, K., and O. Eriksson. 2005. Fleshy fruits: origins, niche shifts, and diversification. Oikos 109: 255–272.

Brady, S. G., T. R. Schultz, B. L. Fisher, and P. S. Ward. 2006. Evaluating alternative hypothesis for the early evolution and diversification of ants. Proc. Nat'l. Acad. Sci. U.S.A. 103: 18172–18177.

Bucheli, S., J.F. Landry, and J. Wenzel. 2002. Larval case architecture and implications of host-plant associations for North American *Coleophora* (Lepidoptera; Coleophoridae). Cladistics 18: 71–93.

Burckhardt, D. 2005. Biology, ecology, and evolution of gall-inducing psyllids (Hemiptera: Psylloidea), pp. 144–157. In A. Raman, C. W. Schaefer, and T.M. Withers (eds.), Biology, ecology, and evolution of gall-inducing arthropods. Science Publishers, Enfield, NH.

Cattin, M.F., L.F. Bersier, C. Banašek-Richter, R. Baltensperger, and J. P. Gabriel. 2004. Phylogenetic constraints and adaptation explain food-web structure. Nature 427: 835–839.

Chesser, P.T., and R.M. Zink. 1994. Modes of speciation in birds: a test of Lynch's method. Evolution 48: 490–497.

Cho, S. 1997. Molecular phylogenetics of the Heliothinae (Lepidoptera: Noctuidae) based on the nuclear genes for elongation factor-1α and dopa decarboxylase. Ph.D. dissertation, University of Maryland, College Park.

Condon, M.A., and G.J. Steck. 1997. Evolution of host use in fruit flies of the genus *Blepharoneura* (Diptera: Tephritidae): cryptic species on sexually dimorphic host plants. Biol. J. Linn. Soc. 60: 443–466.

Connor, E.F., and M.P. Taverner. 1997. The evolution and adaptive significance of the leaf-mining habit. Oikos 79: 6–25.

Cook, L.G., and P.J. Gullan. 2004. The gall-inducing habit has evolved multiple times among the eriococcid scale insects (Sternorrhyncha: Coccoidea: Eriococcidae). Biol. J. Linn. Soc. 83: 441–452.

Cook, J.M., A. Rokas, M. Pagel, and G.N. Stone. 2002. Evolutionary shifts between host oak sections and host-plant organs in *Andricus* gallwasps. Evolution 56: 1821–1830.

Coyne, J.A., and H.A. Orr. 2004. Speciation. Sinauer, Sunderland, MA.

Crespi, B.J., and C.P. Sandoval. 2000. Phylogenetic evidence for the evolution of ecological specialization in *Timema* walking-sticks. J. Evol. Biol. 13: 249–262.

Davis, C.C., C.D. Bell, S. Mathews, and M.J. Donoghue. 2002. Laurasian migration explains Gondwanan disjunctions: evidence from Malpighiaceae. Proc. Natl. Acad. Sci. USA 99: 6833–6837.

Denno, R.F., M.S. McClure, and J.R. Ott. 1995. Interspecific interactions in phytophagous insects: competition reexamined and resurrected. Annu. Rev. Entomol. 40: 297–331.

Dethier, V. 1941. Chemical factors determining the choice of food plants by *Papilio* larvae. Am. Nat. 75: 61–73.

Dietrich, C.H. 1999. The role of grasslands in the diversification of leafhoppers (Homoptera: Cicadellidae): a phylogenetic perspective, pp. 44–49. In C. Warwick (ed.), Proceedings of the 15th North American Prairie Conference. Natural Areas Association, Bend, OR.

DiMichele, W.A., A.K. Behrensmeyer, T.D. Olszewski, C.C. Labandeira, J.M. Pandolfi, S.L. Wing, and R. Bobe. 2004. Long-term stasis in ecological assemblages: evidence from the fossil record. Annu. Rev. Ecol. Syst. 35: 285–322.

Dobler, S. 2001. Evolutionary aspects of defense by recycled plant compounds in herbivorous insects. Basic Appl. Ecol. 2: 15–26.

Dobler, S., P. Mardulyn, J.M. Pasteels, and M. Rowell-Rahier. 1996. Host-plant switches and the evolution of chemical defense and life history in the leaf beetle genus *Oreina*. Evolution 50: 2373–2386.

Dorchin, N., A. Freidberg, and O. Mokady. 2004. Phylogeny of the Baldratiina (Diptera: Cecidomyiidae) inferred from morphological, ecological and molecular data sources, and evolutionary patterns in plant-galler relationships. Mol. Phylogenet. Evol. 30: 503–515.

Ehrlich, P.R., and P.H. Raven. 1964. Butterflies and plants: a study in coevolution. Evolution 18: 586–608.

Fagan, W.F., E. Siemann, C. Mitter, R.F. Denno, A.F. Huberty, H.A. Woods, and J.J. Elser. 2002. Nitrogen in insects: implications for trophic complexity and species diversification. Am. Nat. 160: 784–802.

Faith, D.P., and P.S. Cranston. 1991. Could a cladogram this short have arisen by chance alone? On permutation tests for cladistic structure. Cladistics 7: 1–28.

Fang, Q.Q., S. Cho, J.C. Regier, C. Mitter, M. Matthews, R.W. Poole, T.P. Friedlander, and S.W. Zhao. 1997. A new nuclear gene for insect phylogenetics: dopa decarboxylase is informative of relationships within Heliothinae (Lepidoptera: Noctuidae). Syst. Biol. 46: 269–283.

Farrell, B.D. 1998. Inordinate fondness explained: why are there so many beetles? Science 281: 555–559.

Farrell, B.D. 2001. Evolutionary assembly of the milkweed fauna: cytochrome oxidase I and the age of *Tetraopes* beetles. Mol. Phylogenet. Evol. 18: 467–478.

Farrell, B., and C. Mitter. 1990. Phylogenesis of insect-plant interactions: have *Phyllobrotica* leaf beetles and Lamiales diversified in parallel? Evolution 44: 1389–1403

Farrell, B.D., and C. Mitter. 1993. Phylogenetic determinants of insect/plant community diversity, pp. 253–266. In R. E. Ricklefs and D. Schluter (eds.), Species diversity: historical and geographical perspectives. University of Chicago Press, Chicago.

Farrell, B.D., and C. Mitter. 1998. The timing of insect/plant diversification: might *Tetraopes* (Coleoptera : Cerambycidae) and *Asclepias* (Asclepiadaceae) have co-evolved? Biol. J. Linn. Soc. 63: 553–577.

Farrell, B.D., and A.S. Sequeira. 2004. Evolutionary rates in the adaptive radiation of beetles on plants. Evolution 58: 1984–2001.

Farrell, B.D., D. Dussourd, and C. Mitter. 1991. Escalation of plant defense: Do latex/resin canals spur plant diversification? Am. Nat. 138: 881–900

Farrell B.D., A.S. Sequeira, B.C. O'Meara, B.B. Normark, J.H. Chung, and B. H. Jordal. 2001. The evolution of agriculture in beetles (Curculionidae: Scolytinae and Platypodinae). Evolution 55: 2011–2027.

Favret, C., and D.J. Voegtlin. 2004. Speciation by host-switching in pinyon *Cinara* (Insecta: Hemiptera: Aphididae). Mol. Phylogenet. Evol. 32: 139–151.

Feeny, P. 1991. Chemical constraints on the evolution of swallowtail butterflies, pp. 315–339. In P.W. Price, T.M. Lewinsohn, G.W. Fernandes, and W.W. Benson (eds.), Plant-animal interactions: evolutionary ecology in tropical and temperate regions. John Wiley and Sons, New York.

Filchak, K.E., J.B. Roethele, and J.L. Feder. 2000. Natural selection and sympatric divergence in the apple maggot *Rhagoletis pomonella*. Nature 407: 739–742.

Fitzpatrick, B.M., and M. Turelli. 2006. The geography of mammalian speciation: mixed signals from phylogenies and range maps. Evolution 66: 601–615.

Funk, D.J., K.E. Filchack, and J. L. Feder. 2002. Herbivorous insects: model systems for the comparative study of speciation ecology. Genetica 116: 251–267.

Futuyma, D.J. 1989. Macroevolutionary consequences of speciation, pp. 557–578. In D. Otte and J. A. Endler (eds.), Speciation and its consequences. Sinauer, Sunderland, MA.

Futuyma, D.J., and G. Moreno. 1988. The evolution of ecological specialization. Annu. Rev. Ecol. Syst. 12: 207–233.

Futuyma, D.J., M.C. Keese, and D.J. Funk. 1995. Genetic constraints on macroevolution: the evolution of host affiliation in the leaf beetle genus *Ophraella*. Evolution 49: 797–809.

Gagné, R.J. 1989. The plant-feeding gall midges of North America. Cornell University Press, Ithaca, NY.

Gassman, A.J., A. Levy, T. Tran, and D.J. Futuyma. 2006. Adaptations of an insect to a novel host plant: a phylogenetic approach. Funct. Ecol. 20: 478–485.

Gaston, K.J., D. Reavey, and G.R. Valladares. 1992. Intimacy and fidelity: internal and external feeding by the British microlepidoptera. Ecol. Entomol. 17: 86–88.

Gillespie, J.J., K.M. Kjer, C.N. Duckett, and D.W. Tallamy. 2003. Convergent evolution of cucurbitacin feeding in spatially isolated rootworm taxa (Coleoptera: Chrysomelidae; Galerucinae, Luperini). Mol. Phylogenet. Evol. 29: 161–175.

Gillespie, J.J., K.M. Kjer, E.G. Riley, and D.W. Tallamy. 2004. The evolution of cucurbitacin pharmacophagy in rootworms: insight from Luperini paraphyly, pp. 37–57. In P. Jolivet, J. A. Santiago-Blay, and M. Schmitt (eds.), New developments in the biology of Chrysomelidae. SPB Academic Publishing, The Hague, The Netherlands.

Graham, A. 1999. Late cretaceous and cenozoic history of North American vegetation north of Mexico. Oxford University Press, Oxford.

Grimaldi, D., and M.S. Engel. 2005. Evolution of the insects. Cambridge University Press, Cambridge, MA.

Hansen, T.F., and S.H. Orzack. 2005. Assessing current adaptation and phylogenetic inertia as explanations of trait evolution: the need for controlled comparisons. Evolution 59: 2063–2072.

Harry, M., M. Solignac, and D. Lachaise. 1998. Molecular evidence for parallel evolution of adaptive syndromes in fig-breeding *Lissocephala* (Drosophilidae). Mol. Phylogenet. Evol. 9: 542–551.

Heie, O.E. 1996. The evolutionary history of aphids and a hypothesis on the coevolution of aphids and plants. Boll. Zool. Agr. Bachic. 28: 149–155.

Holloway, J.D., and E.S. Nielsen. 1998. Biogeography of the Lepidoptera, pp. 423–461. In N. P. Kristensen (ed.), Handbook of zoology, Lepidoptera. vol. 1: Systematics and evolution. W. de Gruyter, Berlin.

Hsiao, T.H., and J.M. Pasteels. 1999. Evolution of host-plant affiliation and chemical defense in *Chrysolina-Oreina* leaf beetles as revealed by mtDNA phylogenies, pp. 321–342. In M. L. Cox (ed.), Advances in Chrysomelidae biology. Backhuys, Leiden, The Netherlands.

Hsiao, T.H., and D.M. Windsor. 1999. Historical and biological relationships among Hispinae inferred from 12S mtDNA sequence data, pp. 39–50. In M.L. Cox (ed.), Advances in Chrysomelidae biology. Backhuys, Leiden, The Netherlands.

Isaac, N.J.B., P.M. Agapow, P. H. Harvey, and A. Purvis. 2003. Phylogenetically nested comparisons for testing correlates of species richness: a simulation study of continuous variables. Evolution 57: 18–26.

Janz, N., and S. Nylin. 1998. Butterflies and plants: a phylogenetic study. Evolution 52: 486–502.

Janz, N., K. Nyblom, and S. Nylin. 2001. Evolutionary dynamics of host-plant specialization: a case study of the tribe Nymphalini. Evolution 55: 783–796.

Janz, N., S. Nylin, and N. Wahlberg. 2006. Diversity begets diversity: host expansions and the diversification of plant-feeding insects. BMC Evol. Biol. 6: 4.

Jervis, M. A., C.L. Boggs, and P.N. Ferns. 2005. Egg maturation strategy and its associated trade-offs: a synthesis focusing on Lepidoptera. Ecol. Entomol. 30: 359–375.

Joy, J. B., and B. J. Crespi. 2007. Adaptive radiation of gall-inducing insects within a single host-plant species. Evolution 61: 784–795.

Karban, R., and A.A. Agrawal. 2002. Herbivore offense. Annu. Rev. Ecol. Syst. 33: 641–664.

Kawakita, A, and M. Kato. 2006. Assessment of the diversity and species specificity of the mutualistic association between *Epicephala* moths and *Glochidion* trees. Mol. Ecol. 15: 2567–2581.

Kawakita, A., A. Takimura, T. Terachi, T. Sota, and M. Kato. 2004. Cospeciation analysis of an obligate pollination mutualism: have *Glochidion* trees (Euphorbiaceae) and pollinating *Epicephala* moths (Gracillariidae) diversified in parallel? Evolution 58: 2201–2214.

Kergoat, G.J., A. Delobel, G. Fédière, B. Le Rü, and J.-F. Silvain. 2005. Both host-plant phylogeny and chemistry have shaped the African seed-beetle radiation. Mol. Phylogenet. Evol. 35: 602–611.

Kristensen, N.P. 1984. Studies on the morphology and systematics of primitive Lepidoptera (Insecta). Steenstrupia 10: 141–191.

Kristensen, N.P. 1997. Early evolution of the Lepidoptera + Trichoptera lineage: phylogeny and the ecological scenario. Mem. Mus. Natl. Hist. Nat. 173: 253–271.

Kuhn, J., E.M. Pettersson, B.K. Feld, A. Burse, A. Termonia, J.M. Pasteels, and W. Boland. 2004. Selective transport systems mediate sequestration of plant glucosides in leaf beetles: a molecular basis for adaptation and evolution. Proc. Natl. Acad. Sci. USA 101: 13808–13813.

Labandeira, C.C., 2002. The history of associations between plants and animals, pp. 26–74. In C. Herrera and O. Pellmyr (eds.), Plant-animal interactions: an evolutionary approach. Blackwell Science, Oxford.

Lewinsohn, T.M., V. Novotny, and Y. Basset. 2005. Insects on plants: diversity of herbivore assemblages revisited. Annu. Rev. Ecol. Evol. Syst. 36: 597–620.

Lopez-Vaamonde, C., H.C.J. Godfray, and J.M. Cook. 2003. Evolutionary dynamics of host-plant use in a genus of leaf-mining moths. Evolution 57: 1804–1821.

Losos, J.B., and R.E. Glor. 2003. Phylogenetic comparative methods and the geography of speciation. Trends Ecol. Evol. 18: 220–227.

Lynch, J.D. 1989. The gauge of speciation: on the frequencies of modes of speciation, pp. 527–553. In D. Otte and J.A. Endler (eds.), Speciation and its consequences. Sinauer, Sunderland, MA.

Machado, C.A., N. Robbins, M.T.P. Gilbert, and E.A. Herre. 2005. Critical review of host specificity and its coevolutionary implications in the fig/fig-wasp mutualism. Proc. Natl. Acad. Sci. USA 102: 6558–6565.

Mackenzie, A., and A.F.G. Dixon. 1990. Host alternation in aphids: constraint versus optimization. Am. Nat. 136: 132–134.

Magallón, S. 2004. Dating lineages: molecular and paleontological approaches to the temporal framework of clades. Int. J. Plant. Sci. 165: S7-S21.

Magallón, S., and M.J. Sanderson. 2001. Absolute diversification rates in angiosperm clades. Evolution 55: 1762–1780.

Marvaldi, A.E., A.S. Sequeira, C.W. O'Brien, and B.D. Farrell. 2002. Molecular and morphological phylogenetics of weevils (Coleoptera, Curculionoidea): do niche shifts accompany diversification? Syst. Biol. 51: 761–785.

Mayhew, P.J. 2002. Shifts in hexapod diversification and what Haldane could have said. Proc. R. Soc. Lond. B 269: 969–974.

McKenna D.D., and B.D. Farrell. 2006. Tropical forests are both evolutionary cradles and museums of leaf beetle diversity. Proc. Natl. Acad. Sci. USA 103: 10947–10951.

Miller, D.G., and B. Crespi. 2003. The evolution of inquilinism, host-plant use and mitochondrial substitution rates in *Tamalia* gall aphids. J . Evol. Biol. 16: 731–743.

Miller, J.S. 1987. Host-plant relationships in the Papilionidae (Lepidoptera): parallel cladogenesis or colonization? Cladistics 3: 105–120.

Mitchell, A., C. Mitter, and J.C. Regier. 2006. Systematics and evolution of the cutworm moths (Lepidoptera: Noctuidae): evidence from two protein-coding nuclear genes. Syst. Entomol. 31: 21–46.

Mitter, C., and D.R. Brooks. 1983. Phylogenetic aspects of coevolution, pp. 65–98. In D.J. Futuyma and M. Slatkin (eds.), Coevolution. Sinauer, Sunderland, MA.

Mitter, C., and B. Farrell. 1991. Macroevolutionary aspects of insect-plant relationships, pp. 35–78. In E.A. Bernays (ed.), Insect/plant interactions, vol. 3. CRC Press, Boca Raton, FL.

Mitter, C., B. Farrell, and B. Wiegmann.1988. The phylogenetic study of adaptive zones: has phytophagy promoted insect diversification? Am. Nat. 132: 107–128.

Mitter, C., R.W. Poole, and M. Matthews. 1993. Biosystematics of the Heliothinae (Lepidoptera: Noctuidae). Annu. Rev. Entomol. 38: 207–225.

Moore, B.R., K.M.A. Chan, and M.J. Donoghue. 2004. Detecting diversification rate variation in supertrees, pp. 487–533. In O.R.P. Bininda-Emonds (ed.), Phylogenetic supertrees: combining information to reveal the tree of life. Kluwer Academic Publishers, Dordrecht, The Netherlands.

Moran, N.A. 1988. The evolution of host plant alternation in aphids: evidence for specialization as a dead end. Am. Nat. 132: 681–706.

Moran, N.A. 1990. Aphid life cycls: two evolutionary steps. Am. Nat. 136: 135–138.

Moran, N.A. 1992. The evolution of aphid life cycles. Annu. Rev. Entomol. 37: 321–348.

Moreau, C.S., C.D. Bell, R. Vila, S.B. Archibald, and N.E. Pierce. 2006. Phylogeny of the ants: diversification in the age of angiosperms. Science 312: 101–104.

Morris, D.C., L.A. Mound, M.P. Schwarz, and B.J. Crespi. 1999. Morphological phylogenetics of Australian gall-inducing thrips and their allies: the evolution of host-plant affiliations, domicile use and social behaviour. Syst. Entomol. 24: 289–299.

Morse, G.E., and B.D. Farrell. 2005. Ecological and evolutionary diversification of the seed beetle genus *Stator* (Coleoptera: Chrysomelidae: Bruchinae). Evolution 59: 1315–1333.

Murphy, S.M. 2004. Enemy-free space maintains swallowtail butterfly host shift. Proc. Natl. Acad. Sci. USA 101: 18048–18052.

Nee, S., A.O. Mooers, and P. H. Harvey, 1992. Tempo and mode of evolution revealed from molecular phylogenies. Proc. Natl. Acad. Sci. USA 89: 8322–8326.

Nee, S., T.G. Barraclough, and P. H. Harvey. 1996. Temporal changes in biodiversity: detecting patterns and identifying causes, pp. 230–252. In K.J. Gaston (ed.), Biodiversity: a biology of numbers and difference. Blackwell Scientific, Oxford.

Nickel, H. 2003. The leafhoppers and planthoppers of Germany (Hemiptera, Auchenorrhyncha): patterns and strategies in a highly diverse group of phytophagous insects. Pensoft Series Faunistica No. 28. Pensoft Publishers, Sofia, Bulgaria.

Normark, B.B. 2000. Molecular systematics and evolution of the aphid family Lachnidae. Mol. Phylogenet. Evol. 14: 131–140.

Nosil, P. 2002. Transition rates between specialization and generalization in phytophagous insects. Evolution 56: 1701–1706.

Nosil, P., and A.O. Mooers. 2005. Testing hypotheses about ecological specialization using phylogenetic trees. Evolution 59: 2256–2263.

Nosil, P., B.J. Crespi, and C.P. Sandoval. 2003. Reproductive isolation driven by the combined effects of ecological adaptation and reinforcement. Proc. R. Soc. Lond. B 270: 1911–1918.

Nylander, J.A.A. 2004. Bayesian phylogenetics and the evolution of gall wasps. Ph.D. dissertation, University of Uppsala, Uppsala, Sweden.

Nylander, J., F. Ronquist, J.P. Huelsenbeck, and J.L. Nieves-Aldrey. 2004. Bayesian phylogenetic analysis of combined data. Syst. Biol. 53: 47–67.

Nyman, T., A. Widmer, and H. Roinnen. 2000. Evolution of gall morphology and host-plant relationships in willow-feeding sawflies (Hymenoptera: Tenthredinidae). Evolution 54: 526–533.

Nyman, T., B.D. Farrell, A.G. Zinovjev, and V. Vikberg. 2006. Larval habits, host-plant associations, and speciation in nematine sawflies (Hymenoptera: Tenthredinidae). Evolution 60: 1622–1637.

Ortiz-Rivas, B., A. Moya, and D. Martínez-Torres. 2004. Molecular systematics of aphids (Homoptera: Aphididae): new insights from the long-wavelength opsin gene. Mol. Phylogenet. Evol. 30: 24–37.

Page, R.D.M. (ed.). 2003. Tangled trees: phylogeny, cospeciation and coevolution. University of Chicago Press, Chicago.

Paradis, E. 1997. Assessing temporal variations in diversification rates from phylogenies: estimation and hypothesis testing. Proc. R. Soc. Lond. B 264: 1141–1147.

Pellmyr, O. 2003. Yuccas, yucca moths and coevolution: a review. Ann. Missouri Bot. Gard. 90: 35–55.

Percy, D.M. 2003. Radiation, diversity, and host-plant interactions among island and continental legume-feeding psyllids. Evolution 57: 2540–2556.

Percy, D.M., R.D.M. Page, Q.C.B. Cronk. 2004. Plant-insect interactions: double-dating associated insect and plant lineages reveals asynchronous radiations. Syst. Biol. 53: 120–127.

Plantard, O., J.D. Shorthouse, and J. Y. Rasplus. 1997. Molecular phylogeny of the genus *Diplolepis* (Hymenoptera: Cynipidae), pp. 247–260. In G. Csoka, W.J. Mattson, G.N. Stone, and P.W. Price (eds.), Biology of gall inducing arthropods. Gen. Tech. Rep. NC-199. U.S. Department of Agriculture, Forest Service, North Central Forest Experiment Station, St. Paul, MN.

Powell, J.A. 1980. Evolution of larval food preferences in microlepidoptera. Annu. Rev. Entomol. 25: 133–159.

Powell, J.A., C. Mitter, and B. Farrell. 1998. Evolution of larval feeding habits in Lepidoptera, pp. 403–422. In N.P. Kristensen (ed.), Handbook of zoology, Lepidoptera. Vol. 1: Systematics and evolution. W. de Gruyter, Berlin.

Pratt, G.F. 1994. Evolution of *Euphilotes* (Lepidoptera: Lycaenidae) by seasonal and host shifts. Biol. J. Linn. Soc. 51: 387–416.

Price, P.W. 1980. Evolutionary biology of parasites. Monographs in population biology, vol. 15. Princeton University Press, Princeton, NJ.

Purvis, A. 1996. Using interspecies phylogenies to test macroevolutionary hypotheses, pp. 153–168. In P. H. Harvey, A.J.L. Brown, J.M. Smith, and S. Nee (eds.), New uses for new phylogenies. Oxford University Press, Oxford.

Ree, R.H. 2005. Detecting the historical signature of key innovations using stochastic models of character evolution and cladogenesis. Evolution 59: 257–265.

Rodman, J.E., P.S. Soltis, D.E. Soltis, K.J. Sytsma, and K.G. Karol. 1998. Parallel evolution of glucosinolate biosynthesis inferred from congruent nuclear and plastid gene phylogenies. Am. J. Bot. 85: 997–1006.

Ronquist, F., and J. Liljeblad. 2001. Evolution of the gall wasp-host plant association. Evolution 55: 2503–2522.

Rudgers, J.A., S.Y. Strauss, and J.F. Wendel. 2004. Trade-offs among anti-herbivore resistance traits: insights from Gossypieae (Malvaceae). Am. J. Bot. 91: 871–880.

Rundle, H.D., and P. Nosil. 2005. Ecological speciation. Ecol. Lett. 8: 336–352.

Sachet, J.M., A. Roques, and L. Després. 2006. Linking patterns and processes of species diversification in the cone flies *Strobilomyia* (Diptera: Anthomyiidae). Mol. Phylogenet. Evol. 41: 606–621.

Sanderson, M.J., and M.J. Donoghue. 1994. Shifts in diversification rate with the origin of angiosperms. Science 264: 1590–1593.

Sargent, R.D. 2004. Floral symmetry affects speciation rates in angiosperms. Proc. R. Soc. Lond. B 271: 603–608.

Schick, K., Z.W. Liu, and P. Goldstein. 2003. Phylogeny, historical biogeography, and macroevolution of host use among *Diastrophus* gall wasps (Hymenoptera: Cynipidae). Proc. Entomol. Soc. Wash. 105: 715–732.

Schluter, D. 1987. Character displacement and the adaptive divergence of finches on islands and continents. Am. Nat. 131: 799–824.

Schluter, D. 1988. Character displacement and the adaptive divergence of finches on islands and continents. Am. Nat. 131: 799–824.

Schluter, D. 2000. The ecology of adaptive radiation. Oxford University Press, Oxford.

Schulmeister, S. 2003. Simultaneous analysis of basal Hymenoptera (Insecta): introducing robust-choice sensitivity analysis. Biol. J. Linn. Soc. 79: 245–275.

Simpson, G.G. 1953. The major features of evolution. Columbia University Press, New York.

Singer, M.S., and J.O. Stireman. 2005. The tri-trophic niche concept and adaptive radiation of phytophagous insects. Ecol. Lett. 8: 1247–1255.

Singer, M.S., Y. Carriere, C. Theuring, and T. Hartmann. 2004. Disentangling food quality from resistance against parasitoids: diet choice by a generalist caterpillar. Am. Nat. 164: 423–429.

Smith, A.R., K.M. Pryer, E. Schuettpelz, P. Korall, H. Schneider, and P. G. Wolf. 2006. A classification for extant ferns. Taxon 55: 705–731.

Spencer, K.A. 1990. Host specialization in the world Agromyzidae (Diptera). Kluwer Academic Publishers, Dordrecht, The Netherlands.

Stireman, J.O. 2005. The evolution of generalization? Parasitoid flies and the perils of inferring host range evolution from phylogenies. J. Evol. Biol. 18: 325–336.

Strong, D.R., J.H. Lawton, and R. Southwood. 1984. Insects on plants: community patterns and mechanisms. Harvard University Press, Cambridge, MA.

NINETEEN

Evolution of Insect Resistance to Transgenic Plants

BRUCE E. TABASHNIK AND YVES CARRIÈRE

"If Darwin were alive today the insect world would delight and astound him with its impressive verification of his theories of survival of the fittest. Under the stress of intensive chemical spraying the weaker members of the insect populations are being weeded out" (Carson 1962). When Rachel Carson wrote that insightful passage in *Silent Spring*, evolution of insecticide resistance had been documented in about 100 species of pests. In the ensuing 30 years, the number jumped to more than 500 species (Georghiou and Lagunes-Tejeda 1991). This remarkable ability of insects to adapt quickly to toxins used to control them threatens agriculture and human health worldwide (Roush and Tabashnik 1990; Denholm and Rowland 1992; Hemingway and Ranson 2000).

The quantity and variety of examples of pesticide resistance also offer opportunities for determining how response to selection is affected by various factors, including behavior, dominance, fitness trade-offs, founder events, gene flow, genetic constraints, haplodiploidy, life-history traits, major and minor genes, multitrophic interactions, and population dynamics (e.g., Gould 1984; Roush and McKenzie 1987; Rosenheim and Tabashnik 1991, 1993; Carrière and Roff 1995; Carrière et al. 1995; Rosenheim et al. 1996; Bourguet and Raymond 1998; ffrench-Constant et al. 1998; Peck et al. 1999; Groeters and Tabashnik 2000; Carrière 2003; Mitchell and Onstad 2005). While study of resistance can provide fundamental insights about evolution, efforts to manage resistance enable application and testing of evolutionary theories. In particular, can strategies based on evolutionary principles delay evolution of pest resistance to insecticidal transgenic crops?

Transgenic Crops with *Bacillus thuringiensis* (Bt) Toxins

Genetic engineering creates possibilities for defending plants from herbivory with a diverse array of toxins from plants, animals, and microbes (Schuler et al. 1998; Moar 2003; Ferry et al. 2004, 2006; Cohen 2005). However, so far insecticidal crystal (Cry) proteins from the bacterium *Bacillus thuringiensis* (Bt) are the basis for nearly all genetically engineered crop protection against herbivorous insects in commercial agriculture. Transgenic Bt crops that kill some key pests can reduce reliance on insecticide sprays, thereby providing economic, health, and environmental benefits (Shelton et al. 2002; Carrière et al. 2003; Cattaneo et al. 2006). Evolution of resistance by pests, however, would cut short the efficacy of Bt crops and the associated benefits.

First planted on a large scale in 1996, crops genetically modified to produce Bt toxins covered more than 20 million hectares worldwide during 2005, with a cumulative total of more than 120 million hectares from 1996 to 2005 (James 2005; Lawrence 2005). Although the variety of Bt toxins used in transgenic crops has increased recently (ISB 2004), cotton producing Bt toxin Cry1Ac and corn producing either Cry1Ab or Cry1Ac accounted for nearly all of the acreage of insect-protected transgenic crops grown during the past decade. These toxins kill some key lepidopteran pests of cotton and corn. Cry1Ab and Cry1Ac are so similar that evolution of resistance to one usually confers cross-resistance to the other (Tabashnik et al. 1996; Ferré and Van Rie 2002). Thus, from the standpoint of herbivore resistance, these two toxins can be considered as one type of toxin. Whereas chemically inducible Bt toxin production is feasible (Bates et al. 2005), commercially grown Bt crops produce toxin continuously. In effect, the first generation of Bt crops exposed pest populations over vast areas to a single type of Bt toxin throughout the growing season.

The widespread and prolonged exposure to Bt toxins in transgenic crops represents one of the largest, most sudden selections for resistance ever seen in herbivorous insects (Tabashnik et al. 2003). Therefore, evolution of resistance by target pests is considered the primary threat to the continued success of Bt crops (Gould 1998; US EPA 2001; Ferré and Van Rie 2002; Griffitts and Aroian 2005). Resistance to a Bt toxin is a genetically based decrease in the frequency of individuals susceptible to the toxin caused by exposure of the population to the toxin (Tabashnik 1994).

Many pests have quickly evolved resistance to Bt toxins in the laboratory (Tabashnik 1994; Ferré and Van Rie 2002). Furthermore, evolution of resistance to Bt sprays is documented for two lepidopteran pests of crucifers, with evidence from greenhouse populations of cabbage looper, *Trichoplusia ni* (Hübner) (Janmaat and Myers 2003), and field populations of diamondback moth, *Plutella xylostella* (L). (Tabashnik et al. 1990, 2003). Based on pervasive resistance to conventional insecticides, rapid responses to lab selection with Bt toxins, and modeling results, worst-case scenarios yielded predictions that pests would evolve resistance to Bt crops in as little as three years (Gould et al. 1997; Roush 1997).

Contrary to these predictions, extensive monitoring studies have not detected field-evolved increases in the frequency of resistance during the first decade of Bt crop use (Tabashnik et al. 2003, 2005a). To better understand this phenomenon, we focus here on evolution of insect resistance to the Bt toxins in transgenic crops. The sections below summarize information about Bt toxins and their mode of action, the genetic basis of Bt resistance, the refuge strategy for delaying pest resistance to Bt crops, and a case study of pink bollworm resistance to Bt cotton in Arizona. The chapter ends with conclusions about what has been learned about insect resistance to Bt crops and a look to the future.

Bt Toxins and Their Mode of Action

Many Bt toxins kill certain key pests, yet unlike broad-spectrum insecticides, they have little or no toxicity to most nontarget organisms, including people, wildlife, and most other insects (Mendelsohn et al. 2003; Naranjo et al. 2005; O'Callaghan et al. 2005; Cattaneo et al. 2006). Sprays of Bt toxins have been used safely and effectively in organic and conventional agriculture and forestry for decades. Collectively, the more than 140 known Bt toxins can kill a wide variety of insects, including moths, beetles, and flies, yet the spectrum of toxicity for each toxin is usually narrow (Schnepf et al. 1998; Griffits and Aroian 2005).

The specificity of Bt toxins arises from their mode of action (Schnepf et al. 1998). Bt toxins are eaten, dissolved in the alkaline insect midgut, and activated by insect proteases from full-length protoxin to active toxin. The next step—specific binding of active toxin to midgut membrane receptors—is a key determinant of specificity. Much evidence implies that this specific binding causes pores in midgut membranes that ultimately lead to cell lysis and insect death (Schnepf et al. 1998; Griffits and Aroian 2005). Some recent work with cell lines, however, suggests that specific binding initiates a magnesium-dependent cellular signaling pathway that ultimately kills the insect (Zhang et al. 2005).

Genetic Basis of Resistance to Bt Toxins

Disruption of any of the steps required for toxicity could cause insect resistance to Bt toxins. Thus, in principle, many mechanisms of resistance could occur in insect populations (Heckel 1994). Do evolutionary constraints result in the same adaptation to Bt toxins across diverse insect taxa, as seen with resistance to some synthetic insecticides (ffrench-Constant et al. 1998, 2000)? Or do multiple modes of resistance arise? Even though the same gene is linked with resistance to Cry1Ac in some strains of at least three species of moths, insects can achieve resistance to Bt toxins by various evolutionary solutions.

Resistance to Bt toxins has been studied mostly in lepidopteran pests that are primary targets of Bt sprays and Bt crops. The most common type of Bt resistance in Lepidoptera, called "mode 1," entails strong resistance to one or more Bt toxins in the Cry1A family, limited cross-resistance to Cry1C and most other Bt toxins, recessive inheritance, and reduced binding of one or more Cry1A toxins to midgut membrane target sites (Tabashnik et al. 1998). Mode 1 resistance is documented for some strains of five pests: diamondback moth; Indianmeal moth, *Plodia interpunctella* (Hübner); tobacco budworm, *Heliothis virescens* (F.); pink bollworm, *Pectinophora gossypiella* (Saunders); and cotton bollworm, *Helicoverpa armigera* (Hübner) (Tabashnik et al. 1998; Akhurst et al. 2003; González-Cabrera et al. 2003; Xu et al. 2005).

The molecular genetic basis of mode 1 resistance is known only in the latter three species. In these three cotton pests, laboratory-selected mode 1 resistance is tightly linked with a gene encoding a cadherin protein that binds Cry1Ac in susceptible insects (Gahan et al. 2001; Morin et al. 2003; Xu et al. 2005). So far, at least five different mutations of the cadherin gene that interfere with production of a full-length protein are associated with resistance to Cry1Ac in these three species. Evidence that the resistance-associated mutations block binding of Bt toxin is reported for *Heliothis virescens* (Jurat-Fuentes et al. 2004) and suspected in the others. Although the locations of the mutations within the cadherin gene are unique to each species, each species harbors at least one mutation that introduces a premature stop codon. Whereas only one resistance-associated cadherin mutation has been identified in tobacco budworm and cotton bollworm, pink bollworm has at least three, including two deletions that do not introduce a premature stop codon.

Many cases of resistance to Bt toxins in Lepidoptera do not fit the mode 1 pattern (Tabashnik et al. 1998; Ferré and Van Rie 2002; Griffits and Aroian 2005). Indeed, several examples show that mode 1 resistance and other modes of resistance can occur in the same species of insect (Tabashnik et al. 1998). At least two alternatives to cadherin-based resistance to Bt toxins have been identified at the molecular genetic level. One appears to be another version of target site resistance involving an aminopeptidase N rather than cadherin. A strain of beet armyworm, *Spodoptera exigua* (Hübner), with resistance to Bt toxin Cry1C lacks an aminopeptidase N that may be a receptor for Cry1C (Herrero et al. 2005). Unlike the aforementioned examples involving altered or absent target sites, resistance to Bt toxin Cry1Ab in a strain of European corn borer, *Ostrinia nubilalis* (Hübner), was due primarily to reduced trypsin-like activity of a proteinase that activates Bt protoxin to active toxin (Li et al. 2005). The major locus conferring field-evolved mode 1 resistance in several strains of diamondback moth is not linked with cadherin, several aminopeptidases N, or other candidate genes tested so far (Baxter 2005; Baxter et al. 2005).

The Refuge Strategy for Delaying Pest Resistance to Bt Crops

Although many strategies have been proposed to delay pest resistance to Bt crops (Tabashnik 1994; Roush 1997; Gould 1998; Bates et al. 2005; Zhao et al. 2005), we focus here on the one that is most widely used: the refuge strategy. To discuss the theory underlying the refuge strategy, we begin with assumptions about the genetic basis of resistance to Bt crops. We assume the simplest genetic model, namely, resistance to Bt crops is controlled by a single locus with two alleles, *r* for resistance and *s* for susceptibility. Although this is oversimplified, it is a reasonable starting point because mutations at single loci do confer resistance to Bt toxins in several well-studied cases (Tabashnik et al. 1997; Gahan et al. 2001; Morin et al. 2003; Baxter et al. 2005; Herrero et al. 2005; Li et al. 2005; Xu et al. 2005). Furthermore, resistance to the intense selection imposed by Bt crops is most likely to involve loci with major effects (McKenzie 1996; Groeters and Tabashnik 2000).

The refuge strategy is mandated in most countries where Bt crops are grown, including the United States, which accounts for more than half of the world's Bt crop acreage (US EPA 2001; James 2005; Lawrence 2005). The refuge strategy is based on evolutionary theory elaborated in dozens of papers (e.g., Comins 1977; Georghiou and Taylor 1977; Curtis et al. 1978; Tabashnik and Croft 1982; Gould 1998; Peck et al. 1999; Caprio 2001; Carrière and Tabashnik 2001; Onstad et al. 2002) and on small-scale experiments with diamondback moth (Liu and Tabashnik 1997a; Shelton et al. 2000; Tang et al. 2001). The theory underlying the refuge strategy is to reduce heritability of resistance by (1) providing refuges of non-Bt host plants that produce susceptible adults, (2) promoting mating between resistant and susceptible adults, and (3) decreasing the dominance of resistance.

To implement the refuge strategy, farmers grow refuges of non-Bt host plants near Bt crops to promote survival of susceptible pests. This strategy is expected to work best if resistance is conferred by rare, recessive alleles and if most of the extremely rare resistant adults emerging from Bt crops mate with susceptible adults from refuges. The theory predicts that such conditions will greatly delay evolution of resistance.

Refuge Size

Although rigorous large-scale tests of the refuge strategy are difficult and have not been conducted yet, modeling results and small-scale experiments show that resistance is expected to evolve slower as the area of refuges relative to Bt crops increases. The regulations implemented in the field represent a compromise between the conflicting goals of delaying resistance (which favors large refuges) and minimizing constraints on growers (which favors minimal regulation and small refuges).

In the United States, the percentage of cotton acreage required for refuges depends on whether refuges are planted inside or outside of Bt cotton fields. For refuges outside of Bt cotton fields, refuge size ranges from 5% non-Bt cotton if the refuge is not sprayed with insecticides effective against pests targeted by Bt cotton, to 20% non-Bt cotton if the refuge is sprayed with such insecticides (US EPA 2001). Non-Bt cotton refuges inside Bt cotton fields must be at least 5% of the area of the field; they can be treated with any insecticides except Bt sprays as long as the refuge and Bt cotton are treated simultaneously. The percentage of corn acreage required for non-Bt corn refuges ranges from 20% in regions with little or no Bt cotton to 50% in regions with substantial amounts of Bt cotton (US EPA 2001).

Insect Gene Flow

The success of the refuge strategy depends on insect gene flow between refuges and Bt crops. The extent and impact of this gene flow is affected by many factors, including pest movement and mating patterns, and the spatial and temporal distribution of refuges and Bt crops (Carrière et al. 2004a; Sisterson et al. 2004, 2005). For example, because refuges are sources of insect pests and Bt crop fields are sinks, movement between refuges and Bt crops can reduce population size in refuges and regionally (Riggin-Bucci and Gould 1997; Onstad and Guse 1999; Caprio et al. 2004; Carrière et al. 2003, 2004a, 2004b). Moreover, movement from Bt crops to refuges can bring resistance alleles to refuges, which may increase heritability of resistance when the rare resistant adults emerging from Bt crops mate with adults from refuges bearing resistance alleles (Comins 1977; Caprio and Tabashnik 1992; Sisterson et al. 2004).

For simplicity, many models of insecticide resistance evolution assume that insect movement is sufficient to achieve random mating between adults from refuges and areas treated with insecticide (e.g., Georghiou and Taylor 1977). This apparently led to the claim that random mating between adults from refuges and Bt crops is crucial for the refuge strategy (e.g., Roush 1997; Liu et al. 1999; Glaser and Matten 2003). Although the refuge strategy does require mating between resistant and susceptible adults, results from many modeling studies show that random mating is not essential or even optimal (e.g., Tabashnik 1990; Caprio and Tabashnik 1992; Gould 1998; Peck et al. 1999; Caprio 2001; Ives and Andow 2002; Onstad et al. 2002; Carrière et al. 2004b; Sisterson et al. 2005). Nonetheless, models have not yielded consistent conclusions about how much movement is best for delaying resistance. In particular, recent modeling studies have reported that resistance evolves slowest when adult movement between refuges and Bt crops is either high (Peck et al. 1999; Storer 2003), low (Ives and Andow 2002), or intermediate (Caprio 2001).

Because models used in various studies differ in many ways, the precise cause of such contradictory conclusions among studies is not obvious. Conversely, sensitivity analyses in which only one or a few assumptions are varied systematically in a single study enable testing of the hypothesis that interactions among factors can alter the effect of movement on the rate of resistance evolution (Tabashnik and Croft 1982). Sensitivity analyses show that interactions among the relative abundance, spatial distribution, and temporal distribution of refuges and Bt crop fields can alter the effects of movement on resistance evolution (Sisterson et al. 2005). Furthermore, it appears that differences in conclusions among studies can be explained by differences in assumptions about the relative abundance and distribution of refuges and Bt crop fields (Sisterson et al. 2005). One robust result is that resistance can be delayed effectively by fixing locations of refuges across years and distributing refuges uniformly to ensure that Bt crop fields are not isolated from refuges. However, rotating fields between refuges and Bt crops between years may provide better insect control and reduce the need for insecticide sprays.

TABLE 19.1
Dominance (*h*) of Resistance to *Bacillus thuringiensis* (Bt) Toxins as a Function of Toxin Concentration

Bt toxin concn. (μg per ml diet)	Survival (%) of Larvae Exposed to Bt Toxin: *ss*	*rs*	*rr*	Dominance (*h*)
Heliothis virescens vs. Cry1Ab (Gould et al. 1995)				
0.32	47	100	100[a]	1.00
1.6	16	75	100	0.70
8.0	0	31	97	0.32
40.0	0	0	97	0.00
Pectinophora gossypiella vs. Cry1Ac (Tabashnik et al. 2002a)				
0.32	37	93	100[a]	0.89
1.0	4	52	100[a]	0.50
3.2	0	8	100[a]	0.08
10.0	0[b]	0.5	100	0.005

NOTE: Adapted from Tabashnik et al. 2004. We calculated *h* as (survival of *rs* − survival of *ss*)/(survival of *rr* − survival of *ss*) (Liu and Tabashnik 1997b). Values of *h* vary from 0 (recessive) to 1 (dominant).

[a]Survival inferred to be 100% based on 100% survival at a higher toxin concentration.

[b]Survival inferred to be 0% based on 0% survival at a lower toxin concentration.

Dominance of Resistance

In the early stages of resistance evolution, alleles conferring resistance are expected to be rare. Hence, their dominance is crucial. For example, before commercialization of Bt cotton, the estimated frequency of alleles conferring resistance to Cry1Ac in *H. virescens* from four U.S. states was 0.0015, based on tests of progeny from single-pair matings between wild males and *rr* females from a laboratory-selected strain (Gould et al. 1997). When resistance alleles are rare, they occur mostly in heterozygotes, and resistant homozygotes are extremely rare. Assuming Hardy-Weinberg equilibrium in the example above, the expected frequency of *rs* (3×10^{-3}) is more than a thousand times greater than that of *rr* homozygotes (2.2×10^{-6}). Therefore, the response of heterozygotes—which is determined by dominance—governs the early trajectory of resistance evolution.

Although dominance is sometimes considered an invariant genetic property, the refuge strategy exploits the principle that dominance can depend on the environment (Curtis et al. 1978; Bourguet et al. 1996; Tabashnik et al. 2004). In particular, the dominance of resistance depends on the dose of toxin. The refuge strategy is sometimes called the "high-dose refuge strategy" because results from many modeling studies show that refuges are most effective if the dose of toxin received by insects eating Bt plants is high enough to kill all or nearly all *rs* individuals (e.g., Curtis et al. 1978; Tabashnik and Croft 1982; Roush 1997; Gould 1998). In other words, refuges are predicted to work best if the toxin concentration in Bt plants is high enough to make resistance functionally recessive.

Bioassay results from some key lepidopteran pests show that the dominance of their resistance to Bt toxins decreases as toxin concentration increases (Tabashnik et al. 1992, 2002a; Gould et al. 1995; Liu and Tabashnik 1997b; Tang et al. 1997; Zhao et al. 2000; Liu et al. 2001b; Alves et al. 2006). Two examples illustrate why this occurs (Table 19.1). At low toxin concentrations, survival is low to moderate for susceptible homozygotes (*ss*), and relatively high for heterozygotes (*rs*) and resistant homozygotes (*rr*). The similarity between *rs* and *rr* at low toxin concentrations yields dominant resistance. Conversely, at sufficiently high toxin concentrations, survival is low for *ss* and *rs* relative to *rr*, which yields recessive resistance. Responses of *H. virescens* to Cry1Ab incorporated in artificial diet show that inheritance of resistance varied from completely dominant to completely recessive as the concentration of Bt toxin increased from low to high (Table 19.1). Although the variation is not as extreme, the trend is similar in responses of pink bollworm to Cry1Ac (Table 19.1).

To achieve high concentrations of Bt toxins in transgenic crop plants, Bt toxin genes have been modified for improved expression in plants (Mendelsohn et al. 2003). For some but not all targeted pests, the toxin concentrations in Bt crops are high enough to render resistance functionally recessive (Tabashnik et al. 2000b, 2003). Results from the noctuid moth *Helicoverpa armigera* demonstrate that the dominance of resistance to Bt cotton can vary as the concentration of Cry1Ac changes during the growing season (Bird and Akhurst 2004, 2005). Whereas resistance was recessive ($h = 0$) on young cotton plants with relatively high Cry1Ac concentration, it was additive ($h = 0.49$) on older cotton plants with 75% lower Cry1Ac concentration.

Fitness Costs

Fitness costs occur when fitness in refuges is lower for individuals with resistance alleles than for individuals without resistance alleles. Fitness costs reflect antagonistic pleiotropy causing a trade-off across environments; resistance alleles increase fitness on Bt crops but decrease fitness in the absence of Bt toxins. From the perspective of mutation-selection balance theory, fitness costs keep resistance alleles rare before populations are exposed extensively to Bt toxins. From a management standpoint, fitness costs select against resistance in refuges. In principle, evolution of resistance can be delayed substantially or even reversed if fitness costs and refuges are sufficiently large (Lenormand and Raymond 1998; Carrière and Tabashnik 2001; Carrière et al. 2002, 2004b, 2005a; Tabashnik et al. 2005a).

In the initial stages of resistance evolution, the dominance of fitness costs is crucial, just as the dominance of resistance is vital. As noted above, when Bt crops are first introduced, most *r* alleles occur in *rs* individuals. If resistance is recessive, these *rs* individuals are killed by Bt crops and survive only in refuges. Thus, the fitness of *rs* relative to *ss* in refuges is a key determinant of resistance evolution. Nonrecessive fitness costs make fitness in refuges lower for *rs* than *ss*. Accordingly, nonrecessive fitness costs can strongly favor a decrease in resistance through selection in refuges, even though the extremely rare *rr* individuals are favored by selection in Bt crop fields (Carrière and Tabashnik 2001; Carrière et al. 2002; Tabashnik et al. 2005a).

Fitness costs are usually associated with resistance to Bt toxins (e.g., Groeters et al. 1994; Tabashnik 1994; Alyokhin and Ferro 1999; Oppert et al. 2000; Carrière et al. 2001a, 2001b, 2005a; Ferré and Van Rie 2002; Bird and Akhurst 2004, 2005; Higginson et al. 2005; but also see Tang et al. 1997; Huang et al. 2005). Fitness costs associated with Bt resistance were recessive in most of the few cases in which their dominance was examined (Carrière et al. 2001a, 2001b, 2005a; Janmaat and Myers 2005). However, nonrecessive fitness costs affecting overwintering survival and other life-history traits on some host plants occurred in *H. armigera* (Bird and Akhurst 2004, 2007).

Like the dominance of resistance, fitness costs associated with resistance to insecticides are affected by genotype-environment interactions (McKenzie 1990; Foster et al. 2003; Agnew et al. 2004; Bourguet et al. 2004). In particular, fitness costs associated with Bt resistance are increased by stress imposed by suboptimal host plants (Carrière et al. 2004c, 2005a; Janmaat and Myers 2005; Bird and Akhurst 2007), competition for mates (Higginson et al. 2005), crowding (Raymond et al. 2005), and natural enemies (Gassmann et al. 2006). This creates the opportunity to manipulate fitness costs to enhance the success of the refuge strategy (Carrière et al. 2001b, 2004a, 2005a; Pittendrigh et al. 2004; Gassmann et al. 2006).

Incomplete Resistance

Many early models assumed that *rr* individuals are completely resistant to Bt crops, that is, fitness of *rr* is equal on Bt and non-Bt host plants (Gould 1998). Although this is true in some special cases, such as resistant diamondback moth on experimental Bt crucifers versus non-Bt crucifers (Ramachandran et al. 1998; Tang et al. 1999), it is not generally applicable. In particular, the fitness of *rr* is often lower on Bt crop plants than on their non-Bt counterparts (Liu et al. 1999, 2001a; Bird and Akhurst 2004). This disadvantage suffered by resistant insects on transgenic plants relative to their conventional nontransgenic counterparts is called incomplete resistance (Carrière and Tabashnik 2001). Unlike fitness costs, incomplete resistance cannot cause reversal of resistance evolution. However, lower fitness of *rr* in a Bt crop field relative to a refuge weakens selection for resistance and thus can help to delay resistance.

Pink Bollworm versus Bt Cotton in Arizona

We are part of a team that has been studying pink bollworm resistance to Bt cotton in Arizona since 1997, the second year of large-scale use of Bt crops. This project, which is based at the University of Arizona, has benefited greatly from collaboration with the U.S. Department of Agriculture's Western Cotton Research Laboratory and the Arizona cotton growers. Pink bollworm is the major lepidopteran pest of cotton in the southwestern United States; it also attacks cotton in many other nations (Henneberry and Naranjo 1998). Females lay eggs on cotton plants and larvae bore into cotton bolls where they eat cotton seeds, with the potential to completely destroy yield.

The risk of pink bollworm resistance to Bt cotton producing Cry1Ac was considered high in Arizona for the following reasons: (1) lab selection with Cry1Ac in artificial diet produced strains resistant to Bt cotton in a few generations; (2) since 1997, Bt cotton has accounted for more than half of Arizona's cotton; (3) Bt cotton kills close to 100% of susceptible pink bollworm larvae that eat it; (4) pink bollworm

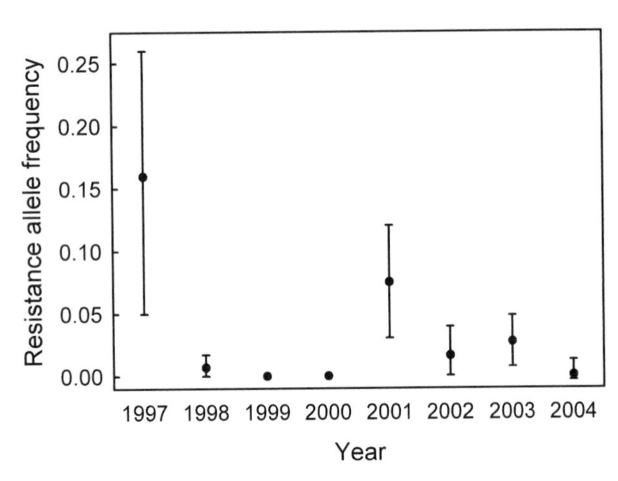

FIGURE 19.1. The mean resistance allele frequency (with 95% confidence intervals) for pink bollworm in Arizona from 1997 to 2004 (from Tabashnik et al. 2005a; copyright 2005, National Academy of Sciences, U.S.A.). The frequency of resistance to Cry1Ac, the toxin in Bt cotton, was estimated with laboratory diet bioassays testing an average of 2680 pink bollworm larvae per year from 10 to 17 sites per year. The resistance (*r*) allele frequency was calculated for each site assuming Hardy-Weinberg equilibrium. In 1999 and 2000, no larvae survived on diet treated with a diagnostic concentration of Cry1Ac (10 μg toxin per ml diet).

is an ecological specialist on cotton in Arizona; and (5) pink bollworm has up to five generations per year in Arizona (Henneberry and Naranjo 1998; Tabashnik et al. 2000a; Carrière et al. 2005b). Despite these concerns, and to our surprise, monitoring of field populations revealed that the frequency of resistance to Bt cotton showed no net increase from 1997 to 2004 (Fig. 19.1). The lack of field-evolved resistance despite extensive use of Bt cotton and rapid resistance evolution in the laboratory suggests that the refuge strategy has helped to delay pink bollworm resistance to Bt cotton. Below, we summarize the monitoring evidence and an explanation for the observed pattern based on incorporation of data from the field and greenhouse into a simple population genetic model.

Monitoring Results

To monitor resistance to Cry1Ac, we sampled pink bollworm by collecting bolls from 10 to 17 cotton fields in Arizona per year. The progeny of field-collected individuals from each site were reared separately and tested using bioassays. Neonates were tested individually on artificial diet with 10 μg Cry1Ac per ml diet, which kills *ss* and *rs,* but not *rr* (Tabashnik et al. 2000a, 2002a, 2005b). Additional neonates were tested on diet without toxin so that mortality on treated diet could be corrected for control mortality.

Unexpectedly, the highest frequency of resistance to Cry1Ac occurred in 1997, the first year of monitoring with bioassays (Fig. 19.1). The statewide mean *r* allele frequency in 1997 was 0.16 with a 95% confidence interval (CI) of 0.05 to 0.26. To put this in perspective, the estimated mean frequency of *rr* was 0.026 (0.16^2) for pink bollworm in 1997, which is more than 10,000 times higher than the estimated frequency of *rr* (2.2×10^{-6}) for *Heliothis virescens* (Gould et al. 1997). Five of 10 cotton fields sampled in 1997 yielded one or more resistant individuals that survived exposure to the diagnostic concentration of Cry1Ac (Tabashnik et al. 2000a). Contrary to our expectations, the *r* allele frequency dropped in 1998 to 0.0070 (95% CI = 0–0.002). Whereas survival at the diagnostic concentration of Cry1Ac was 3.2% (16 of 500) in 1997, it was only 0.09% (1 of 1,100) in 1998. The frequency of resistance showed some variation among years, but no net increase from 1998 to 2004 (2004 mean = 0.004, 95% CI = 0–0.01).

Rapid responses to laboratory selection confirmed the relatively high *r* allele frequency detected in field populations in 1997. Two rounds of laboratory selection with survivors pooled from statewide samples yielded a resistant strain (AZP-R) with larvae capable of survival on Bt cotton (Tabashnik et al. 2000a). Subsequently, independent laboratory selections were conducted with two of the strains derived from the field in 1997 that had contributed individuals to AZP-R. These selections produced a resistant strain (MOV97-R) from a single site in western Arizona and another resistant strain from a single site in eastern Arizona (SAF97-R) (Tabashnik et al. 2005b). The quick response to selection in each of the three strains suggests that individuals sampled from the field in 1997 harbored *r* alleles. Although recessive mutations in a cadherin gene were associated with resistance in all three strains, the frequency of the three identified resistance alleles (*r1, r2,* and *r3*) varied among strains. AZP-R had all three alleles (Morin et al. 2003), which is consistent with the multiple origins of individuals pooled to create this strain. MOV97-R had only alleles *r1* and *r3*, while SAF97-R had only *r1* and *r2* (Tabashnik et al. 2005b).

Several independent lines of evidence support the conclusion that pink bollworm resistance to Bt cotton in Arizona did not increase from 1997 to 2004. Unlike the rapid response to lab selection achieved with strains derived from the field in 1997, similar selections did not yield resistant strains from Arizona in subsequent years. Field efficacy of Bt cotton has remained >99% (Dennehy et al. 2004), and no cases of reduced efficacy caused by resistance have been documented, despite a widely publicized grower hotline for reporting potential problems. DNA-based screening shows that cadherin *r* alleles were rare in field populations from 2001 to 2005 (Morin et al. 2004; Tabashnik et al. 2006). Furthermore, long-term regional declines of pink bollworm populations have occurred in areas of Arizona with high use of Bt cotton (Carrière et al. 2003).

A Simple Population Genetic Model

To better understand the observed outcome in the field, we used a simple population genetic model to explore conditions expected to cause increases, decreases, or stability of

Gould, F., A. Martínez-Ramírez, A. Anderson, J. Ferré, F. J. Silva, and W. J. Moar. 1992. Broad-spectrum resistance to *Bacillus thuringiensis* in *Heliothis virescens*. Proc. Natl. Acad. Sci. USA 89: 7986–7990.

Gould, F., A. Anderson, A. Reynolds, L. Bumgarner, and W. Moar. 1995. Selection and genetic analysis of a *Heliothis virescens* (Lepidoptera: Noctuidae) strain with high levels of resistance to *Bacillus thuringiensis* toxins. J. Econ. Entomol. 88: 545–1559.

Gould, F., A. Anderson, A. Jones, D. Sumerford, D. G. Heckel, J. Lopez, S. Micinski, R. Leonard, and M. Laster. 1997. Initial frequency of alleles for resistance to *Bacillus thuringiensis* toxins in field populations of *Heliothis virescens*. Proc. Natl Acad. Sci. USA 94: 3519–3523.

Griffitts, J. S., and R. V. Aroian. 2005. Many roads to resistance: how invertebrates adapt to Bt toxins. BioEssays 27: 614–624.

Groeters, F. G., and B. E. Tabashnik. 2000. Roles of selection intensity, major genes, and minor genes in evolution of insecticide resistance. J. Econ. Entomol. 93: 1580–1587.

Groeters, F. R., B. E. Tabashnik, N. Finson, and M. W. Johnson. 1994. Fitness costs of resistance to *Bacillus thuringiensis* in the diamondback moth *(Plutella xylostella)*. Evolution 48: 197–201.

Hartl, D. L., and A. G. Clark. 1989. Principles of population genetics, 2nd ed. Sinauer, Sunderland, MA.

Heckel, D. G. 1994. The complex genetic basis of resistance to *Bacillus thuringiensis* in insects. Biocontrol Sci. Technol. 4: 405–417.

Hemingway, J., and H. Ranson. 2000. Insecticide resistance in insect vectors of human disease. Annu. Rev. Entomol. 45: 371–391.

Henneberry, T. J., and S. E. Naranjo. 1998. Integrated management approaches for pink bollworm in the southwest United States. Integr. Pest Manag. Rev. 3: 31–52.

Herrero, S., T. Gechev, P. L. Bakker, W. J. Moar, and R. A. de Maagd. 2005. *Bacillus thuringiensis* Cry1Ca-resistant *Spodoptera exigua* lacks expression of one of four aminopeptidase N genes. BMC Genomics 6: 96.

Higginson, D. M., S. Morin, M. Nyboer, R. Biggs, B. E. Tabashnik, and Y. Carrière. 2005. Evolutionary trade-offs of insect resistance to Bt crops: fitness costs affecting paternity. Evolution 59: 915–920.

High, S. M., M. B. Cohen, Q. Y. Shu, and I. Altosaar. 2004. Achieving successful deployment of Bt rice. Trends Plant Sci. 9: 286–292.

Huang, F, L. L. Buschman, and R. A. Higgins. 2005. Larval survival and development of susceptible and resistant *Ostrinia nubilalis* (Lepidoptera: Pyralidae) on diet containing *Bacillus thuringiensis*. Agric. Forest. Entomol. 7: 45–52.

ISB (Information Systems for Biotechnology). 2004. Crops no longer regulated by USDA. Virginia Tech, Blacksburg. Available at www.isb.vt.edu/cfdocs/biopetitions1.cfm.

Ives, A. R., and D. A. Andow. 2002. Evolution of resistance to Bt crops: directional selection in structured environments. Ecol. Lett. 5: 792–801.

James, C. 2005. Global status of biotech/GM Crops in 2005. 2005 ISAAA briefs no. 34. International Service for the Acquisition of Agri-biotech Applications, Ithaca, NY.

Janmaat, A. F., and J. H. Myers. 2003. Rapid evolution and the cost of resistance to *Bacillus thuringiensis* in greenhouse populations of cabbage loopers, *Trichoplusia ni*. Proc. R. Soc. Lond. B 270: 2263–2270.

Janmaat, A. F., and J. H. Myers. 2005. The cost of resistance to *Bacillus thuringiensis* varies with the host plant of *Trichoplusia ni*. Proc. R. Soc. Lond. B 272: 1031–1038.

Jurat-Fuentes, J. L., L. J. Gahan, F. L. Gould, D. G. Heckel, and M. J. Adang. 2004. The HevCaLP protein mediates binding specificity of the Cry1A class of *Bacillus thuringiensis* toxins in *Heliothis virescens*. Biochem. 43: 14,299–14,305.

Labbe, P., T. Lenormand, and M. Raymond. 2005. On the worldwide spread of an insecticide resistance gene: a role for local selection. J. Evol. Biol. 18: 1471–1484.

Lawrence, S. 2005. Agbio keeps growing. Nat. Biotechnol. 23: 281.

Lenormand, T., and M. Raymond. 1998. Resistance management: the stable zone strategy. Proc. R. Soc. Lond. B 265: 1985–1990.

Li, H., B. Oppert, R. A. Higgins, F. Huang, L. L. Buschman, J.-R. Gao, and K. Y. Zhu. 2005. Characterization of cDNAs encoding three trypsin-like proteinases and mRNA quantitative analysis in Bt-resistant and-susceptible strains of *Ostrinia nubilalis*. Insect Biochem. Mol. Biol. 35: 47–860.

Liu, Y. B., and B. E. Tabashnik. 1997a. Experimental evidence that refuges delay insect adaptation to *Bacillus thuringiensis* toxin Cry1C in diamondback moth. Appl. Environ. Microbiol. 63: 2218–2223.

Liu, Y. B., and B. E. Tabashnik. 1997b. Inheritance of resistance to *Bacillus thuringiensis* toxin Cry1C in diamondback moth. Appl. Environ. Microbiol. 63: 2218–2223.

Liu, Y. B., B. E. Tabashnik, T. J. Dennehy, A. L. Patin, and A. C. Bartlett. 1999. Development time and resistance to *Bt* crops. Nature 400: 519.

Liu, Y. B., B. E. Tabashnik, T. J. Dennehy, A. L. Patin, M. A. Sims, S. K. Meyer, and Y. Carrière. 2001a. Effects of Bt cotton and Cry1Ac toxin on survival and development of pink bollworm (Lepidoptera: Gelechiidae). J. Econ. Entomol. 94: 1237–1242.

Liu, Y. B., B. E. Tabashnik, S. K. Meyer, Y. Carrière, and A. C. Bartlett. 2001b. Genetics of pink bollworm resistance to *Bacillus thuringiensis* toxin Cry1Ac. J. Econ. Entomol. 94: 248–252.

McKenzie, J. A. 1990. Selection at a dieldrin resistance locus in overwintering populations of *Lucilia cuprina* (Wiedemann). Aust. J. Zool. 38: 493–501.

McKenzie, J. A. 1996. Ecological and evolutionary aspects of insecticide resistance. R. G. Landes Co. and Academic Press, Austin, TX.

Mendelsohn, M., J. Kough, Z. Vaitizis, and K. Matthews. 2003. Are Bt crops safe? Nat. Biotech. 21: 1003–1009.

Mitchell, P. D., and D. W. Onstad. 2005. Effect of extended diapause on evolution of resistance to transgenic *Bacillus thuringiensis* corn by northern corn rootworm (Coleoptera: Chrysomelidae). J. Econ. Entomol. 98: 2220–2234.

Moar, W. J. 2003. Breathing new life into insect-resistant plants. Nat. Biotechnol. 21: 1152–1154.

Morin, S., R. W. Biggs, M. S. Sisterson, L. Shriver, C. Ellers-Kirk, D. Higginson, D. Holley, J. J. Gahan, D. G. Heckel, Y. Carrière, T. J. Dennehy, J. K. Brown, and B. E. Tabashnik. 2003. Three cadherin alleles associated with resistance to *Bacillus thuringiensis* in pink bollworm. Proc. Natl. Acad. Sci. USA 100: 5004–5009.

Morin, S., S. Henderson, J. A. Fabrick, Y. Carrière, T. J. Dennehy, J. K. Brown, and B. E. Tabashnik. 2004. DNA-based detection of Bt resistance alleles in pink bollworm. Insect Biochem. Mol. Biol. 34: 1225–1233.

Naranjo, S. E., G. Head, and G. P. Dively. 2005. Field studies assessing arthropod nontarget effects in Bt transgenic crops: Introduction. Environ. Entomol. 34: 1178–1180.

O'Callaghan, M., T. R. Glare, E. P. J. Burgess, and L. A. Malone. 2005. Effects of plants genetically modified for insect resistance on nontarget organisms. Annu. Rev. Entomol. 50: 271–292.

Onstad, D. W., and G. A. Guse. 1999. Economic analysis of the use of transgenic crops and nontransgenic refuges for management of European corn borer (Lepidoptera: Pyralidae). J. Econ. Entomol. 92: 1256–1265.

Onstad, D. W., G. A. Guse, P. Porter, L. L. Buschman, R. A. Higgins, P. E. Sloderbeck, F. B. Peairs, and G. B. Gronholm. 2002. Modeling the development of resistance by stalk-boring lepidopteran insects (Crambidae) in areas with transgenic corn and frequent insecticide use. J. Econ. Entomol. 95: 1033–1043.

Oppert, B., R. Hammel, J. E. Thorne, and K. J. Kramer. 2000. Fitness cost of resistance to *Bacillus thuringiensis* in the Indianmeal moth, *Plodia interpunctella*. Entomol. Exp. Appl. 96: 281–287.

Peck, S., F. Gould, and S. P. Ellner. 1999. Spread of resistance in spatially extended regions of transgenic cotton: implications for management of *Heliothis virescens* (Lepidoptera: Noctuidae). J. Econ. Entomol. 92: 1–16.

Pittendrigh, B. R., P. J. Gaffney, J. E. Huesing, D. W. Onstad, R. T. Roush, and L. L. Murdock. 2004. "Active" refuges can inhibit the evolution of resistance in insects towards transgenic insect-resistant plants. J. Theor. Biol. 231: 461–474.

Ramachandran, S., G. D. Buntin, J. N. All, B. E. Tabashnik, P. L. Raymer, M. J. Adang, D. A. Pulliam, and C. N. Stewart Jr. 1998. Survival, development, and oviposition of resistant diamondback moth (Lepidoptera: Plutellidae) on transgenic canola producing a *Bacillus thuringiensis* toxin. J. Econ. Entomol. 91: 1239–1244.

Raymond, B., A. H. Sayyed, and D. J. Wright. 2005. Genes and environment interact to determine the fitness costs of resistance to *Bacillus thuringiensis*. Proc. R. Soc. Lond. B 272: 1519–1524.

Raymond, M., and M. Marquine. 1994. Evolution of insecticide resistance in *Culex pipiens* populations: the Corsican paradox. J. Evol. Biol. 7: 315–337.

Riggin-Bucci, T. M., and F. Gould. 1997. Impact of intraplot mixtures of toxic and non-toxic plants on population dynamics of diamondback moth (Lepidoptera: Plutellidae) and its natural enemies. J. Econ. Entomol. 90: 241–251.

Rosenheim, J. A., and B. E. Tabashnik. 1991. Influence of generation time on the rate of response to selection. Am. Nat. 137: 527–541.

Rosenheim, J. A., and B. E. Tabashnik. 1993. Generation time and evolution. Nature. 365: 791–792.

Rosenheim, J. A., M. W. Johnson, R. F. L. Mau, S. C. Welter, and B. E. Tabashnik. 1996. Biochemical preadaptations, founder events, and the evolution of resistance in arthropods. J. Econ. Entomol. 89: 263–273.

Roush, R. T. 1997. Bt-transgenic crops: just another pretty insecticide or a new start for resistance management? Pestic. Sci. 51: 328–344.

Roush, R. T., and J. A. McKenzie. 1987. Ecological genetics of insecticide and acaricide resistance. Annu. Rev. Entomol. 32: 361–380.

Roush, R. T., and B. E. Tabashnik (eds.). 1990. Pesticide resistance in arthropods. Chapman and Hall, New York.

Schnepf, E., N. Crickmore, J. van Rie, D. Lereclus, J. Baum, J. Feitelson, D. R. Zeigler, and D. H. Dean. 1998. *Bacillus thuringiensis* and its pesticidal crystal proteins. Microbiol. Mol. Biol. Rev. 62: 775–806.

Schuler, T. H., G. M. Poppy, B. R. Kerry, and I. Denholm. 1998. Insect-resistant transgenic plants. Trends Biotechnol. 16: 168–175.

Shelton, A. M., J. D. Tang, R. T. Roush, T. D. Metz, and E. D. Earle. 2000. Field tests on managing resistance to *Bt*-engineered plants. Nat. Biotechnol. 18: 339–342.

Shelton, A. M., J.-Z. Zhao, and R. T. Roush. 2002. Economic, ecological, food safety, and social conseqauences of the deployment of Bt transgenic plants. Annu. Rev. Entomol. 47: 845–881.

Sisterson, M. S., L. Antilla, Y. Carrière, C. Ellers-Kirk, and B. E. Tabashnik. 2004. Effects of insect population size on evolution of resistance to transgenic crops. J. Econ. Entomol. 97: 1413–1424.

Sisterson, M. S., Y. Carrière, T. J. Dennehy, and B. E. Tabashnik. 2005. Evolution of resistance to transgenic crops: interactions between insect movement and field distribution. J. Econ. Entomol. 98: 1751–1762.

Storer, N. P. 2003. A spatially explicit model simulating western corn rootworm (Coleoptera: Chrysomelidae) adaptation to insect-resistant maize. J. Econ. Entomol. 96: 1530–1547.

Tabashnik, B. E. 1990. Modeling and evaluation of resistance management tactics, pp. 153–182. In R. T. Roush and B. E. Tabashnik (eds.), Pesticide resistance in arthropods. Chapman and Hall, New York.

Tabashnik, B. E. 1994. Evolution of resistance to *Bacillus thuringiensis*. Annu. Rev. Entomol. 39: 47–79.

Tabashnik, B. E., and B. A. Croft. 1982. Managing pesticide resistance in crop-arthropod complexes: interactions between biological and operational factors. Environ. Entomol. 11: 1137–1144.

Tabashnik, B. E., N. L. Cushing, N. Finson, and M. W. Johnson. 1990. Field development of resistance to *Bacillus thuringiensis* in diamondback moth (Lepidoptera: Plutellidae). J. Econ. Entomol. 83: 1671–1676.

Tabashnik, B. E., J. M. Schwartz, N. Finson, and M. W. Johnson. 1992. Inheritance of resistance to *Bacillus thuringiensis* in diamondback moth (Lepidoptera: Plutellidae). J. Econ. Entomol. 85: 1046–1055.

Tabashnik, B. E., T. Malvar, Y. B. Liu, D. Borthakur, B. S. Shin, S. H. Park, L. Masson, R. A. de Maagd, and D. Bosch. 1996. Cross-resistance of diamondback moth indicates altered interactions with domain II of *Bacillus thuringiensis* toxins. Appl. Environ. Microbiol. 62: 2839–2844.

Tabashnik, B. E., Y. B. Liu, N. Finson, L. Masson, and D. G. Heckel. 1997. One gene in diamondback moth confers resistance to four *Bacillus thuringiensis* toxins. Proc. Natl. Acad. Sci. USA 94: 1640–1644.

Tabashnik, B. E., Y. B. Liu, T. Malvar, D. G. Heckel, L. Masson, and J. Ferré. 1998. Insect resistance to *Bacillus thuringiensis*: uniform or diverse? Philos. Trans. R. Soc. Lond. B 353: 1751–1756.

Tabashnik, B. E., A. L. Patin, T. J. Dennehy, Y. B. Liu, Y. Carrière, M. A. Sims, and L. Antilla. 2000a. Frequency of resistance to *Bacillus thuringiensis* in field populations of pink bollworm. Proc. Natl. Acad. Sci. USA 97: 12980–12984.

Tabashnik, B. E., R. T. Roush, E. D. Earle, and A. M. Shelton. 2000b. Resistance to Bt toxins. Science 287: 42.

Tabashnik, B. E., Y. B. Liu, T. J. Dennehy, M. A. Sims, M. S. Sisterson, R. W. Biggs, and Y. Carrière. 2002a. Inheritance of resistance to Bt toxin Cry1Ac in a field-derived strain of pink bollworm (Lepidoptera: Gelechiidae). J. Econ. Entomol. 95: 1018–1026.

Tabashnik, B. E., T. J. Dennehy, M. A. Sims, K. Larkin, G. P. Head, W. J. Moar, and Y. Carrière. 2002b. Control of resistant pink bollworm by transgenic cotton with *Bacillus thuringiensis* toxin Cry2Ab. Appl. Environ. Microbiol. 68: 3790–3794.

Tabashnik, B. E., Y. Carrière, T. J. Dennehy, S. Morin, M. Sisterson, R. T. Roush, A. M. Shelton, and J.-Z. Zhao. 2003. Insect resistance to transgenic Bt crops: lessons from the laboratory and field. J. Econ. Entomol. 96: 1031–1038.

Tabashnik, B. E., F. Gould, and Y. Carrière. 2004. Delaying evolution of insect resistance to transgenic crops by decreasing dominance and heritability. J. Evol. Biol. 17: 904–912.

Tabashnik, B. E., T. J. Dennehy, and Y. Carrière. 2005a. Delayed resistance to transgenic cotton in pink bollworm. Proc. Natl. Acad. Sci. USA 102: 15389–15393.

Tabashnik, B. E., R. W. Biggs, D. M. Higginson, S. Henderson, D. C. Unnithan, G. C. Unnithan, C. Ellers-Kirk, M. S. Sisterson, T. J. Dennehy, Y. Carrière, and S. Morin. 2005b. Association between resistance to Bt cotton and cadherin genotype in pink bollworm. J. Econ. Entomol. 98: 635–644.

Tabashnik, B.E., J.A. Fabrick, S. Henderson, R.W. Biggs, C.A. Yafuso, M.E. Nyboer, N.M. Manhardt, L.A. Coughlin, J. Sollome, Y. Carrière, T. J. Dennehy, and S. Morin. 2006. DNA screening reveals pink bollworm resistance to Bt cotton remains rare after a decade of exposure. J. Econ. Entomol. 99: 1525–1530.

Tang, J.D., S. Gilboa, R.T. Roush, and A.M. Shelton. 1997. Inheritance, stability, and lack-of-fitness costs of field-selected resistance to *Bacillus thuringiensis* in diamondback moth (Lepidoptera: Plutellidae) from Florida. J. Econ. Entomol. 90: 732–741.

Tang, J.D., H.L. Collins, R.T. Roush, T.D. Metz, E.D. Earle, and A.M. Shelton. 1999. Survival, weight gain, and oviposition of resistant and susceptible *Plutella xylostella* (L.) on broccoli expressing Cry1Ac toxin of *Bacillus thuringiensis*. J. Econ. Entomol. 92: 47–55.

Tang, J.D., H.L. Collins, T.D. Metz, E.D. Earle, J.Z. Zhao, R.T. Roush, and A.M. Shelton. 2001. Greenhouse tests on resistance management of Bt transgenic plants using refuge strategies. J. Econ. Entomol. 94: 240–247.

US EPA (U.S. Environmental Protection Agency). 2001. Biopesticides registration action document: *Bacillus thuringiensis* plant-incorporated protectants. Available at www.epa.gov/pesticides/biopesticides/pips/bt_brad.htm.

US EPA. 2002. Biopesticides registration action document – *Bacillus thuringiensis* Cry2Ab2 protein and its genetic material necessary for its production in cotton. Available at www.epa.gov/pesticides/biopesticides/ingredients/factsheets/factsheet_006487.htm.

Wu, K., Y. Guo, and S. Gao. 2002a. Evaluation of the natural refuge function for *Helicoverpa armigera* within *Bacillus thuringiensis* transgenic cotton growing areas in North China. J. Econ. Entomol. 95: 832–837.

Wu, K., Y. Guo, N. Lv, J.T. Greenplate, and R. Deaton. 2002b. Resistance monitoring of *Helicoverpa armigera* to *Bacillus thuringiensis* insecticidal protein in China J. Econ. Entomol. 95: 826–831.

Xu, X., L. Yu, and Y. Wu. 2005. Disruption of a cadherin gene associated with resistance to Cry1Ac -endotoxin of *Bacillus thuringiensis* in *Helicoverpa armigera*. Appl. Environ. Microbiol. 71: 948–954.

Zhang, X., M. Candas, N.B. Griko, L. Rose-Young, and L.A. Bulla Jr. 2005. Cytotoxicity of *Bacillus thuringiensis* Cry1Ab toxin depends on specific binding of the toxin to the cadherin receptor BT-R1 expressed in insect cells. Cell Death Differ. 12: 1407–1416.

Zhao, J.Z. , H.L. Collins, J.D. Tang, J. Cao, E.D. Earle, R.T. Roush, S. Herrero, B. Escriche, J. Ferré, and A.M. Shelton. 2000. Development and characterization of diamondback moth resistance to transgenic broccoli expressing high levels of Cry1C. Appl. Environ. Microbiol. 66: 3784–3789.

Zhao, J.-Z., J. Cao, H.L. Collins, S.L. Bates, R.T. Roush, E.D. Earle, and A.M. Shelton. 2005. Concurrent use of transgenic plants expressing a single and two *Bacillus thuringiensis* genes speeds insect adaptation to pyramided plants. Proc. Natl. Acad. Sci. USA 102: 8426–8430.

TWENTY

Exotic Plants and Enemy Resistance

JOHN L. MARON AND MONTSERRAT VILÀ

The increasing movement of organisms to new regions by humans is enabling species to breach natural dispersal barriers that normally constrain their geographic distribution. Oddly enough, despite being introduced to areas that may be very different from their home region, some exotics become spectacularly more successful in evolutionarily novel environments than in areas in which they evolved. How some exotics come to dominate these new habitats, despite being often inconspicuous members of their native community, is one of ecology's central mysteries. Unraveling this mystery involves understanding how introduced organisms faced with novel abiotic or biotic conditions make accommodations to their new environments. In this chapter, we consider one dimension of this accommodation process: how exotic plants respond to the altered assemblage of natural enemies they face within their introduced ranges.

Where they are native, plants are attacked by a diverse group of pests, including both specialists and generalists. Together, these enemies can impose diffuse, conflicting, or shifting selection pressures on plants (Hare and Futuyma 1978; Fox 1988; Marquis 1990; Rausher 1992; Pilson 1996; Juenger and Bergelson 1998; but see Maddox and Root 1990). In contrast, exotics face a greatly simplified natural-enemy landscape. Human-mediated transport to new areas instantaneously liberates exotics from their coevolved specialist fauna (Mitchell and Power 2003; Blair and Wolfe 2004; Hinz and Schwarzlaender 2004; Vilà et al. 2005). Indeed, this escape from specialist natural enemies has both been a leading hypothesis for exotic success and a founding principle on which biological control is based (Williams 1954; Maron and Vilà 2001; Keane and Crawley 2002). In addition to escaping from specialists, exotics that lack taxonomic relatives within recipient communities and therefore possess defenses that are chemically unique to the community they invade may also escape from generalists. Host switching by specialists from natives to exotics is also less likely for exotics with no close taxonomic relatives in the recipient community.

Although many exotics escape from consumers, some may actually face greater selection pressure by enemies in their introduced than in their native ranges. The most powerful example of this flip side of enemy escape is weed biological control. When a biocontrol insect is introduced, almost instantaneously, plants that have for decades grown free from specialists can suddenly face devastating attack by these herbivores. Since biocontrol insects are often freed from trophic control themselves, they can build to high numbers and potentially impose greater selection on exotic target plants than they might on their native hosts. Moreover, since some exotics represent a new and underutilized food source for generalists, they may actually accumulate generalist pests in their introduced range (Strong 1974; Auerbach and Simberloff 1988; Jobin et al. 1996; Memmott et al. 2000; Graves and Shapiro 2003; Carroll et al. 2005). This may particularly be the case if exotics have reduced genetic diversity and hence reduced genetic diversity of polymorphic defenses against generalists (Colautti et al. 2004).

Historically, ecologists have emphasized phenotypic plasticity as a primary mechanism by which exotic plants coped with novel abiotic or biotic circumstances in their introduced range (Baker 1974; Wu and Jain 1978; Rice and Mack 1991; Williams et al. 1995). High levels of adaptive plasticity (which might be present in particular species prior to their introduction) have often been often posited as a predictor of invasiveness (Baker 1974; Rejmanek and Richardson 1996; Mal and Lovett-Doust 2005) and important in allowing exotics to cope with a range of heterogeneous environments. Genetic impoverishment from founder effects (Baker 1974; Morgan and Marshall 1978; Barrett and Richardson 1986), the perceived long time span over which it took evolution to operate, and the relatively short invasion history of many exotics have appeared to make evolution an unlikely mechanism that could account for rapid phenotypic adaptation to new conditions. In terms of response to natural enemies, it was assumed that nonnatives that faced reduced selection by enemies should lower levels of defense in a plastic manner, enabling reallocation of resources to growth (Bazzaz et al. 1987). Recently, however, these assumptions have met with reconsideration.

Rapid Evolution of Exotics in Response to Enemy Pressure

Recently, evolution has been considered more seriously within the ecological context of invasions (Huey et al. 2000; Bone and Fares 2001; Lee 2002; Rice and Emory 2003; Stockwell et al. 2003). A growing number of studies show that organisms faced with novel abiotic or biotic conditions can rapidly evolve adaptations to these new conditions (Reznick et al. 1997; Huey et al. 2000; Grant and Grant 2002; Maron et al. 2004a). Indeed, some of the best evidence of rapid evolutionary change has come from exotic plants (Reznick and Ghalambor 2001). Although the greater attention paid to evolutionary processes in invasion biology by ecologists has been fairly recent (Hänfling and Kollman 2002; Lee 2002; Rice and Emory 2003), it is important to note that evolutionary biologists were long ago interested in how exotics adapted to new areas (Baker and Stebbins 1965; Brown and Marshall 1981). Classic studies in the 1960s and 1970s focused on the evolutionary potential of weeds and demonstrated that some exotic populations possessed surprising amounts of genetic variation (Clegg and Allard 1972; Jain and Martins 1979; Brown and Marshall 1981). Other work demonstrated that weeds were capable of undergoing genetically based adaptation to conditions within their introduced ranges (Baker and Stebbens 1965; Baker 1974), as evidenced by the formation of locally adapted races or ecotypes (Hodgson 1964). Only recently, however, have both the ecological and evolutionary aspects of invasions seen a greater coupling (Webber and Schmid 1998; Neuffer and Hurka 1999; Hänfling and Kollman 2002; Maron et al. 2004a; Phillips and Shine 2004; Callaway et al. 2005a, 2005b). This likely reflects a broad trend for greater fusion between ecologists and evolutionary biologists in considering the evolutionary dimension of species interactions.

Here we focus on the specific case of how exotic plants may evolve in response to an altered enemy landscape in the introduced range. We discuss some current hypotheses that make predictions about how enemy defense should evolve in exotic plants and review the empirical tests of these hypotheses. We then summarize results from our own work, where we have compared various aspects of enemy defense in native European and exotic North American genotypes of the short-lived perennial St. John's wort *(Hypericum perforatum).*

Exotics as Substrates for Studying the Evolutionary Response of Plants to Natural Enemies

Since entire assemblages of consumers are often eliminated when plants colonize new regions, species introductions serve as large biogeographical experiments that can allow inferences about how changes in enemy pressure may influence the evolution of plant defense. In native plant-consumer systems, understanding this issue has been challenging. Native plants are usually attacked by tens if not hundreds of herbivores. Surgically removing entire groups of species to examine how they affect the evolution of a particular resistance trait is logistically daunting. Ironically, where this approach has been performed most successfully, it has been on exotic species that have naturalized to their recipient community, rather than on native plants (e.g., North American *Pastinaca sativa* [Berenbaum et al. 1986], North American *Ipomoea purpurea* [Simms and Rausher 1989], North American *Arapidopsis thaliana* [Maurico and Rausher 1997]). Moreover, many native plant–native herbivore systems may be at evolutionary equilibrium, making assessments of directional selection more difficult. However, when plants are introduced to new regions, the identity of consumers, the intensity of their attack, and therefore their overall selective effects often change in predictable ways. For example, exotics are often liberated from their entire suite of specialist enemies in the native range. They may acquire new generalist enemies with different selective effects than those encountered in the native range. Furthermore, for many exotics, since the approximate date of introduction is known, one can conservatively determine the time span over which evolution may have occurred.

By placing native and introduced genotypes in common gardens in both their introduced and native ranges, it is possible to explore how changes in consumer pressure may have influenced exotic plant defense. In common gardens in the native range, one can determine whether exotic plants that are brought home have lost resistance to their native enemies. In the introduced range, by placing biocontrol agents on native and exotic genotypes, one can ask whether exotic plants have lost or gained resistance to specialist herbivores.

Hypotheses and Evidence for Plant Defenses

Recent theories make several predictions about how plant resistance traits should evolve in exotic plants. The "evolution of increased competitive ability" hypothesis predicts that exotic plants that are liberated from specialists should rapidly evolve reduced resistance to this group of consumers (Blossey and Nötzold 1995). This prediction is based on the assumption that plant defenses against specialists are energetically costly, and therefore there is a selective advantage to saving these costs. Although there is certainly some evidence that constitutive defenses might incur allocation costs (Bergelson and Purrington 1996), increasing evidence suggests that costs might be more ecological than energetic (Koricheva 2002). For example plant defense traits can negatively influence plant attractiveness to pollinators (Strauss et al. 1999). Furthermore, allocation costs are expected to be highest for plants growing in low-resource environments. In contrast, successful exotics may be successful because they fill a novel functional role in recipient communities that enables them to tap into "free" or unused resources (Holmes and Rice 1996; Dyer and Rice 1999; Shea and Chesson 2002; Fargione et al. 2003). If this is so, costs of maintaining existing defenses may be minimal.

In addition to losing resistance to specialists, it has also been posited that exotics might rapidly acquire enhanced resistance to generalists (Müller-Schärer et al. 2004; Joshi and Vrieling 2005). For natives, levels of chemical defense are thought to represent a trade-off between two opposing forces (van der Meijden 1996). On one hand, specialists are thought to select for lower levels of qualitative (toxic) defenses because these compounds are often used as host-finding or oviposition cues or even as feeding stimulants by adapted specialists (Rees 1969). Generalists, on the other hand, may select for higher levels of these compounds as greater concentrations provide increased resistance to these consumers (van der Meijden 1996). Thus, if specialists that limit directional selection for increased resistance against generalists are lost in the introduced range, and if exotics accumulate generalists where they are introduced, then selection should favor greater defenses against generalists in recipient communities compared to where they are native (van der Meijden 1996; Müller-Schärer et al. 2004; Joshi and Vrieling 2005).

Evolution in Exotics in Response to Biocontrol

What might be the evolutionary response of exotics to attack by biocontrol agents? As the above arguments suggest, prior to the introduction of biocontrol agents, biocontrol targets may have lost defenses to specialists and therefore be more vulnerable to these agents than they would be at home (Burdon et al. 1981; Müller-Schärer and Steinger 2004). Alternatively, the initial increased vulnerability of some genotypes to biocontrol agents might render populations initially highly variable in their resistance to biocontrol agents (Burdon et al. 1981; Garcia-Rossi et al. 2003). After the introduction of biocontrols, however, one might expect plants to rapidly evolve increased resistance to biocontrol agents. This might occur because selection imposed by biocontrol can be intense. That is, in cases where biocontrol is successful and sustained it is often because biocontrol agents kill plants, just as with chemical control of weeds. It is well known that agricultural weeds have rapidly evolved resistance to a variety of herbicides (Georghiou 1986; Powles and Holtum 1994; Cavan et al. 1998). Native plants in recipient communities may also rapidly evolve resistance to allelopathic chemicals produced by invaders (Callaway et al. 2005a).

Yet, although weeds have great potential to quickly evolve resistance to their biocontrol agents, several factors may hinder this process. For example, if there are negative genetic correlations between resistance and other ecologically important traits, this can retard directional selection toward increased resistance. This was observed by Henter and Via (1995) in their study of pea aphid–parasitoid interactions. They found that despite ample genetic variation for resistance of pea aphids to their common parasitoid, resistance did not increase after several generations of selection. In plant-pathogen systems, disease-resistant genotypes can actually *decrease* in the presence of a pathogen (Parker 1991; Burdon and Thompson 1995), also presumably because of negative genetic correlations. Recent work by Etterson and Shaw (2001) showed that adaptation to climate change by the prairie plant *Chamaecrista fasciculata* was constrained due to among-trait genetic correlations that were antagonistic to the direction of selection.

As well, population limitation produced by biocontrol does not necessarily imply strong selection. Theory predicts that spatial heterogeneity, which provides refuges for plants under attack, can decouple strong population limitation from strong selection (Rohani et al. 1994; Alstad and Andow 1995). Gene flow from populations not under control can erode selection in populations that are under attack. Temporal heterogeneity in selection has been shown to have similar effects. Empirical work has shown that boom-bust dynamics can characterize interactions between biocontrol agents and weeds (McEvoy et al. 1993). In theoretical models, these volatile population dynamics can weaken selective responses by weeds to biocontrol agents (Holt and Hochberg 1997). Holt and Hochberg (1997) argue that heterogeneity in selection may explain why there is limited evidence for the breakdown of evolutionary stability in the case of biological control and yet so many examples of rapid evolution of resistance to chemical control, where control is more geographically uniform. Alternatively, the lack of evidence for evolutionary instability in biological control programs may simply reflect a paucity of adequate research (Kraaijeveld et al. 1998). Since studies are rarely undertaken to understand how the interaction between biocontrol agents and their host plants is played out (Simberloff and Stiling 1996), whether weeds rapidly evolve resistance to biocontrol agents remains a mystery.

Despite the potential for biocontrol targets to rapidly evolve resistance to biocontrol agents, biocontrol theory makes the implicit assumption that interactions between biocontrol agents and target plants are evolutionarily stable (Huffaker et al. 1971). Any evolutionary change on the part of biocontrol targets and controlling agents are assumed to be counterbalancing (i.e., a coevolutionary arms race). What little discussion there has been regarding evolutionary change in target weeds usually concerns how hybridization and introgression among weeds might alter compatibility between hosts and potential control agents (Ehler 1998; but see Pimentel 1986). As Newman et al. (1996) wrote, "because most weed biological control agents are specialists and thought to be adapted to overcome their host's defensive systems, the role of resistance has rarely been considered in weed biological control" (p. 382).

Experimental Evidence for Evolution of Exotic Plant Defense

Enemy resistance has now been compared among native and exotic genotypes for 14 plant species in 18 separate studies (Table 20.1). In these studies, plant resistance to herbivory is defined as either effects of plants on herbivore performance (growth, survival) or the amount of herbivore damage imposed on plants. Eleven studies explored whether exotic genotypes have lost resistance to specialist herbivores; 12 studies compared native and introduced conspecifics for resistance to generalist pests. Of those studies that quantified resistance to generalists, all but the study by Stastny et al. (2005) did so by comparing herbivore performance (larval growth or development time) or herbivore damage to plants in laboratory feeding trials. In contrast, over half of the studies on resistance to specialists occurred in the field, in common gardens. Four of the 12 studies that examined exotic plant resistance to generalists included both common garden and laboratory bioassay studies, while none of the studies on specialists included both laboratory and field tests.

Of the specialist herbivores that have been tested, 7 out of 11 (64%) showed greater performance or higher levels of damage on exotic versus native genotypes. These results, while still limited, suggest that exotics may more commonly lose resistance to specialist herbivores than their native counterparts. In contrast, of the 12 studies on generalists, only 2 (Siemann and Rogers 2003; Maron et al. 2004b) found that exotics had lost resistance to this group of pests. Two studies (Joshi and Vrieling 2005; Leger and Forister 2005) found that exotics actually evolved greater resistance to generalists, as predicted by Müller-Schärer et al. (2004). The majority of studies, however, found no evidence that exotics have altered defense against generalists.

Few studies have examined whether weeds that are biocontrol targets evolve resistance to their agents. The best example of this phenomenon comes from the classic study by Burden et al. (1981). They found that a less-resistant form of skeleton weed *(Chondrilla juncea)* to a rust *(Puccinia chondrillina)* was controlled but replaced through time by two more resistant forms (forms B and C) (Burdon et al. 1981; Cullen and Groves 1981). More recently, Garcia-Rossi et al. (2003) examined differences in herbivore resistance between populations of *Spartina alterniflora* that had never been separated from this species' specialist herbivore, the planthopper *Prokelisia marginata*, and those that had been separated for over 100 years. They found that *Spartina* that had been liberated from its planthopper herbivore was much less resistant and more variable in its resistance than were plants from populations that had always been exposed to *P. marginata*. Garcia-Rossi et al. (2003) predict that biocontrol should rapidly eliminate the less-resistant genotypes, leaving populations of plants with high resistance to *P. marginata*. Finally, the work by Berenbaum and Zangerl is relevant to the issue of whether plants rapidly evolve resistance to specialists once they are introduced. Over the last 125 years, introduced wild parsnip *(Pastinaca sativa)* has evolved increased production of the secondary defensive compound sphondin in response to the introduction (in 1883) of its specialist herbivore, the parsnip webworm, *Depressaria pastinacella* (Berenbaum and Zangerl 1998; Zangerl and Berenbaum 2003).

Response of St. John's Wort to an Altered Enemy Landscape

Background and History

St. John's wort has several attributes that make it an attractive plant to explore how the addition or subtraction of enemies in the introduced range influences the evolution of enemy defense. First, the introduction history of St. John's wort in North America is well documented. *Hypericum perforatum* was first introduced into the eastern United States in 1793 (first reported in Lancaster, Pennsylvania); it was first found in Oregon between 1840 and 1850 and in California by 1900 (Campbell and Delfosse 1984). In the West, plants quickly became established in overgrazed rangelands, spread rapidly, and grew in dense monocultures. Livestock that ate portions of the plant became sick; thus, successful invasion took rangeland out of production. Second, different populations of St. John's wort have experienced divergent histories of herbivore pressure in North America. In western North America, in 1945, Huffaker spearheaded the introduction of a chrysomelid beetle, *Chrysolina quadrigemina* (Coleoptera: Chrysomelidae), to control *H. perforatum* (Huffaker and Holloway 1949; Holloway and Huffaker 1951). This was the first introduction of a biocontrol insect to control an exotic plant in North America. The results were stunning. *Chrysolina quadrigemina* populations established quickly and grew rapidly. Within five years of their introduction, biocontrol beetles had markedly reduced *H. perforatum* populations. The most dramatic effect was in California, where *H. perforatum* was reduced to less than 1% of its former range (McCaffrey et al. 1995; Ritcher 1996).

Although St. John's wort has been present for centuries in central and eastern North America (Sampson and Parker

TABLE 20.1

Outcome of Studies That Have Compared Resistance to Generalist and/or Specialist Herbivores and Pathogens among Native and Introduced Populations

Plant Studied	Resistance to Generalists	Resistance to Specialists	Experimental Approach	Number of Populations	Authors
Lythrum salicaria	ND	–/ND	FB	1/1	Blossey and Nötzold 1995
		ND	FB	6/6	Willis et al. 1999
Spartina alternaflora		–	FB	5/6[a]	Daehler and Strong 1997
Sapium sebiferum	–		FB	Pooled[b]	Siemann and Rogers 2003
			FB	Pooled[b]	Lankau et al. 2004
Silene latifolia		–	CG-introduced range	20/20	Blair and Wolfe 2004
		–	CG-native range	20/20	Wolfe et al. 2004
Alliaria petiolata	ND	–	FB	8/7	Bossdorf et al. 2004
Hypericum perforatum	–	ND	CG-native range	15/15	Maron et al. 2004b
			CG-introduced range	18/30	
Barbarea vulgaris	ND		CG-native range + FB	3/3	Buschmann et al. 2005
Bunias vulgaris	ND		CG-native range + FB	3/3	Buschmann et al. 2005
Cardaria draba	ND		CG-native range + FB	3/3	Buschmann et al. 2005
Rorippa austriaca	ND		CG-native range + FB	3/3	Buschmann et al. 2005
Senecio jacobaea	+	–	CG-native range	15/16	Joshi and Vrieling 2005
		–	CG-introduced range	4/4	Stastny et al. 2005
Eschscholzia californica	+		FB	4/7	Leger and Forister 2005
Solidago gigantean		–	CG-native range	10/10	Meyer et al. 2005
Ambrosia artemisiifolia	ND	ND[c]	CG-native range CG-introduced range	2/1	Genton et al. 2005

NOTE: –, within a common environment, exotic genotypes have lower resistance than do native conspecifics; +, exotic genotypes have greater resistance than do natives; ND, no difference in resistance; CG, common garden; FB, feeding bioassay; blank cells reflect cases where data were not collected; numbers separated by slash indicate number of native/number of introduced populations sampled.

[a]Clones rather than populations sampled.

[b]Seeds pooled from collections made across one (Lankau et al. 2004) or two (Siemann and Rogers 2003) provinces in China (native range) and from the Huston metropolitan area (introduced range) in the United States.

[c]Genton et al. (2005) do not report whether the herbivores that attacked plants in the native range were generalists or specialists.

1930; Voss 1985), until recently plants have never been exposed to biocontrol. Only a small number of *C. quadrigemina* were released into eastern Ontario in 1969. These insects have since spread in Ontario and to Minnesota, but populations remain at low density and their impacts on St. John's wort populations appear minimal (Harris and Maw 1984; Fields et al. 1988; Hoebeke 1993; Julien and Griffiths 1998).

Control of St. John's wort in portions of the West was successful because beetles killed their host plants. Larvae feeding on the new leaves and stems that are produced in winter and spring have the greatest impact on plant survival. Although adult beetles can completely defoliate plants and destroy flowers and seeds, it is usually larval feeding that kills plants (Holloway et al. 1957).

Interestingly, plants growing in the shade gained a refuge from biocontrol, because the beetles performed more poorly in the shade. Biocontrol in other regions in the West (as well as Australia and other countries) has not always been as universally successful as in California. Poor control is due to several factors, including reduced efficacy of *C. quadrigemina* in colder climates, the presence of more extensive shaded habitat, and summer rain, which allows plants with a less extensive root system to survive. Summer rain also appears important because it enables *H. perforatum* to recover from defoliation (Huffaker 1957; Harris and Maw 1984; Williams 1984).

Additional biocontrol agents have been introduced subsequent to the initial establishment of *C. quadrigemina*. These have included a buprestid root-boring beetle, *Agrilus hyperici*, a gall-forming midge, *Zeuxidiplosis giardi*, and a second chrysomelid beetle, *Chrysolina hyperici*. In Canada, the aphid *Aphis chloris* has also been introduced. These agents add to biocontrol effectiveness in areas where performance by *C. quadrigemina* has been poor (i.e., cool, wet areas). However, since *C. quadrigemina* has been the most common and also the most effective biocontrol agent throughout most of the West, we focus on whether plants have evolved resistance to this herbivore in particular.

Another attribute of St. John's wort that makes it an attractive plant to study the evolution of defense is that the secondary chemistry is reasonably well understood. Individuals produce several toxic defensive compounds (i.e., qualitative defenses), chief of which are hypericin and pseudohypericin. These polycyclic naphthodianthrones (Trifunovic et al. 1998) are produced by glands located along the outer edge of leaves. These powerful photo-oxidants are light activated and cause cell damage. Like many allelochemicals, these compounds appear to have different effects on adapted and nonadapted enemies (Feeny 1992). Hypericins have been shown to be effective at deterring generalist insect herbivores (Arnason et al. 1983, 1992; Fields et al. 1990; Mitch 1994; Sirvent et al. 2003), they can be toxic to large grazing mammals (Giese 1980), and they have potent antimicrobial properties that provide defense against generalist pathogens (Arnason et al. 1983). While generalist insect herbivores avoid consuming portions of leaves containing glands that sequester hypericin (Guillet 1997), specialists use hypericin as a host-finding cue (Rees 1969). Guillet et al. (2000) found that a specialist noctuid (Lepidoptera) caterpillar had higher rates of ingestion when fed plant tissue high in hypericin compared to when fed tissue low in hypericin.

Experimental Approach

We have compared levels of genetically based enemy resistance between introduced and native genotypes of St. John's wort by growing plants in common gardens. Common garden studies enable one to control for environmental effects on phenotypes and examine the degree to which phenotypic variation may be genetically based (assuming no substantial maternal effects are present). By placing the progeny of plants from native European populations and exotic western and central North American populations into common gardens in North America (Washington and California) we have examined (1) if plants from central North America that have not been exposed to biocontrol for the last century have lost resistance to the specialist biocontrol beetle, *C. quadrigemina,* and (2) whether plants from western North America that have been exposed to this agent since the mid- to late 1940s have rapidly evolved resistance to *C. quadrigemina*. By growing exotic and native genotypes in common gardens in Europe (Spain), we explored (1) whether exotics have lost resistance to generalist soil pathogens from the native range, and (2) whether exotics have altered resistance to a native specialist aphid *(A. chloris)* that has been introduced into North America and Australia as a biocontrol agent. In addition, by examining the defensive chemistry of plants in multiple common gardens, we quantified levels of constitutive defenses in plants across gardens, asking whether exotic and native genotypes consistently differ in their levels of qualitative defenses against generalists.

COMMON GARDEN EXPERIMENTS IN THE INTRODUCED RANGE

We collected seed capsules of *H. perforatum* from 18 populations across Europe, 18 populations from western North America, and 14 populations across central North America (seed collection methods are outlined by Maron et al. [2004a]). We established common gardens in Snohomish, Washington, in May 2000 and in Pope Valley, California, in March 2001. The Washington garden contained plants from all 50 populations, whereas (due to space and logistical considerations) the California garden contained plants from only 36 populations. In each garden we created 10 (Washington) or 9 (California) experimental blocks, with each block containing six plots. Plants in half of these plots were exposed to herbivory by *C. quadrigemina*; and plants in the remaining plots were protected from herbivory. Control plants were kept free of larval beetles by spraying individuals once in spring with an insecticide (Isotox). Any adult beetles found on control plants were regularly removed by hand throughout summer.

Within each treatment type, plots within a block contained plants from one of the three regions. One individual from each population within a region was represented in each plot. Different blocks contained unique individuals collected from different maternal plants in each source population. Plants from the same region that were in "exposed" or "protected" herbivory treatments within the same block were maternal sibs and likely clones since St. John's wort produces over 90% of its seeds apomicticly (Arnholdt-Schmitt 2000; Mayo and Langridge 2003). Plant, plot, and block spacings are given elsewhere (Maron et al. 2004a).

We quantified variation in resistance to *C. quadrigemina* in two ways. First, we counted the number of beetles that accumulated on plants in the year after introduction. Since beetles were free to move between plants in common gardens, the cumulative number of beetles that accumulated on particular genotypes in control plots across a season provided one metric of resistance (Maddox and Root 1987). We censused beetles on plants in Washington in 2001 and 2002. In 2002, the beetle population crashed and beetle numbers were extremely low. Second, we determined the effects of biocontrol beetles on plant fecundity by comparing seed production of plants exposed and protected from herbivory. We estimated plant fecundity by harvesting, drying, and weighing seed capsules at the end of summer and then converting capsule weight to capsule number based on regressions of relationships between these two variables.

COMMON GARDEN EXPERIMENTS IN THE NATIVE RANGE

In an additional common garden, in an old field at the Universitat Autònoma de Barcelona campus field station in Bellaterra, Spain, we determined whether North American and European genotypes differed in their resistance to several native enemies that naturally colonized plants. We assessed resistance to three generalist soil pathogens: *Colletotrichum* sp. (Coelomycetes), *Alternaria* sp. (Hyphomycetes), and *Fusarium oxysporum* (Hyphomycetes). These generalist soil fungal pathogens are dispersed by water (Andrés et al. 1989) and cause necrosis (or anthracnose, in the case of *Colletotricum*). We also assessed resistance to the specialist aphid *A. chloris*. In this garden we quantified resistance to these pests in two ways. For soil pathogens, on control plants we compared how the number of plants attacked by pathogens differed among European and North American populations. We also compared the survival of plants from both regions that were either exposed or protected from pathogen attack. For aphids, on control plants we determined how the number of aphid colonies per plant differed among plants from Europe and North America.

In 2002, we planted seedlings from 30 source populations into this garden, 15 each from Europe and North America. Due to space limitations we omitted central North American populations. Plots contained different individuals from each western North American or European source population. Half of the plots were sprayed with fungicide to suppress native fungal pathogens; plants in the remaining plots were exposed to natural levels of pathogen attack (see Maron [2004b] for a full description of the plot layout). In midsummer of 2002 and 2003, when aphid numbers were at their peak, we censused the number of aphids on plants by counting the number of aphid clumps per plant. This metric is a good estimator of aphid load per plant. During their first summer in Spain, experimental plants remained pathogen free. However, this changed in 2003, and starting in January of this year, we censused plants every two weeks, noting pathogen infection.

DEFENSIVE CHEMISTRY

To determine how defensive chemistry differed between native and introduced genotypes, we sampled leaf tissue of plants that we grew in Washington, and in a second common garden in Spain (at the Mas Badia Experimental Field Station near Girona, Spain (latitude 42° 19′). Sampling and methods for chemical analyses are detailed elsewhere (Maron et al. 2004b).

Results

RESISTANCE TO SPECIALIST BIOCONTROL AGENTS

We found no consistent evidence that St. John's wort from introduced populations had either lost or gained resistance to the specialist biocontrol agent, *C. quadrigemina*. In the common garden in California, in the year following their introduction into the garden, there were no significant differences in the number of adult beetles on plants from the native versus introduced region (Fig. 20.1A) (repeated measures nested ANOVA, $P = 0.44$). Instead, larger plants had more beetles (plant volume and beetle numbers log transformed; $R^2 = 0.25$, $P < 0.0001$). Although beetle addition significantly reduced seed capsule production (Fig. 20.1B) (ANOVA, $P < 0.0001$), there was no difference in the negative effects of beetle herbivory among plants from the three regions (i.e., no significant region by treatment interaction, ANOVA, $P = 0.15$).

In Washington in 2001, again one year after beetles had been introduced to the common garden, we found significantly more beetles on western North American genotypes than on European or central North American plants (Fig. 20.1A) (repeated measures nested ANOVA, $P < 0.03$). However, since in 2001 plants from western North America were larger than plants from the other two regions (Maron et al. 2004a) it is likely that, as in California, larger plants attracted more beetles. In both common gardens, plants from different populations also differed significantly in the number of beetles found on them (repeated measures nested ANOVA, $P < 0.008$ and $P < 0.003$ for California and Washington, respectively).

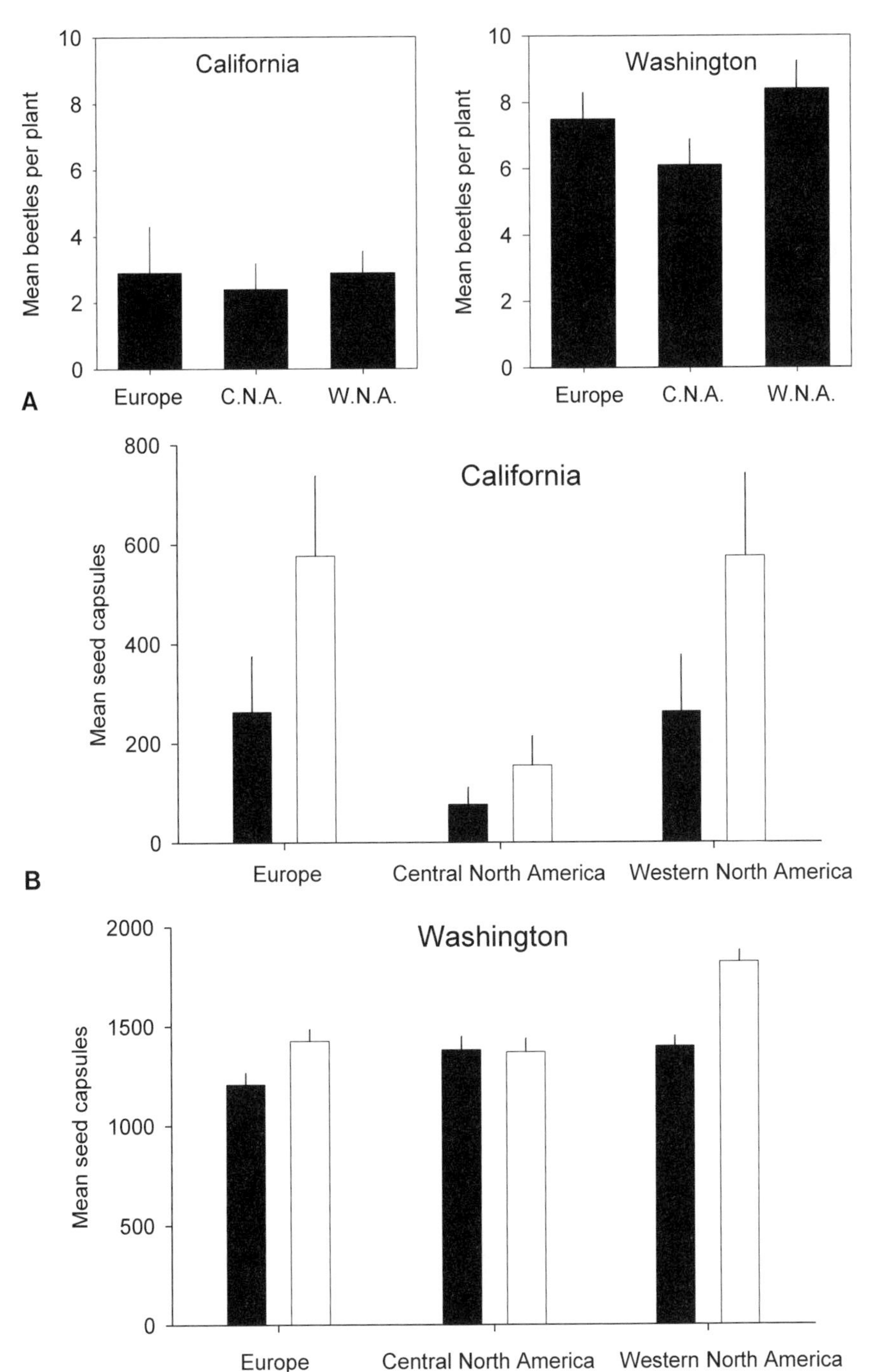

FIGURE 20.1. A. Mean (+SEM) number of biocontrol beetles (*Chrysolina quadrigemina*) found on plants from different regions (C.N.A., Central North America; W.N.A., Western North America) in common gardens in California in 2002 (*upper left*) and in Washington in 2001 (*upper right*). Values are from the census in which beetles were at their seasonal peak. B. Mean effect of beetle (*Chrysolina quadrigemina*) herbivory on seed capsule production in California and Washington. Black bars, plants exposed to bioncontrol beetles; clear bars, plants protected from herbivory. Regional means are the average of population means in each region.

In Washington, herbivory by beetles reduced seed capsule production; overall, control plants produced significantly more seed capsules than those exposed to biocontrol (Fig. 20.1B) (ANOVA, $P < 0.0001$). The impact of beetle herbivory on per capita seed production varied by region of plant origin (significant treatment by region interaction; ANOVA, $P < 0.004$); the magnitude reduction in seed capsule production scaled positively with beetle numbers (ANOVA, $P < 0.002$). Since plants from Europe and central North America supported fewer beetles than did plants from western North America, beetles had less impact on their per capita seed production than for genotypes from western North America. Herbivory on plants from European and central North American populations reduced their seed production by an average of 15% and 0.8%, respectively. In contrast, herbivory resulted in a 23% drop in the per capita seed production of plants from western North American populations. After statistically controlling for differences among plants in beetle numbers, western North American plants still suffered significantly greater reductions in seed production than did plants from the other two regions (ANCOVA, $P < 0.009$).

In 2002, the beetle population across the Washington garden was much reduced compared to 2001. On average, there were 1.7 (± SEM 0.26), 1.4 (± SEM 0.19), and 0.92

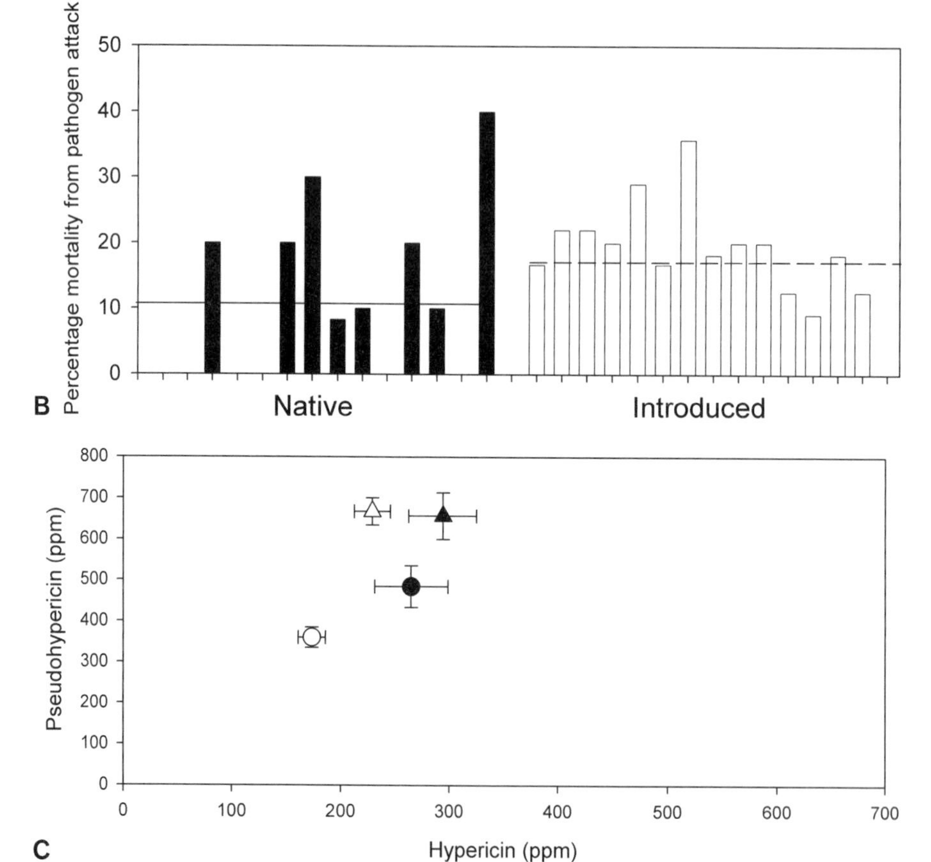

FIGURE 20.2. In common garden in Spain, percentage of individuals within native European (closed bars) and introduced North American (open bars) populations that were infected by pathogens (A) or died after infection (B). Horizontal lines are the mean of population means from European or North American populations. Lines that differ in style (i.e., solid versus dashed lines) indicate statistically significant differences between regions. (C) Relationship between mean of population means (±SEM) hypericin and mean of population means (±SEM) pseudohypericin among St. John's wort grown in Washington (circles) and Spain (triangles) common gardens. Open symbols, nonnative populations; closed symbols, native populations.

(± SEM 0.16) beetles per plant for plants from Europe, central North America, and western North America, respectively. In contrast to results from 2001, the number of adult beetles on plants from the three regions was not significantly different (ANOVA, $P = 0.16$).

In Spain, resistance to the specialist native aphid, *A. chloris*, also was not different between plants from Europe and North America. There was no significant difference in the number of aphid colonies that naturally colonized plants from Europe and North America (nested ANOVA, $P = 0.44$).

RESISTANCE TO GENERALIST PATHOGENS

In contrast to these varied results on resistance against specialists, we found strong support for St. John's wort from exotic populations having reduced resistance to generalist consumers. Introduced genotypes of St. John's wort from western North America that were grown in Europe had lower resistance to generalist soil pathogens than did native St. John's wort. Levels of pathogen attack (Fig. 20.2A), and mortality due to this attack (Fig. 20.2B) were both higher among plants from western North American populations compared to native European populations.

DEFENSIVE CHEMISTRY

In common gardens in North America and Europe, exotic North American genotypes produced significantly less hypericin compared to European genotypes (Fig. 20.2C). There was no difference in mean hypericin values between western and central North American genotypes (post hoc comparison, $P = 0.19$), despite their different presumed histories of herbivore pressure. Exotic genotypes in Washington (the only garden where it was measured) also produced, on average, 19% less hypericide than did European genotypes, a significant difference (Maron et al. 2004b). We found no significant differences in levels of pseudohypericin between native and exotic genotypes across gardens in Washington and Spain.

Explanation of Results from Work on St. John's Wort

Our results from work on St. John's wort provide an intriguing counterpoint to the studies we reviewed. We found that exotic St. John's wort populations (1) have reduced resistance to generalists (pathogens), (2) lower chemical defenses (which are likely most effective against generalists) compared to plants from native populations, but (3) no loss or gain in resistance to specialist insects. In contrast, among the studies we reviewed, there was more support for a loss of resistance in exotics to specialist enemies, and there was generally no difference among native and introduced plants in resistance to generalist enemies (Table 20.1).

Why might our results differ from those of other studies? One possibility is that we studied generalist pathogens, whereas other studies have investigated resistance to generalist herbivores. We think it more likely, however, that the answer may lie in how selection operates on St. John's wort defense in the native range. We speculate that generalist pathogens may be a potent source of selection on native St. John's wort. In natural populations in Spain, we have witnessed the wholesale die-off of dense St. John's wort populations from pathogen attack. Moreover, in a survey of 43 different populations across Western Europe, an average of 18% of the plants sampled showed evidence of attack by pathogens. Nearly one-third of sampled populations had at least one plant that was dead, apparently from pathogen assault (Vilà et al. 2005). While many specialist insects also attack *H. perforatum* in Europe (Wilson 1943), many of these species possess behavioral or biochemical adaptations to overcome the photoactivated defenses of *Hypericum* (Rees 1969; Arnason et al. 1992; Guillet et al. 2000). Thus, toxic defenses against generalist pathogens may be favored in Europe (sensu van der Meijden 1996). In North America, St. John's wort clearly escapes from its specialist herbivores (except where plants have been exposed to biocontrol). But because of its rather novel chemistry (we know of no abundant and widespread native *Hypericum* species that grow in identical habitats as *H. perforatum*) it may also escape from generalist herbivores. In fact, in the introduced range only 4% of the plants have signs of herbivore damage compared to 23% in the native European range (Vilà et al. 2005). As well, in our common gardens in North America, plants attracted few generalist herbivores. Given both the limited level of attack in North America by generalist enemies at large, and the fact that costs of pathogen resistance are often much higher than those of herbivore resistance (Bergelson and Purrington 1996), selection in North America may favor an overall decline in defense investment to generalist pathogens as opposed to specialist insects.

But why has St. John's wort from western North America not rapidly evolved resistance to specialist biocontrol agents? With regard to the biocontrol beetle *C. quadrigemina*, it may be that in western North America, although local populations are suppressed by this beetle, disturbed areas where populations have undergone local suppression may be recolonized by propagules from nearby areas where plants have escaped control. Since St. John's wort gains a refuge from beetle herbivory in the shade (Holloway 1957), plants escaping control in the shade could provide propagules that recolonize sunny areas. This process could decouple population regulation from an evolutionary response to strong selection (Holt and Hochberg 1997). It would be interesting to test whether there are differences in resistance between genotypes from western North America that have long grown in the shade with those that have grown in the sun. Alternatively, it may be that (by chance) populations we sampled have not faced a long history of repeated biocontrol. To our knowledge, western North America plants used in our common garden in Spain have never been exposed to the biocontrol agent *A. chloris*, so this might explain why we found no evidence that plants from western North America had greater resistance to this specialist.

Of course, the reasoning outlined above is speculative. What it suggests, however, is that a key to interpreting work in this area is an increased understanding of the relative magnitude of selective effects of generalists versus specialists in native and introduced populations. Future studies would benefit from a tighter coupling of observational data from souce populations (levels and identity of enemy attack) and common gardens (genetically based differences in defense between exotics and natives in progeny from source populations).

Among-Population Variation in Enemy Resistance

An important ancillary benefit of the growing interest in comparative studies of resistance among exotic and native genotypes is the focus on population-level comparisons. In contrast, research on the evolution of defense in native systems has almost exclusively examined the causes and consequences of variation in resistance within single populations, since individuals are the unit upon which selection acts. As such, our understanding of sources of variation in resistance across larger spatial scales is limited. Even for populations in close proximity we are mostly ignorant about what drives among-population variation in resistance.

A robust result shared among many of the studies we reviewed is that there are substantial levels of among-population variation in resistance, even among introduced populations. In our work, we found 3.6-fold differences in levels of hypericin among introduced populations, and 4.2-fold differences among native populations for plants grown in our Washington common garden. We have found similarly high levels of genetically based variation in other traits (seed production, plant size) among native and introduced populations as well (Maron et al. 2004a). For natives, if genetically based levels of resistance reflect past selective effects of consumers (Berenbaum et al. 1986; Marquis 1992), then these data strongly suggest that enemy pressure and selection on resistance traits vary dramatically, even among populations from within a small portion of the distribution of a species. This in turn provides interesting support for coevolutionary "hotspots" and "coldspots" (sensu Thompson 1999a, 1999b), which for chemical defense has been best documented in poisonous snake-newt interactions (Brodie and Brodie 1999; Geffeney et al. 2002). A future challenge for those interested in native plant–consumer dynamics will be to understand what the sources and consequences of this variation are for plant fitness, abundance, and dynamics. For those interested in the evolutionary trajectory of exotics, it suggests that the view of plants universally going from being well defended where they are native to poorly defended where they are introduced is overly simplistic.

Are Differences in Exotic and Native Phenotypes the Product of Evolution?

Throughout this chapter, we have implied that differences in resistance between exotics and natives in common environments suggest evidence for rapid evolutionary effects. It is important to stress, however, that this need not be the case. Several crucial pieces of evidence are needed before concluding that differences in phenotypes between natives and exotics are driven by evolution. First, one must rule out founder effects. If exotic populations are, by chance, founded by individuals with low resistance, differences between natives and exotics have nothing to do with evolution. Genetic data can be used to shed light on invasion history. For example, for St. John's wort we analyzed variation in neutral amplified fragment length polymorphism (AFLP) makers among introduced and native genotypes. These data showed substantial neutral genetic variation and suggested that St. John's wort has been introduced multiple times, from multiple source populations, into North America (Maron et al. 2004a).

Second, one should rule out maternal effects. Maternal effects occur when differences in maternal environment influence how seeds are provisioned, which then affects trait values of juvenile or adult plants. Thus, if the maternal environment is universally different between native and introduced populations, this could potentially produce a consistent difference in phenotype between natives and exotics, even when they are grown in a constant environment. One indirect approach to examine whether maternal effects might be important is to compare seed weight between native and introduced populations. We have done this for St. John's wort and found no significant differences in seed weights between populations from Europe, western North America, or central North America (J.L.M., unpublished data). Of course a more direct and powerful approach involves growing plants for two or more generations in a common environment, collecting seeds from these plants, and using them as the source of material for common garden experiments.

Third, since an assumption behind comparative studies of exotic and natives is that enemy defense can evolve in the introduced range, ideally one should measure the heritability of putative defensive traits to ensure that this assumption is correct. Where heritability of resistance traits has been assessed for native plants, it is usually the case that heritabilities are greater than zero, implying they can evolve (Maddox and Root 1987; McCrea and Abrahamson 1987; Fritz and Price 1988; Kennedy and Barbour 1992). Fourth, a sufficient number of populations should be examined so that levels of resistance across a reasonable portion of the native and introduced range can be determined (Bossdorf et al. 2005). Of the 19 comparisons made between exotic and native genotypes from the papers we reviewed, only an average of 8 populations were tested from the native range and 9 were tested from the native range. Since these tests assume that sampled populations represent the situation across an entire range, future studies should sample a greater number of populations. In our experience, adding populations often increases the among-region variance. Thus, somewhat counterintuitively, sampling only a few populations may actually increase the probability of finding differences between regions.

Two aspects of how exotics respond to altered enemy pressure in the introduced range have thus far received relatively scant attention. The first concerns tolerance. It is not clear how selection on tolerance might operate in the introduced range. The answer to this will of course depend on whether tolerance has significant costs (Agrawal et al. 1999), how tolerance and resistance trade off with each other, if at all (Strauss and Agrawal 1999), and the prevalence of generalist herbivores that graze on exotics in the introduced range (Müller-Schärer et al. 2004). Only two studies have explicitly compared tolerance between exotic and native genotypes, and neither of these studies found significant differences in tolerance (van Kleunen and Schmid 2003; Bossdorf et al. 2004). In our work on St. John's wort, we have similarly found no evidence for differences in tolerance to simulated herbivory between native and exotic plants (R.B. and J.L.M., unpublished data). The second issue concerns whether induced defenses differ between native and exotic genotypes. We know of no explicit comparison that has been made in levels of induced defenses between native and exotic conspecifics.

Finally, common garden studies should ideally be conducted in multiple locations (sensu Clausen et al. 1940). As Rice and Mack (1991) have stated, "although the potential for ecotypic variation and evolutionary change can be studied within a common garden or glasshouse environment, its demonstration requires a reciprocal transplant experiment in the field" (p. 98). Our previous results serve to highlight this fact. When we compared seed capsule production of native and exotic genotypes of St. John's wort, results from one or even two gardens were not necessarily mirrored across all gardens (Maron et al. 2004a). For example, there were strong differences in fecundity between western North American and European plants in common gardens in Washington and Sweden but not in California and Spain. Similarly, we found that exotic St. John's wort populations produced significantly less of the secondary compound, pseudohypericin, in our common garden in Spain. However, this was not the case in Washington (Maron et al. 2004a). Had we established only one common garden, we would have come to erroneous conclusions regarding trait differences between native and exotic genotypes.

Conclusions

While a growing body of work shows that exotic plants often have genetically based differences in herbivore defense compared to their native conspecifics, ascribing these differences to rapid evolution will require more detailed study than has occurred to date. In particular, future studies will need to sample a greater number of populations across the native and introduced range and couple common garden results with data from which the invasion history of particular species can be inferred. As well, a greater congruence of methodology will aid in the synthesis of results. To date, studies have used different approaches to measuring resistance. Some studies have estimated resistance by measuring herbivore performance on plants, while others have examined the fitness effects of the consumers on plants in the field—a more direct estimate of resistance. Finally, many studies have assayed resistance by measuring levels of chemical defense. Future studies would be bolstered by making a more direct link between levels of particular secondary chemicals that are assayed and effects on consumers (sensu Berenbaum and Zangerl 1998).

More generally, future interpretation of comparative biogeographic studies of plant defense will be strengthened by resolution of several issues. The first involves understanding the selective impacts of specialists versus generalists in native populations. Current theories regarding how plant defense should evolve in exotic plants predict different outcomes for defenses against specialists versus generalists (Blossey and Nötzold 1995; van der Meijden 1996; Müller-Schärer et al. 2004; Joshi and Vrieling 2005). These are predicated on the notion that exotics escape from specialists, whereas this may or may not be the case for generalists. But for exotics to truly escape from specialists, the selective impacts of these enemies must be relatively strong across the native range. At the very least, specialists must impose stronger selective effects than generalists where plants are native. If this is not the case, exotics may have little to escape from. The second issue, related to the first, concerns how one classifies consumers that attack plants in their native range. Consumers do not always neatly fit into a tidy specialist-generalist dichotomy. Although some specialists are highly restricted in their host use, others are likely more polyphagous than is appreciated (Novotny et al. 2002). By the same token, although particular species of generalists may feed on many taxonomically diverse species, some individuals may "specialize" on particular host plants. Thus, making accurate predictions about how enemy defense may evolve in exotic plants will require more detailed natural history data on the selective pressures imposed by consumers across native populations that are sampled. Finally, there is the question of what plant traits confer resistance to particular types of consumers. The most frequently used organizing hypothesis is that toxic "qualitative" defenses deter generalists and that less-toxic, digestibility-reducing "quantitative" defenses deter specialists (Feeny 1976; Rhoades and Cates 1976). Thus, escape from specialists is thought to result in the loss of quantitative defenses but not qualitative defenses (Müller-Schärer et al. 2004). Many resistance traits, however, may not be so easily assigned to one of these two discrete groups (Stamp 2003). Surely toxic qualitative defenses provide some measure of defense against specialists, just as less-toxic quantitative defenses must provide some resistance to generalists. Similarly, rather than assigning consumers into the discrete groups of generalist or specialist, greater natural history information is needed on the particular consumers that plants escape from when introduced, their feeding proclivities, and the traits that confer resistance to them. In the future, a more nuanced approach to defense and enemy specialization will be required to make better sense of biogeographic comparisons among plant populations.

Finally, we close this chapter by making a plea that was first made by Harper in 1977. Biocontrol represents an excellent opportunity to infer how particular specialist herbivores may influence the evolution of resistance in plants that have been naïve to specialist herbivory for decades. New biocontrol introductions are increasing, and it would be extremely valuable, and easy, to save seeds from a variety of populations of exotics prior to when they become targets for biocontrol so that future researchers could directly compare phenotypes of ancient versus contemporary genotypes to directly determine whether biocontrol is a potent selective force on exotics.

Acknowledgments

We greatly thank D. Ewing and P. Matas for greenhouse assistance in Washington and Spain, respectively. R. Bommarco, B. Clifton, S. Elmendorf, D. Grosenbacher, J. Jones, T. Hirsch, T. Huettner, J. Pijoan, and M. Wolven helped establish and maintain the Washington common garden and K. Parker graciously allowed us to establish research plots on her land. I. Gimeno and L. Marco helped to establish and maintain the Spanish common gardens. We thank A. Angert, D. Ayers, J. Combs, U. Gamper, S. Gardner, D. Greiling, F. Grevstad, J. Hess, R. Keller, P. Kittelson, E. Knapp, E. Ogheri, P. Pysek, A. Sears, R. Sobhian, A. Stanley, J. Taft, E. Weber, A. Weis, and A. Wolf for helping with seed collection. John Aranson collaborated with work on plant chemistry; students and staff at the University of Ottawa who assisted with the phytochemcial analysis included Andrew Knox, Alex Turcotte, Andrew Burt, and John Livesey. This work was supported by grants to J.L.M. from NSF grants DEB-0296175 and OPP-0296208 and from a fellowship from the Catalunya PIV. The Institut de Ciència i Technologia Ambientals, Universitat Autònoma de Barcelona kindly provided working space. M.V. was supported by MCYT grant REN2000-0361 GLO.

References Cited

Agrawal, A. A., S. Y. Strauss, and M. J. Stout. 1999. Costs of induced responses and tolerance to herbivory in male and female fitness components of wild radish. Evolution 53: 1093–1104.

Alstad, D. N., and D. A. Andow. 1995. Managing the evolution of insect resistance to transgenic plants. Science 268: 1894–1896.

Andrés de, M. F., F. García-Arenal, M. M. López, and P. Megarejo. 1998. Patógenos de plantas descritas en España. Editorial de Ministerio de Agricultura, Pesca y Alimentacíon, Madrid.

Arnason, J. T., G. H. N. Towers, B. J. R. Philogene, and J. D. H. Lambert. 1983. The role of natural photosensitizers in plant resistance to insects, pp. 139–151. In A. Hedin (ed.), Plant resistance to insects. ACS Symposium Series 208. American Chemical Society, Washington, DC.

Arnason, J. T., Philogene, B. J. R., and G. H. N. Towers. 1992. Phototoxins in plant-insect interactions, pp. 317–341. In G. A. Rosenthal and M. R. Berenbaum (eds.), Herbivores: their interactions with secondary plant metabolites. Academic Press, New York.

Arnholdt-Schmitt, B. 2000. RAPD analysis: a method to investigate aspects of the reproductive biology of *Hypericum perforatum* L. Theor. Appl. Genet. 100: 906–911.

Auerbach, M., and D. Simberloff. 1988. Rapid leaf-miner colonization of introduced trees and shifts in sources of herbivore mortality. Oikos 52: 41–50.

Baker, H. G. 1974. The evolution of weeds. Annu. Rev. Ecol. Syst. 5: 1–24.

Baker, H. G., and G. L. Stebbins (eds.). 1965. The genetics of colonizing species. Academic Press, New York.

Barrett, S. C. H., and B. J. Richardson. 1986. Genetic attributes of invading species, pp. 21–33. In R. H. Groves and J. J Burdon (eds.), Ecology of biological invasions. Cambridge University Press, Cambridge, UK.

Bazzaz, F. A., N. Chiariello, P. D. Coley, and L. Pitelka. 1987. The allocation of resources to reproduction and defense. Bioscience 37: 58–67.

Berenbaum, M. R., and A. R. Zangerl. 1998. Chemical phenotype matching between plant and its insect herbivore. Proc. Natl. Acad. Sci. USA 95: 13743–13748.

Berenbaum, M. R., A. R. Zangerl, and J. K. Nitao. 1986. Constraints on chemical coevolution: wild parsnips and the parsnip webworm. Evolution 40: 1215–1228.

Bergelson, J., and C. B. Purrington. 1996. Surveying patterns in the cost of resistance in plants. Am. Nat. 148: 536–558.

Blair, A. C., and L. M. Wolfe. 2004. The evolution of an invasive plant: an experimental study with *Silene latifolia*. Ecology 85: 3035–3042.

Blossey, B., and R. Nötzold. 1995. Evolution of increased competitive ability in invasive nonindigenous plants: a hypothesis. J. Ecol. 83: 887–889.

Bone, E., and A. Farres. 2001. Trends and rates of microevolution in plants. Genetica 112–113: 165–182.

Bossdorf, O., S. Schröder, D. Prati, and H. Auge. 2004. Palatability and tolerance to simulated herbivory in native and introduced populations of *Alliaria petiolata* (Brassicaceae). Am. J. Bot. 91: 856–862.

Bossdorf, O., H. Auge, L. Lafuma, W. E. Rogers, E. Siemann, and D. Prati. 2005. Phenotypic and genetic differentiation between native and introduced plant populations. Oecologia 144: 1–11.

Brodie, E. D. III, and E. D. Brodie Jr. 1999. Predator-prey arms races. BioScience 49: 557–568.

Brown, A. H. D., and D. R. Marshall. 1981. Evolutionary changes accompanying colonization in plants, pp. 73–98. In G. G. E. Scudder and J. L. Reveal (eds.). Evolution today. Proceedings of the Second International Congress of Systematics and Evolutionary Biology. Hunt Institute for Botanical Documentation, Carnegie-Mellon University, Pittsburgh, PA.

Burdon, J. J., and J. N. Thompson. 1995. Changed patterns of resistance in a population of *Linum marginale* attacked by the rust pathogen *Melampsora lini*. J. Ecol. 83: 199–206.

Burdon, J. J., Groves, R. H., and J. M. Cullen. 1981. The impact of biological control on the distribution and abundance of *Chondrilla juncea* in south-eastern Australia. J. Appl. Ecol. 18: 957–966.

Buschmann, H. P. J. Edwards, and H. Dietz. 2005. Variation in growth pattern and response to slug damage among native and invasive provenances of four perennial Brassicaceae species. J. Ecol. 93: 322–334.

Callaway, R. M., W. M. Ridenour, T. Laboski, T. Weir, and J. M. Vivanco. 2005a. Natural selection for resistance to the allelopathic effects of invasive plants. J. Ecol. 93: 567–583.

Callaway, R. M., J. L. Hierro, and A. S. Thorpe. 2005b. Evolutionary trajectories in plant and soil microbial communities: *Centauria* invasions and the geographic mosaic of coevolution, pp 341–363. In D. F. Sax, S. D. Gaines, and J. J. Stachowicz (eds.) Exotic species, bane to conservation and boon to understanding: ecology, evolution and biogeography. Sinauer, Sunderland, MA.

Campbell, M. H., and E. S. Delfosse. 1984. The biology of Australian weeds. 13. *Hypericum perforatum*. J. Aust. Inst. Ag. Sci. 50: 63–73.

Carroll, S. P., J. E. Loye, H. Dingle, M. Mathieson, T. R. Famula, and M. P. Zaluck. 2005. And the beak shall inherit: evolution in response to invasion. Ecol. Lett. 8: 944–951.

Cavan, G., Biss, P., and S. R. Moss. 1998. Herbicide resistance and gene flow in wild oats *(Avena fatua* and *Avena sterilis* ssp. *ludoviciana)*. Ann. Appl. Biol. 133: 207–217.

Clausen, J., D. D. Keck, and W. M. Hiesey. 1940. Experimental studies on the nature of species. I. The effects of varied environments on North American plants. Carn. Inst. Wash. pub. no. 520. Carnegie Institute of Washington, Washington, DC.

Clegg, M. T., and R. W. Allard. 1972. Pattern of genetic differentiation in the slender wild oat species *Avena barbata*. Proc. Natl. Acad. Sci. USA 69: 1820–1824.

Coluatti, R., A. Ricciardi, I. A. Grigorovich, and H. J. MacIsaac. 2004. Is invasion success explained by the enemy release hypothesis? Ecol. Lett. 7: 721–733.

Cullen, J. M., and R. H. Groves. 1981. The population biology of *Chondrilla juncea* L. in Australia. Proc. Ecol. Soc. Aust. 10: 121–134.

Daehler, C. C., and D. R. Strong. 1997. Reduced herbivore resistance in introduced smooth cordgrass *(Spartina alterniflora)* after a century of herbivore-free growth. Oecologia 110: 99–108.

Dyer, A. R., and K. J. Rice. 1999. Effect of competition on resource availability and growth of a California bunchgrass. Ecology 80: 2697–2710.

Ehler, L. E. 1998. Invasion biology and biological control. Biol. Control 13: 127–133.

Elton, C. S. 1958. The ecology of invasions by animals and plants. Metheun, London.

Etterson, J. R., and R. G. Shaw. 2001. Constraint to adaptive evolution in response to global warming. Science 294: 151–154.

Fargione, J., C. S. Brown, and D. Tilman. 2003. Community assembly and invasion: an experimental test of neutral versus niche processes. Proc. Nat. Acad. Sci. USA 100: 8916–8920.

Feeny, P. 1976. Plant apparency and chemical defense, pp. 1–40. In J. W. Wallace and R. L. Mansell (eds.), Recent advances in phytochemistry, vol. 10. Plenum Press, New York.

Feeny, P. 1992. The evolution of chemical ecology: contributions from the study of herbivorous insects, pp. 1–44. In G. A. Rosenthal and M. R. Berenbaum (eds.), Herbivores, their interactions with secondary plant metabolites, 2nd ed., vol. 2. Academic Press, San Diego.

Fields, P. G., J. T. Arnason, and B. J. R. Philogene. 1988. Distribution of *Chrysolina* spp. (Coleoptera: Crysomelidae) in eastern Ontario, 18 years after their initial release. Can. Enotomol. 120: 937–938.

Fields, P. G., J. T. Arnason, and B. J. R. Philogene. 1990. Behavioural and physical adaptations of three insects that feed on the phototoxic plant *Hypericum perforatum*. Can. J. Bot. 68: 339–346.

Fox, L. R. 1988. Diffuse coevolution within complex communities. Ecology 69: 906–907.

Fritz, R. S., and P. W. Price. 1988. Genetic variation among plants and insect community structure: willows and sawflies. Ecology 69: 845–856.

Garcia-Rossi, D., N. Rank, and D. R. Strong. 2003. Potential for self-defeating biological control? Variation in herbivore vulnerability among invasive *Spartina* genotypes. Ecol. Appl. 13: 1640–1649.

Geffeney, S., E. D. Brodie Jr., and P. C. Ruben. 2002. Mechanisms of adaptation in a predator-prey arms race: TTX-resistant sodium channels. Science 297: 1336–1339.

Genton, B. J., P. M. Kotanen, P. O. Cheptou, C. Adolphe, and J. A. Shykoff. 2005. Enemy release but no evolutionary loss of defence in a plant invasion: an inter-continental reciprocal transplant experiment. Oecologia 146: 404–414.

Georghiou, G. P. 1986. The magnitude of the resistance problem, pp. 14–43. In Pesticide resistance: strategies and tactics for management. National Academy Press, Washington, DC.

Giese, A. C. 1980. Hypericism, pp. 229–255. In K. C. Smith (ed.), Photochemical and photobiological reviews. Plenum Press, New York.

Grant, P. R., and B. R. Grant. 2002. Unpredictable evolution in a 30-year study of Darwin's finches. Science 296: 707–711.

Graves, S. D., and A. M. Shapiro. 2003. Exotics as host plants of the California butterfly fauna. Biol. Conserv. 110: 413–433.

Guillet, G. 1997. Ecophysical importance of phototoxins in plant-insect relationships. Ph.D. dissertation, University of Ottawa, Ottawa, Canada.

Guillet, G., C. Podesfinski, C. Regnault Roger, J. T. Arnason, and B. J. R. Philogene. 2000. Biochemical and behavioral adaptations of generalist and specialist herbivorous insects feeding on *Hypericum perforatum* (St. John's wort). Environ. Entomol. 29: 135–139.

Hänfling, B., and J. Kollman. 2002. An evolutionary perspective on invasions. Trends Ecol. Evol. 17: 545–546.

Hare, J. D., and D. J. Futuyma. 1978. Different effects on variation in *Xanthium strumarium* L. (Compositae) on two insect seed predators. Oecologia 37: 109–12.

Harper, J. L. 1977. Population biology of plants. Academic Press, London.

Harris, P., and M. Maw. 1984. *Hypericum perforatum* L., St. John's wort (Hypericaceae), pp. 171–177. In J. S. Kelleher and M. A. Hulme (eds.), Biological control programmes against insects and weeds in Canada 1969–1980. Commonwealth Agricultural Bureaux, Slough, UK.

Henter, H. J., and S. Via. 1995. The potential for coevolution in a host-parasitoid system. I. Genetic variation within an aphid population in susceptibility to a parasitic wasp. Evolution 49: 427–438.

Hinz, H. L., and M. Schwarzlaender. 2004. Comparing invasive plants from their native and exotic range: what can we learn from biological control? Weed Technol. 18: 1533–1541.

Hodgson, J. M. 1964. Variations in ecotypes of Canada thistle. Weeds 12: 167–171.

Hoebeke, E. R. 1993. Establishment of *Uropora quadrifasciata* (Diptera: Tephritidae) and *Chrysolina quadrigemina* (Coleoptera: Chrysomelidae) in portions of the eastern United States. Entomol. News 104: 143–152.

Holloway, J. K. 1957. Weed control by insect. Sci. Am. 197: 56–62.

Holloway, J. K., and C. B. Huffaker. 1951. The role of *Chrysolina gemellata* in the biological control of Klamath weed. J. Econ. Entomol. 44: 244–247.

Holmes, T. H., and K. J. Rice. 1996. Patterns and growth and soil-water utilization in some exotic annuals and native perennial bunchgrasses of California. Ann. Bot. 78: 233–243.

Holt, R. D., and M. E. Hochberg. 1997. When is biological control evolutionarily stable (or is it)? Ecology 78: 1673–1683.

Huey, R. B., G. W. Gilchrist, M. L. Carlson, D. Berrigan, and L. Serra. 2000. Rapid evolution of a geographic cline in size in an introduced fly. Science 287: 308–309.

Huffaker, C. B. 1957. Fundamentals of biological control of weeds. Hilgardia 27: 101–157.

Huffaker, C. B., and J. K. Holloway. 1949. Changes in range plant populations structure associated with feeding of imported enemies of Klamath weed (*Hypericum perforatum* L.). Ecology 30: 167–175.

Huffaker, C. B., P. S. Messenger, and P. DeBach. 1971. The natural enemy component in natural control and the theory of biological control, pp. 16–67. In C. B. Huffaker (ed.), Biological Control. Plenum Press, New York.

Jain, S. K., and P. S. Martins. 1979. Ecological genetics of the colonizing ability of rose clover (*Trifolium hirtum* All.). Am. J. Bot. 66: 361–366.

Jobin, A., U. Schaffner, and W. Nentwig. 1996. The structure of the phytophagous insect fauna on the introduced weed *Solidago altissima* in Switzerland. Entomol. Exp. Appl. 79: 33–42.

Johshi, J., and K. Vrieling. 2005. The enemy release and EICA revisited: incorporating the fundamental difference between specialist and generalist herbivores. Ecol. Lett. 8: 704–714.

Juenger, T., and J. Bergelson. 1998. Pairwise versus diffuse natural selection and the multiple herbivores of scarlet gilia, *Ipomopsis aggregata*. Evolution 52: 1583–1592.

Julien, M. H., and M. W. Griffiths (eds.). 1998. Biological control of weeds. CABI Publishing, New York.

Keane, R.M., and M.J. Crawley. 2002. Exotic plant invasions and the enemy release hypothesis. Trends Ecol. Evol. 17: 164–170.

Kennedy, G.G., and J.D. Barbour. 1992. Resistance variation in natural and managed systems, pp. 13–41. In R.S. Fritz and E.L. Simms (eds.), Plant resistance to herbivores and pathogens. University of Chicago Press, Chicago.

Koricheva, J. 2002. Meta-analysis of sources of variation in fitness costs of plant antiherbivore defenses. Ecology 93: 176–190.

Kraaijeveld, A.R., J.J.M. van Alphen, and H.C.J. Godfray. 1998. The coevolution of host resistance and parasitoid virulence. Parasitology 116: S29–S45.

Lankau, R.A., W.E. Rogers, and E. Siemann. 2004. Constraints on the utilisation of the invasive Chinese tallow tree *Sapium sebiferum* by generalist native herbivores in coastal prairies. Ecol. Entomol. 29: 66–75.

Lee, C.E. 2002. Evolutionary genetics of invasive species. Trends Ecol. Evol. 17: 386–391.

Leger, E.A., and M.L. Forister. 2005. Increased resistance to generalist herbivores in invasive populations of California poppy *(Eschscholzia californica)*. Divers. Distrib. 11: 311–317.

Maddox, G.D., and R.B. Root. 1987. Resistance to 16 diverse species of herbivorous insects within a population of goldenrod, *Solidago altissima*: genetic variation and heritability. Oecologia 72: 8–14.

Maddox, G.D., and R.B. Root. 1990. Structure of the encounter between goldenrod *(Solidago altissima)* and its diverse insect fauna. Ecology 71: 2115–2124.

Mal, T.K., and J. Lovett-Doust. 2005. Phenotypic plasticity in vegetative and reproductive traits in the invasive weed, *Lythrum salicaria* (Laythraceae), in response to soil moisture. Am. J. Bot. 92: 819–825.

Maron, J.L., and M. Vilà. 2001. Do herbivores affect plant invasion? Evidence for the natural enemies and biotic resistance hypotheses. Oikos 95: 363–373.

Maron, J.L., M. Vilà, R. Bommarco, S. Elmendorf, and P. Beardsley. 2004a. Rapid evolution of an invasive plant. Ecol. Monogr. 74: 261–280.

Maron, J.L., M. Vilà, and J. Arnason. 2004b. Loss of enemy resistance among introduced populations of St. John's wort, *Hypericum perforatum*. Ecology 85: 3243–3253.

Marquis, R.J. 1990. Genotypic variation in leaf damage in *Piper arieianum* (Piperaceae) by a multispecies assemblage of herbivores. Evolution 44: 104–120.

Marquis, R.J. 1992. The selective impacts of herbivores, pp. 301–325. In R.S. Fritz and E.L. Simms (eds.), Plant resistance to herbivores and pathogens. University of Chicago Press, Chicago.

Maurico, R., and M.D. Rausher. 1997. Experimental manipulation of putative selective agents provides evidence for the role of natural enemies in the evolution of plant defense. Evolution 51: 1435–1444.

Mayo, G.M., and P. Langridge. 2003. Modes of reproduction in Australian populations of *Hypericum perforatum* L. (St. John's wort) revealed by DNA fingerprinting and cytological methods. Genome 46: 573–579.

McCaffrey, J.P., C.L. Campbell, and L.A. Andres. 1995. Biological control in the western United States: accomplishments and benefits of regional research, pp. 281–285. In J.R. Nechols, L.A. Andres, J.W. Beardsley, R.D. Goeden, and C.G. Jackson (eds.), Univ. Calif. Div. Agric. Nat. Res. pub. 3361 . Davis, CA.

McCrea, K.D., and W.G. Abrahamson. 1987. Variation in herbivore infestation: historical vs. genetic factors. Ecology 68: 822–827.

McEvoy, P.B., N.T. Rudd, C.S. Cox, and M. Huso. 1993. Disturbance, competition and herbivory effects on ragwort *Senecio jacobaea* populations. Ecol. Monogr. 63: 55–75.

Memmott, J., S.V. Fowler, Q. Paynter, A.W. Sheppard, and P. Syrett. 2000. The invertebrate fauna on broom, *Cytisus scoparius*, in two native and two exotic habitats. Acta Oecol. 21: 213–222.

Meyer, G.R., R. Clare, and E. Weber. 2005. An experimental test of the evolution of increased competitive ability hypothesis in goldenrod, *Solidego gigantean*. Oecologia 144: 299–307.

Mitchell, C.G., and A.G. Power. 2003. Release of invasive plants from fungal and viral pathogens. Nature 421: 625–627.

Mitich, L.W. 1994. Intriguing world of weeds: Common St. Johns wort. Weed Technol. 8: 658–661.

Morgan, G.F., and D.R. Marshall. 1978. Allozyme uniformity within and variation between races of the colonizing species *Xanthium strumarium* L. (Noogoora burr). Aust. J. Biol. Sci. 31: 284–292.

Müller-Schärer, H., and T. Steinger. 2004. Predicting evolutionary change in invasive, exotic plants and its consequences for plant-herbivore interactions, pp. 137–162. In , L.E. Ehler, R. Sforza, and T. Mateille (eds.), Genetics, evolution and biological control. CABI Publishing, Wallingford, UK.

Müller-Schärer, H., U. Schaffner, and T. Steinger. 2004. Evolution in invasive plants: implications for biological control. Trends Ecol. Evol. 19: 417–422.

Neuffer, B., and H. Hurka. 1999. Colonization history and introduction dynamics of *Capsella bursa-pastoris* (Brassicaceae) in North America: isozymes and quantitative traits. Mol. Ecol. 8: 1667–1681.

Newman, R.I., D. Thompson, and D. Richman. 1998. Conservation strategies for the biological control of weeds, pp. 371–396. In P. Barbosa (ed.), Conservation biological control. Academic Press, New York.

Novotny, V., Y. Basset, S.E. Miller, G.D. Weiblen, B. Bremer, L. Cizek, and P. Drozd. 2002. Low host specificity of herbivorous insects in a tropical forest. Nature 416: 841–844.

Parker, M.A. 1991. Nonadaptive evolution of disease resistance in an annual legume. Evolution 54: 1209–1217.

Phillips, B.L., and R. Shine. 2004. Adapting to an invasive species: toxic cane toads induce morphological change in Australian snakes. Proc. Natl. Acad. Sci. USA 101: 17150–17155.

Pilson, D. 1996. Two herbivores and constraints on selection for resistance in *Brassica rapa*. Evolution 50: 1492–1500.

Pimentel, D. 1986. Population dynamics and the importance of evolution in successful biological control, pp. 3–18. In , J.M. Franz (ed.), Biological plant and health protections. Springer-Verlag, New York.

Powles, S.B., and J.A.M. Holtum. 1994. Herbicide resistance in plants. Lewis Publishers, Boca Raton, FL.

Rausher, M. 1992. Natural selection and the evolution of plant-insect interactions, pp. 20–88. In B.D. Roitberg and M.B. Isman (eds.), Insect chemical ecology: an evolutionary approach. Chapman and Hall, New York.

Rees, C. J.C. 1969. Chemoreceptor specificity associated with choice of feeding site by the beetle, *Chrysolina brunsvicensis* on its food-plant, *Hypericum hirsutum*. Entomol. Exp. Appl. 12: 565–583.

Rejmánek, M., and D.M. Richardson. 1996. What attributes make some plant species more invasive? Ecology 77: 1655–1661.

Reznick, D.N., and C.K. Ghalambor. 2001. The population biology of contemporary adaptations: what empirical studies reveal about the conditions that promote adaptive evolution. Genetica 112–113: 183–198.

Reznick, D.N, F.H. Shaw, R.H. Rodd, and R.G. Shaw. 1997. Evaluation of the rate of evolution in natural populations of guppies *(Poecilia reticulata)*. Science 275: 1934–1937.

Rhoades, D.F., and R.G. Cates. 1976. Towards a general theory of plant antiherbivore chemistry, pp. 168–213. In J.W. Wallace and R.L. Mansell (eds.), Recent advances in phytochemistry, vol. 10. Plenum Press, New York.

Rice, K.J., and N.C. Emory. 2003. Managing microevolution: restoration in the face of global change. Front. Ecol. 1: 469–478.

Rice K.J., and R.N. Mack. 1991. Ecological genetics of *Bromus tectorum*. III. The demography of reciprocally sown populations. Oecologia 88: 91–101.

Ritcher, P.O. 1996. Biological Control of Insects and Weeds in Oregon. Ag. Ex. Sta. Oregon State Univ. Tech. Bull. 90: 1–39.

Rohani, P., H. C.J. Godfray, and M.P. Hassell. 1994. Aggregation and the dynamics of host-parasitoid systems: a discrete generation model with within-generation redistribution. Am. Nat. 144: 491–509.

Sampson, A.W., and K.W. Parker. 1930. St. Johns-wort on rangelands of California. Calif. Agric. Exp. Sta. Bull. 503.

Shea, K., and P. Chesson. 2002. Community ecology theory as a framework for biological invasions. Trends Ecol. Evol. 17: 170–176.

Siemann, E., and W.E. Rogers. 2003. Herbivory, disease, recruitment limitation, and success of alien and native tree species. Ecology 84: 1489–1505.

Simberloff, D., and P. Stiling. 1996. How risky is biological control? Ecology 77: 1964–1974.

Simms, E. L., and M.D. Rausher. 1989. The evolution of resistance to herbivory in *Ipomoea purpurea* II. Natural selection by insects and costs of resistance. Evolution 43: 573–585.

Sirvent, T.M., S.B. Krasnoff, and D.M. Gibson. 2003. Induction of hypericins and hyperforins in *Hypericum perforatum* in response to damage by herbivores. J. Chem. Ecol. 29: 2667–2681.

Stamp, N. 2003. Out of the quagmire of plant defense hypotheses. Q. Rev. Biol. 78: 23–55.

Stastny, M., U. Schaffner, and E. Elle. 2005. Do vigour of introduced populations and escape from specialist herbivores contribute to invasiveness? J. Ecol. 93: 27–37.

Stockwell, C.A., A.P. Hendry, and M. T. Kinnison. 2003. Contemporary evolution meets conservation biology. Trends Ecol. Evol. 18: 94–101.

Strauss, S.Y., and A.A. Agrawal. 1999. The ecology and evolution of plant tolerance to herbivory. Trends Ecol. Evol. 14: 179–185.

Strauss, S.Y., D.H. Siemens, M.B. Decher, and T. Mitchell-Olds. 1999. Ecological costs of plant resistance in the currency of pollination. Evolution 53: 1105–1113.

Strong, D.R. 1974. Rapid asymptotic species accumulation in phytophagous insect communities: the pests of Cacao. Science 185: 164–1066.

Thompson, J.N. 1999a. The raw material for coevolution. Oikos 84: 5–16.

Thompson, J.N. 1999b. Specific hypotheses on the geographic mosaic of coevolution. Am. Nat. 153: S1–S14.

Trifunovic, S., V. Vajs, S. Macura, N. Juranic, Z. Djarmati, R. Jankov, and S. Milosavljevic. 1998. Oxidation products of hyperforin from *Hypericum perforatum*. Phytochemistry 49: 1305–1310.

van der Meijden, E. 1996. Plant defense, an evolutionary dilemma: contrasting effects of (specialist and generalist) herbivores and natural enemies. Entomol. Exp. Appl. 80: 307–310.

van Kleunen, M., and B. Schmid. 2003. No evidence for an evolutionary increased competitive ability (EICA) in the invasive plant, *Solidago canadensis*. Ecology 84: 2816–2823.

Vilà, M., J.L. Maron, and L. Marco. 2005. Evidence for the enemy release hypothesis in *Hypericum perforatum* L. Oecologia 142: 474–479.

Voss, E.G. 1985. Michigan flora, part II: Dicots (Saururaceae-Cornaceae). Cranbrook Institute of Science, Cranbrook, MI.

Weber, E., and B. Schmid. 1998. Latitudinal population differentiation in two species of *Solidago* (Asteraceae) introduced into Europe. Am. J. Bot. 85: 1110–1121.

Williams, D.G., R.N. Mack, and R.A. Black. 1995. Ecophysiology of introduced *Pennisetum setaceum* on Hawaii: the role of phenotypic plasticity. Ecology 76: 1569–1580.

Williams, J.R. 1954. The biological control of weeds, pp. 95–98. In Report of the Sixth Commonwealth Entomological Congress, London.

Williams, K.S. 1984. Climatic influences on weeds and their herbivores: biological control of St. John's wort in British Columbia, pp. 127–132. In E.S. Delfosse (ed.), Proceedings of the VI international symposium for the biological control of weeds. Agriculture Canada, Ottawa, ON.

Willis, A.J., M.B. Thomas, and J.H. Lawton. 1999. Is the increased vigour of invasive weeds explained by a trade-off between growth and herbivore resistance? Oecologia 120: 275–283.

Wilson, F. 1943. The entomological control of St. John's wort (*Hypericum perforatum* L.). Council Sci. Ind. Res. Austr. Bull. 169: 1–88.

Wolfe, L.M., J. A. Elzinga, and A. Biere. 2004. Increased susceptibility to enemies following introduction in the invasive plant *Silene latifolia*. Ecol. Lett. 7: 813–820.

Wu, K.K., and S.K. Jain. 1978. Genetic and plastic responses in geographic differentiation of *Bromus rubens* populations. Can. J. Bot. 56: 873–879.

Zangerl, A.R., and M.R. Berenbaum. 2003. Phenotype matching in wild parsnip and parsnip webworms: causes and consequences. Evolution 57: 806–815.

TWENTY-ONE

Life-History Evolution in Native and Introduced Populations

ROBERT F. DENNO, MERRILL A. PETERSON,
MATTHEW R. WEAVER, AND DAVID J. HAWTHORNE

A life-history strategy encompasses an integrated suite of traits associated with reproduction (e.g., fecundity, age to first reproduction, offspring size, and voltinism) and the placement of offspring on suitable resources in space (e.g., dispersal) and time (e.g., diapause) (Southwood et al. 1974; Denno et al. 1996; Roff 2002). Notably, there is tremendous variation, both within and among species, in particular life-history traits (Denno et al. 1991, 1996; Roff 1992, 2002; Schmidt et al. 2005), and the challenge is to explain how such trait diversity arises and what factors influence the evolution of particular traits (Roff 1992, 2002).

Despite the clear environmental drawbacks of invasive species (Hobbs 1989; Mack et al. 2000; Perrings et al. 2000; Pimentel et al. 2000; Crooks 2002; Bertness et al. 2004; Normile 2005; Williamson 2005), such species incursions do provide a unique opportunity to explore how novel selective regimes influence the evolution of particular life-history traits (Ellstrand and Schierenbeck 2000; Quinn et al. 2000; Sakai et al. 2001). In fact, many of the best examples of rapid directional selection on such traits involve invasive species (Thompson 1998; Reznick and Ghalambor 2001; Sakai et al. 2001). The evolution of life-history traits in introduced populations may be rapid in part because some life-history traits, such as dispersal and high reproductive potential, may promote the colonization of novel habitats (MacArthur and Wilson 1967; Pianka 1970; Southwood et al. 1974), whereas a different suite of traits may be favored in the newly colonized region. For example, habitat features such as persistence, stability, and heterogeneity, which are thought to play a major role in the evolution of life-history strategies (Southwood et al. 1974; Stearns 1992; Roff 1994, 2002; Denno et al. 1996, 2001), are likely to differ substantially among regions. These habitat differences can in turn favor new trait optima following colonization (Pianka 1970; Lee 2002; Roff 2002). Thus, with invading species, it is possible to determine a species' life-history strategy (e.g., traits associated with dispersal, reproduction and voltinism) in its native range, to document traits associated with successful colonization of novel habitats, to assess changes in life-history traits in introduced populations, and most notably to elucidate differences in selective regimes (e.g., habitat persistence) between native and introduced populations that underlie life-history change in the newly colonized habitat (see Crawley 1986; Ehrlich 1986; Lawton and Brown 1986; Denno et al. 1996; Thompson 1998; Sakai et al. 2001).

The model system we employ here to explore life-history evolution in native and introduced populations involves *Prokelisia* planthoppers (Hemiptera: Delphacidae), small phloem-feeding herbivorous insects, and their only host plants, *Spartina* cordgrasses that grow as expansive monocultures in intertidal salt marshes (Daehler and Strong 1996; Denno et al. 1996; Bertness et al. 1999). Both *Prokelisia* planthoppers and their *S. alterniflora* host plant occur natively along the Atlantic and Gulf coasts of North America (Mobberley 1956; Chapman 1977; Denno et al. 1996). However, *Spartina* cordgrasses have been introduced both accidentally and for restoration purposes and now threaten native wetlands and intertidal salt marshes throughout the world (Ranwell 1967; Aberle 1993; Daehler and Strong 1996). Moreover, the planthopper *P. marginata* has been accidentally introduced into western North America (California) and southern Europe (Portugal, Slovenia, and Spain) and for the biological control of alien *Spartina* populations that have invaded salt marshes in Washington State (Grevstad et al. 2003a, 2003b; J. Quartau and M.R. Wilson, 2004, personal communication). Accidental introductions of *P. marginata* have occurred anthropogenically, perhaps as embedded eggs in *Spartina* plants used for marsh restoration, as eggs in discarded *Spartina* used as oyster packing material, or as eggs in *Spartina* rhizomes in ship ballast (Sayce 1988; Callaway and Josselyn 1992; Aberle 1993; Daehler and Strong 1996).

Because selection experiments and field data suggest that the life-history traits of planthoppers respond rapidly to selection (Denno et al. 1991; Marooka and Tojo 1992), we argue that the life-history strategy of an introduced planthopper has the potential to evolve quickly in response to local habitat conditions. In particular, dispersal characters (flight capability versus flightlessness) respond quickly to selection in planthoppers (Marooka and Tojo 1992). Moreover, the relationship between habitat persistence and the incidence of dispersal (percent flight-capable adults) is established for planthopper populations in general, and specifically for *P. marginata* (Denno et al. 1991, 1996, 2001). Thus, habitat persistence can be quantified in source and novel habitats and be used to predict any change the incidence of dispersal (Denno et al. 1991, 1996).

From a life-history perspective, our major objectives were to (1) provide background information on the life-history strategies and dispersal ecology of planthoppers, (2) identify variation in the incidence of dispersal and associated reproductive traits (fecundity and voltinism) in native and introduced populations of *P. marginata*, (3) provide evidence that variation in dispersal ability is genetically based, (4) verify that habitat persistence is the underlying ecological determinant of variation in dispersal capability across populations, (5) identify life-history traits that promote successful colonization and establishment, and (6) show that following colonization, selection resulting from habitat characteristics (persistence and heterogeneity) can promote change in the incidence of dispersal. Overall, we predict that following colonization, the incidence of dispersal in introduced populations of *P. marginata* will rapidly evolve to reflect the persistence characteristics of the colonized habitat and not those prevailing in the source area. With our assessment we aim to highlight how invasive species can be used to determine the effects of novel selective regimes on life-history evolution.

Life-History Strategy and Dispersal Ecology of *Prokelisia* Planthoppers

Life Cycle and Natural History

Prokelisia marginata and its noninvasive congener *P. dolus* are the most common herbivores on their *Spartina* host plants (Wilson et al. 1994; Denno et al. 1996; Grevstad et al. 2003a). Both *Prokelisia* species can be extremely abundant, with densities often exceeding 10,000 individuals per m^2 (Denno et al. 1996, 2000b). Female planthoppers insert their eggs in living *Spartina* blades where they remain for approximately two weeks before hatching (Denno et al. 1989, 1996). Nymphs (five instars) and adults feed on the phloem sap of their *Spartina* hosts, and development from egg to adult takes approximately six weeks (Denno et al. 1989, 1996). In the absence of significant predation, planthopper populations can grow rapidly, frequently outbreak (Denno and Peterson 2000; Denno et al. 2005), and can significantly diminish the growth of *Spartina* (Daehler and Strong 1995, 1997; Wu et al. 1999; Denno et al. 2002; Finke and Denno 2004). Overwintering takes place on the high marsh, away from the water's edge, as active nymphs nestled into leaf litter, and protective leaf litter is critical for their winter survival (Denno et al. 1996). All stages of *Prokelisia* are able to withstand short-term tidal inundation (Vince et al. 1981; Roderick 1987; Throckmorton 1989).

Flight Polymorphism, Dispersal, and Constraints on Reproduction

Before examining geographic patterns of dispersal and associated life-history traits in *P. marginata*, it is important to know how such traits are integrated in planthoppers. In general, planthoppers are ideal for the study of life-history evolution because they exhibit wing dimorphism, which facilitates the detection of flight-capable from flightless individuals in a population (Denno et al. 1991). For example, populations of most planthopper species including *P. marginata* contain two wing forms: flightless adults (brachypters with reduced wings) and flight-capable adults (macropters) that possess fully developed wings and can disperse long distances (Denno 1994). However, the proportion of each wing form can vary tremendously among different species (Denno et al. 1991; Denno 1994) and geographically among populations of the same species (Iwanaga et al. 1987; Iwanaga and Tojo 1988; Denno et al. 1996). Importantly, the fraction of the macropterous wing form in a population can be used as a reliable index of the level of potential dispersal in the population (Denno et al. 1991).

Wing form in planthoppers is determined by a developmental switch that responds to environmental cues (Denno 1994; Denno et al. 2001). However, the sensitivity of the switch is heritable and under polygenic control (Denno 1994; Zera and Denno 1997). Of all the proximate cues known to affect wing form in planthoppers, population density is by far the most important for most species (Denno and Roderick 1990). The density that triggers the production of macropterous forms, however, can differ among species and among populations of the same species (Denno et al. 1991, 1996; Denno 1994). In general, variation in the wing-form response of individuals in a population to density is strongly related to the incidence of dispersal in that population such that selection for increased dispersal results in a more sensitive developmental switch, the triggering of macropters production at low densities, and a high proportion of flight-capable morphs in the population (Denno et al. 1991).

Although dispersal is advantageous for colonizing temporary and novel habitats, it does not occur without costs in planthoppers, costs that are often leveled against reproduction (Denno et al. 1989; Denno 1994). For instance, although macropters are the colonizing morph, they are both less fecund and reproduce later in life compared to brachypterous females (Denno et al. 1989; Denno 1994). As a consequence, the net replacement rate of macropters is only half as high as that for brachypters (Denno et al. 1989). The reproductive delay and reduced fecundity observed in the macropterous form of female planthoppers support the view that flight capability is costly and that phenotypic trade-offs between flight and reproduction exist (Roff 1986; Denno et al. 1989; Roff and Fairbairn 1991; Zera and Denno 1997). Thus, dispersive planthopper species, characterized by a high incidence of macroptery in populations, likely suffer from reduced fecundity and may be less able to establish in novel habitats (Denno 1993). This said, planthopper life histories are extremely flexible and as such may allow for both successful colonization and establishment, at least to some degree. For instance, their wing-dimorphic life history allows for both colonization and reproduction to be enhanced by partitioning these two functions between the two adult wing forms (Denno and Peterson 2000). Just how this situation might arise is illustrated by the typical population-dynamic scenario that occurs repeatedly among mobile species exploiting temporary habitats (Denno and Roderick 1990). In a newly developing habitat (e.g., an agricultural crop), macropters colonize at very low densities, mate soon after arrival, leave offspring that develop mostly into fecund brachypterous adults due to the initially low-density conditions, and rapid population growth ensues (Denno and Peterson 2000). In the generations that follow, nymphs molt increasingly more into macropterous adults that then escape deteriorating conditions by dispersing to other habitats. Thus, the dimorphic life history exhibited by *Prokelisia* and other planthoppers allows for effective colonization of new habitats by macropters and enhanced establishment in the novel habitat by the more prolific brachypters (Denno and Peterson 2000).

Moreover, both selection experiments and field data suggest that the life-history traits of planthoppers respond rapidly to selection, and thus the possibility for rapid adaptation in new habitats. For instance, macropterous and brachypterous lines of the planthopper *Nilaparvata lugens* can be developed after only 30 generations of selection in the laboratory on an initial population consisting of an equal mix of both wing forms (Marooka and Tojo 1992). Moreover, planthoppers existing in relative young habitats where fewer than 50 generations have been achieved are highly brachypterous (>85%), suggesting a rapid loss of dispersal ability since colonization (Denno et al. 1991).

Habitat Characteristics and the Incidence of Dispersal in Planthoppers

Three habitat characteristics, namely persistence, dimensionality, and isolation are considered central in shaping the evolution of insect dispersal strategies (Roff 1990; Denno et al. 1991, 1996, 2000a, 2001; Wagner and Liebherr 1992; Denno 1994; Travis and Dytham 1999). In general, theory predicts reductions in flight capability for species exploiting habitats that are very persistent, low-profile in structural dimension, or isolated (Roff 1990; Denno et al. 2000a, 2001). Specifically for planthoppers, habitat persistence and dimensionality appear to influence the evolution of dispersal more than does habitat isolation (Denno et al. 2000a, 2001).

In persistent habitats, selection should favor reduced levels of dispersal due to the reproductive costs associated with flight capability (Roff 1986, 1990; Denno et al. 1989, 1991). In temporary habitats, however, wings should facilitate the tracking of changing resources (Roff 1990; Denno et al. 1991). Consequently, as habitats become more ephemeral, the expectation is to observe a higher incidence of dispersal. Indeed, an interspecific comparison among 38 different planthopper species found a significant negative relationship between habitat persistence (the maximum number of planthopper generations achievable given the longevity of the habitat) and the incidence of dispersal (percent macroptery) in a population (Denno et al. 1991). The highest levels of dispersal (>50% macroptery) were recorded for species inhabiting short-lived agricultural crops (e.g., *N. lugens*) or natural disturbed habitats (e.g., *P. marginata*) that persist for less than one year. In contrast, the lowest incidence of dispersal (≤1%) occurred in species occupying persistent habitats such as bogs, freshwater marshes, and salt marshes that have existed in North America for 2000 to 12,000 years (Denno et al. 1991).

Although wings facilitate escape from temporary habitats and the colonization of new ones, wings also function in habitat negotiation, particularly in complex, three-dimensional vegetation (Waloff 1983; Denno 1994). For example, mate finding and rediscovery of an optimal feeding site following escape from a predator may prove difficult for flightless brachypters in arboreal habitats. By contrast, the consequences of releasing from a host plant in low-profile vegetation (e.g., grasses) are likely to be minimal because resources can be easily found by walking. Thus, selection should favor the retention of flight capability in arboreal species even though their habitats are persistent. This prediction was borne out by a comparative study of 150 delphacid species, which revealed a positive relationship between host-plant height and incidence of macroptery (Denno 1994). Even though both habitat persistence and dimensionality bear on the evolution of dispersal strategies of planthoppers at large, only habitat persistence is likely to influence the incidence of dispersal in *Prokelisia* planthopper populations due to the low-profile nature (<1 m in height) of the *Spartina* host plants they exploit. Thus, for the remainder of this chapter we focus on habitat persistence as it influences dispersal and other associated traits in native and introduced populations of *Prokelisia*.

Geographic Distribution of the Insect and Host

Prokelisia marginata feeds and develops on four closely related cordgrass species: *S. alterniflora*, *S. foliosa*, *S. anglica*, and *S. maritima* (Wilson et al. 1994; Denno et al. 1996;

Grevstad et al. 2003a; J. Quartau and M.R. Wilson, 1998, personal communication). *Spartina alterniflora* occurs natively along the Atlantic and Gulf coasts of North America but has been introduced and is invasive in Pacific-coast estuaries in California, Oregon, and Washington (Mobberley 1956; Aberle 1993; Daehler and Strong 1996, 1997; Denno et al. 1996). *Spartina foliosa* grows natively in isolated marshes along the Pacific coast from the San Francisco Bay area south through southern California and into Baja California (Mobberley 1956; Denno et al. 1996). *Spartina anglica*, an amphidiploid hybrid between *S. alterniflora* and the European *S. maritima*, has been introduced into the San Francisco Bay and Puget Sound estuaries (Raybould et al. 1991; Aberle 1993; Daehler and Strong 1996; Wu et al. 1999). Finally, *S. maritima* grows natively in European estuaries in the Netherlands and England, along the southern Iberian Peninsula in Portugal and Spain, and in the Gulf of Venice in Italy and Slovenia (Mobberley 1956; Padinha et al. 2000; Sánchez et al. 2001; Nieva et al. 2005).

Prokelisia marginata is native to much of the Atlantic and Gulf coasts of North America, where it feeds and develops only on *S. alterniflora* (Denno et al. 1996). In the San Francisco Bay marshes, accidentally introduced populations of *P. marginata* (~1970) feed on the native cordgrass *S. foliosa*, on populations of *S. alterniflora*, also introduced ca. 1970, and on hybrids between the two (Roderick 1987; Denno et al. 1996; Daehler and Strong 1997; Wu et al. 1999; Ayers and Strong 2002). *Prokelisia marginata* was introduced into Washington State from a San Francisco Bay source population in 2000 for the biological control of invasive populations of *S. alterniflora*, and it also develops in the greenhouse on *S. anglica* (Wu et al. 1999; Grevstad et al. 2003a; Garcia-Rossi et al. 2003). In southern Portugal and Spain, introduced populations of *P. marginata* were first collected in 1997 and 1998, respectively, feeding on *S. maritima* (J. Quartau and M. R. Wilson, 1998, personal communication). Even more recently, introduced populations of *P. marginata* have been reported in Slovenia (Seljak 2004) in association with *S. maritima* that occurs abundantly throughout the region (Mobberley 1956).

Geographic Source of Introduced Populations

To compare life-history characteristics between native and introduced populations of *P. marginata*, it is critical to accurately identify the geographic source of the introduced population(s). To do so, populations of *P. marginata* were sampled throughout its native and introduced geographic range (Table 21.1), and samples were prepared for DNA analysis. Frozen planthoppers (one to nine individuals per population) were prepared for DNA extraction by first grinding them in a 1.7 ml microfuge tube using a disposable pestle. DNA was subsequently extracted using a QIAamp Tissue Kit (Qiagen, Inc. Valencia, CA), eluted into 200 μl of the supplied buffer, and stored at −20°C. In all, 11 populations were sampled within its native range (7 from Atlantic coastal locations and 4 Gulf coast sites), and 4 populations were sampled from putative introduced populations (2 from California and 2 from Portugal). Samples of two additional delphacid planthopper species, the congener *P. dolus* and the more distantly related *Toya venilia*, were taken to establish the outgroup for phylogenetic analysis. In all, 53 individuals of *P. marginata* and 2 individuals each of *P. dolus* and *T. venilia* were sequenced (Table 21.1). Approximately 450 bp of the planthopper mitochondrial *cytochrome-oxidase I* (COI) gene were amplified using the primers C1-J-1751 (5′-GGATCACCT-GATATAGCATTCCC) and C1-N-2197 (5′-CCCGGTAAAAT-TAAAATATAAACTTC) (Simon et al. 1994). Amplification products were treated with shrimp alkaline phosphatase and exonuclease I to remove unincorporated primer (USB Corp. Cleveland, OH) and were cycle-sequenced on both strands using dideoxy-terminator chemistry and the amplification primers. Labeled sequencing reactions were analyzed on an ABI 3100 capillary sequencer using a 54 cm capillary array and POP-6 electrophoresis matrix (Applied Biosystems, Foster City, CA). Sequences were edited and aligned using the software Sequencher (GeneCodes Corp., Ann Arbor, MI). Editing included truncating all sequences to a common length of 335 bp. Variation in mitochondrial DNA has proved useful for inferring intraspecific relationships (Harrison 1989), and mtDNA sequence data from COI has been used specifically to elucidate relationships over a wide taxonomic range of delphacid planthoppers (Dijkstra et al. 2003).

Subsequently, DNA sequence data from COI were used to construct a haplotype phylogeny. From this phylogeny it was possible to identify broad patterns of phylogeographic structure, assess the probable geographic origin of introduced populations, and confirm the suspected recent timing of invasions. The software Modeltest was used to identify the best model for maximum likelihood analysis for the set of unique *P. marginata* haplotypes (Posada and Crandall 1998). Relationships among mtDNA sequences were estimated in PAUP* V4.10b (Swofford 1996), using neighbor joining on maximum likelihood distances under the HKY + G model (Posada and Crandall 1998) on 1000 bootstrap replicates of the data set. Trees were rooted at *T. venilia*. Individuals from all populations of *P. marginata* and the two outgroup species were included in the analysis (Table 21.1).

Phylogenetic analysis revealed extensive geographic structure among native North American populations of *P. marginata* (Fig. 21.1). Haplotypes of *P. marginata* clustered largely by geographic region with two well-supported sister clades: one comprising haplotypes originating from populations along the mid-Atlantic coast (Virginia north to New York), and another consisting of haplotypes that were found in populations along the south Atlantic coast and Gulf coast. Within the south Atlantic–Gulf coast clade were two other sister lineages, one well-supported group with haplotypes from the western Gulf (Louisiana and Texas) and another unresolved polytomy that includes haplotypes largely from the south Atlantic (South Carolina south to

TABLE 21.1
Locations of *Prokelisia marginata* Populations and Outgroup Planthopper Species

Location	*Sample Date*	*No. Adults Sequenced*
Prokelisa marginata Populations		
Pacific coast		
Skaggs Island, Napa Co., CA	September 26, 1995	3
East Palo Alto, San Mateo Co., CA	September 19, 1995	3
Gulf coast		
S. Padre Island, Cameron Co., TX	May 16, 1994	3
Matagorda, Matagorda Co., TX	May 14, 1994	4
Oak Grove, Cameron Pa., LA	May 12, 1994	2
Pass Christian, Harrison Co., MS	May 12, 1994	4
Atlantic coast		
Captree Island, Suffolk Co., NY	July 8, 1999	2
Tuckerton, Ocean Co., NJ	July 29, 1994	7
Ocean City, Worcester Co., MD	August 15, 1994	2
Virginia Beach, Chesapeake Co., VA	September 24, 1994	9
White Hall Terrace, Berkeley Co., SC	May 23, 1994	3
Fernandina Beach, Nassau Co., FL	September 19, 1993	3
Crescent Beach, St. Johns Co., FL	September 19, 1993	2
Europe		
Olhao, Algarve, Portugal	May 7, 1998	3
Tavira, Algarve, Portugal	October 8, 1997	3
Outgroup Species		
Prokelisia dolus populations		
Tuckerton, Ocean Co., NJ	July 29, 1994	1
Fernandina Beach, Nassau Co., FL	September 19, 1993	1
Toya venilia populations		
Flamingo Pond, Anegada, BVI	October 19, 1997	1
Great Harbor, Jost Van Dyck, BVI	October 18, 1998	1

northern Florida), the eastern Gulf (Mississippi), and even one Virginia haplotype.

Notably, all haplotypes from California and Portugal were identical to each other and to a frequent haplotype found only in populations from the mid-Atlantic clade (New Jersey and Virginia), thus implicating mid-Atlantic populations in North America as the source of introduced planthoppers (Fig. 21.1). Thus, based on this single gene, there was no documented variation among haplotypes from the putative introduced populations.

Genetic Variation in Native and Introduced Populations

An important factor that may influence the ability of introduced populations to evolve in response to the conditions imposed by novel environments is the degree to which the introduction process diminishes population-genetic variation (Hopper et al. 1993; Lee 2002). Both founder effects and postcolonization bottlenecks (Nei et al. 1975) may diminish variation in recently established populations (e.g., Suarez et al. 1999; Hufbauer et al. 2004). As a result, those populations may harbor little variation that could serve as the raw material on which natural selection might act (Endler 1986; Lindholm et al. 2005).

Compared to the mid-Atlantic source populations from which they originated, California populations of *P. marginata* show marked reductions in allelic diversity at a suite of allozyme loci (Peterson and Denno 1997). This result was mirrored by the lack of mitochondrial DNA haplotype variation in populations in both California and Portugal, compared to those populations in the mid-Atlantic source area (Fig. 21.1). Thus, it appears that adaptive evolution in the introduced populations in both regions would have to have occurred in spite of substantial reductions in genetic variation. Interestingly,

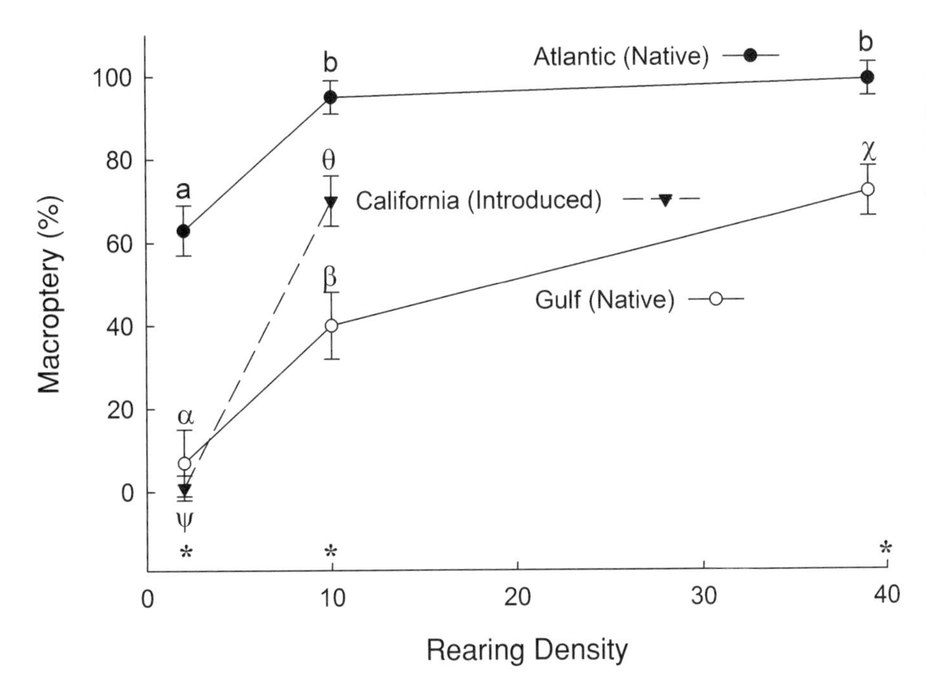

FIGURE 21.3. Effect of rearing density on the frequency of emerging percent macropterous adults of *Prokelisia marginata*. The density-wing-form response is compared between native populations collected on the Atlantic (Tuckerton, NJ) and Gulf (Cedar Key, FL) coasts and from an introduced population in California (Bolinas). Within a population, means (± SEM) with different letters (Roman or Greek) are significantly different ($P < 0.05$). Significant differences between populations (within-density comparisons) are indicated along the abscissa (* = $P < 0.05$; ANOVA followed by Sidak's adjustment for multiple comparisons). Data extracted from Denno et al. (1996) and Roderick (1987).

and 40 individuals per cage) (Denno et al. 1996), whereas individuals from the introduced population in California were raised at densities of 1 and 10 in similar sized cages and on the same number of *Spartina* plants (Roderick 1987).

Results show that Atlantic and Gulf coast populations of *P. marginata* differ dramatically in their tendency to produce macropters (Denno et al. 1996) (Fig. 21.3). For the two native populations of *P. marginata*, more adults molted into macropters as rearing density increased, but the overall incidence of macroptery was significantly higher in the Atlantic population (84%) than in the Gulf coast population (38%). The production of the flight-capable wing form was density dependent in the introduced California population as well (Fig. 21.3). Moreover, the incidence of percent macropter production can be compared among the three populations under common rearing conditions at low (1 to 3 individuals per cage) and moderate (10 individuals per cage) densities. When this was done, the incidence of macroptery was significantly higher under low-density conditions for individuals from the Atlantic population (63 ± 6%) than for those from either the Gulf (7 ± 8%) or California (1 ± 3%) populations (Fig. 21.3). At the moderate rearing density of 10 individuals per cage, the fraction of macropterous adults produced was again significantly higher for the Atlantic coast population (95 ± 4%) than the California population (70 ± 6%), which was significantly more macropterous than the Gulf population (40 ± 8%).

These data suggest that selection for dispersal capability has acted differentially between the Atlantic and Gulf coast populations, resulting in a greater production of macropters at a given density in the Atlantic populations. Notably, density-wing-form responses paralleled those of the field survey, whereby Atlantic coastal populations have a much higher incidence of macroptery (92 ± 2%) than Gulf coast populations (17 ± 5%). Moreover, 35 years after its introduction into California, *P. marginata* shows evidence of reduced dispersal, whereby the same level of crowding results in a lower incidence of macroptery than in a representative Atlantic source population (Fig. 21.3). We do not have comparable density-macroptery data for the introduced populations of *P. marginata* in Portugal. However, a relatively low incidence of macroptery occurs in Portuguese populations (31 ± 20%) (Fig. 21.2) despite moderately high field densities (>750 individuals per m^2) (M. R. Wilson, 1998, personal communication). Field densities such as these (>750 per m^2) in Atlantic coastal marshes are associated with a much higher incidence of macroptery (74%) (Denno and Roderick 1992). Overall, data suggest that the geographic variation in macroptery observed in native populations of the *Prokelisia* planthoppers likely reflects genetically based differences in macropter production and thus dispersal capability (Fig. 21.3). Moreover, data from introduced populations of *P. marginata* in California, and to some extent in Portugal, suggest that the observed reduction in the incidence of dispersal following colonization has a genetic basis. Apparently, in spite of relatively low genetic variation in the introduced populations, selection acting against dispersal has reduced the sensitivity of the developmental switch to crowding, resulting in a lower proportion of flight-capable adults in populations in both regions.

Habitat Factors Underlying Geographic Variation in Dispersal

At a regional spatial scale, the incidence of percent macroptery in native populations of *Prokelisia* planthoppers is inversely related to habitat persistence, indexed as the proportion of each species' population able to endure through winter in its primary habitat for development

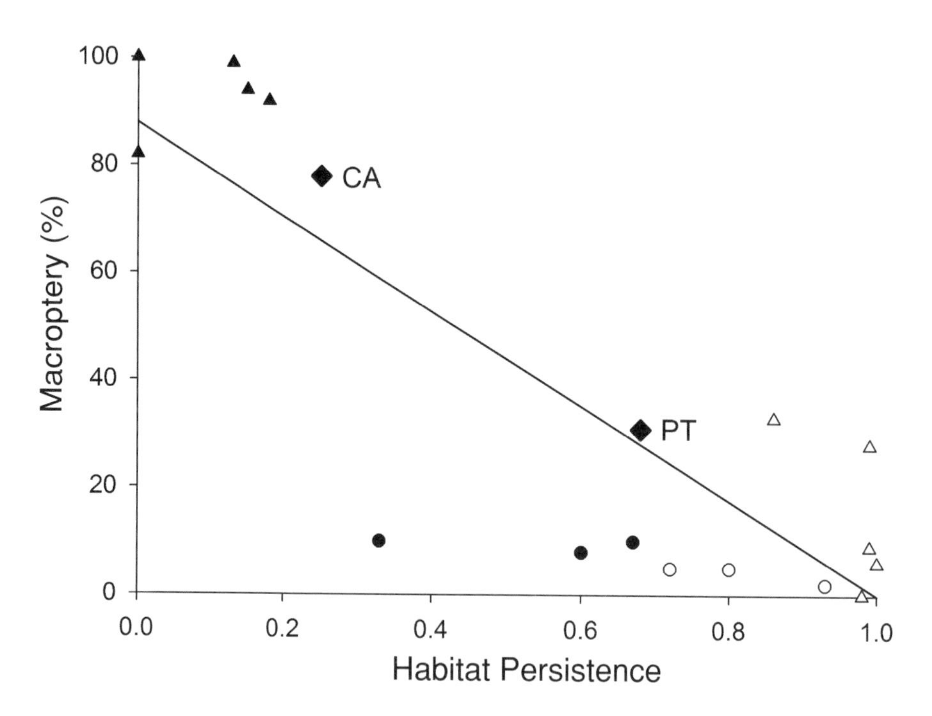

FIGURE 21.4. Relationship between the incidence of percent macroptery observed in populations of *Prokelisia marginata* and *P. dolus* and habitat persistence (proportion of the population persisting through winter in the primary habitat for development: $Y_{\text{Macroptery}} = 89.8 - 90.8X$, $R^2 = 0.75$, $P < 0.001$). *Prokelisia marginata*: native populations on the Atlantic (▲) and Gulf (●) coasts and introduced populations (◆) from California (CA) and Portugal (PT). *Prokelisia dolus*: native populations from the Atlantic (Δ) and Gulf (○) coasts (modified from Denno et al. 1996).

(Denno et al. 1996; Denno and Peterson 2004) (Fig. 21.4). At locations along the Atlantic coast, only a small proportion (<20%) of the *P. marginata* population remains through winter in the primary habitat for development (low marsh), and associated levels of macroptery during the growing season are high (>75%). A combination of ice shearing and the tidal removal of protective leaf litter in low-marsh habitats in the mid-Atlantic states makes persisting through winter in this disturbed habitat difficult (Denno et al. 1996). Thus, from a planthopper's perspective, low-marsh habitats are temporary and exploitable only during the growing season. In this region of heterogeneous marsh structure and harsh winter climate, dispersal is retained because it allows for effective interhabitat migration between overwintering habitats on the high marsh and optimal sites for development during summer that lie hundreds of meters away in low-marsh *Spartina* (Denno 1988; Denno et al. 1996). At Gulf coast locations, a much higher proportion (30 to 70%) of the *P. marginata* population persists through winter in the low marsh, and associated levels of macroptery are relatively low (<15%). The low levels of macroptery in most Gulf coast populations of *P. marginata* are promoted because this species can remain in low-marsh habitats year-round in this climatically equitable region.

For its native congener *P. dolus*, the proportion of individuals remaining through winter in its primary habitat for development (meadows of high-marsh *Spartina*) is high (>70%) at all locations along the Atlantic and Gulf coasts where levels of macroptery are relatively low (<35%) (Fig. 21.4). Because *P. dolus* can remain throughout the year in its primary habitat for development on the less-disturbed high marsh, a low incidence of dispersal is promoted (Denno et al. 1996). Thus, *P. dolus* provides independent confirmation that a low incidence of dispersal is associated with persistent habitats.

One can now ask if the lower incidence of dispersal in the introduced populations of *P. marginata* in California and Portugal compared to the mid-Atlantic source area (91 ± 3%) reflects selection for reduced dispersal extending from more persistent habitats in the invaded habitat. Determining the relationship between habitat persistence and ambient macroptery was possible for the introduced population at Bolinas, California, because both the incidence of macroptery in the field (78%) and the proportion of the population remaining in the low marsh during winter (0.25) are known (Roderick 1987; Denno et al. 1996). This point (CA) is predicted well by the established relationship between percent macroptery and habitat persistence for native populations of *P. marginata* (Fig. 21.4). We conclude from this analysis that the reduced incidence of dispersal in seen in *P. marginata* approximately 35 years after its introduction is attributable in part to the more persistent nature of its optimal habitat for development (low-marsh *Spartina*) in California compared to that along the mid-Atlantic coast. The milder winters in Bolinas (average winter low temperature during December, January and February is 5.9°C) and reduced mean tidal range there (0.91 m) compared to the more severe winters (average winter lows range from −1.4 to −6.3°C) and higher tidal ranges (0.93 to 2.13 m) in the mid-Atlantic states likely contribute to the greater persistence of *P. marginata* in low-marsh habitats in California (Conway and Liston 1974; Ruffner 1980; Pearce and Smith 1990; Tide Tables 1992, 1993; Denno et al. 1996). Note that the incidence of macroptery at Bolinas is greater than the average for populations throughout the San Francisco Bay area (29 ± 4%) (Denno et al. 1996). Thus, it is likely that this population yields an underestimate of the degree to which the tendency to produce macropters in this region has declined over the 35 years following introduction.

diversity of introduced populations of guppies *Poecilia reticulata* in Australia. Mol. Ecol. 14: 3671–3682.

MacArthur, R., and E.O. Wilson. 1967. The theory of island biogeography. Princeton University Press, Princeton, NJ.

Mack, R.N., D. Simberloff, W.M. Lonsdale, H. Evans, M. Clout, and F.A. Bazzaz. 2000. Biotic invasions: causes, epidemiology, global consequences and control. Ecol. Appl. 10: 689–710.

Mann. R. 1979. Exotic species in mariculture. MIT Press, Cambridge, MA.

Marooka, S., and S. Tojo. 1992. Maintenance and selection of strains exhibiting specific wing form and body colour under high density conditions in the brown planthopper, *Nilaparvata lugens* (Homoptera: Delphacidae). Appl. Entomol. Zool. 27: 445–454.

Mattos Machado, T., A. Solé-Cava, J.R. David, and B.C. Bitner-Mathé. 2005. Allozyme variability in an invasive drosophilid, *Zaprionus indianus* (Diptera: Drosophilidae): comparison of a recently introduced Brazilian population with Old World populations. Ann. Soc. Entomol. France (NS) 41: 7–13.

Meffert, L.M. 2000. The evolutionary potential of morphology and mating behavior: the role of epistasis in bottlenecked populations, pp. 177–193. In J.B. Wolf, E.D. Brodie III, and M.J. Wade (eds.), Epistasis and the evolutionary process. Oxford University Press, Oxford, UK.

Mobberley, D.G. 1956. Taxonomy and distribution of the genus *Spartina*. Iowa St. Coll. J. Sci. 30: 471–574.

Nei, M., T. Maruyama, and R. Chakraborty. 1975. The bottleneck effect and genetic variability in populations. Evolution 29: 1–10.

Newton, A., and S.M. Mudge. 2003. Temperature and salinity regimes in a shallow, mesotidal lagoon, the Ria Formosa, Portugal. Estuar. Coast. Shelf Sci. 56: 1–13.

Nieva, F.J. J., E.M. Castellanos, J.M. Castillo, and M.E. Figueroa. 2005. Clonal growth and tiller demography of the invader cordgrass *Spartina densiflora* Brongn. at two contrasting habitats in SW European salt marshes. Wetlands 25: 122–129.

Normais Climatológicas. 2005. Normais Climatológicas 1965–1999. Faro, Intituto de Meterologia, Lisbon.

Normile, D. 2005. Expanding trade with China creates ecological backwash. Science 306: 968–969.

Padinha, C., R. Santos, and M.T. Brown. 2000. Evaluating environmental contamination in Ria Formosa (Portugal) using stress indexes of *Spartina maritima*. Mar. Environ. Res. 49: 67–78.

Pearce, E.A., and G. Smith. 1990. Times Books world weather guide. Random House, New York.

Perrings, C., M. Williamson, and S. Dalmazzone (eds.). 2000. The economics of biological invasions. Edward Elgar Publishing, Cheltenham, UK.

Peterson, M.A., and R.F. Denno. 1997. The influence of intraspecific variation in dispersal strategies on the genetic structure of planthopper populations. Evolution 51: 1189–1206.

Pianka, E. 1970. On *r* and *K* selection. Am. Nat. 104: 592–597.

Pimentel, D., L. Lach, R. Zuniga, and D. Morrison. 2000. Environmental and economic costs of nonindigenous species in the United States. BioScience 50: 53–65.

Posada, D., and K.A. Crandall. 1998. MODELTEST: testing the model of DNA substitution. Bioinformatics 14: 817–818.

Quinn, T.P., M.J. Unwin, and M.T. Kinnison. 2000. Evolution of temporal isolation in the wild: Genetic divergence in timing of migration and breeding by introduced chinook salmon populations. Evolution 54: 1372–1385.

Ranwell, D.S. 1967. World resources of *Spartina townsendii* (sensu lato) and economic use of *Spartina* marshland. J. Appl. Ecol. 4: 239–256.

Raybould, A.F., A.J. Gray, M.J. Lawrence, and D.F. Marshall. 1991. The evolution of *Spartina anglica* C.E. Hubbard (Gramineae): the origin of genetic variability. Biol. J. Linn Soc. 43: 111–126.

Reed, D.H., and R. Frankham. 2001. How closely correlated are molecular and quantitative measures of genetic variation? A meta-analysis. Evolution 55: 1095–1103.

Reznick, D.N., and C.K. Ghalambor. 2001. The population ecology of contemporary adaptations: what empirical studies reveal about the conditions that promote adaptive evolution. Genetica 112/113: 183–198.

Roderick, G.K. 1987. Ecology and evolution of dispersal in California populations of a salt marsh insect, *Prokelisia marginata*. Ph.D. dissertation, University of California, Berkeley.

Roff, D.A. 1986. The evolution of wing dimorphism in insects. Evolution. 40: 1009–1020.

Roff, D.A. 1990. The evolution of flightlessness in insects. Ecol. Monogr. 60: 389–421.

Roff, D. 1992. The Evolution of life histories. Chapman and Hall, New York.

Roff, D.A. 1994. Habitat persistence and the evolution of wing dimorphism in insects. Am. Nat. 144: 772–798.

Roff, D. 2002. Life history evolution. Sinauer, Sunderland, MA.

Roff, D.A., and D.J. Fairbairn. 1991. Wing dimorphisms and the evolution of migratory polymorphisms among the Insecta. Am. Zool. 31: 243–251.

Rosenberg L. J., and J. I. Magor. 1987. Predicting windborne displacements of the brown planthopper, *Nilaparvata lugens* from synoptic weather data. 1. long-distance displacements in the north-east monsoon. Journal of Animal Ecology 56: 39–51.

Ruano, F. 1997. Fisheries and farming of important marine bivalves in Portugal. U.S. Department of Commerce, NOAA Technical Report NMFS 129: 191–200.

Ruffner, J.A. 1980. Climates of the states. Gale Research Company, Detroit, MI.

Sakai, A.K., F.W. Allendorf, J.S. Holt, D.M. Lodge, J. Molofsky, K.A. With, S. Baughman, R.J. Cabin, J.E. Cohen, N.C. Ellstrand, D.E. McCauley, P. O'Neill, I.M. Parker, J.N. Thompson, and S.G. Weller. 2001. The population biology of invasive species. Annu. Rev. Ecol. Syst. 32: 305–332.

Sánchez, J.M., D.G. SanLeon, and J. Izco. 2001. Primary colonization of mudflat estuaries by *Spartina maritima* (Curtis) Fernald in northwest Spain: vegetation structure and sediment accretion. Aquat. Bot. 69: 15–25.

Sayce, K. 1988. Introduced cordgrass *Spartina alterniflora* Loisel. in saltmarshes and tidelands of Willapa Bay, Washington, pp. 1–70. U.S. Fish and Wildlife Service report, USFWS FWSI-87058 TS. Willapa Bay National Wildlife Refuge, Ilwaco, WA.

Schmidt, P.S., L. Matzkin, M. Ippolito, and W.F. Eanes. 2005. Geographic variation in diapause incidence, life history traits, and climatic adaptation in *Drosophila melanogaster*. Evolution 59: 1721–1732.

Seljak, G. 2004. *Prokelisia marginata* (Van Duzee, 1897): a Nearctic planthopper new to Slovenia and Europe (Auchenorrhyncha: Delphacidae). Acta Entomol. Slovenica 12: 260–263.

Shigesada, N., and K. Kawasaki. 1997. Biological invasions: theory and practice. Oxford University Press, New York.

Simon, C., F. Frati, A. Beckenbach, B. Crespi, H. Liu, and P. Flook. 1994. Evolution, weighting, and phylogenetic utility of mitochondrial gene sequences and a compilation of conserved polymerase chain reaction primers. Ann. Entomol. Soc. Am. 87: 651–701.

Southwood, T.R. E., R.M. May, M.P. Hassell, and G.R. Conway. 1974. Ecological strategies and population parameters. Am. Nat. 108: 791–804.

Stearns, S. 1992. The evolution of life histories. Oxford University Press, Oxford, England.

Suarez, A.V., Tsutsui, N.D., Holway, D.A., and T.J. Case. 1999. Behavioral and genetic differentiation between native and introduced populations of the Argentine ant. Biol. Invas.. 1: 43–53.

Swofford, D.L. 1996. PAUP: phylogenetic analysis using parsimony and other methods, version 4.0. Sinauer, Sunderland, MA.

Swofford D. L. 1998. "PAUP* 4.0 beta version." Sinauer, Sunderland, MA.

Thompson, J.N. 1998. Rapid evolution as an ecological process. Trend Ecol. Evol. 13: 329–332.

Throckmorton, A.E. 1989. The effects of tidal inundation, spider predation, and dispersal on the population dynamics of *Prokelisia marginata* (Homoptera: Delphacidae) in north Florida salt marshes. Biological Science FSU: 223.

Tide Tables. 1992. East coast of North and South America. U.S. Department of Commerce, NOAA, National Ocean Service, Washington, DC.

Tide Tables. 1993. West coast of North and South America. U.S. Department of Commerce, NOAA, National Ocean Service, Washington, DC.

Travis, J.M., and C. Dytham. 1999. Habitat persistence, habitat availability and the evolution of dispersal. Proc. R. Soc. Lond. B 266: 723–728.

Turner, R.E. 1976. Geographic variation in salt marsh macrophyte production: a review. Contrib. Mar. Sci. 20: 47–68.

Vince, S.W., I. Valiela, and J.M. Teal. 1981. An experimental study of the structure of herbivorous insect communities in a salt marsh. Ecology 62: 1662–1678.

Wagner, D.L., and J.K. Liebherr. 1992. Flightlessness in insects. Trends Ecol. Evol. 7: 216–220.

Waloff, N. 1983. Absence of wing polymorphism in the arboreal, phytophagous species of some taxa of temperate Hemiptera: an hypothesis. Ecol. Entomol. 8: 229–232.

Webster, D. 2003. Maryland's oyster fishery. Maryland Aquafarmer 3: 5–8.

Williamson, M. 2005. Biological invasions. Kluwer Academic Publishers, Norwell, MA.

Wilson, S.W., C. Mitter, R.F. Denno, and M.R. Wilson. 1994. Evolutionary patterns of host plant use by delphacid planthoppers and their relatives, pp. 7–113. In R.F. Denno and T.J. Perfect (eds.), Planthoppers: their ecology and management. Chapman and Hall, New York.

Wu, M.-Y., S. Hacker, D. Ayers, and D.R. Strong. 1999. Potential of *Prokelisia* spp. as biological control agents of English cordgrass, *Spartina angelica*. Biol. Control 16: 267–273.

Zera, A.J., and R.F. Denno. 1997. Physiology and ecology of dispersal polymorphisms in insects. Annu. Rev. Entomol. 42: 207–231.

criteria that were not affected by learning (Parmesan et al. 1995). After alighting, most of them showed chemical and physical preferences for *Pedicularis* over *Collinsia* (Singer 1983; M.C.S., 1979–2006 personal observations). Those that discriminated among individual *Pedicularis* plants did so adaptively (Ng 1988). After deciding to oviposit, Rabbit butterflies showed positive geotropism and laid clusters averaging 48 eggs per cluster at the base of the plant (Moore 1989).

Target Condition

The Rabbit metapopulation spent around 10 years, from 1979 to 1989, evolving toward greater use of *C. torreyi* as an oviposition host (Singer et al. 1993). We can judge the target condition from the suites of adaptation seen in metapopulations living in the same habitat type as Rabbit, but where *C. torreyi* is the sole and presumably traditional host. Two such metapopulations are shown in Fig. 22.1; of these the one named on the map is Tamarack, about 50 km north of Rabbit. The second, Leek Springs, is the northernmost *Collinsia*-feeding pie in column 6 of Fig. 22.1.

Both Leek Springs and Tamarack are at the same elevation as Rabbit, and all three habitats have the same mixture of potential host species. However, at Tamarack and Leek, *Pedicularis* is not used for oviposition by *E. editha*. All butterflies from these sites that we have tested (>300) preferred *Collinsia* over *Pedicularis*; none were without preference. The preference for *Collinsia* was very strong, with females accepting *Pedicularis* for oviposition only after several days without encountering *Collinsia*. Further, Leek and Tamarack females alighted on *Collinsia* more frequently than did Rabbit butterflies when released in the same area (M.C.S. and D. Boughton, unpublished data).

After alighting, Tamarack females discriminated adaptively among individual *Collinsia* plants, preferring young over old. Having chosen a plant for oviposition they laid small egg clusters (mean of five to six eggs) in the upper half of the plant, not at the base. These observed differences between Rabbit (*Pedicularis*-adapted) and Tamarack (*Collinsia*-adapted) butterflies in oviposition preference, geotropism, and egg cluster size have been maintained in laboratory culture with all larvae raised on a common host (M.C.S., unpublished data).

Rabbit Evolution Phase I

Humans create a novel patch type, which butterflies colonize throughout a network of novel and ancestral patches. The stage for these events was set when logging by the U.S. Forest Service between 1967 and 1982 created a series of habitat patches of a novel type that were available for colonization by *E. editha*. Thus, humans inadvertently created a patch network with two patch types: unchanged patches and novel patches where the insects' host was removed and an alternate host provided. In these novel, logged patches the traditional host of the insects, *P. semibarbata*, had been killed because it is a hemiparasite on trees. At the same time, in these clearings the abundant native annual, *C. torreyi*, was rendered suitable to the insects because its lifespan was extended by the fertilization that followed the logging and burning (Singer 1983; Moore 1989; Boughton 1999).

In undisturbed patches, classified in previous publications as "outcrop" patches because they often contained large boulders (which is why they were not logged), *Pedicularis* was still available, and oviposition on *Collinsia* resulted in almost certain death of offspring from rapid host senescence (Moore 1989; Boughton 1999). We never observed more than minimal use of *Collinsia* in outcrops; in some years we found a single egg cluster, but in most years we found none. Each year from 1979 to 2005 outcrop butterflies have continued to use principally *Pedicularis*, with occasional use of *Castilleja*.

In the early 1980s fitness was highest on *Collinsia* in clearings, lower on *Pedicularis* in outcrops, and negligible on *Collinsia* in outcrops. Consequently, natural selection favored oviposition on *Collinsia* in clearings and avoidance of this host in outcrops (Singer and Thomas 1996). Granted this starting condition, it is not surprising that the Rabbit Meadow butterflies did indeed incorporate *Collinsia* into their diet in clearings. The first clearing had been colonized before our studies began in 1979, and all the large clearings in the patch network were occupied by 1984, across an area of 8 by 10 km.

TABLE 22.2

Preferences of *E. Editha* Captured Emerging from *Collinsia* at Rabbit Meadow, Showing Change Over Time

	Preferring Pedicularis	*No preference*	*Preferring Collinsia*
Years 1979–1984	35	17	3
Year 1989	10	23	10

Rabbit Evolution Phase II, 1980s

High insect densities on novel host in anthropogenic patch type, traditional patches unchanged, rapid evolution of preference, source-sink set up. Artificial enhancement of the suitability of *Collinsia* led to larval densities becoming an order of magnitude higher in clearings than in outcrops, and clearings became net exporters of butterflies (Thomas et al. 1996; Boughton 1999). Initially, butterflies in the clearing populations retained preferences for their traditional host and traditional habitat. In 1979 to 1982 most butterflies emerging in clearings preferred to oviposit on *Pedicularis* despite having developed on *Collinsia* from eggs naturally laid there (Singer 1983; Singer et al. 1992a; Singer and Thomas 1996). However, response to selection did not take long. Table 22.2 shows results of oviposition preference tests performed on butterflies captured as mating pairs immediately after

TABLE 22.3
Preferences of *E. editha* Captured Ovipositing at Rabbit Meadow, 1979–1981

	Preferring Pedicularis	*No preference*	*Preferring Collinsia*
Ovipositing on *Pedicularis*	26	4	0
Ovipositing on *Collinsia*	2	8	12

emerging from *Collinsia* in a single large clearing in 1979 to 1984 and in 1989. Even though all these butterflies had developed on *Collinsia*, the change between the two periods was highly significant. By 1989 preference for *Collinsia* was as frequent as preference for *Pedicularis*, among butterflies developing on *Collinsia*. Table 22.3 shows that these tested preferences were strongly related to actual oviposition choices made in the field. It is clear that evolution of preference for *Collinsia* proceeded rapidly in the clearing habitats during the 1980s (Singer et al. 1993).

Rabbit Evolution Phase III, 1990s

Anthropogenic patch type abandoned, recolonizations from traditional patches obstructed by phenological barrier. Following the rapid evolution toward use of *Collinsia* in the clearing habitat type in the 1980s, the clearing populations suffered a series of climatically driven catastrophes in 1989, 1990, and 1992, resulting in their extirpation in 1992 to 1993 (Singer et al. 1994; Singer and Thomas 1996; Thomas et al. 1996). In 1989 and 1990, mild winters caused early (April) eclosion of clearing adults, compared to the normal flight at 2350 m (June to July). In 1989 this caused starvation of adults that emerged before their nectar sources were in bloom, while in 1990 they were hit by heavy snowfall (normal for April). In 1992 a summer freeze of −5 C on June 15 killed *Collinsias* in the clearings and starved the feeding larvae (Singer et al. 1994; Thomas et al. 1996; Singer and Thomas 1996). Throughout these events, the naturally later-emerging outcrop populations were unaffected.

Recolonization of the clearings did not occur for several years because of a phenological barrier. Insect development was faster in clearings than in outcrops by an average of 10 days, due to hotter surface air temperatures in clearings (Boughton 1999). Rapid development allowed clearing butterflies to use their annual host without heavy mortality from host senescence (Singer 1983; Boughton 1999). In most years, immigrants from outcrops to clearings arrived too late for successful reproduction. Boughton (1999) used insecticide to extirpate patches of the outcrop habitats (containing *Pedicularis*) and found that they were colonized more than 100 times faster than the empty clearing patches.

Boughton concluded that in the 1990s the system had two alternative stable states. One such state had substantial use of both host species but higher densities of insects on *Collinsia* in the clearings, and net export of insects from clearing to outcrop. In this state clearings acted as sources and outcrops as "pseudosinks" (differing from true sinks because they persisted after immigration from clearings ceased). The alternate state had established populations on *Pedicularis* in outcrops and none in clearings, with clearings acting as true sinks. That is, postextinction clearings absorbed immigrants that had no surviving offspring because of a phenological mismatch (adults arrived too late in the season). Switching between these stable states affected natural selection on diet at the metapopulation scale. When clearings were occupied (the 1980s), selection averaged across the system favored acceptance of *Collinsia*; when they were unoccupied (post-1992), selection opposed *Collinsia* preference (Singer and Thomas 1996; Thomas et al. 1996; Boughton 1999). The entire system had been flipped from one of these stable states to the other by a single extreme weather event: the summer freeze of 1992.

Rabbit Evolution Phase IV, 2000s

Novel host abandoned even though it would still support highest fitness if insects had completed their adaptation to it. Just as the spatial constriction of the Schneider population to areas without *Collinsia* in 1989 failed to eliminate preference for *Collinsia* from the population (see above), the extirpation of the clearing populations at Rabbit in 1992 did not eliminate postalighting preferences for *Collinsia* from the population. The proportion of *Collinsia* preferences declined steeply with the reversal of selection, but we still found some insects preferring *Collinsia* in the mid-1990s (Singer and Thomas 1996). Temporary recolonization of clearing patches occurred, starting in 1995, but all were extinct again by 1999 (Boughton 1999).

Since 1999 there have been no established populations in the clearings. Why not? The answer is not simple, but we believe that there is one overriding cause. The clearings themselves have undergone considerable succession since our study at Rabbit began in 1979. Some no longer have suitable *Collinsia* populations at all, but most have *Collinsia* that appears suitable for insects adapted to use this host. We have stated above that butterflies adapted to use *Collinsia* (i.e., from populations monophagous on *Collinsia*, such as Tamarack) made adaptive discriminations among individual *Collinsias*, preferring young over old individuals. Likewise, butterflies adapted to use *Pedicularis* made adaptive discriminations among individual *Pedicularis* plants (Ng 1988).

However, there is a related finding that poses perhaps the most severe problem for host switching. Butterflies adapted to use *Pedicularis*, such as Rabbit butterflies, made *maladaptive* discriminations among *Collinsias*, preferring older plants (that had just finished blooming) over younger, still-blooming *Collinsias*. In another study, Mackay (1985b), working at Rabbit Meadow in the newly colonized clearing, measured size, density, and phenology of a sample of *Collinsias* that bore naturally laid eggs and of a sample that did not. He constructed an algorithm to determine whether a particular *Collinsia* was likely to receive eggs. He then chose a random sample of *Collinsia* plants and placed neonate larvae on them as well as on a sample chosen by his algorithm. Plants chosen by the algorithm as oviposition substrates supported lower larval survival than those chosen randomly (judged less likely to receive eggs). These studies indicate that Rabbit (*Pedicularis*-adapted) butterflies are in a bad position when they are confronted with a phenologically diverse *Collinsia* population—they do not just make poor choices, they make the *worst* choices.

When the clearings were freshly made, the fact that Rabbit butterflies tended to choose the most mature plants was not important because the newly fertilized plants were almost all phenologically suitable for the insects. This is clear from data on the fates of naturally laid eggs on *Collinsia* at Rabbit, that were documented in three years by Moore (1989) who made complete life tables for the butterflies on their two principal hosts. By 2003, however, the clearing *Collinsias* were a mixed bunch, with some still showing strong effects of the original anthropogenic fertilization and some reverting to their original small size and short lifespans—under which circumstances the insects had evolved their local preferences for *Pedicularis*.

In this recent situation, with *Collinsia* habitat declining in quality, and individuals showing very diverse phenology, Rabbit butterflies were experimentally forced to oviposit on *Collinsia* in formerly occupied clearings. Each insect was offered a series of staged encounters with *Collinsia* in the field and allowed to oviposit when it chose to do so. As expected, most Rabbit butterflies chose plants that were destined to cause offspring starvation by early senescence. In contrast, Tamarack (*Collinsia*-adapted) butterflies imported to the Rabbit site and given equivalent manipulated host encounters chose plants that survived much longer.

We (M.C.S., M.B., and B.W., unpublished) followed the fates of these eggs and of eggs laid by Rabbit insects on *Pedicularis*. This experiment was done in 2003 and 2004 in two patch-pairs per year (a patch-pair is an adjacent clearing and outcrop patch). In each of the four patch-pair/year combinations the result was the same: the highest survival was of imported *Collinsia*-adapted insects on *Collinsia*, the second highest fitness was that of Rabbit insects on *Pedicularis,* and the lowest was of Rabbit insects on *Collinsia*. Censuses of the condition of the *Collinsia* 10 days prior to larval diapause indicated that survival of Rabbit larvae would have been extremely low even if they had developed from eggs laid 10 days earlier, as would have been the case if they had been laid by butterflies emerging in clearing populations.

Our conclusion is that the condition described by Moore (1989), Thomas et al. (1996), and Boughton (1999) no longer applied. In 2003 to 2004 it was no longer true that the fitness of Rabbit insects was higher on their novel than on their traditional host. It would not even have been true had the insects developed on *Collinsia* and emerged 10 days earlier. Fitness was higher on the traditional host, *Pedicularis*; the direction of natural selection on diet had been reversed and the insects had responded by abandoning *Collinsia* entirely.

At the same time, the highest fitness in the habitat could still be obtained by feeding on *Collinsia* if the butterflies were well adapted to using it. This was exemplified clearly by the performance of imported *Collinsia*-adapted Tamarack insects. Rabbit insects have abandoned the host that could confer the highest fitness by failing to evolve fast enough the adaptations necessary to maintain populations on that host. Particularly important would be ability to cope with environmental fluctuations that alter the proportion of phenologically suitable plants from year to year.

Over the past 20 years, experimental comparisons between Rabbit butterflies and those from the presumed ancestral state at Colony have shown that, since extinction of clearing habitats in 1992, differences in female host preferences between ancestral (Colony) and disturbed (Rabbit) populations have diminished steadily. In 2004, for the first time, the preference estimates were identical at Rabbit and Colony. We would like to make one more comparison before drawing a conclusion, but we tentatively conclude that the Rabbit metapopulation has returned to its evolutionary starting point before the impact of logging caused colonization of *Collinsia.* In some sense, by returning to their ancestral state, the Rabbit insects have fallen off an adaptive peak.

Mechanics of Evolution of Oviposition Preference at Rabbit: A Plethora of Phenomena

By applying postalighting preference tests in different years to butterflies captured in different patches of the human-impacted Rabbit metapopulation and in the undisturbed Colony metapopulation (representing the ancestral condition at rabbit) we drew the following conclusions.

1. Genetic variation of preference between adjacent patches of different type (clearing/outcrop) superficially appeared to be driven by local selection but was actually driven by assortative flight behavior. Differences between patches took the form that would be expected if it were a response to patch-specific natural selection: butterflies in the clearings were more accepting of the clearing host, *Collinsia* (Singer 1983; Singer and Thomas 1996). However, these differences were not principally responses to selection at all. Instead, they were caused by the insects behaviorally assorting themselves among the patches according to their preference genotypes. This conclusion stems from multiple studies:
   a. Prior experiments had shown preference-associated movement between an adjacent clearing and outcrop patch. Thomas and Singer (1987) had performed

mark-release-recapture and applied postalighting preference tests to butterflies that stayed in their habitats and those that moved between patches. The insects least likely to move were those that preferred the host used in the habitat where they were marked, next were those without preference, while the most likely to move were those preferring the host used in the other habitat.

b. As metapopulation-wide selection shifted from favoring *Collinsia* preference to *Collinsia* rejection, preferences in adjacent clearing and outcrop patches changed in parallel. The parallel nature of the shifts are shown by contingency table analyses, which documented effects of patch type and year, but no significant interaction (Singer and Thomas 1996). After the clearing populations had been extirpated, in 1994, we again applied preference tests to butterflies captured in the Rabbit clearing and outcrop, although the butterflies captured in the clearing had not emerged there but had immigrated from nearby outcrops. These tests showed much reduced acceptance of *Collinsia* compared to the 1980s but exactly the same difference in preference between the patches that we had observed when both patches had contained strong butterfly populations (Singer and Thomas 1996).

c. Subsequently the same effect was confirmed in the much larger metapopulation of a close relative, *Melitaea cinxia,* in Finland (Hanski and Singer 2001; Hanski and Heino 2003). These observations, taken together, suggested that preference-biased movement was maintaining evolutionary divergence between patches of different host composition.

2. Variation of preference among patches of the same type was caused by gene flow. During the 1980s both butterfly density and preference in outcrop patches were associated with isolation from populations in clearings. Not surprisingly, acceptance of the clearing host, *Collinsia*, was highest in outcrop patches that were least isolated from clearing butterflies (i.e., clearing populations were both close and large). After the extirpation of the clearing populations by frost in 1992 there was no longer any association between isolation from clearings and either butterfly density or preference (Thomas et al. 1996). In fact, the patches in the Rabbit metapopulation that had been most different from each other in preference pre-1992 showed identical preference profiles in 1994 (M.C.S. and D. Vasco, unpublished data). Once the biased gene flow from clearings into outcrops had been cut off, preferences rapidly became homogeneous among the outcrop patches.

3. The difference in preference between the disturbed Rabbit metapopulation and the undisturbed Colony metapopulation was due to natural selection. The effects of logging had caused natural selection for acceptance of *Collinsia* in some patches of the Rabbit system, but not at Colony in the undisturbed National Park. After the extirpation of the clearing populations at Rabbit, we compared preferences between insects from an outcrop patch at Rabbit and freshly caught butterflies from Colony. The Rabbit butterflies were significantly more accepting of *Collinsia*, despite that fact that they and the Colony butterflies had developed from eggs laid naturally on *Pedicularis* (Singer and Thomas 1996). This result suggested an evolutionary legacy of gene flow from clearings into outcrops at Rabbit, a legacy that lasted after the clearing populations had been extirpated and *Collinsia* was no longer used.

Third Host Shift, at Sonora Junction

The third host shift represents a natural population extinction and recolonization followed by a diet expansion. By chance, we have observed this natural extinction and recolonization in a population that had been subjected to considerable study. At Sonora Junction in the 1980s, the preferred host of *E. editha* was *Castilleja pilosa*; the second-preferred was *Collinsia parviflora*, and the third-preferred was *Penstemon rydbergii*. With the exception of a single insect that preferred *Penstemon* over *Collinsia*, we found no deviation from these ranks in repeated samples from the population (Singer et al. 1989; Singer and Parmesan 1993; M.C.S., unpublished data). Butterflies have either ranked *Castilleja* > *Collinsia* > *Penstemon* or they have shown no preference. Despite the number-one ranking of *Castilleja*, in some years it has received only 20% of the eggs laid and *Collinsia* has received 80% (Singer et al. 1989). This was because *Castilleja* has been sufficiently rare that searching insects failed to find it before reaching the level of ovipositional motivation at which they would accept the much more abundant *Collinsia*.

Frenchman Lake (Fig. 22.1) is a site with similar plant abundances to Sonora, but with mirror-image host use: at Frenchman, the insects laid most of their eggs on *Penstemon* and only one egg cluster has ever been found on *Collinsia*. A detailed comparison was made between the Sonora and Frenchman populations of the butterfly *(E. editha)* and two host species *(Penstemon rydbergii* and *Collinsia parviflora)*. Our conclusions were that the two populations of *Collinsia* were not very different, but that Frenchman *Penstemon* was genetically more acceptable than Sonora *Penstemon* to butterflies from either site, and that Frenchman butterflies were genetically more *Penstemon*-preferring than those from Sonora (Singer and Parmesan 1993). Sonora butterflies preferred *Penstemon* over *Collinsia* about half the time, when presented with plant pairs from Frenchman rather than from their own site. Thus, the difference between the two sites in plant-insect interaction (observed diet) emerged equally from genetic variation among plants and among insects.

With this background, we turn to the observed diet shift at Sonora. The population became naturally extinct in the late 1990s, was absent for about four years, and the site was recolonized in 2001 or 2002. After recolonization the majority of eggs were laid on *Penstemon* (B.W., R. Plowes, and M.C.S., unpublished data), which, prior to 1998, had been the least-preferred of the three hosts and had only one observed oviposition over several years of censusing (Singer and Parmesan 1993).

Preference tests conducted on the recolonized Sonora population have revealed a dramatic change. While preferences were previously homogeneous (Singer et al. 1989; Singer and Parmesan 1993), they are now extremely diverse. By controlling for variation within plant species by offering all insects the same individual plants, we now find all possible rank orders for the three hosts, *Penstemon, Collinsia,* and *Castilleja*. Any one of these may now be ranked at either the top or the bottom of the preference hierarchy.

Speed of Evolution and Diversity of Preference

Our original study of the Sonora population compared it with Schneider and showed that the Schneider population achieved diversity of diet by diversity of preference rank, while the Sonora population achieved diet diversity by weakness rather than diversity of preference (Singer et al. 1989). At Schneider there were *Plantago*-preferring insects and *Collinsia*-preferring insects, and these preferences strongly influenced the choices that the butterflies actually made in the field. At Sonora the historically high use of *Collinsia* (preextinction) occurred in the absence of butterflies that preferred this host because their preferences for their favored host, *Castilleja*, were weak.

We subsequently compared preference diversity in a series of 10 populations in which diet was not rapidly evolving, and 2 (Rabbit and Schneider) where rapid evolution was occurring. Our conclusion was that preference ranks were significantly more diverse (by rank test) in the rapidly evolving populations (Singer et al. 1994; Thomas and Singer 1998). For reasons that we did not attempt to guess, bouts of rapid diet evolution were accompanied by unusual diversity of preference rank.

We can now add the recent events at Sonora to this set of observations. Diet at Sonora has changed dramatically, and this change has been accompanied by the appearance of preference rank diversity at this site with formerly homogeneous preferences. It is hard to say in what way these events are evolutionary. It is possible that the source for the colonization was a population with host composition and host preferences just like those now present at Sonora. However, we have searched without success for a local *Penstemon*-feeding population from which this recolonization could have occurred. The three closest populations (2, 5, and 7 km) from which recolonization could have occurred were monophagous on *Collinsia*. A more distant population (about 20 km) was monophagous on *Castilleja*. None were oligophagous, as was the extinct Sonora population. At the single site where *Penstemon* occurred, it was rejected in preference trials by the local insects. However, in the two *Collinsia*-feeding populations where both *Penstemon* and *Castilleja* plants were absent, *Penstemon* was relatively quickly accepted (though never most-preferred) in experimental preference trials.

Thus, our physical exploration of the landscape makes us think it most likely that the source population for the recolonization of Sonora was monophagous on, and had preference for, *Collinsia* but contained females that would oviposit on *Penstemon* before dying. The lines of evidence indicate that the Sonora population is now evolving rapidly in response to strong natural selection stemming from differences between the plant communities at Sonora and in the source site.

Types of Anthropogenic Effect

Our results highlight the impacts of humans on evolution in "natural" systems. These examples are part of a widespread and diverse literature on human impacts. When we raise salmon to supplement wild stocks, we inadvertently affect selection on life-history traits and then alter the wild populations into which our cultured fish are released (Heath et al. 2003). When we domesticate plants, we inadvertently apply selection on their defensive chemistry and alter the interactions between those plants and the insects that encounter them (Benrey et al. 1998). When we move plants and herbivorous insects around the world, whether deliberately or not, we expose the plants to novel insects and the insects to novel plants (chapters in this volume: Feder and Forbes, Maron and Vila, Boggs and Ehrlich). Native insects can accumulate rapidly on exotic plants (Strong 1974; Agrawal and Kotanen 2003; Graves and Shapiro 2003). Conversely, insects introduced as agents for biological control of weeds have notoriously attacked plants other than their intended targets (Louda et al. 1997). Which insects feed on which plant species in these dynamic and complex circumstances may depend on unexpected features of their biology. For example, Fox et al. (1997) describe how *Stator* beetles alter their egg size adaptively in response to feeding on native hosts but fail to do so in response to exotic hosts. In consequence, they fail to colonize an exotic host that requires large egg size except where that host grows sympatrically with a native host that induces the insects to mature large eggs.

Range shifts of plants or insects caused by introduction of exotic species can establish novel plant-insect associations without evolutionary change; but, as our study systems shows, evolution is likely to quickly follow. In both the anthropogenic host shifts that we describe here we had access to the presumed ancestral condition and ascertained that at least some insects would have accepted the novel host immediately. However, from the moment that we first observed these novel associations, rapid evolution of host preference was already occurring. Depending on both genetic and ecological factors, rapid evolution may occur both of invasive species and of species that interact with them (Lee 2002; van Klinken and Edwards 2002). Carroll et al. (2005) describe how rapid evolution of beak length in an Australian soapberry bug adapts the bug to feeding on seeds of an exotic balloon vine. Siemann and Rogers (2003) conclude that invasive *Sapium sebiferum* have lost resistance to generalist herbivores.

Cryptically Anthropogenic Effects

When the pattern of insect-host associations changes because of range shifts caused by artificial introduction of exotic insects or exotic plants, the anthropogenic influence is clear. However, sometimes more subtle anthropogenic forces are at work. Both geographic ranges and phenologies of plants and insects are currently being affected by climate change (Grabherr et al. 1994; Parmesan 1996; Fitter and Fitter 2002; Parmesan and Yohe 2003; Root et al. 2003, 2004). However, insect ranges are showing stronger magnitude of response and shorter lag times than are plant ranges (Parmesan 2001). Further, closely related insects show considerable variation in their response. Some insect ranges are shifting poleward to the extent predicted if they are to maintain their climate envelopes, while others are not moving at all, even when suitable habitat appears to exist (Parmesan et al. 1999). Because not all species within a taxon share identical range-shift responses, and because insects and plants are showing systematic differences, anthropogenic climate change will break apart established associations between species and make new ones.

There are strong interactions between insect diet and climate. Nylin (1988) and Scriber and Lederhouse (1992) independently pointed out that insects in seasonal environments must fit an integral number of generations into the available season. When there is only just enough time to do this, insects should be specialists and confine themselves to the host on which they can mature fastest. When there is ample time for X generations but not for $X + 1$, then the diet can be broader and can include plants on which growth rates are not maximal.

As poleward range limits expand under the influence of warming climate, the interactions between host affiliation, growth rate, temperature, and voltinism are played out across physical landscapes with warm and cool microhabitats, north-facing and south-facing slopes, and different species of potential hosts. An educational example is that of the British butterfly *Aricia agestis*, studied by Thomas et al. (2001), in which populations near the northern range boundary had previously adapted to cool conditions by specializing on the host genus, *Helianthemum*, that grows in hot microclimates and hence supports fast larval growth. Climate warming did not initially cause range expansion because *Helianthemum* was absent immediately to the north of the range limit. However, warming did permit rapid evolution of broader diet at the range limit, to a host previously used only by more southern populations, *Geranium*, which grows in cooler microclimates. Once this local diet evolution had occurred, the boundary did expand northward across the band from which *Helianthemum* was absent but *Geranium* was present (Thomas et al. 2001). This example shows how responses to climate change may involve complex interplay between ecological processes such as patch colonization and evolutionary processes, in this case the evolution of diet breadth.

Without very detailed knowledge, the diet evolution of *A. agestis* would not be classified as anthropogenic. We could perhaps classify it as cryptically anthropogenic, in the same class as the evolution of *E. editha* at Rabbit Meadow that we documented above, that involved no exotic species and that would not have been recognized as anthropogenic without a time-series of detailed field data.

A second type of cryptic human influence could be impacting wild species in a "natural" state, stemming from human influence on population extinction and colonization rates. Another way to describe the events at Sonora would be to state that the diet breadth of the newly founded population at this site is greater than that of the older population that it has replaced. This may not be an accident. We have circumstantial evidence that it might be a more general phenomenon. We have noticed a novel relationship in the mtDNA gene tree published by Radtkey and Singer (1995) for a series of *E. editha* populations in California. Among populations, there is a significant inverse relationship between diet breadth and genetic diversity. Populations with broader diets tend to contain fewer mitochondrial genotypes.

A possible cause of this relationship would be an effect of population age on both genetic and diet diversities. Young populations may have low genetic diversity from bottlenecks involved in population founding, and high diet diversity as the insects begin adapting to the local plant community. Older populations may acquire genetic diversity from immigration and mutation while becoming more specialized to feed on the host that is most suitable locally. Janz and Nylin (this volume), looking at phylogenetic evidence across species, deduced that an alternation between specialization and generalization occurs across the large butterfly family Nymphalidae, to which our study insects belong. This phenomenon appears mirrored at a very different scale in our study system. If our line of reasoning is true, diet breadth *within* a species expands and contracts in parallel with population founding and population aging.

Humans have obviously affected population extinction rates, with most human activities either directly or indirectly leading to the demise of many populations. However, habitat alteration can also create suitable habitat (as at Rabbit Meadow), and introduced exotics can become new hosts (as at Schneider's Meadow). It is possible, then, that human-driven alterations of population extinction and colonization rates could, via the mechanisms discussed above, alter genetic diversity and diet breadth, even in apparently natural populations.

Relevance to Conservation

Working with a single butterfly species, we have described three host shifts: one of them apparently natural, one clearly anthropogenic involving an exotic host, and one that is cryptically anthropogenic. From a conservation perspective, what are the practical consequences of the types of anthropogenic change we have observed? The events at Rabbit seem innocuous; those at Schneider do not. At Rabbit, humans altered the landscape and the alteration was followed by succession and a return of the habitat to a condition resembling its initial state. Insect diet changed first to include a novel host and then to exclude it as succession rendered it less suitable. These changes occurred through a combination of plasticity and evolution in response to selection (Singer et al. 1989; 1992b). Once the novel host had been colonized, rapid evolution of preference occurred and then, equally rapidly, reversed itself. (If this happened repeatedly we might call it "peristaltic" evolution.) Despite the fact that the metapopulation does not now stand on the highest available adaptive peak, it is not significantly worse off than prior to human intervention, and it seems no less able to adapt to future changes.

In contrast, the Schneider population has become both more fully adapted to the anthropogenic changes and more dependent on the nature of future human activities. In our original account (Singer et al. 1993) we suggested that, if this population were to complete the host shift to *Plantago* (as it has now effectively done), it would become vulnerable to the same types of change that have doomed so many European butterfly populations that used to feed on this same host. Specifically, abandonment of traditional farming methods in Europe has, by allowing succession and cooling the microclimate, harmed the insects much more than their hosts.

Evolutionary athlete though it be, Edith's checkerspot is not likely to be able to change as fast as humans can alter the landscape at Schneider's Meadow. We shall see, in fact, because its ability to do this is about to be put to the test. Schneider's Meadow lies within the city limits of Carson City, the capital city of Nevada. The meadow has been sold, and the developer plans to render it into a golf course and upscale housing development. Legal challenges to this plan are ongoing as we write. The almost-abandoned traditional host, *Collinsia,* grows mostly among sagebrush, in drier microhabitats than the *Plantago* that grows around the edges of the spring-fed wet meadow. The golf course is likely to remove more *Plantago* than *Collinsia*. If so, the insects would have better prospects now if they had not abandoned their traditional host.

Consequences for a Species Adopting an Exotic Host

The incorporation of *P. lanceolata* into the diet of North American species is not surprising, since in Europe this plant is a frequent host to Melitaeine butterflies (Tolman and Lewington 1997). In the United States at least four independent colonizations of *P. lanceolata* by native butterflies have been observed. The butterflies include *Junonia coenia* (Bowers 1984), *Euphydryas phaeton* (Bowers et al. 1992), and *E. editha* (Singer et al. 1993; P. Severns, 2006, personal communication). However, use of the exotic *P. lanceolata* has not helped threatened species recover.

Plantago lanceolata is used by at least two threatened subspecies of *E. editha: E. e. monoensis* at Schneider's Meadow has been known to oviposit on *P. lanceolata* since at least the 1950s, and *E. e. taylori* in the vicinity of Corvallis has used *P. lanceolata* since at least the 1960s (D.V. McCorkle, 1970, personal communication). Yet in neither case has the ability to feed on this widely distributed exotic weed led to perceptible mitigation of these insects' threatened status. *Euphydryas e. taylori* is now thought to be extinct in Canada, even though it had evolved to the point where *P. lanceolata* was its only recorded host, and this exotic plant is still abundant throughout the region.

In fact, to date, this evolution of using *P. lanceolata* in existing populations has never been reported to have led to the founding of new populations in habitats containing this host. For example, regional censuses suggest that *E. editha* existed at Schneider's Meadow prior to the arrival of *P. lanceolata*. Schneider's Meadow habitat closely resembles others in the region where *P. lanceolata* does not (yet) grow and where the local *E. editha* feed on *Collinsia parviflora* and/or *Penstemon rydbergii*. So, the probable sequence of events is that *P. lanceolata* was introduced by humans to Schneider's Meadow and the insects simply included it in their diet without changing their spatial distribution. This population of *E. editha* is classed in the subspecies *E. e. monoensis*, which occurs along the eastern slopes of the Sierra Nevada and which is currently known from only five or six populations. It is a puzzle that this threatened taxon has apparently not founded new populations on its newly acquired, widely distributed exotic host, since a high proportion of female emigrants from the Schneider's Meadow population prefer to oviposit on *P. lanceolata* and would therefore tend to colonize habitat patches containing this host (Hanski and Singer 2001).

The continued rarity of *E. e. taylori* is equally puzzling. The historical range of *E. e. taylori* extended from central Oregon to Vancouver Island (British Columbia) at low elevations. It is currently thought to be extinct in Canada but is still known from two or three small populations in Washington (D. Grossboll, 2004, personal communication). One of the Washington populations gives a clue to the original diet: it feeds on a native (now rare) *Castilleja*. *Euphydryas e. taylori* had been presumed extinct from Oregon until 1998, when it was rediscovered at Corvallis by A. Warren. This rediscovered population inhabits a meadow dominated by exotic herbs where detailed censuses show that it is currently monophagous on *P. lanceolata* (P. Severns, 2006, personal communication). The exotic *P. lanceolata* cannot be the traditional host, but almost all populations recorded in the 1950s and 1960s in Oregon and Canada had never been observed to use any host other than *Plantago*. Yet this apparent complete evolutionary shift to an abundant and widespread exotic did not prevent extinction of nearly all of these populations, and subsequent candidacy for listing of *E. e. taylori* under the U.S. Endangered Species Act.

In our original account of the rapid evolution at Schneider's Meadow (Singer et al. 1993) we pointed out that the insects were both adapting to human land management techniques and becoming dependent on them. *Plantago lanceolata* is only a suitable host for Melitaeine butterflies where it grows in short sward and larvae can thermoregulate in early spring in the hot microclimate close to the ground, by basking on soil or rocks. This situation is maintained in European habitats by grazing or mowing for hay, and Melitaeine butterflies are often abundant in traditionally maintained hay-meadows. When these meadows are abandoned, the vegetation becomes denser and the butterflies die out, even though their hosts may still be abundant. Widespread local extinctions of European melitaeines have resulted from abandonment of pasture and hay-meadows as well as from conversion of land to more intensive agriculture. Although many North-American populations of *P. lanceolata* currently appear physically suitable for butterfly larvae, the European experience shows that insects that use those populations thereby become vulnerable to rapid changes in land use. If they were to look to the very long term, North American checkerspots might be advised to beware the immediate appeal of the exotic *P. lanceolata*.

Conclusions

In conclusion, we know a great deal about anthropogenic evolution caused by our direct attempts to control weeds or pathogens and by deliberate genetic modification of crops. However, we know much less about anthropogenic effects that are caused by inadvertent rather than by deliberate human intervention. Our discussion raises several questions. Can we quantify the role that humans are currently playing in evolution of "natural" systems and specifically in plant-insect interactions? How often does it happen that we inadvertently cause evolution, fail to notice this, and then fail to predict the effects of evolved dependence on our activities that threatened taxa have evolved? These are important questions but little studied as yet. We hope that our discussion here will help motivate readers to go and find some of the answers.

Acknowledgments

We are much grateful to Gwen Gage for the artwork; to Carol Boggs, Paul Ehrlich, Niklas Janz, Sören Nylin, and Kelley Tilmon for their careful critiques of the manuscript; to Nikhil Advani and Rob Plowes for permission to cite their unpublished data, and to Daniel Bolnick, Davy Boughton, Jim Bull, Tom Juenger, Mark Kirkpatrick, Jim Mallet, Sasha Mikheyev, and Chris Thomas for their mostly cheerful comments and advice over the course of this project. Camille Parmesan helped greatly with the final draft. The National Science Foundation provided funding (DEB-0215436) to M.C.S. and B.W.

References Cited

Agrawal, A. A., and P. M. Kotanen. 2003. Herbivores and the success of exotic plants: a phylogenetically controlled experiment. Ecol. Lett. 6: 712–715

Bale J. S., G. J. Masters, I. D. Hodkinson, C. Awmack, T. M. Bezemer, V. K. Brown, J. Butterfield, A. Buse, J. C. Coulson, J. Farrar, J. E. G. Good, R. Harrington, S. Hartley, T. H. Jones. R. L. Lindroth, M. C. Press, I. Symrnioudis, A. D. Watt, and J. B. Whittaker. 2002. Herbivory in global climate change research: direct effects of rising temperature on insect herbivores. Global Change Biol. 8: 1–16.

Benrey, B., A. Callejas, L. Rios, K. Oyama, and R. F. Denno. 1998. Effects of domestication of *Brassica* and *Phaseolus* on the interaction between phytophagous insects and parasitoids. Biol. Control 11: 130–140.

Boughton, D. A. 1999. Empirical evidence for complex source-sink dynamics with alternative states in a butterfly metapopulation. Ecology 80: 2727–2739.

Bowers, M. D. 1984. Iridoid glycosides and hostplant specificity in larvae of the buckeye butterfly, *Junonia coenia* (Nymphalidae). J. Chem. Ecol. 10: 1567–1577.

Bowers, M. D., N. E. Stamp, and S. Collinge. 1992. Early stage of host-range expansion by a specialist herbivore, *Euphydryas phaeton* (Nymphalidae). Ecology 73: 526–536.

Bradshaw, W. E., and C. M. Holzapfel. 2001. Genetic shift in photoperiodic response correlated with global warming. Proc. Natl. Acad. Sci. USA 98: 14509–14511.

Carroll, S. P., J. E. Loye, H. Dingle, M. Mathieson, T. R. Famula, and M. P. Zalucki. 2005. And the beak shall inherit: evolution in response to invasion. Ecol. Lett. 8: 944–951.

Ehrlich, P. R. 1965. The population biology of the butterfly *Euphydryas editha*. II. The structure of the Jasper Ridge Colony. Evolution 19: 327–336.

Ehrlich, P. R., and I. Hanski. 2004. On wings of checkerspots: a model system for population biology. Oxford University Press, New York.

Ehrlich, P. R., R. R. White, M. C. Singer, S. W. McKechnie, and L. E. Gilbert. 1980. Extinction, reduction, stability and increase: the responses of checkerspot butterflies *(Euphydryas editha)* to the California drought. Oecologia 46: 101–105.

Fitter, A. H., and R. S. R. Fitter. 2002. Rapid changes in flowering time in British plants. Science 296: 1689–1691.

Ford, E. B. 1945. Butterflies. Collins, London.

Fox, C. W., J. A. Nilsson, and T. A. Mousseau. 1997. The ecology of diet expansion in a seed-feeding beetle: pre-existing variation, rapid adaptation and maternal effects? Evol. Ecol. 11: 183–194.

Grabherr, G., M. Gottfried, and H. Pauli. 1994. Climate effects on mountain plants. Nature 369: 448.

Graves, S. D., and A. M. Shapiro. 2003. Exotics as host plants of the California butterfly fauna. Biol. Conserv. 110: 413–433.

Haag, C. R., M. Saastamoinen, J. H. Marden, and I. Hanski. 2005. A candidate locus for variation in dispersal rate in a butterfly metapopulation. Proc. R. Soc. Lond. B 272: 2449–2456.

Hanski, I. 1999. Metapopulation ecology. Oxford University Press, New York.

Hanski, I., and M. Heino. 2003. Metapopulation-level adaptation of insect host plant preference and extinction-colonization dynamics in heterogeneous landscapes. Theor. Popul. Biol. 64: 281–290.

Hanski, I., and M. C. Singer. 2001. Extinction-colonization dynamics and host-plant choice in butterfly metapopulations. Am. Nat. 158: 341–353.

Harrison, S. D., D. D. Murphy, and P. R. Ehrlich. 1988. Distribution of the Bay Checkerspot, *Euphydryas editha bayensis*: evidence for a metapopulation model. Am. Nat. 132: 360–382.

Heath, D. D., J. W. Heath, C. A. Bryden, R. M. Johnson, and C. W. Fox. 2003. Rapid evolution of egg size in captive salmon. Science 299: 1738–1740.

Hill, J. K., C. D. Thomas, and D. S. Blakeley. 1999. Evolution of flight morphology in a butterfly that has recently expanded its geographic range. Oecologia 121: 165–170.

Huey, R. B., G. W. Gilchrist, M. L. Carlson, D. Berrigan, and L. Serra. 2000. Rapid evolution of a geographic cline in size in an introduced fly. Science 287: 308–309.

Jefferies, R. L., and J. L. Maron. 1997. The embarrassment of riches: atmospheric deposition of nitrogen and community and ecosystem processes. Trends Ecol. Evol. 12: 74–78.

Lee, C. E. 2002. Evolutionary genetics of invasive species. Trends Ecol. Evol. 17: 386–391.

Louda, S. M., D. Kendall, J. Connor, and D. Simberloff. 1997. Ecological effects of an insect introduced for the biological control of weeds. Science 277: 1088–1090.

Mackay, D. A. 1985a. Prealighting search behavior and host plant selection by ovipositing *Euphydryas editha* butterflies. Ecology 66: 142–151.

Mackay, D. A. 1985b. Conspecific host discrimination by ovipositing *Euphydryas editha* butterflies: its nature and its consequences for offspring survivorship. Res. Popul. Ecol. 27: 87–98.

Maron, J. L., and M. Vila. 2001. When do herbivores affect plant invasion? Evidence for the natural enemies and biotic resistance hypotheses. Oikos 95: 361–373.

Moore, S. D. 1989. Patterns of of juvenile mortality within an oligophagous insect population. Ecology 70: 1726–1737.

Ng, D. 1988. A novel level of interactions in plant-insect systems. Nature 334: 611–612.

Nylin, S. 1988. Host plant specialization and seasonality in a polyphagous butterfly, *Polygonia c-album* (Nympalidae). Oikos 53: 381–386.

Parmesan, C. 1996. Climate and species' range. Nature 382: 765–766.

Parmesan, C. 2001. Coping with modern times? Insect movement and climate change. In I. Woiwod, D. R. Reynolds, and C. D. Thomas (eds.), Insect movement: mechanisms and consequences. CAB International, Wallingford, UK

Parmesan, C. 2005. Range and abundance changes, ch. 12. In T. Lovejoy and L. Hannah (eds.), Climate change and biodiversity. Yale University Press, New Have, CT.

Parmesan, C., and G. Yohe. 2003. A globally coherent fingerprint of climate change impacts across natural systems. Nature 421: 37–42.

Parmesan, C., M. C. Singer, and I. Harris. 1995. Absence of adaptive learning from the oviposition foraging behaviour of a checkerspot butterfly. Anim. Behav. 50: 161–175.

Parmesan, C., N. Ryrholm, and C. Stefanescu. 1999. Poleward shifts in geographical ranges of butterflies associated with regional warming. Nature 399: 579–583.

Pounds, J. A., M. P. L. Fogden, and J. H. Campbell. 1999. Biological response to climate change on a tropical mountain. Nature 398: 611–615.

Radtkey, R. R., and M. C. Singer. 1995. Repeated reversals of host-preference evolution in a specialist insect herbivore. Evolution 49: 351–359.

Rausher, M. D. 1982. Population differentiation in *Euphydryas editha*: larval adaptation to different hosts. Evolution 36: 581–590.

Root, T. L., J. T. Price, K. R. Hall, S. H. Schneider, C. Rosenzweig, and A. Pounds. 2003. "Fingerprints" of global warming on wild animals and plants. Nature 421: 57–60.

Root, T. L., D. P. MacMynowski, M. D. Mastandrea, and S. S. Schneider. 2004. Human-modified temperatures induce species changes: joint attribution. Proc. Natl. Acad. Sci. USA 102: 7465–7469.

Schlaepfer, M.A., P. W. Sherman, B. Blossey, and M. C. Runge. 2005. Introduced species as evolutionary traps. Ecol. Lett. 8: 241–246.

Scriber, J. M., and R. C. Lederhouse. 1992. The thermal environment as a resource dictating geographic patterns of feeding specialization of insect herbivores, pp 429–466. In M. R. Hunter, T. Ohgushi, and P. W. Price (eds.), Effects of resource distribution on plant-animal interactions. Academic Press, New York.

Siemann, E., and W. E. Rogers. 2003. Reduced resistance of invasive varieties of the alien tree *Sapium sebiferum* to a generalist herbivore. Oecologia 135: 451–457.

Singer, M. C. 1971. Evolution of food-plant preference in the butterfly *Euphydryas editha*. Evolution 25: 383–389.

Singer, M. C. 1983. Determinants of multiple host use in a phytophagous insect population. Evolution 37: 389–403.

Singer, M. C. 1984. Butterfly—host plant relationships. In R. I. Vane-Wright and P. R. Ackery (eds.), The biology of butterflies. Symposia of the Royal Entomological Society XIII, London.

Singer, M. C., and C. Parmesan. 1993. Sources of variation in patterns of plant-insect interaction. Nature 361: 251–253.

Singer, M. C., and C. D. Thomas. 1996. Evolutionary responses of a butterfly metapopulation to human and climate-caused environmental variation. Am. Nat. 148: S9–S39.

Singer, M. C., D. Ng, and C. D. Thomas. 1988. Heritability of oviposition preference and its relationship to offspring performance within a single insect population. Evolution 42: 977–985.

Singer, M. C., C. D. Thomas, H. L. Billington, and C. Parmesan. 1989. Variation among conspecific insect populations in the mechanistic basis of diet breadth. Anim. Behav. 37: 751–759.

Singer, M. C., D. Ng, D. Vasco, and C. D. Thomas. 1992a. Rapidly evolving associations among oviposition preferences fail to constrain evolution of insect diet. Am. Nat. 139: 9–20.

Singer, M. C., D. A. Vasco, C. Parmesan, C. D. Thomas, and D. Ng. 1992b. Distinguishing between "preference" and "motivation" in food choice: an example from insect oviposition. Anim. Behav. 44: 463–471.

Singer, M. C., C. D. Thomas, and C. Parmesan. 1993. Rapid human-induced evolution of insect-host associations. Nature 366: 681–683.

Singer, M. C., C. D. Thomas, H. L. Billington, and C. Parmesan. 1994. Correlates of speed of evolution of host preference in a set of twelve populations of the butterfly *Euphydryas editha*. Ecoscience 1: 107–114.

Spiller, D. A., and A. A. Agrawal. 2001. Intense disturbance enhances plant susceptibility to herbivory: natural and experimental evidence. Ecology 84: 890–897.

Strong, D. R. 1974. Rapid asymptotic species accumulation in phytophagous insect communities: the pests of cacao. Science 185: 1064–1066.

Thomas C.D. 2001. Ecological and evolutionary processes at expanding range margins. Nature 411: 577–581.

Thomas, C.D. , E.J. Bodsworth, R.J. Wilson, A.D. Simmons, Z.G. Davies, M. Musche, and L. Conradt. 2001. Ecological and evolutionary processes at expanding range margins. Nature 411: 577–581.

Thomas, C.D., D. Ng, M.C. Singer, J.L.B. Mallet, C. Parmesan, and H.L. Billington. 1987. Incorporation of a European weed into the diet of a North American herbivore. Evolution 41: 892–901.

Thomas, C.D., and M.C. Singer. 1987. Variation in host preference affects movement patterns within a butterfly population. Ecology 68: 1262–1267.

Thomas, C.D., and M.C. Singer. 1998. Scale-dependent evolution of specialization in a checkerspot butterfly, pp. 343–374. In S. Mopper and S.Y. Strauss (eds.), Genetic structure and local adaptation in natural insect populations. Chapman and Hall, London.

Thomas, C.D., M.C. Singer, and D.A. Boughton. 1996. Catastrophic extinction of population sources in a butterly metapopulation. Am. Nat. 148: 957–975.

Tolman, T., and R. Lewington. 1997. Butterflies of Britain and Europe. Harper Collins, London.

van Klinken, R.D., and O.R. Edwards. 2002. Is host-specificity of weed biological control agents likely to evolve rapidly following establishment? Ecol. Lett. 5: 1088–1090.

Wahlberg, N. 2000. The ecology and evolution of Melitaeine butterflies. Ph.D. dissertation, University of Helsinki.

Warren, M.S., and J.K. Hill. 2001. Rapid responses of British butterflies to opposing forces of climate and habitat change. Nature 414: 65–69.

TWENTY-THREE

Conservation of Coevolved Insect Herbivores and Plants

CAROL L. BOGGS AND PAUL R. EHRLICH

It goes without saying that one cannot consider the conservation of plants without thinking about their relationship with the creatures that eat them—and herbivorous insects have long been among the worst enemies of the global flora (Becerra 1997, 2003, 2005). And as anyone who has worked extensively with butterflies or other herbivorous insects can tell you, the distribution, abundance, and phenology of food plants are absolutely key to understanding the dynamics and conservation of their populations. It is no accident that the entire field of coevolution sprung from a study of the reciprocal evolutionary interaction of butterfly caterpillars and their photosynthesizing victims (Ehrlich and Raven 1964). The prominence and popularity with amateurs of butterflies makes them the best-known large group of insect herbivores and an important model system in population biology (Boggs et al. 2003). Similarly, many of the examples in what follows will be drawn from work on that system.

Thus, conserving either herbivorous insects or plants automatically requires consideration of the other group, their coevolutionary partners. The dramatic effect that insect herbivores can have on plant populations was demonstrated years ago by the control of a vast *Opuntia* cactus invasion in Australia by the introduction of a coevolved predator, the moth *Cactoblastis cactorum* Berg (Mooney and Hobbs 2000). So was the powerful evolutionary impacts that even a small insect herbivore could have on relatively large plants, as when the tiny lycaenid, *Glaucopsyche lygdamus* Doubleday, was shown to dramatically reduce seed set of lupine plants (Breedlove and Ehrlich 1968). The other side of the coin is even more obvious—known to every butterfly collector who observes populations disappearing as larval food plants are extirpated by development (e.g., Ehrlich and Hanski 2004, p. 60). The scale of human activities is such that many (and potentially all) populations of plants and herbivorous insects are being influenced by those activities, and thus so are patterns of their coevolution.

The drivers that endanger coevolved populations are, of course, the same ones that endanger organisms in general. Humanity probably influences interactions of all organisms by brute destruction of habitats, and insect herbivores–plant interactions are no exception. Invasive plants can disrupt coevolutionary complexes by arriving in a new location without their native predators, outcompeting the native host plants of specialist insect herbivores or creating dangerous fire-prone thatch (Powell and Parker 1993). In the process, they can disrupt ecosystem services. For instance, in places where Latin American asters of the genus *Eupatorium* invade, many problems are exacerbated when the plant is no longer imbedded in complexes with coevolved insect herbivores such as the tephritid gall fly *Procecidochares utilis* Stone (Groves 1989): the stock carrying capacity of grazing lands is often reduced, and succession in milpa agriculture is slowed, among other problems.

Changes in disturbance regimes due either directly to human activities or indirectly to introductions as previously noted can threaten host plants, their coevolved herbivorous insects, and the evolutionary dynamic between them. Climate change demonstrably can affect herbivore-plant complexes, and again likely affect the evolutionary dynamic within such complexes. For example, overharvesting of plants has had devastating effects on some cactus populations (Hernandez 1995; Ortega-Baes and Godinez-Alvarez 2006), and this clearly alters their coevolution with specialist insect herbivores. It reduces the food resources available to the insect populations and puts a selective premium on individuals with a tendancy to oviposition "errors" and/or less specialized adaptation to host defenses. In contrast, no butterfly is likely to have become globally extinct from overharvesting (although populations of some lycaenid species such as *Lycaena dispar* Haworth and *Plebejus sephirus kovacsi* Szabo (Balint 1993a, 1993b) have been hurried toward extinction by butterfly collectors).

Not only do these threats affect the population dynamics of species involved in plant-herbivore interactions, they also affect the evolutionary dynamics underlying the coevolutionary interaction. Much less is known about the effects of anthropogenic threats on these dynamics, yet they are equally critical to conservation of biodiversity in the longer run. These dynamics shape species interactions at multiple scales, producing biogeographic mosaics of types, strengths, and identities of interacting species pairs at the landscape scale as a result of adaptation and coevolution at the regional to local scale (Thompson 2005). The resulting spatial patterns, which are readily disrupted by human activity, are an important component of global-scale biodiversity.

Below we address in more detail each of the sources of conservation threat, both to coevolved populations of plants and their insect herbivores, and to the coevolutionary dynamics themselves. Where possible, we put our analysis in the context of spatial scale and complexity. Our aim is to provide a conceptual overview of current knowledge of the conservation problems associated with plant–insect herbivore coevolved interactions.

Habitat Destruction

The coevolutionary linkages between insects and their host plants partially determine the conservation impacts of habitat destruction. In the simplest analysis, habitat destruction carried out at a large enough scale results in loss of host plants, and subsequent loss of their dependent insects. As expected, broad-brush diversity studies on native butterflies in California (Hawkins and Porter 2003) and rare herbivorous insects in Great Britain (Hopkins et al. 2002) indicate a positive correlation between plant diversity and herbivorous insect diversity, although in California much of this relationship is accounted for by environmental variables (evapotranspiration and topography) (Hawkins and Porter 2003).

Examples of Direct Effects of Habitat Destruction

Although little studied, the sparse distributions of some of the few relatively well-known herbivorous species in Amazonia, such as riodinine butterflies (K. S. Brown Jr. 1993), suggest that even small areas of deforestation can exterminate long-coevolved associations. Much better documented are the fates of a series of San Francisco area populations of the Bay checkerspot butterfly (*Euphydryas editha* Boisduval) that have gone extinct as their habitats have disappeared under homes and freeways (Ehrlich and Hanski 2004). Habitat destruction has also been responsible for the extirpation of many European populations of lycaenid butterflies (Munguira et al. 1993). Wetland drainage is a classic example of a form of habitat destruction that wipes out plant–insect herbivore complexes. An example is the drainage of British fenlands (wet meadows) for agriculture in recent centuries. Drying and subsequent scrub invasion of the well-studied Wicken Fen wet meadows in Cambridgeshire suggests the fate of the plants and herbivorous insects of British fenlands. Thirty-five species of plants have gone extinct, as have four butterfly species, more than four moths, and a weevil (Colston and Friday 1999). The flagship butterfly species of wetland loss is the large copper, *Lycaena dispar*. It went extinct in Britain and was reestablished with Dutch stock, a marginal, heavily managed population that persists. Elsewhere in central and western Europe it is declining because of wetland loss and reduction of growth and nutrient quality of its coevolutionary "partner," *Rumex hydrolapathum* Hudson, the great water dock (Duffy 1993). Whether that change in nutrient quality was selected for in the dock's coevolutionary "race" (Ehrlich 1970) with *L. dispar* is unknown.

Note that a plant does not need to be forced to extinction for its coevolutionary relationship with an herbivorous insect to be shattered. For example, today one of the main oviposition plants for *E. editha*, *Plantago erecta* Morris, occurs frequently in sparse populations on nonserpentine soils. But in the San Francisco Bay area it is only on serpentine soils where it develops the thick "lawns" required to nourish larvae and support a checkerspot population. As a result, the butterfly is restricted to serpentine habitat and is threatened with extinction due to habitat loss.

The Role of Host Range

Koh et al. (2004c) developed a model to predict species coextinction rates of tropical butterflies in response to extinction of their hosts. Coextinction rates depended on the degree of polyphagy by the insects. Estimates for coextinction of butterflies, for example, increase exponentially with increasing extinction of host plants. Small numbers of plant extinctions are predicted to have relatively little overall impact on butterfly diversity, since many butterflies are polyphagous. However, butterfly extinctions increase quite rapidly at higher levels of plant extinctions, as additional monophagous and now polyphagous butterfly species are affected. The analysis of Koh et al. (2004c) suggests that less than 0.1% of beetles and butterflies are now extinct due to loss of their host plants (primarily to habitat loss), whereas 0.4% (4672 species) of beetles and 0.8% (142 species) of butterflies are at risk, given current plant endangerment records. In another analysis, Koh et al. (2004a) predicted expected numbers of butterfly extinctions in Singapore based on plant extinctions. Observed butterfly extinctions were significantly greater than predicted, suggesting that herbivores may go extinct before their host plants due to a necessary host-plant density threshold and/or to other factors.

Human disturbance can also create new types of habitat, with novel hosts, for some species. If variation exists for host-plant preference by ovipositing females, insects can respond to disturbance by switching to use of a new alternate plant whose availability is associated with disturbance, potentially enhancing the insects' population sizes and survival probabilities. Singer et al. (1993) demonstrate this for two populations of the butterfly *E. editha* in the Sierra Nevada. Habitat disturbance was in the form of logging in one case, and introduction of a new species via grazing activities in the second. In both instances, butterflies remained in the disturbed habitat and began utilizing hosts that were within the species' repertoire but not previously used by the population under study. The frequencies of butterfly's host-plant preferences evolved after the disturbances, suggesting that dependence by the butterfly on human disturbance regimes, and the plants fostered by those regimes, could result. (See also M. C. Singer, this volume.) However, if the population retains genetic diversity for host-plant use, loss of the disturbed habitat will not be lethal. For the *E. editha* population in logged habitat, Thomas et al. (1996) showed that the logged areas served as a source and the old habitat as a sink due to the success of the butterfly in the logged area. When a later frost killed the new host in the logged (source) habitat, the butterfly retracted back to, and persisted in, its original area. This case study also demonstrates the relative nature of source and sink populations, providing a cautionary tale for conservationists focused on preserving only current source populations.

These studies suggest the more general hypothesis that host-plant breadth determines the ease of conservation of particular insects. Monophagous insects are hypothesized to be more prone to endangerment due to habitat destruction, and harder to conserve than are poly- or oligophagous insects. This is due both to lack of substitutability of host plants for insects with narrow host ranges and to more restricted habitat requirements. Working on the urbanized island of Singapore, Koh and Sodhi (2004) showed that butterflies found in urban, disturbed habitats were more likely to be polyphagous, while those in remaining forested regions were more likely to be monophagous. Larval host-plant specificity was one of the most important factors determining extinction (Koh et al. 2004b). Likewise, polyphagous butterfly species dominated smaller calcareous grassland fragments in German agricultural landscapes (Tscharntke et al. 2002). These data are useful for conservation triage, indicating that more attention needs to be paid to insect-plant pairs in which the insect has a restricted host range.

Spatial Ecology: Habitat Fragmentation and Heterogeneity

Looked at from a large enough scale, all habitats are patchy, often with diffuse boundaries or ecotones between them. Habitat destruction alters the scale and pattern of these patches and may replace ecotones with sharp transitions and new barriers to movement. This poses several types of conservation problems. First, the remaining habitat may be distributed as smaller pieces within a hostile matrix, leading to difficulty maintaining viable population sizes of plants and/or insects. Second, such fragments may also be sufficiently isolated as to preclude effective metapopulation dynamics. Spatial patterns of gene flow may also be altered, affecting local and regional coevolutionary interactions and the biogeographic mosaics that result (Thompson and Cunningham 2002; Thompson 2005). Gene flow within a larger population or among demes in a metapopulation may alter the outcome of selection on members of insect-plant pairs, through effects both on a migration-selection balance and on available variation for traits under selection or drift (e.g., Via 1991).

The construction of corridors is one method explored by conservation biologists to address problems of small fragments, although their efficacy is still not well understood (Debinski and Holt 2000). Nonetheless, a series of elegant experimental studies at the landscape level have shown that corridors can increase long-distance movement between fragments for butterflies (Haddad 1999). In addition, corridors can facilitate some plant-animal interactions, including seed dispersal and pollination (Tewksbury et al. 2002), indicating that plant movement as well as insect movement can be assisted. However, plants and insects presumably do not have the same movement rate between fragments, even in the presence of corridors, which may affect the coevolutionary dynamics between insects and their host plants over time. This remains as a profitable future area for research.

Species Introductions

Coevolved plant–insect herbivore interactions may be particularly vulnerable to disruption due to introduction of nonnative species. Introduced species that interact with a native plant-insect pair can drive changes in the population dynamics of either or both of the native species, as well as drive evolutionary changes in the trophic relationships (e.g., Carroll et al. 2005). Both evolutionary and ecological changes have implications for conservation. Here, we first consider the direct effects of introduced species that "intrude" into the plant-insect relationship, and then the indirect effects of introduced species that affect either the plant or insect through competition, predation, or mutualistic interactions.

Plant Introductions

To understand the possible direct effects of introduced species, suppose that the nonnative is a plant. Several possibilities exist with respect to its interaction with a native insect herbivore. Ovipositing females may ignore the plant, in which case there is no ecological or evolutionary effect on the native species that results from a direct trophic interaction. Second, native female insects may oviposit on the plant, and their larvae successfully develop on the plant (Thomas et al. 1987; Graves and Shapiro 2003). Third, native female insects may oviposit on the plant, yet their larvae cannot successfully develop on the plant. These last two scenarios may yield significant ecological and evolutionary changes, with implications for conservation.

The third scenario, that native insects oviposit on the introduced plant but their larvae cannot develop on it, is in some ways the most insidious, since it is the most likely to go undetected in the wild. Several instances of this are known among butterflies (Straatman 1962; Bowden 1971; Courant et al. 1994; Feldman and Haber 1998). In the case of the native Colorado *Pieris napi mcdunnoughi* Remington and the introduced Eurasian mustard, *Thlaspi arvense* L., the butterfly oviposits on the plant, but young larvae will not eat it (Chew 1975, 1977; C.L.B., unpublished data), although the plant is used as a normal host by *P. napi napi* L. in Sweden, where both are native. The maladaptation in Colorado is a characteristic of the butterfly, rather than the plant, as shown by reciprocal feeding trials using both European and American plants and butterflies (C.L.B. and C. Wiklund, unpublished data). Female oviposition preference is heritable and at least partially sex-linked (C.L.B., R.S. Niell, F. Shaw, V.O. Ezenwa, S. Simmers, K. White, A.K. Leidner, unpublished data). However, there is almost no variation within the population for larval survival on the plant (C.L.B., unpublished data), although hybridization studies between American and Swedish butterflies suggest a heritable basis for larval performance (C.L.B. and C. Wiklund, unpublished data). The opportunity thus exists for evolutionary change in the butterfly population in response to the introduced plant, but the simultaneous effects of population size reduction and selection remain to be explored. To complicate matters further, while the introduced plant is an aggressive invader of disturbed soils, it is out-competed during later succession (C.L.B., personal observation). Invasion/succession dynamics will therefore also affect the ecological and evolutionary fate of the butterfly population. In addition, patterns of gene flow into the butterfly population from other populations not exposed to the plant will affect the evolutionary dynamic. This case study again reflects the importance of biogeographic mosaics at several scales, from continental to local, which must be accounted for in dealing with conservation problems brought on by host-plant introductions.

Insect Introductions

Alternatively, rather than a plant introduction, an insect may be introduced. In some cases, such introductions are done intentionally in the context of biocontrol of introduced pest plants and may be species that feed on the pest plant in its native region. Nonetheless, the overall conceptual context is the same as for introductions of plants. Ovipositing introduced female insects may expand their host range to include both native and introduced pest plants (Louda et al. 2003). As for plant introductions, the outcome also depends on whether the resulting larvae can develop on these new hosts.

Such insect introductions may have serious implications for both plant and plant-insect conservation and must be considered when evaluating control methods for introduced plants. In a series of detailed studies on weevils introduced for biocontrol of thistles in the United States, Louda and coworkers have explored the effects of such introduced flower-feeding insects on both native and introduced thistles (Louda et al. 2003). Among other findings, introduced biocontrol insects can alter selection regimes, with the potential to drive native plants to extinction. Studying the introduced weevil *Rhinocyllus conicus* Frölich. on a native thistle (*Cirsium canescens* Nuttall), Rose et al. (2005) modeled the population dynamics of the thistle-native herbivorous insect–introduced weevil system. Native floral herbivorous insects significantly decreased the population growth rate of the native thistle, but not below persistence values, while leaf-feeding herbivorous insects had no detectable effect on the thistle's growth rate. However, the introduced weevil *R. conicus* imposed strong selection on the thistle's size at flowering, which in turn strongly affected seed set and population growth, with the potential to drive the thistle to extinction. To make matters worse, increasing density of introduced thistles (*Carduus nutans* L.) significantly increased the proportion of plants attacked and egg load in a native thistle [*Cirsium undulatum* Nuttall (Spreng.)] by *R. conicus*. These data indicate that introduced plants can maintain large enough population sizes to support a large enough population of introduced biocontrol insect herbivores to result in significant damage to native nontarget plants (Rand and Louda 2004). The ecological context of release of biocontrol agents is thus as important as the host preferences documented in preliminary studies prior to release (Louda et al. 2003) and must be taken into consideration when defining a control strategy for nonnative plants.

Shifts in an introduced plant's palatability and tolerance to herbivory by insects from its native range may also occur after introduction. This reflects evolutionary change at the local scale, as envisioned by the geographic mosaic theory of coevolution (Thompson 2005). It also poses serious challenges to control of introduced plants, since the frequency of either newly unpalatable or tolerant plant genotypes will increase under selection if insects from the native range are released for biocontrol (Garcia-Rossi et al. 2003). Populations of an introduced cordgrass (*Spartina alterniflora* Loisel.) in estuaries on the Pacific coast that had not been exposed to a planthopper (*Prokelisia marginata* Van Duzee) from their native range had less resistance overall to the planthopper than did either native cordgrass populations or introduced cordgrass populations with long exposure to the planthopper (Garcia-Rossi et al. 2003). One introduced cordgrass population, which was not resistant to the planthopper, showed virtually no ill effects from the high densities of planthoppers under experimental greenhouse conditions.

Indirect Effects of Introductions

Introductions of organisms not directly involved with a plant–insect herbivore interaction can still affect the relationship through effects on one of the pair. For example, in the San Francisco Bay area of California, invasive European weeds are gradually occupying serpentine grasslands as fallout from air pollution increases the soils' nitrogen content and the weeds gradually evolve tolerance to their high concentrations of toxic metals (Ehrlich and Hanski 2004). It is likely that these trends will exterminate the *E. editha– Plantago–Castilleja* interaction there within this century. This can be viewed as the continuation on a local scale of the regional-scale gross alteration of the entire California flora (and its associated insect herbivores) by the importation of European weeds by the conquistadors centuries ago.

In South Africa, populations of the lycaenid *Argyrocupha malagrida* Wallengren are threatened by heavy invasions of their habitat by alien vegetation. Interestingly, the invasive Argentine ant *Linepithema humile* Shattuck has recently appeared in the fynbos and is threatening lycaenid and other phytophagous insects there in the longer term, since they are not seed dispersers of the local flora as are native ants (Sameways 1993). This invader thus will potentially disrupt a series of plant–insect herbivore complexes. In addition, it threatens lycaenids directly by exterminating the native ants with which the native butterflies have a mutualistic coevolutionary relationship. At much larger scales of time and space, the ultimate invaders affecting plant–insect herbivore coevolution must have been *Homo erectus* Dubois and *H. sapiens* L. "invading" the rest of Eurasia from their African homeland and changing fire regimes. Then *H. sapiens* moved into the New World and likely exterminated the megafauna (Burney and Flannery 2005), dramatically altering the large herbivore fauna that transformed the flora of the Western Hemisphere and doubtless its coevolutionary relationships with insects. The rapid cultural evolution that has recently so magnified human impacts, of course, promises to make those alterations pale by comparison.

Changes in Disturbance Regimes

Changes in disturbance regimes can have far-reaching effects on species interactions and population viabilities, both through direct effects and through effects on habitat availability (Singer et al. 1993). Fire is one example of a crucial factor in the evolution of many ecosystems. Changes in fire regimes can have dramatic effects on plant–insect herbivore systems, in some ways playing a role parallel to that of a megaherbivore (Bond and Keeley 2005), and often must be of central concern to a conservation manager. While increasing fire frequency due to anthropogenic causes can have ecological effects on populations (J. W. Brown 1993; Prince 1993), fire suppression can also lead to the disappearance of the early successional host plants that could be crucial to some insects, for example, the lycaenids *Eumaeus atala* Poey in Florida (Emmel and Minno 1993) and the Natal endemic *Orachrysops ariadne* Butler (Henning et al. 1993).

Grazing by vertebrates can also have significant effects on plant–insect herbivore systems. In some cases, grazing is beneficial and can be used as a management tool to maintain habitat for threatened plant–insect herbivore pairs. For the checkerspot butterfly *E. editha* in the San Francisco Bay area, for example, grazing of serpentine grasslands helps maintain the native host plants necessary for the butterfly's survival, in the face of encroaching introduced Eurasian plants (Ehrlich and Hanski 2004).

Climate Change

Climate change, a portion of which is anthropogenic (Barnett et al. 2005), has implications for the conservation of plant–insect herbivore systems through both changes in mean climate parameters and in their variances (Boyce et al. 2006). Such changes are commonly thought of as a driver of changes in species' ranges. However, the impacts can be more locally direct, through effects on phenology and species interactions. As one example, climate change may pull apart coevolved plant-insect phenologies, as happened in the Jasper Ridge populations of the butterfly *E. editha* and contributed to their extinction (McLaughlin et al. 2002a, 2002b). Here, increasing climate variability that affected the phenology of both the butterfly and its host plants drove population dynamic fluctuations that contributed to the extinction of two butterfly populations. The potential of anthropogenic climate change to disrupt coevolved relationships is also suggested by the discovery that among several host plants that satisfactorily support development of *E. editha* in Colorado, the insect feeds solely on *Castilleja linariifolia* Benth., the most drought resistant (Holdren and Ehrlich 1982). This suggests that the butterfly is already living on the edge. Further exacerbation of the periodic regional droughts might remove even *C. linariifolia* as a potential host in some or all years and wipe out the complex.

The possible influences of climate change on coevolved relationships are implicit in Becerra's fine work on the distribution and diversification trees of the genus *Bursera*, characteristic of Mexico's tropical dry forest (Becerra 2005). Presumably the *Bursera-Blepharida* (flea beetle) coevolutionary system (Becerra 2003) diversified in response to climate change caused by mountain building. At first glance, one might expect some coevolutionary complexes to diversify in response to anthropogenic alteration of climate, and indeed they might. But there likely will be substantial differences from the climate-change situation described by Becerra and that facing Earth's biota now. Changes now entrained may be much, much more rapid, reducing time for adaptation for all actors in a complex. And the opportunities for migration in response to changing environments will, for many organisms, be constrained by human-created barriers of farmland, grazed terrain, areas occupied by exotic species, and concrete-coated roads and urban areas. A most likely result will be increased isolation of ever smaller populations that are failing to evolve rapidly enough to adjust to altered temperature and humidity regimes. When plant populations go extinct locally, populations of insect herbivores will follow them unless they can adapt to another local host. If the insect goes extinct first, that may enhance the possibilities of survival for plants freed from one source of reduced fitness, but for the plant population to persist would require an extremely fortuitous sequence of events. In mutualistic relationships, such as in tight plant-pollinator coevolution, there would be a sort of "law of the minimum." If one partner goes extinct, the other is almost certainly doomed. In plant–insect herbivore complexes, extinction of the herbivore means the plant "wins" the coevolutionary "race," even if change eventually dooms the plant to a pyrrhic victory.

Complex Interactions: Conservation Implications

Our discussion so far has assumed that insects eat plants and that the interaction is thus antagonistic, from the plant's viewpoint. Real life is more complex. Many insect herbivores also have mutualistic interactions with the plants, pollinating flowers as well as laying eggs (e.g., Pellmyr et al. 1996; Thompson and Cunningham 2002; Adler and Bronstein 2004; Yu et al. 2004). In some cases, such as for the plant *Lithophragma parviflorum* (Hook.) and the moth *Greya politella* Walsingham, both the presence and the strength of the antagonistic and mutualistic interactions can also vary over space and time (e.g., Thompson and Cunningham 2002; Thompson 2005). Maintenance of diverse populations of plants and insects encompassing variation in antagonist/mutualist pattern is thus critical to conservation of the plant-insect system over the longer run.

Of course, coevolutionary interactions of plants and insect herbivores can also have implications in other parts of the food chain. The most obvious place where this occurs is when the insects evolve the ability sequester plant defensive compounds to employ as defenses against their own predators. Milkweed (*Asclepias* spp.) populations vary greatly in their content of cardenolide defensive chemicals (Brower et al. 1982), and monarch butterflies acquire their famous distastefulness from their larval foodplants (Brower 1984). Milkweed populations are being exterminated in many areas, and the cardenolide content of the surviving ones will influence the predator protection of monarchs, the efficacy of the defenses of the monarch's co-mimic, *Limenitis archippus* Cramer (Ritland and Brower 1991), and populations of avian predators that potentially attack both. And coevolution will continue in the entire system until and unless some of the actors become globally extinct. The same can be said of the lycaenid-lupine system that also features *Lupinus* populations with very different alkaloidal defenses (Dolinger et al. 1973), and likely many other systems as well.

Conclusion

A theme emerging from our discussion is that spatial patterns are important. Paralleling the significant role that spatial environmental heterogeneity and/or metapopulation structure plays in the ecology and evolution of single species (e.g., Hanski and Gaggiotti 2004), the geographic mosaic of population interactions arising out of the vagaries of species interactions across space and time affects the ecological and evolutionary fate of those interactions (Thompson 2005). Conservationists have exploited tools of metapopulation viability to help manage single populations. We now need an equivalent concept structure relating to "mosaic viability" to deal with species interactions, including insect herbivores and their plants. What number of mosaic elements is needed to maintain both insect and plant populations, and also their genetic diversity? How do the geographic structure of the mosaic, the strength of interactions among species, and other biological vagaries affect the required overall size of the mosaic of interacting populations? What are the implications of barriers to movement or other human-derived alterations, which may act differentially on plants and insects? While information is rapidly accumulating, we are not yet able to answer these questions with the type of conceptual models now available for metapopulations, or now starting to be available for food webs (e.g., Brose et al. 2004).

Conservation of plants and herbivores are two sides of the same coin, and our major conclusion is as disturbing as it is unsurprising. When human activities dramatically affect the other organisms that are the working parts of our life-support systems, deleterious effects are likely to be propagated throughout ecosystems. The history of plant invasions in which insect herbivores have been left behind provides dramatic evidence of the potential threats (Williamson 1996). Ecosystem services ranging from supplying weed control, genetic diversity, and pollination for agriculture to provision of esthetic values can be reduced by such invasions. The alteration of coevolutionary interactions, both direct and diffuse, is part of a general trend of unthinking modification of the future of evolutionary processes on our planet (Myers 1991). Since the vast majority of coevolutionary interactions are unknown, the known ones are at best partially understood, and the general theory of plant-herbivore coevolution is still evolving (Boege and Marquis 2005), those interactions can only rarely be taken into consideration when contemplating conservation dilemmas. This puts even more pressure on biologists in general and conservation biologists in particular to attempt to change the major drivers of environmental destruction—the "three horsemen of IPAT": overpopulation, overconsumption, and the employment of environmentally malign technologies and socio-political-economic institutions and attitudes (Ehrlich and Ehrlich 2005). Biologists must help guide cultural evolution to avoid unnecessary and potentially dangerous alteration of interactions and foreclosure of evolutionary and coevolutionary options (Ehrlich 2001, 2002, 2003).

Acknowledgments

We thank J.N. Thompson for helpful comments on the manuscript.

References Cited

Adler, L.S., and J.L. Bronstein. 2004. Attracting antagonists: does floral nectar increase leaf herbivory? Ecology 85: 1519–1526.

Balint, Z. 1993a. The threatened lycaenids of the Carpathian Basin, east-central Europe, pp. 105–111. In T.R. New (ed.), Conservation biology of Lycaenidae (butterflies). IUCN, Gland, Switzerland.

Balint, Z. 1993b. The pannonian zephyr blue, *Plebejus sephirus kovacsi* Szabo, pp. 103–104. In T.R. New (ed.), Conservation biology of Lycaenidae (butterflies). IUCN, Gland, Switzerland.

Barnett, T., F. Zwiers, G. Hegerl, M. Allen, T. Crowley, N. Gillet, K. Hasselmann, P. Jones, B. Santer, R. Schnur, P. Scott, K. Taylor, and S. Tett. 2005. Detecting and attributing external influences on the climate system: a review of recent advances. J. Climate 18: 1291–1314.

Becerra, J.X. 1997. Insects on plants: chemical trends in host use. Science 276: 253–256.

Becerra, J.X. 2003. Synchronous coadaptation in an ancient case of herbivory. Proc. Natl. Acad. Sci. USA 100: 12804–12807.

Becerra, J.X. 2005. Timing the origin and expansion of the Mexican tropical dry forest. Proc. Natl. Acad. Sci. USA 102: 10919–10923.

Boege, K., and R.J. Marquis. 2005. Facing herbivory as you grow up: the ontogeny of resistance in plants. Trends Ecol. Evol. 20: 441–448.

Boggs, C.L., W.B. Watt, and P.R. Ehrlich (eds.). 2003. Butterflies: ecology and evolution taking flight. University of Chicago Press, Chicago.

Bond, W.J., and J.E. Keeley. 2005. Fire as a global "herbivore": the ecology and evolution of flammable ecosystems. Trends Ecol. Evol. 20: 387–394.

Bowden, S.R. 1971. American white butterflies and English foodplants. J. Lepidopt. Soc. 25: 6–12.

Boyce, M.S., C.V. Haridas, C. Lee, C.L. Boggs, E.M. Bruna, T. Coulson, D. Doak, J.M. Drake, J.-M. Gaillard, C.C. Horvitz, S. Kalisz, B.E. Kendall, T. Knight, E.S. Menges, W.F. Morris, C.A. Pfister, and S.D. Tuljapurkar. 2006. Demography in an increasingly variable world. Trends Ecol. Evol. 21: 141–148.

Breedlove, D.E., and P.R. Ehrlich. 1968. Plant-herbivore coevolution: lupines and lycaenids. Science 162: 671–672.

Brose, U., A. Ostling, K. Harrison, and N.D. Martinea. 2004. Unified spatial scaling of species and their trophic interactions. Nature 428: 167–171.

Brower, L.P. 1984. Chemical defence in butterflies, pp. 109–134. In R.I. Vane-Wright and P.R. Ackery (eds.), The biology of butterflies. Academic Press, Orlando, FL.

Brower, L.P., J.N. Seiber, C.J. Nelson, P. Tuskes, and S.P. Lynch. 1982. Plant-determined variation in the cardenolide content, thin layer chromatography profiles, and emetic potency of monarch butterflies, *Danaus plexippus*, reared on the milkweed, *Asclepias eriocarpa* in California. J. Chem. Ecol. 8: 579–633.

Brown, J.W. 1993. Thorne's hairsteak, *Mitoura thornei* Brown, pp. 116–119. In T.R. New (ed.), Conservation biology of Lycaenidae (butterflies). IUCN, Gland, Switzerland.

Brown, K.S., Jr. 1993. Riodininae: Amazonian genera with most species very rare or local, pp. 151. In T.R. New (ed.), Conservation biology of Lycaenidae (butterflies). IUCN, Gland, Switzerland.

Burney, D.A., and T.F. Flannery. 2005. Fifty millennia of catastrophic extinctions after human contact. Trends Ecol. Evol. 20: 395–401.

Carroll, S.P., J.E. Love, H. Dingle, M. Mathieson, T.R. Famula, and M.P. Zalucki. 2005. And the beak shall inherit: evolution in response to invasion. Ecol. Lett. 8: 944–951.

Chew, F.S. 1975. Coevolution of pierid butterflies and their cruciferous foodplants. I. The relative quality of available resources. Oecologia 20: 1117–11127.

Chew, F.S. 1977. Coevolution of pierid butterflies and their cruciferous foodplants. II. The distribution of eggs on potential foodplants. Evolution 31: 568–579.

Colston, A., and L. Friday. 1999. Wicken Fen: 100 years either side of the millennium. Nat. Cambridgeshire 41: 46–58.

Courant, A.V., A.E. Holbrook, E.D. Van der Reijden, and F.S. Chew. 1994. Native pierine butterfly (Pieridae) adapting to naturalized crucifer? J. Lepidopt. Soc. 48: 168–170.

Debinski, D.M., and R.D. Holt. 2000. A survey and overview of habitat fragmentation experiments. Conserv. Biol. 14: 342–355.

Dolinger, P.M., P.R. Ehrlich, W.L. Fitch, and D.E. Breedlove. 1973. Alkaloid and predation patterns in Colorado lupine populations. Oecologia 13: 191–204.

Duffy, E. 1993. The large copper (Dutch: grote vuurvlinder), *Lycaena dispar* Haworth, pp. 81–82. In T.R. New (ed.), Conservation biology of Lycaenidae (butterflies). IUCN, Gland, Switzerland.

Ehrlich, P.R. 1970. Coevolution and the biology of communities, pp. 1–11. In K.L. Chambers (ed.), Biochemical coevolution. Oregon State University Press, Corvallis.

Ehrlich, P.R. 2001. Intervening in evolution: ethics and actions. Proc. Natl. Acad. Sci. USA 98: 5477–5480.

Ehrlich, P.R. 2002. Human natures, nature conservation, and environmental ethics. Bioscience 52: 31–43.

Ehrlich, P.R. 2003. Bioethics: are our priorities right? Bioscience 53: 1207–1216.

Ehrlich, P.R., and A.H. Ehrlich. 2005. One with Nineveh: Politics, consumption, and the human future, (with new afterword). Island Press, Washington, DC.

Ehrlich, P., and I. Hanski (eds.). 2004. On the wings of checkerspots: a model system for population biology. Oxford University Press, Oxford, UK.

Ehrlich, P.R., and P.H. Raven. 1964. Butterflies and plants: a study in coevolution. Evolution 18: 586–608.

Emmel, T.C., and M.C. Minno. 1993. The Atala butterfly, *Eumaeus atala florida* (Röber), pp. 120–130. In T.R. New (ed.), Conservation biology of Lycaenidae (butterflies). IUCN, Gland, Switzerland.

Feldman, T.S., and W.A. Haber. 1998. Oviposition behavior, host plant use, and diet breadth of *Anthanassa* butterflies (Lepidoptera: Nymphalidae) using plants in the Acanthaceae in a Costa Rican community. Fl. Entomol. 81: 396–406.

Garcia-Rossi, D., N. Rank, and D.R. Strong. 2003. Potential for self-defeating biological control? Variation in herbivore vulnerability among invasive *Spartina* genotypes. Ecol. Appl. 13: 1640–1649.

Graves, S.D., and A.M. Shapiro. 2003. Exotics as host plants of the California butterfly fauna. Biol. Conserv. 110: 413–433.

Groves, R.H. 1989. Ecological control of invasive terrestrial plants, pp. 437–461. In A. Drake, H.A. Mooney, F. d. Castri, R.H. Groves, F.J. Kruger, M. Rejmnek, and M. Williamson (eds.), Biological invasions: a global perspective. Wiley, Chichester, UK.

Haddad, N.M. 1999. Corridor and distance effects on interpatch movements: a landscape experiment with butterflies. Ecol. Appl. 9: 612–622.

Hanski, I., and O.E. Gaggiotti [eds.]. 2004. Ecology, genetics, and evolution of metapopulations. Elsevier Academic Press, Amsterdam.

Hawkins, B.A., and E.E. Porter. 2003. Does herbivore diversity depend on plant diversity? The case of California butterflies. Am. Nat. 161: 40–49.

Henning, S.F., G.A. Henning, and M.J. Samways. 1993. *Orachrysops (Lepidochrysops) ariadne* (Butler); subfamily Polyommatinae, tribe Polyommatini, pp. 159. In T.R. New (ed.), Conservation biology of Lycaenidae (butterflies). IUCN, Gland, Switzerland.

Hernandez, H.B., RT. 1995. Endangered cacti in the Chihuahuan Desert. I. Distribution patterns. Conserv. Biol. 9: 1176–1188.

Holdren, C.E., and P.R. Ehrlich. 1982. Ecological determinants of food plant choice in the checkerspot butterfly *Euphydryas editha* in Colorado. Oecologia 52: 417–423.

Hopkins, G.W., J.I. Thacker, A.F. G. Dixon, P. Waring, and M.G. Telfer. 2002. Identifying rarity in insects: the importance of host plant range. Biol. Conserv. 105: 293–307.

Koh, L.P., and N.S. Sodhi. 2004. Importance of reserves, fragments, and parks for butterfly conservation in a tropical urban landscape. Ecol. Appl. 14: 1695–1708.

Koh, L.P., N.S. Sodhi, and B.W. Brook. 2004a. Co-extinctions of tropical butterflies and their hostplants. Biotropica 36: 272–274.

Koh, L.P., N.S. Sodhi, and B.W. Brook. 2004b. Ecological correlates of extinction proneness in tropical butterflies. Conserv. Biol. 18: 1571–1578.

Koh, L.P., R.R. Dunn, N.S. Sodhi, R.K. Colwell, H.C. Proctor, and V.S. Smith. 2004c. Species coextinctions and the biodiversity crisis. Science 305: 1632–1634.

Louda, S.M., R.W. Pemberton, M.T. Johnson, and P.A. Follett. 2003. Nontarget effects: The Achilles' heel of biological control? Retrospective analyses to reduce risk associated with biocontrol introductions. Ann. Rev. Entomol. 48: 365–396.

Louda, S.M., R.W. Pemberton, M.T. Johnson, and P.A. Follett. 2003. Nontarget effects: The Achille's heel of biological control? Retrospective analysis to reduce risk associated with biocontrol introductions. Annu. Rev. Entomol. 48: 365–396.

McLaughlin, J.F., J.J. Hellmann, C.L. Boggs, and P.R. Ehrlich. 2002a. Climate change hastens population extinctions. Proc. Natl. Acad. Sci. USA 99: 6070–6074.

McLaughlin, J.F., J.J. Hellmann, C.L. Boggs, and P.R. Ehrlich. 2002b. The route to extinction: population dynamics of a threatened butterfly. Oecologia 132: 538–548.

Mooney, H.A., and R.J. Hobbs. 2000. Invasive species in a changing world. Island Press, Washington, DC.

Munguira, M.L., J. Martin, and E. Balletto. 1993. Conservation biology of the Lycaenidae: A European overview, pp. 23–34. In T.R. New (ed.), Conservation biology of Lycaenidae (butterflies). IUCN, Gland, Switzerland.

Myers, N. 1991. The biotic crisis and the future of evolution. Proc. Natl. Acad. Sci. USA 98: 538992.

Ortega-Baes, P., and H. Godinez-Alvarez. 2006. Global diversity and conservation priorities in the Cactaceae. Biodivers. Conserv.15: 817–827.

Pellmyr, O., J.N. Thompson, J.M. Brown, and R.G. Harrison. 1996. Evolution of pollination and mutualism in the yucca moth lineage. Am. Nat. 148: 827–847.

Powell, J.A., and M.W. Parker. 1993. Lange's metalmark, *Apodemia mormo langei* Comstock, pp. 116–119. In T.R. New (ed.), Conservation biology of Lycaenidae (butterflies). IUCN, Gland, Switzerland.

Prince, G.B. 1993. The Australian hairstreak, *Pseudalmenus chlorinda* (Blanchard), pp. 159. In T.R. New (ed.), Conservation Biology of Lycaenidae (Butterflies). IUCN, Gland, Switzerland.

Rand, T.A., and S.M. Louda. 2004. Exotic weed invasion increases the susceptibility of native plants ot attack by a biocontrol herbivore. Ecology 85: 1548–1554.

Ritland, D.B., and L.P. Brower. 1991. The viceroy butterfly is not a batesian mimic. Nature 350: 497–498.

Rose, K.E., S.M. Louda, and M. Rees. 2005. Demographic and evolutionary impacts of native and invasive insect herbivores on *Cirsium canescens*. Ecology 86: 453–465.

Sameways, M.J. 1993. Threatened Lycaenidae of South Africa. pp. 62–69. In T.R. New (ed.), Conservation biology of Lycaenidae (butterflies). IUCN, Gland, Switzerland.

Singer, M.C., C.D. Thomas, and C. Parmesan. 1993. Rapid human-induced evolution of insect-host associations. Nature 366: 681–683.

Straatman, R. 1962. Notes on certain Lepidoptera ovipositing on plants which are toxic to their larvae. J. Lepidopt. Soc. 16: 99–103.

Tewksbury, J.J., D.J. Levey, N.M. Haddad, S. Sargents, J.L. Orrocks, A. Weldon, B.J. Danielson, J. Brinkerhoff, E.I. Damschen, and P. Townsend. 2002. Corridors affect plants, animals and their interactions in fragmented landscapes. Proc. Natl. Acad. Sci. USA 99: 12923–12926.

Thomas, C.D., M.C. Singer, and D.A. Boughton. 1996. Catastrophic extinction of population sources in a butterfly metapopulation. Am. Nat. 148: 957–975.

Thomas, C.D., D. Ng, M.C. Singer, C. Parmesan, and H.L. Billington. 1987. Incorporation of a European weed into the diet of a North American herbivore. Evolution 41: 892–901.

Thompson, J.N. 2005. The geographic mosaic of coevolution. University of Chicago Press, Chicago.

Thompson, J.N., and B.M. Cunningham. 2002. Geographic structure and dynamics of coevolutionary selection. Nature 417: 735–738.

Tscharntke, T., I. Staffan-Dewenter, A. Kruess, and C. Thies. 2002. Contribution of small habitat fragments to conservation of insect communities of grassland-cropland landscapes. Ecol. Appl. 12: 354–363.

Via, S. 1991. The genetic structure of host plant adaptation in a spatial patchwork: Demographic variability among reciprocally transplanted pea aphid clones. Evolution 45: 827–852.

Williamson, M. 1996. Biological invasions. Chapman and Hall, London.

Yu, D.W., J. Ridley, E. Jousselin, E.A. Herre, S.G. Compton, J.M. Cook, J.C. Moore, and G.D. Weiblen. 2004. Oviposition strategies, host coercion, and the stable exploitation of figs by wasps. Proc. R. Soc. Biol. Sci. Ser. B 271: 1185–1195.

INDEX

Text: 8.25/12 ITC Stone Serif Medium
Display: Akzidenz Grotesk family
Composition: Aptara
Indexer: Live Oaks Indexing
Printer and Binder: Sheridan Books, Inc.